Topics in Biostatistics

METHODS IN MOLECULAR BIOLOGY™

John M. Walker, SERIES EDITOR

419. **Post-Transcriptional Gene Regulation,** edited by *Jeffrey Wilusz, 2008*
418. **Avidin–Biotin Interactions:** *Methods and Applications,* edited by *Robert J. McMahon, 2008*
417. **Tissue Engineering, Second Edition,** edited by *Hannsjörg Hauser and Martin Fussenegger, 2007*
416. **Gene Essentiality:** *Protocols and Bioinformatics,* edited by *Andrei L. Osterman, 2008*
415. **Innate Immunity,** edited by *Jonathan Ewbank and Eric Vivier, 2007*
414. **Apoptosis in Cancer:** *Methods and Protocols,* edited by *Gil Mor and Ayesha Alvero, 2008*
413. **Protein Structure Prediction, Second Edition,** edited by *Mohammed Zaki and Chris Bystroff, 2008*
412. **Neutrophil Methods and Protocols,** edited by *Mark T. Quinn, Frank R. DeLeo, and Gary M. Bokoch, 2007*
411. **Reporter Genes for Mammalian Systems,** edited by *Don Anson, 2007*
410. **Environmental Genomics,** edited by *Cristofre C. Martin, 2007*
409. **Immunoinformatics:** *Predicting Immunogenicity In Silico,* edited by *Darren R. Flower, 2007*
408. **Gene Function Analysis,** edited by *Michael Ochs, 2007*
407. **Stem Cell Assays,** edited by *Vemuri C. Mohan, 2007*
406. **Plant Bioinformatics:** *Methods and Protocols,* edited by *David Edwards, 2007*
405. **Telomerase Inhibition:** *Strategies and Protocols,* edited by *Lucy Andrews and Trygve O. Tollefsbol, 2007*
404. **Topics in Biostatistics,** edited by *Walter T. Ambrosius, 2007*
403. **Patch-Clamp Methods and Protocols,** edited by *Peter Molnar and James J. Hickman, 2007*
402. **PCR Primer Design,** edited by *Anton Yuryev, 2007*
401. **Neuroinformatics,** edited by *Chiquito J. Crasto, 2007*
400. **Methods in Lipid Membranes,** edited by *Alex Dopico, 2007*
399. **Neuroprotection Methods and Protocols,** edited by *Tiziana Borsello, 2007*
398. **Lipid Rafts,** edited by *Thomas J. McIntosh, 2007*
397. **Hedgehog Signaling Protocols,** edited by *Jamila I. Horabin, 2007*
396. **Comparative Genomics,** *Volume 2,* edited by *Nicholas H. Bergman, 2007*
395. **Comparative Genomics,** *Volume 1,* edited by *Nicholas H. Bergman, 2007*
394. **Salmonella:** *Methods and Protocols,* edited by *Heide Schatten and Abe Eisenstark, 2007*
393. **Plant Secondary Metabolites,** edited by *Harinder P. S. Makkar, P. Siddhuraju, and Klaus Becher, 2007*
392. **Molecular Motors:** *Methods and Protocols,* edited by *Ann O. Sperry, 2007*
391. **MRSA Protocols,** edited by *Yinduo Ji, 2007*
390. **Protein Targeting Protocols, Second Edition,** edited by *Mark van der Giezen, 2007*
389. ***Pichia*** **Protocols, Second Edition,** edited by *James M. Cregg, 2007*
388. **Baculovirus and Insect Cell Expression Protocols, Second Edition,** edited by *David W. Murhammer, 2007*
387. **Serial Analysis of Gene Expression (SAGE):** *Digital Gene Expression Profiling,* edited by *Kare Lehmann Nielsen, 2007*
386. **Peptide Characterization and Application Protocols,** edited by *Gregg B. Fields, 2007*
385. **Microchip-Based Assay Systems:** *Methods and Applications,* edited by *Pierre N. Floriano, 2007*
384. **Capillary Electrophoresis:** *Methods and Protocols,* edited by *Philippe Schmitt-Kopplin, 2007*
383. **Cancer Genomics and Proteomics:** *Methods and Protocols,* edited by *Paul B. Fisher, 2007*
382. **Microarrays, Second Edition:** *Volume 2, Applications and Data Analysis,* edited by *Jang B. Rampal, 2007*
381. **Microarrays, Second Edition:** *Volume 1, Synthesis Methods,* edited by *Jang B. Rampal, 2007*
380. **Immunological Tolerance:** *Methods and Protocols, edited by Paul J. Fairchild, 2007*
379. **Glycovirology Protocols,** edited by *Richard J. Sugrue, 2007*
378. **Monoclonal Antibodies:** *Methods and Protocols,* edited by *Maher Albitar, 2007*
377. **Microarray Data Analysis:** *Methods and Applications,* edited by *Michael J. Korenberg, 2007*
376. **Linkage Disequilibrium and Association Mapping:** *Analysis and Application,* edited by *Andrew R. Collins, 2007*
375. **In Vitro Transcription and Translation Protocols:** *Second Edition,* edited by *Guido Grandi, 2007*
374. **Quantum Dots:** *Applications in Biology,* edited by *Marcel Bruchez and Charles Z. Hotz, 2007*
373. **Pyrosequencing® Protocols,** edited by *Sharon Marsh, 2007*
372. **Mitochondria: Practical Protocols,** edited by *Dario Leister and Johannes Herrmann, 2007*
371. **Biological Aging:** *Methods and Protocols,* edited by *Trygve O. Tollefsbol, 2007*
370. **Adhesion Protein Protocols,** *Second Edition,* edited by *Amanda S. Coutts, 2007*
369. **Electron Microscopy:** *Methods and Protocols, Second Edition,* edited by *John Kuo, 2007*
368. **Cryopreservation and Freeze-Drying Protocols,** *Second Edition,* edited by *John G. Day and Glyn Stacey, 2007*
367. **Mass Spectrometry Data Analysis in Proteomics,** edited by *Rune Matthiesen, 2007*
366. **Cardiac Gene Expression:** *Methods and Protocols,* edited by *Jun Zhang and Gregg Rokosh, 2007*
365. **Protein Phosphatase Protocols:** edited by *Greg Moorhead, 2007*
364. **Macromolecular Crystallography Protocols:** *Volume 2, Structure Determination,* edited by *Sylvie Doublié, 2007*
363. **Macromolecular Crystallography Protocols:** *Volume 1, Preparation and Crystallization of Macromolecules,* edited by *Sylvie Doublié, 2007*
362. **Circadian Rhythms:** *Methods and Protocols,* edited by *Ezio Rosato, 2007*
361. **Target Discovery and Validation Reviews and Protocols:** *Emerging Molecular Targets and Treatment Options, Volume 2,* edited by *Mouldy Sioud, 2007*

METHODS IN MOLECULAR BIOLOGY™

Topics in Biostatistics

Edited by

Walter T. Ambrosius

Department of Biostatistical Sciences
Wake Forest University Health Sciences
Winston-Salem, NC

HUMANA PRESS ✻ TOTOWA, NEW JERSEY

999 Riverview Drive, Suite 208
Totowa, New Jersey 07512

www.humanapress.com

This publication is printed on acid-free paper. ∞
ANSI Z39.48-1984 (American Standards Institute)

Permanence of Paper for Printed Library Materials.

For additional copies, pricing for bulk purchases, and/or information about other Humana titles, contact Humana at the above address or at any of the following numbers: Tel.: 973-256-1699; Fax: 973-256-8341; E-mail: orders@humanapr.com; or visit our Website: www.humanapress.com

10 9 8 7 6 5 4 3 2 1

Library of Congress Control Number: 2006940292

ISBN: 978-1-58829-531-6 e-ISBN: 978-1-59745-530-5

Preface

The primary purpose of this book is to provide scientists with a broad survey of biostatistical methods. This book is intended for use by scientists in all disciplines. To this end, we present a series of biostatistical techniques illustrated with examples. Many of the examples are from biology and medicine. Elementary methods include descriptive statistics, study design, statistical inference, categorical variables, evaluation of diagnostic tests, comparison of means (*t*-test, nonparametric tests, and analysis of variance), linear regression, and logistic regression. All of these methods can be performed with paper, pencil, and a calculator in simple cases. The secondary purpose of this book is to introduce more complicated statistical methods requiring either collaboration with a biostatistician or use of a statistical package. Our goal is not to teach the reader how to use these methods but rather to teach the "language of statistics" so that collaborating with a biostatistician or deciphering a software manual is more productive.

Walter T. Ambrosius

Acknowledgments

I would like to thank the authors; without their excellent chapters, this book would never have come to fruition. I thank Stephen W. Looney for his recommendation of me to Humana Press. I appreciate the confidence in me Humana Press and the Series Editor, John M. Walker, showed by selecting me as editor of this volume. I thank the production staff at Chernow Editorial Services. I gratefully acknowledge support of the National Institutes of Health (M01-RR07122) and the Wake Forest University Health Sciences General Clinical Research Center. Finally, I thank my wife, Leslie A. Underwood, who proofread every chapter, and our daughter, Emma G. Ambrosius, who assisted with hugs and kisses.

Walter T. Ambrosius

Contents

Contributors

TODD A. ALONZO, PhD • *Children's Oncology Group, University of Southern California, Arcadia, CA*

WALTER T. AMBROSIUS, PhD • *Department of Biostatistical Sciences, Wake Forest University Health Sciences, Winston-Salem, NC*

NANCY BERMAN, PhD • *Statistical Research Associates, Washington, DC*

KATHLEEN M. CAMPBELL, MD • *Cincinnati Children's Hospital, Cincinnati, OH*

L. DOUGLAS CASE, PhD • *Department of Biostatistical Sciences, Wake Forest University Health Sciences, Winston-Salem, NC*

LANG CHEN, MS • *Department of Biostatistics, University of Alabama at Birmingham, Hoover, AL*

RALPH B. D'AGOSTINO, JR., PhD • *Department of Biostatistical Sciences, Wake Forest University School of Medicine, Winston-Salem, NC*

LYNN E. EBERLY, PhD • *Division of Biostatistics, University of Minnesota, Minneapolis, MN*

JASON P. FINE, PhD • *Department of Statistics and Biostatistics, University of Wisconsin—Madison, Madison, WI*

TASHA E. FINGERLIN, PhD • *University of Colorado, Health Sciences at Denver, Aurora, CO*

MARK E. GLICKMAN, PhD • *Department of Health Services, Boston University School of Public Health, Bedford, MA*

JAMES J. GRADY, PhD • *University of Texas Medical Branch, Galveston, TX*

CHRISTINA GULLÍON, PhD • *The Center for Health Research/Kaiser Permanente, Portland, OR*

RAYMOND G. HOFFMANN, PhD • *Division of Biostatistics, Medical College of Wisconsin, Milwaukee, WI*

EDWARD H. IP, PhD • *Department of Biostatistical Sciences, Wake Forest University Health Sciences, Winston-Salem, NC*

HONGYU JIANG, PhD • *Department of Biostatistics, Harvard School of Public Health, Boston, MA*

KYOUNGMI KIM, PhD • *Department of Biostatistics, University of Alabama at Birmingham, Hoover, AL*

CARL D. LANGEFELD, PhD • *Department of Biostatistical Sciences, Wake Forest University Health Sciences, Winston-Salem, NC*

Hyun Ja Lim, PhD • *Department of Community Health and Epidemiology, University of Saskatchewan College of Medicine, Saskatoon, Saskatchewan, Canada*

Douglas W. Mahoney, PhD • *Division of Biostatistics, Health Sciences Research, Mayo Clinic College of Medicine, Rochester, MN*

Tapan Mehta, MS • *Department of Biostatistics, University of Alabama at Birmingham, Hoover, AL*

Josyf C. Mychaleckyj, PhD • *Center for Public Health Genomics, University of Virginia, Charlottesville, VA*

Todd G. Nick, PhD • *Cincinnati Children's Hospital Medical Center, Cincinnati, OH*

Ann L. Oberg, PhD • *Division of Biostatistics, Health Sciences Research, Mayo Clinic College of Medicine, Rochester, MN*

Grier P. Page, PhD • *Department of Biostatistics, University of Alabama at Birmingham, Hoover, AL*

Curtis A. Parvin, PhD • *Department of Pathology and Immunology, Washington University School of Medicine, St. Louis, MO*

Margaret Sullivan Pepe, PhD • *Department of Statistics, Fred Hutchinson Cancer Research Center, University of Washington, Seattle, WA*

Susan M. Perkins, PhD • *Division of Biostatistics, Indiana University School of Medicine, Indianapolis, IN*

Jonathan J. Shuster, PhD • *Department of Health Policy/Epidemiology, University of Florida, Gainesville, FL*

Wanzhu Tu, PhD • *Division of Biostatistics, Indiana University School of Medicine, Indianopolis, IN*

David A. van Dyk, PhD • *Department of Statistics, University of California, Irvine, CA*

Stanislav O. Zakharkin, PhD • *Department of Biostatistics, University of Alabama at Birmingham, Hoover, AL*

Kui Zhang, PhD • *Department of Biostatistics, University of Alabama at Birmingham, Hoover, AL*

1

Study Design: The Basics

Hyun Ja Lim and Raymond G. Hoffmann

Summary

In biomedical research, meaningful conclusions can only be drawn based on data collected from a valid scientific design using appropriate statistical methods. Therefore, the selection of an appropriate study design is important in order to provide an unbiased and scientific evaluation of the research questions. In this chapter, the different kinds of experimental studies commonly used in biology and medicine are introduced. A brief survey of basic experimental study designs, randomization, blinding, possible biases, issues in data analysis, and interpretation of the study results are mainly provided.

Key Words: As-received analysis; bias; blinding; block randomization; carryover effect; cluster design; complete randomization; compliance; crossover design; dropout; experimental study; exploratory analysis; factorial design; group allocation design; historically controlled study; intention-to-treat analysis; masking; per-protocol analysis; randomization; randomized controlled study; stratified randomization; subgroup analysis.

1. Introduction

In biomedical research, meaningful conclusions can only be drawn based on data collected from a valid scientific design using appropriate statistical methods. Therefore, the selection of an appropriate study design is important to provide an unbiased and scientific evaluation of the research questions. Each design is based on a certain rationale and is applicable in certain experimental situations. Before a study design is chosen, some basic design considerations such as goals of the studies, subject or sample selection, randomization and blinding, the selection of controls, and some statistical issues must be considered to justify the use of statistical analyses. In this chapter, the different kinds of experimental

From: *Methods in Molecular Biology, vol. 404: Topics in Biostatistics*
Edited by: W. T. Ambrosius © Humana Press Inc., Totowa, NJ

studies commonly used in biology and medicine will be introduced. The primary purpose of this chapter is to provide scientists with a brief survey of basic experimental study designs, randomization, blinding, and possible biases. Furthermore, issues in data analysis and interpretation of the study results will be considered.

2. Experimental Studies

An *experimental* (or interventional) study is a study in which conditions are controlled and manipulated by the study investigator. This type of study is contrary to an *observational* study in which the investigator does not control conditions, as in surveys and most epidemiologic studies. The major objective of an experimental study is to provide a precise and valid treatment comparison. The design of a study can contribute to this objective by preventing bias, ensuring an efficient comparison, and possessing sufficient simplicity to encourage participation and minimize errors. Experimental studies involve one or more interventions, which an investigator controls, such as treatments, drugs, procedures, technologies, or devices. In general, the purpose of an experimental study is to determine whether a treatment makes a difference or is effective on the study units or samples. Experimental studies in biomedical science are most commonly encountered as laboratory samples, animal experiments, or clinical trials on patients. Experimental units can be human subjects, animals, cells, or materials.

Experimental studies fall into two categories: those with controls and those without controls. Uncontrolled studies are conducted by trying a new treatment on some experimental units. Any benefit or harmful effects seen in the study units will be ascribed to the new treatment. Many uncontrolled studies have suggested that a new treatment was highly effective, only for this apparent benefit to disappear after more careful examination. There are several instances of treatments being investigated and found ineffective after many years. For example, the use of hormonal therapy in menopausal women was believed to reduce the risk of cardiovascular disease ***(1,2)***. A controlled study by the Women's Health Initiative found that it does not protect the heart and may even increase the risk of coronary heart disease ***(3)***.

For controlled studies, the experimental treatment (or intervention) is compared with another treatment. In some studies, a placebo/dummy treatment is used for comparison; in other studies, the existing standard treatment is used. In detecting whether the difference is due to the experimental treatment or to some other factor, studies with controls are more scientific and have greater validity than studies without controls. This section will discuss several controlled experimental designs: randomized, historical, crossover, factorial, and cluster group designs.

2.1. Randomized Controlled Studies

A randomized controlled study (or randomized controlled trial; RCT) is also known as a *parallel* design. *Randomization* refers to the random assignment of experimental units to one of two or more treatments for the purpose of comparing the treatments on some outcome measure. To investigate whether a new treatment is better than other ones, in a randomized controlled study units are allocated into two or more groups and everyone within a group receives the same treatment. The *control* can be a standard treatment or placebo. A placebo is a treatment that appears to be identical to other study treatments, but with no true biological effect. Thus, *placebo-controlled* means experimental units in all study groups receive identical-appearing treatment, but some groups receive an inactive treatment. An example of a randomized controlled study is the Lipids Research Clinics Coronary Primary Prevention Study ***(4)***. To test whether lowering cholesterol can prevent the development of coronary heart disease, in this study a total of 4000 middle-aged males were randomized to receive either the cholesterol-lowering agent cholestyramine or a placebo. The number of experimental units to be allocated to each group in the randomized controlled study is fixed in advance, and experimental units are randomly allocated to one of the groups by a specific randomization scheme detailed in **Sections 3.1–3.3**. A randomized controlled study is not good for a low-prevalence disease because it is difficult to recruit study participants. There is also an ethical issue of using a placebo control group. A randomized controlled study may not be appropriate where disease prognosis is poor with current treatments.

2.2. Historically Controlled Studies

When researchers involved in a controlled study choose comparable study units from a previously existing group of units and compare them with a new intervention, it is called a *historically controlled study*. If history has established the performance of a standard treatment, a new intervention is used for comparison with the outcomes in the historical groups making it a nonrandomized and nonconcurrent study. An example of a historically controlled study is a study of an antibiotic lock (highly concentrated antibiotic solution) for the treatment of catheter-related bacteremia ***(5)***. In this study, a group of patients was given an antibiotic lock. They were compared with previous patients who received routine catheter replacement. Historically controlled studies can be conducted rapidly and at less cost than other types of studies. In addition, there is not an ethical issue of using a control group because the effect of standard treatment is known. Usually, sources of the control groups are from existing laboratory data, published literature, and medical charts or records that were not collected for any study purposes. This can lead to a lack of uniformity in collecting and reporting data and to missing data. Major concerns about historically controlled studies are the quality of data, as well as the accuracy, completeness, and reliability with which control groups were collected. In recent years, outcomes research of large computerized prospective databases, such as cancer registries, have been expanded to provide important information. An improved outcome may be erroneously attributed to the treatment when the improvement may actually be from changes over time in the patient population, patient management, or diagnostic technology. Historically controlled studies are always potentially biased in evaluating whether a new intervention group is better than control groups.

2.3. Crossover Studies

When an experimental unit receives more than one treatment during a study, the design is called a *crossover study*. The word *crossover* has come to be used for studies in which experimental units are given a number of treatments in the same or possibly different number of periods, and each treatment is given at a different time to each experimental unit. In the most common type of crossover study there are two treatments (say **A** and **B**) and two periods. Half of the participants are randomly assigned to start with treatment **A** in the first period and then "cross over" to treatment **B** in the second period; the other half does the opposite. More than two treatments in a crossover study with as many periods as there are treatments can be used. However, these may be difficult to carry out in practice. An example of a crossover study is found in asthma treatments to compare a single dose of formoterol with a single dose of salbutamol *(6)*. Children are randomly allocated to either formoterol followed by salbutamol or salbutamol followed by formoterol. Because all experimental units receive all treatments and only the order of administering the treatments is randomized in a crossover study, randomization itself is less important than in a parallel design.

Both treatments are applied to each experimental unit in the common crossover study, and so each experimental unit forms its own control, and the measured effect of the intervention is the difference in each experimental unit's response to the intervention and control. This allows both between-unit and within-unit analyses. Because the within-unit variation is usually less than the between-unit variation, this leads to the treatment difference being estimated with greater precision. Thus the study requires a smaller sample size. However, it needs a doubling of the study duration compared with a parallel design. To use the crossover design, we need to ensure that the effects from the previous treatment do not carry over into the period of the next treatment. If a carryover effect exists, the analysis is complicated, and the direct comparison of the treatment effects can be invalidated. To remove such carryover effect, a washout period can be imposed between treatment periods with the hope that the prior treatment will cease to affect the experimental unit before starting the next treatment. In many cases, it is difficult to know whether all carryover effects have been eliminated. In crossover studies, the complexity of analysis and interpretation created by the problem of carryover effects are substantial. Patients dropping out of the study also have strong effects and more severe consequences than in a parallel design. For example, if a patient drops out of a crossover study in the second period, a simple analysis cannot use the data from only the first period of treatment.

In general, a crossover design is appropriate for studies of treatments for chronic conditions, such as depression or hypertension. A crossover design is also appropriate for those treatments whose effects can be measured after a short period and the underlying disease condition does not change substantially. It is not appropriate for studies where the treatment changes the patient's underlying disease condition substantially, such as in surgical treatments.

2.4. Factorial Designs

A *factorial design* is a study that tests the effect of more than one treatment in order to reduce time and expenses by looking at two or more factors simultaneously. For example, two treatments (say **A** and **B**) are simultaneously compared with each other and with a control. Experimental units are divided into four groups who receive the control treatment, **A** only, **B** only, and both **A** and **B**. The factorial design, which is a special type of the parallel design, is used to investigate two separate and relatively independent research questions simultaneously in a single study. The factorial design is very efficient when the treatments do not interact with each other. In the presence of a treatment interaction between **A** and **B**, the estimation of the overall effect of **A** or **B** is not straightforward because the effect of **A** differs depending on the presence or absence of **B**. Interaction may be anticipated if the two drugs act on the same response or through the same mechanism. If some interactions are known biologically not to exist or are unimportant, the partial or fractional factorial design can be used to reduce the sample size and complexity of the experiment. This can be done by omitting certain treatment combinations from the design and still estimating all of the other effects. An example of a factorial design is the Physician's Health Study that investigated the roles of aspirin and beta-carotene in reducing the risk of cardiovascular disease and the risk of cancer, respectively *(7)*. Another example of a factorial design is the Women's Health Initiative Study that investigated the role of dietary modification and hormone replacement therapy in reducing the risk of breast and colorectal cancer and the risk of heart disease and osteoporosis, respectively *(8)*.

In general, a factorial design is appropriate to study different diseases such as aspirin on myocardial infarction and beta-carotene on cancer, different mechanisms such as radiotherapy or chemotherapy for tumor, or treatments whose combination may be much better than each individually. However, it usually is not appropriate when the combination of treatments would result in adverse effects.

2.5. Cluster or Group Allocation Designs

To address research questions about the effects of public health programs in the population, applying the intervention to groups of experimental units rather than individual units may be more feasible and cost-effective. When a treatment is randomly allocated to groups or clusters of experimental units, such as a family, school, workplace, or community, the design is called a *cluster* or *group allocation design*. Experimental units within a cluster usually share similar environments or characteristics and thus are correlated. Randomizing by group rather than by individual is easier to conduct. Because randomization is performed at the cluster level rather than at the individual unit level, the randomization unit differs in the analysis. Hence, the standard methods for sample size calculation are not directly applicable. One must consider both intercluster and intracluster variation for data analysis. One example is the study of vitamin A on morbidity and mortality in Indian children, which randomized with villages ***(9)***. Another example is the Hutchinson Smoking Prevention Project, which was a school-based study for evaluation of the long-term effectiveness on the prevention of habitual cigarette smoking among youth, which randomized school districts ***(10)***.

We have described several different study designs. We now turn to an examination of the different methods that can be used within a design.

3. Randomization

The design of an experimental study involves the choice of a method for allocating treatments to experimental units. An investigator's knowledge of treatment allocation might introduce bias by conscious or unconscious selection of experimental units to receive a particular treatment. The most widely used method of unbiased treatment allocation is to use random allocation to assign experimental units to treatment groups. The original concept of randomization was from R.A. Fisher in his classic text, *The Design of Experiments*, in 1935. It was originally used in agricultural experiments, where various plots in a field were randomly assigned to different experimental conditions. Later the idea was adopted for use in other experimental studies and clinical trials.

Randomization is mainly used to avoid possible bias in selection and allocation of experimental units. In addition, randomization is used to prevent the existence of systematic differences among groups that are not due to the intervention effect being compared. Randomization produces balance among groups on known or unknown risk factors and so provides comparability of groups. Randomization is a method to ensure that the intervention and control groups are as similar as possible with respect to all baseline characteristics at the beginning of the study. Randomization ensures the validity of most statistical tests and is thus a fundamental principle of experimental design.

3.1. Complete or Simple Randomization

The simplest randomized scheme is *complete* or *simple randomization*, where experimental units are assigned to a treatment according to a specific constant probability. For example, an investigator assigns each experimental unit into one of two groups, say **A** or **B**, with a specific probability. In the special case of two groups and equal probability of 0.5 to each, complete randomization is represented by a fair coin toss or, equivalently, sampling with replacement of one of two marked cards (**A** and **B**) from a box. More practically, a random number table or random number generating algorithm may be used. This scheme may be applied to more than two groups or applied to two groups with unequal probability. One example of a randomization scheme with an unequal probability is a multicenter chronic hepatitis study that investigated the effectiveness of lamivudine antiviral therapy for patients with chronic hepatitis B and advanced liver disease. In this study, two thirds of patients were randomly assigned to lamivudine treatment group and one third to placebo group *(11)*.

The allocation does not depend on the experimental unit's prognostic factors or on the previous subject's treatment assignment. The distinguishing feature of complete randomization is that the allocation is statistically independent among experimental units, thus there is no special complication in analysis. Complete randomization is simple, easy to implement, and eliminates the possibility of selection bias. However, it has the disadvantage that the number of experimental units in groups may be substantially imbalanced. In addition, the prognostic profile between groups may differ substantially and thus their results perceived as less reliable. Studies exhibiting imbalances may be inefficient and are of a greater concern for small- and medium-sized sample studies.

3.2. Block Randomization

Alternative allocation methods that force balance between treatment groups are called *block randomization,* also known as a *permuted block design*. A block is a group of similar experimental units or characteristics. Blocks can be of any size but are a multiple of the number of the study treatments. Block randomization is used to keep the numbers of experimental units in the different groups closely balanced at all times. For example, if we consider experimental units in blocks of four at a time, there are six ways in which we can allocate treatments so that two experimental units get **A** and two get **B**: **AABB**, **ABAB**, **ABBA**, **BBAA**, **BABA**, **BAAB**. One of these arrangements is selected at random, and the next four experimental units are assigned accordingly. This process is repeated as many times as needed. The method can also be carried out by allocating treatments within each block in the desired proportions such as 3 : 1 or 2 : 1. For each block, a random order of the treatments is used, and this is done independently for each block.

Block randomization provides balance in the numbers in each group throughout the study. Implementation is not complicated, though not as easy as complete randomization. Block randomization also increases the comparability of the treatment groups particularly when experimental unit characteristics may change over time. A disadvantage of block randomization is that treatment assignment is more predictable than complete randomization. Block randomization provides balance in the numbers in each group but does not guarantee balance for important prognostic factors. This can be achieved using stratified randomization.

3.3. Stratified Randomization

Stratified randomization (or *permuted blocks within strata*) is a combined randomization by defining strata based on prognostic factors and performing permuted block randomization within each stratum. The purpose of stratified randomization is to ensure the treatment groups are balanced on important prognostic factors, for example age or disease status/condition, and to provide increased efficiency and power in the analysis. This method requires a separate block randomization list for each stratum. For example, in a study to compare two alternative treatments for heart disease, it would be important to stratify by gender. Two separate lists of random allocation can be generated for males and females. Stratified randomization can be extended to two or more stratifying variables. In this way, the effect of nuisance factors that contribute systematic variation to the differences among experimental units can be eliminated. However, such prognostic factors must be measured prior to randomization. The number of strata depends on sample size. In general, stratifying two or at most three prognostic factors is recommended, unless the sample size is very large.

The problems caused by many strata have motivated alternative adaptive randomization techniques such as the biased coin method ***(12)***, the urn method ***(13,14)***, adaptive stratification ***(15)***, and response randomization.

4. Blinding/Masking

Blinding or *masking* is the purposeful concealment of the treatment assignment (and other relevant information) of the experimental units. Sometimes blinding refers to any attempt to make the various participants in a study unaware of which treatment experimental units have been offered, so that the knowledge cannot cause them to act differently thereby affecting the internal validity of the study. Blinding is needed most when reporting of the outcomes under consideration can be influenced easily by knowledge of treatment, such as a patient's pain or nausea in self-reporting or self-assessment studies.

Blinding can take place at three levels: study units, investigator, and data monitor/analysts. Blinding can be classified into four types depending on the level: unblinded or open-label, single blind, double blind, and triple blind. An unblinded or open-label study is a study in which no blinding is used. The investigator and the experimental units know which treatment the experimental unit receives. When the experimental units are aware of which treatments they receive, they may react in favor of the treatment they receive, which can lead to a serious bias. The investigator may be tempted to look more carefully for outcomes or diagnose the outcome more frequently in a certain group. If the primary study outcome is objective, such as survival, open-label studies are less likely to be biased. However, where the primary outcome is subjective, such as diagnostic measurements, questionnaire scales, or a subject's self-reported opinions, open-label studies are highly susceptible to bias. In general, open-label studies are not considered as adequate as well-controlled blinded studies for providing substantial evidence of treatment effect. However, it is not always possible to blind treatment administration in experimental studies involving treatments requiring different modes of administration, such as a medical device or a form of surgical experiments. In addition, blind treatment administration may not be recommended in experiments where knowledge of treatment assignment is part of the effect being tested, such as dietary or smoking prevention intervention. An open-label study is simple to design and conduct and is less expensive than other designs. The disadvantage is that it has a probable bias introduced by the study units and study evaluators. Thus it is difficult to assess the true effect of the group difference. In the worst case, the study units may decide they don't like the treatment and switch over to the other treatment on their own.

In a single-blinded study, only investigators are aware of which treatment each experimental unit is receiving. The investigator might affect the administration of non–study treatment, data collection, and outcome measurement/assessment. A single-blinded study is simple to carry out and, at times, the investigator's knowledge can help in making judgments for experimental unit care. However, the disadvantage is that it has a potential bias introduced by the investigator, who has the tendency of giving more intensive effort to a particular treatment group.

In a double-blinded study, neither the experimental units nor the investigators are aware of which treatment each experimental unit is receiving. A third party, often a committee, maintains the blindness and monitors data for toxicity and benefit. In a double-blinded study, the risk of bias is reduced, as the investigator's actions to the study groups are equal. This is the recommended design for a clinical trial. However, the disadvantages are that it is more complex, more expensive, and more difficult to administer than other studies. In addition, a double-blinded study needs an effective data-monitoring scheme and an emergency unblinding procedure.

In a triple-blinded study, even the data monitoring committee is not aware of the identity of the groups. The theory is that the committee will evaluate the study results more objectively. The disadvantages are similar to those of a double-blinded study. Additionally, with the added complexity of a triple-blinded study, the decision process in any emergency situation is slow. Sometimes triple blinding refers to not letting the analyst know which treatment a group receives, so that analysis is blinded.

As a part of blinding in drug intervention studies, matching and coding of drugs are used. Matching of drugs means both active and placebo drugs are physically identical in size, shape, taste, color, sheen, and texture. Coding of drugs means the labeling of bottles or vials does not disclose the contents of the drugs. Usually this is done by means of assigning a number to the active and placebo drugs.

5. Biases

An experimental study is conducted to draw inferences about what happens in the study sample and to extend the findings to the population. When investigators design and implement a study, they worry about both random and systematic errors, which might weaken the study inference. Random error is unexplained variability and cannot be attributed to a specific cause. It can be reduced by increasing the number of observations or by training the evaluators to report or score the data in the same way. Systematic error is a deviation that is not a consequence of chance alone. For example, hospital A in a multicenter HIV study may use different assessment criteria of disease progression to AIDS. Systematic error cannot be reduced by simply increasing the number of observations. Only with a good study design and with training of the study team to promote standardization of procedures, such as a laboratory test or X-ray assessment, can systematic error be reduced.

Bias is any deviation from the true value. Minert ***(16)*** defines bias as a preconceived personal preference or inclination that influences the way in which a measurement, analysis, assessment, or procedure is performed or reported. Bias is considered a systematic error that can enter a study at any stage. Thus, bias refers to distortion in the selection of experimental units, collection of data, determination of end points, and final analyses. *Selection bias* results from using an unrepresentative sample of the population from which it comes. In comparative experimental studies, it means that individuals with certain characteristics are more likely to receive particular treatments. *Assessment bias* is caused when study participants (experimental unit or research team) are aware of which treatment experimental units have been offered. *Information bias* results from the information experimental units provide to investigators, which is heavily tainted by their own belief and values. *Observer bias* is caused when the objectivity of the investigators varies. Assessment, information, and observer biases can be reduced by adopting blinding in the study, and selection bias can be reduced by using randomization. It is important in experiments to recognize that although bias can never be completely controlled, the effort to limit biases increases both validity of the study and the ability to detect true differences among treatments.

6. Analyses

Once data have been collected, they will be subjected to a statistical analysis in concordance with the experimental design and its associated model. In simple designs, if the outcome (or end point) measurement is a dichotomous variable, such as yes/no or success/failure, then the proportions between groups can be compared using a chi-square test. If the outcome measurement is continuous, then a t-test can be used to compare the mean difference between two independent groups. When the outcome measurement is continuous but is not normally distributed, nonparametric methods can be used. When a study involves subject recruitment over an extended period of time and the duration of follow-up time through a common calendar time point, survival analysis should be used. The methods of statistical analysis mentioned here will be discussed in later chapters of this book. In this section, however, we will consider the more fundamental issue: which experimental units should be included in the data analysis?

6.1. Compliance

Before statistical methods are applied to a data set, investigators should assess whether the compliance within treatment groups is similar before comparing the outcomes in the groups. When a substantial number of experimental units drop out from the study, are lost to follow-up, do not receive the study intervention, or do not adhere to the study protocol, the investigators should consider the effects of such deviation on the analysis. Here we consider two kinds of dropout: *study dropout* and *treatment dropout*. Study dropout occurs when experimental units do not complete follow-up due to withdrawal or refusal to participate in the study. For a univariate outcome, the outcome measurement might not be obtained from such experimental units, thus they cannot be used in the analysis without some assumptions on missing data (see **Chapter 17**). Treatment dropout is noncompliance or nonadherence by experimental units. This occurs when experimental units do not receive the prescribed study regimen. There are different forms of noncompliance. Experimental units who stop taking or never received the assigned treatment represent one kind of noncompliance. For example, an experimental unit assigned a 6-week course of a twice-daily hypertension drug stops taking the drug after 1 week but remains in the study through the required follow-up time. Another example of noncompliance is departure from the prescribed schedule or dosage or a switch to another treatment. Noncompliance is also used broadly to refer to any instance where the actual treatment received differs from the intended treatment for any reason. Dropout or noncompliance is less serious for many basic science studies using animals, cell lines, or nonliving experimental units than for clinical trials on patients.

There are many ways to measure compliance, including patient and care provider reports, pill counts, blood/urine tests, and electronic monitoring. For example, compliance could be defined as the number of pills taken or the percentage of prescribed pills taken while in the study. Even the most carefully monitored experimental study may fail to achieve perfect compliance. The manner of assessment and the definition of compliance/noncompliance depends on the study objective, study design, type of treatments, expected extent of noncompliance, and planned analysis. Noncompliant patients still provide information about the study outcome. However, poor adherence threatens the validity of the study findings and can diminish the power of the study to detect a difference among the effects of treatment. In some cases, it may cause stopping the trial early. Stratifying analysis by compliance causes serious bias and is rarely performed.

6.2. Intention-to-Treat Analysis

Treatment comparisons must be based on analyses that are consistent with the study design used to generate them. In the case of randomized experimental studies, the analyses of the outcomes of interest must be by assigned treatment. This means that the outcomes are used to judge an experimental unit's assigned treatment regardless of whether or not it followed the assigned treatment when the outcomes are measured. This is called *intention-to-treat (ITT) analysis*.

Intention-to-treat analysis includes all randomized patients in the groups to which they were randomly assigned, regardless of their adherence with the entry criteria, the treatment they actually received, and subsequent withdrawal from treatment ***(17)***. Noncompliant experimental units are included in the analysis as if they had finished the study in compliance with the original treatment assignment. The principle of intention-to-treat is to compare groups as randomized, also known as *as-randomized.* According to this principle, each analysis treatment group should contain all experimental units and only those experimental units randomized to that group, with no postrandomization exclusions. This approach may seem to be illogical at first, as the outcome of a noncompliant experimental unit is counted as a success or failure of a treatment. However, ITT analysis is assured to be unbiased, and it provides valid estimates and tests for the effect on outcome of the assigned treatment in a randomized experimental study.

ITT analysis estimates a parameter of primary interest, namely, treatment *effectiveness*, which is the population effect of prescribing one treatment versus another. The disadvantage of ITT analysis is that noncompliant experimental units will nevertheless be included in the estimates of the effects of that intervention. Thus, ITT analysis cannot estimate treatment *efficacy*, which is the biological effect of the treatment if taken as prescribed. Substantial dropout or nonadherence will also cause the ITT analysis to underestimate the magnitude of the effects of that intervention ***(18)***.

6.3. As-Received Analysis and Per-Protocol Analysis

An alternative to the ITT approach is *as-received* analysis and *per-protocol* (PP) analysis. As-received analysis is based on the particular treatment actually received. Per-protocol or *adherence-only* analysis uses experimental units who remain on assigned treatment and adhere to protocol. The primary analysis of a randomized experimental study should compare experimental units in their randomly assigned treatment groups. However, when a substantial number of experimental units are noncompliant, it is tempting to consider treatment comparisons using only those experimental units with treatment as actually received rather than as prescribed. There are several arguments against this *as-received* approach, such as the prognostic balance by randomization is likely to be disturbed. The sample size will be reduced, or the validity of statistical procedure will be undermined. Results of analysis by treatment as received may suffer a bias introduced by compliance or a factor often related to outcome independent of the treatment received. One example is the Coronary Drug Project, which tested the efficacy of the cholesterol-lowering drug clofibrate on mortality ***(19)***. In this randomized, double-blinded, placebo-controlled study, patients who adhered to the clofibrate regimen had a benefit, whereas those who did not adhere had a death rate similar to the placebo group. The extent and nature of this bias may be related to the definition of compliance in per-protocol analysis. The PP analysis includes all patients who completed the full course of assigned treatment and who had no major protocol violation. Both perprotocol and as-received analyses should be interpreted with caution.

In general, the results of studies are often evaluated with both ITT and per-protocol analyses. Because not all experimental units taking the experimental treatment in the general population will take it for the course as prescribed, the ITT tends to give an estimate of the overall effect that the experimental treatment will have on the population. The PP results estimate the overall effect of the full course of experimental treatment. The results from both ITT and PP analyses are important and should be considered. If both analyses produce similar results, the conclusion of the study is certain and confident. If they differ, results of the ITT analysis are preferred because they preserve the value of randomization.

6.4. Subgroup Analysis

There is often interest in identifying which experimental units do well on a treatment and which do poorly. To answer questions like this, we analyze the data separately for subsets of the data, which is called *subgroup analysis*. In experimental studies, subgroup analyses are defined as comparisons among randomized groups in a subset of the study. The main aim of subgroup analysis is to study consistency of treatment effects among different groups of experimental units and to identify large differences between subgroups. Subgroup analysis may be possible using data from all or some subset of a study. Because it uses a smaller sample than the entire study and so may not be of sufficient power to detect important differences, subgroup analyses can lead to wrong conclusions and are easy to misuse. Thus we should avoid claiming that a treatment is effective (ineffective) in the subgroup population when the differences were observed (unobserved) in a subgroup analysis. The results from subgroup analyses can be useful and may offer hypotheses for subsequent evaluation and future research. As secondary but not confirmatory analysis, subgroup analysis should also be planned in advance before treatment is started. Appropriate secondary analyses in any randomized experimental study should also be based on the intention to treatment principle.

6.5. Exploratory Analyses

Experimental studies should be designed primarily to get precise answers to the main research questions written in the study protocol. However, there is considerable interest in trying to learn something about the underlying biology of the disease during the course of a study beyond the planned study objectives. This is called *exploratory analysis*, which serves to generate and not to prove hypotheses. Usually, data are collected on experimental units in an attempt to understand which variables are useful in predicting subject outcome for use in subsequent studies and in explaining the results of a given study. The general questions in exploratory analyses include: "What are the important prognostic baseline factors?" "How can they be used in the future studies?" "Are there any specific subsets of experimental units that have different outcomes?" Thus, the exploratory analysis can be used as a pilot study to obtain valuable information on optimal experimental conditions, participant sources, and recruitment. From the pilot study, the preliminary result will indicate whether a full-scale study is practical. One should keep in mind that any statistically significant results from any exploratory analyses should be interpreted with caution. Various statistical methods for exploratory analyses and extrapolation will be introduced in the later chapters of this book.

7. Study Interpretation

Interpretation of the results is the final phase of an experimental study. In most cases, the statistical analysis of an experimental study is straightforward, using relatively simple methods such as *t*-test or chi-square tests, and interpretation is also straightforward. However, inferences from a sample to a population depend on the assumption that the experimental units are representative of the population. Because most studies use inclusion/exclusion criterion to select eligible units, extrapolation of results to other units may not be guaranteed. For example, when a study is conducted on middle-aged men, it is not reasonable to assume that the results apply to women or to young or very old men. Because it is quite possible that different groups would respond differently, wider applicability or generalization of the study results should be carefully considered. Before an investigator extrapolates the results from a study to the population in general, there are two aspects that require particular attention. First, the samples studied should be representative of the population of interest. Second, groups being compared should be as alike as possible apart from the features of direct interest. If a study finds a statistically significant difference, the investigators should provide the limitation of the findings and the degree of completeness of the data to evaluate a study. If a study finds no statistically significant difference, then the investigators should provide their understanding of why no difference was found. There might be several possible explanations, such as using inappropriate dosage, a too-small sample size, too many dropouts, lack of adherence, or inadequate outcome measurement. It is desirable to conduct subgroup analysis, which might provide clues about variation in the effectiveness of a treatment for different groups of experimental units. Findings related to the secondary questions may be interesting, but they should be put in the proper perspective.

In summary, we have presented many basic ideas of study design and conduct. Most of these ideas will be expanded upon in later chapters.

References

1. Grodstein, F., and Stampfer, M. (1995) The epidemiology of coronary heart disease and estrogen replacement in menopausal women. *Prog. Cardiovasc. Dis.* **38**, 199–210.
2. Manson, J. E., Hsia, J., Johnson, K. C., Rossouw, J. E., Assaf, A. R., Lasser, N. L., Trevisan, M., Black, H. R., Heckbert, S. R., Detrano, R., Strickland, O. L., Wong, N. D., Crouse, J. R., Stein, E., and Cushman, M. (2003) Estrogen plus progestin and the risk of coronary heart disease. *N. Engl. J. Med.* **349**, 523–534.
3. Women's Health Initiative Study. (2002) Risks and benefits of estrogen plus progestin in healthy postmenopausal women. *JAMA* **288**, 321–333.
4. Lipid Research Clinics Program. (1984) The lipids research clinics coronary primary prevention trials results. I. Reduction in incidence of coronary heart disease. *JAMA* **251**, 351–364.
5. Poole, C. V., Carlton, D., Bimbo, L., and Allon, M. (2004) Treatment of catheter-related bacteraemia with an antibiotic lock protocol: effect of bacterial pathogen. *Nephrol. Dial. Transplant* **19**, 1237–1244.
6. Graff-Lonnevig, V., and Browaldh, L. (1990) Twelve hours bronchodilating effect of inhaled formoteral in children with asthma: a double-blind cross over study versus salbutamol. *Crit. Exp. Allergy* **20**, 429–432.
7. Stampfer, M. J., Buring, J. E., Wilett, W. Rosner, B., Eberleiner, K., and Hennekens, C. H. (1985) The 2×2 factorial design: its application to a randomized trial of aspirin and carotene in US physicians. *Stat. Med.* **4**, 111–116.
8. Women's Health Initiative Study Protocol. (1994) Bethesda, National Institutes of Health.
9. Vijayaraghavan K., Radhaiah, G., Prakasam, B. S., Sarma, K. V. R., and Reddy, V. (1990) Effect of massive dose vitamin A on morbidity and mortality in Indian children. *Lancet* **336**, 1342–1345.
10. Peterson, A. V., Mann, S. L., Kealey, K. A., and Marek, P. M. (2000) Experimental design and methods for school-based randomized trials: experience from the Hutchinson Smoking Prevention Project (HSPP). *Control. Clin. Trials* **21**, 144–165.
11. Liaw, Y., Sung, J., Chow, W. C., Farrell, G., Lee, C., Yuen, H., Tanwandee, T., Tao, O., Shue, K., Keene, O. N., Dixon, J. S., Gray, D. F., and Sabbat, J. (2004) Lamivudine for patients with chronic hepatitis B and advanced liver disease. *N. Engl. J. Med.* **351**, 1521–1531.
12. Efron, B. (1971) Forcing a sequential experiment to be balanced. *Biometrika* **58**, 403–417.
13. Wei, L. J. (1977) A class of designs for sequential clinical trials. *J. Am. Stat. Assoc.* **72**, 382–386.

14. Wei, L. J. (1978) An application of an urn model to the design of sequential controlled clinical trials. *J. Am. Stat. Assoc.* **73**, 559–563.
15. Pocock, S. J., and Simon, R. (1975) Sequential treatment assignment with balancing for prognostic factors in the controlled clinical trials. *Biometrics* **35**, 102–115.
16. Minert, C. L. (1986) *Clinical Trials*. New York, Oxford University Press.
17. Fisher, C. J., Dixon, D. O., Herson, J., Frankowski R. F., Hearron, M. S., and Peace, K. E. (1990) Intend-to-treat in clinical trials. In: Peace, K. E. *Statistical Issues in Drug Research and Development*. New York, Marcel Dekker.
18. Sheiner, L. B., and Rdubin, D. B. (1995) Intention-to-treat analysis and the goals of clinical trials. *Clin. Pharmacol. Ther.* **57**, 6–15.
19. Coronary Drug Project Research Group. (1980) Influence of adherence to treatment and response of cholesterol on mortality in the Coronary Drug Project. *N. Engl. J. Med.* **303**, 1038–1041.

2

Observational Study Design

Raymond G. Hoffmann and Hyun Ja Lim

Summary

Much can be learned about a process by observing changes over time or by comparing two different processes under different conditions. This chapter introduces the major types of observational study designs: the longitudinal or cohort study, the comparative or case-control study, and some of their variants. It also includes examples of the key measures of relationship between factor and outcome in observational studies, the relative risk and the odds ratio. The similarity of the two measures for low incidence outcomes is illustrated, as is the use of attributable risk to assess how much of a binary outcome is due to a single factor.

Key Words: Case-control study; cohort study; cross-sectional study; matched studies; odds ratio; propensity score; prospective cohort; recall bias; retrospective cohort.

1. Introduction

Observational studies are an alternative to experimental studies. An observational study is sometimes termed a *natural experiment*. Instead of being randomized into one group or another to ensure statistical balance, subjects are classified into groups either by the *presence of an exposure*, which is called a *cohort study*, or the *presence or absence of a disease*, which is called a *case-control study*. A subject could be a cell, a bacteria, a specific cell line, a pond of environmental interest, a rat, or a person.

Some examples of the different types of groupings that are used in observational studies are

- Cohort studies (retrospective): having been exposed to asbestos in the workplace or not (or at different levels of asbestos exposure) with lung cancer as the outcome; growing up in an area with high fluoride water compared with growing up in an

From: *Methods in Molecular Biology, vol. 404: Topics in Biostatistics*
Edited by: W. T. Ambrosius © Humana Press Inc., Totowa, NJ

area without much fluoride in the water with dental caries as an outcome; comparing the outcomes of two different treatments for acne based on a registry of clinic patients in the past 5 years; or evaluating the effect of childhood obesity on diabetes using records from 10 years of a pediatric practice.
- Cohort studies (prospective): choosing to smoke or not to smoke with the outcome being the development of lung cancer, emphysema, or heart disease; being part of an ecosystem that is high in volatile organic compounds (VOCs) compared with an ecosystem that is low in VOCs with the outcome being survival of a flora or fauna species; determining whether the apple- or pear-shaped body type (phenotype) leads to an increase in the development of heart disease, hypertension, or diabetes.
- Case-control studies: being part of a group that develops a disease such as lung cancer compared with members of the group that do not develop the disease; comparing HIV polymerase chain inhibitor–resistant HIV to nonresistant HIV in order to identify characteristics differentiating the two groups; comparing patients who have a new highly virulent infectious disease of unknown etiology to subjects without the disease but living in the same neighborhood to identify factors associated with the etiology or cause of the disease; or comparing the results of a microarray analysis applied to cells from cancer patients and noncancer patients or a microarray analysis applied to normal cells and cancer cells from the same subjects.
- Case-control (genetic association) studies: using cases that have high blood levels of methotrexate compared with controls that have low blood levels of methotrexate to identify which alleles of CYP2E1, an enzyme that affects the rate of metabolism of various compounds, relate to this phenotype; comparing severe chronic asthma (cases) to normal children of the same age, gender, and ethnicity to identify genes (or markers) that are associated with the disease; taking blood samples from cases and controls and either using a candidate gene approach or doing a genome-wide scan ***(1,2)***.

The term *subjects* could represent persons, animals, bacteria, or any other kind of experimental unit.

Because the groups in an observational study are *not randomized*, they are not necessarily equivalent for many other factors that in fact may be the real cause of the difference or may be promoters or antagonists of the effect being studied. For example, in a study comparing lung cancer patients and patients without lung cancer, the patients may be representative of different lifestyles so that many risk factors appear to differ between the groups. For example, some patients could belong to a different socioeconomic class that is exposed to some occupational risk factor that differs from the noncancer group but has no relationship to the disease process. A risk factor like this is a potential *confounder* of the relationship. A confounder is a variable that hides either a relationship or a variable that makes a relationship appear strong when it is not.

Risk in an observational study is often stated in terms of the *relative risk* of developing a characteristic based on exposure. If 3 out of 10 mammalian cell cultures exposed to ultraviolet (UV) radiation developed chromosomal abnormalities while 9 out of 10 mammalian cell cultures developed chromosomal abnormalities when exposed to UV radiation plus a common NSAID (nonsteroidal anti-inflammatory drug), then the relative risk (RR) of developing chromosomal abnormalities due to NSAID exposure (in the UV test system) is

$$\mathrm{RR} = \frac{9/10}{3/10} = 3.0.$$

In a study where entities are followed over time, the relative risk is expressed in terms of the time period. For example, suppose 5% of sunbathers develop skin lesions in a year if they use a sunblock of SPF 30 or more, and 10% of sunbathers develop skin lesions in a year if they use a sunblock of only SPF 5. The relative risk of developing skin lesions in a year for using a low-value SPF sunscreen is

$$\mathrm{RR} = \frac{10\% \text{ per year}}{5\% \text{ per year}} = 2.0.$$

Another way of expressing this is that the *protective* effect of using a high-number SPF sunscreen versus a low-number SPF sunscreen is

$$\mathrm{RR} = \frac{5\% \text{ per year}}{10\% \text{ per year}} = 0.5,$$

and sunbathers are only half as likely to develop skin lesions. In **Section 6** of this chapter, we will discuss relative risk in more detail, as well as other measures of risk such as the odds ratio.

2. Cohort Studies

A cohort study is one where two or more groups of subjects are followed over time to see if they develop some disease or if some event occurs. In an exposure study (occupational or environmental), the effect of exposure on multiple outcomes—death, cancer, heart disease—can be observed. There are two types of cohort studies: prospective and retrospective.

2.1. Prospective Cohort Studies

Prospective cohort studies (also known as follow-up studies) follow groups of cells, animals, or patients with different exposures until some point in time where something happens or the study is terminated *(3)*. Usually the outcomes of interest (e.g., death) are specified at the start of the study.

Example: A Prospective Cohort Study

In the British Physician study, a prospective study of smoking, 34,439 male British doctors were invited to participate in a study on the effects of smoking *(5)*. Initially, there were two groups, smokers and nonsmokers. Eventually, a third group, those who quit smoking, was followed for 10 years, then 20 years *(6)*, and recently the 50-year follow-up was reported *(7)*. They were followed to observe what diseases would develop related to smoking status. The risk of lung cancer for smokers was 2.49/1000, whereas the risk of lung cancer for nonsmokers was 0.17/1000. Thus the relative risk of lung cancer for smokers over a 50-year period is

$$\text{RR} = \frac{2.49/1000}{0.17/1000} = 14.7.$$

In the same study, the risk of dying from ischemic heart disease (IHD) in smokers was 10.1/1000, and the risk of IHD in nonsmokers was 6.49/1000, giving an RR for IHD in smokers versus nonsmokers of

$$\text{RR} = \frac{10.1/1000}{64.9/1000} = 1.56.$$

The rarity of lung cancer deaths is the reason that smoking has such an effect on lung cancer. Indeed, we can quantitate how much of the lung cancer mortality is due to smoking by examining the difference in the risk in the smokers. This is called attributable risk (AR) and is a measure of how much of the condition, problem or disease is due to the risk factor.

$$\begin{aligned}\text{AR} &= \frac{\text{Lung cancer mortality due to smoking}}{\text{All lung cancer mortality}} \\ &= \frac{2.49/1000 - 0.17/1000}{2.49/1000} \times 100\% = 98.6\%.\end{aligned}$$

The same calculations give an attributable risk of 35.7% of the IHD mortality in the smokers due to smoking during the 50 years of follow-up.

2.2. Retrospective Cohort Studies

Retrospective cohort studies use historical data to make comparisons based on risk factors or exposures that occurred prior to the event. Historical records of snowfall in different continents can be used to study the effects of global warming. Historical records of bacterial prevalence in different hospitals can be used to study the effects of frequent antibiotic use. Patient records can be used to compare the effect of different treatments. Retrospective cohort designs may also use historical data from prospective cohort studies. For example, the Framingham Heart Study ***(8,9)*** examined the effects of different partitions of the risk factors. Retrospective cohort analyses can be facilitated if the initial design of the cohort study recruits not just 1000 smokers and 1000 nonsmokers but 2000 subjects some of whom will be smokers and some of whom will be nonsmokers. Alternatively, the nonsmoking group can be studied by itself in retrospective cohort studies to examine the effect of other risk factors independent of smoking.

2.3. Analysis of Cohort Studies

Cohort studies are not subject to recall bias (defined as differential recollection of exposure because of the presence of the condition or disease) because the outcome occurs after entry to the study. However, in retrospective cohort studies, missing values for a factor that was not originally one of the primary risk factors can be a severe problem. The term *missing completely at random* means that the probability of an observation being missing does not depend on the observed or unobserved measurements. This type of missing value only affects the magnitude of the effect that the study can detect. Other types of missing values can affect the validity of the estimated risk. For example, if subjects die from a treatment effect that is not one of the primary outcomes (e.g., being hit by a car because of disorientation caused by the treatment), disease-specific mortality will be significantly biased, but all causes of mortality will not be biased (see also **Chapter 17**).

The presence of differences between the groups when the study was started is a problem with either type of cohort study. A study may show that exercise was a protective risk factor against heart disease, but it may be that the entire lifestyle is protective with regular exercise the best indicator for that protective lifestyle. Thus, when analyses of cohort data are performed, methods that group risk factors into similar classes, such as propensity scores, may be used ***(10,11)***. Differences between groups at baseline can be adjusted for by stratification (i.e., putting like hospitals together for studies of bacterial flora or putting experiments performed by the same lab technician together when studying the effect of immunoglobulins on longitudinal measures of inflammation). Regression adjustment is another method for accounting for differences between groups and is discussed in **Chapter 9**.

3. Case-Control Studies

A case-control study compares the characteristics between two groups, usually one that has a condition or disease compared with one that does not have the condition or the disease ***(12)***. These characteristics are termed *risk* *factors* for the development of the disease. Some of the risk factors will be related to the development of the disease, some of them will be due to the presence of the disease but not involved in the development of the disease, and some of them will be due to chance. Statistical analysis is used to assess the probability or odds of the risk factor being related to the disease or condition.

Usually, external evidence for a mechanism of the development of the disease is also used to discriminate risk factors for the development of the disease from markers of the disease presence ***(13)***. Often, a case-control study will be followed by a cohort study to test whether the disease or condition actually develops in subjects with the risk factor.

3.1. Odds Ratios

Because the number of cases and the number of controls is predetermined in a case-control study, the relative risk cannot be used ***(3)***. An alternative way of measuring risk is in terms of the *odds ratio*. The "odds of a disease given a risk factor" is the probability of having the disease with the factor divided by the probability of not having the disease with the factor present. Thus, if the probability is 0.20 or 1 in 5, the odds is 0.2/(1 – 0.2) = 0.2/0.8 = 0.25. It is also described as 1:4 (read as 1 *to* 4 and interpreted as 1 event will occur for every 4 times it does not occur). This is the same type of odds that are given at a racetrack or for a sports team. The *odds ratio* is the ratio of the odds of the disease with the risk factor present divided by the odds of the disease with the risk factor absent. It is used in case-control studies because conditional probability arguments ***(12)*** can be used to show the computation of the odds ratio as the odds of the risk factor. For example, smoking, in lung cancer patients, divided by the odds of smoking in the controls is equivalent to the odds ratio for the disease given the risk factor.

For example, if the proportion of smokers in lung cancer patients is 1 in 10 and the proportion of smokers in controls is 1 in 100, the odds ratio (OR) for lung cancer given smoking is

$$\text{OR} = \frac{\text{Odds of smokers in lung cancer}}{\text{Odds of smokers in ontrols}} = \frac{\left(\frac{1}{10}\right)\Big/\left(\frac{9}{10}\right)}{\left(\frac{1}{100}\right)\Big/\left(\frac{99}{100}\right)}$$

$$= 11.0 = \text{OR of lung cancer in smokers.}$$

If the condition or disease is rare, the odds ratio and the relative risk are almost the same (see **Section 6.2**).

3.2. Choice of Controls

The key design issue in a case-control study is the *choice of the controls*. To obtain an unbiased (correct or appropriate) estimate of the risk, the controls must be comparable with the cases for factors that are not related to the disease (or outcome). For example, in the study of endometrial cancer and estrogens ***(14)***, the controls were women from the same clinic, and some of them had bleeding problems related to exposure to estrogens. Thus the odds ratio was estimated as 1.7. When the controls were not chosen from the same clinic, the odds ratio was estimated as 11.98. Usually, a *community control* is necessary in addition to clinic controls to account for common factors in the controls and cases that may also be associated with the disease. One possibility for a community control is a friend of the same gender, who will most likely be similarly aged and of similar socioeconomic status. Another possibility for a community control is to use controls from the same block, the same census tract, or from within a 1-mile radius of the control. However, if the risk factor is environmental, choosing someone within a 1-mile radius may mean that the control is exposed to the same toxic substance. These same considerations must be taken into account if the study is of the number of mutations observed in a cell selected from ponds near an environmental source compared with ponds that are not near the environmental source. The control ponds must be comparable with the "case" ponds in terms of depth, surface area, and so forth.

Because many disease conditions are rare, one design option for a case-control study is to use 2 or 3 times as many controls as cases to compensate for the shortage of cases. When the disease condition is not rare, a design option to improve the sensitivity of the study is to use 2 or 3 times as many cases as controls so that the effect of a range of exposure to the risk factor in the cases can be compared with the controls.

3.3. Case-Control Genetic Association Studies

The case-control strategy has also been adapted to genetic studies of association. The goal is to either identify the heritability of a trait or to identify the gene or the marker of a gene that is associated with the phenotypic trait. Usually, "cases" represent the presence of some phenotype (e.g., hypertension, curly fruit fly wings, differential pharmacokinetic and pharmacodynamic response, or polymerase inhibitor resistance in the HIV retrovirus). In a genetic association study, the controls are chosen not to have that phenotype. However, the controls again need to be similar to the cases in general genetic background; otherwise false-positive genes will be identified because of admixture (differences between the groups that are unrelated to the outcome of interest). More on genetic association studies will be found in **Chapter 21**.

3.4. Matching and Case-Control Studies

In some studies, the controls are *matched* to the cases to eliminate confounders (e.g., age or gender) that can affect the presence of the disease but are not directly related to the development of the disease. For example, each control may be chosen to be within 2 years of age of the corresponding case. Evaluations of the effects of an intervention on two different cell lines would be done by the same lab technician or with the same batch of chemicals. Another type of matching often found in genetic studies is to match siblings that are specifically chosen to be either affected by the disease or unaffected by the disease. Environmental studies often match on nonenvironmental factors that may predispose to the disease. For example, one might match the cigarette smoking status of parents in a case-control study of leukemia due to exposure to power line radiation.

In another example of a case-control study using matching, researchers examined pemphigus foliaceus, which is an adolescent/early adult autoimmune disease that has both genetic susceptibility and suspected environmental risks (possible insect carriers, etc.). Both family controls and community controls were used in the study ***(15)***. The disease was studied in a remote Indian community in Brazil with one to four age-matched family controls over age 18 matched and one to five age-matched community controls. The goal was to be able to identify differences within the "house," as well as differences in the location of the house, exposures by occupation, pets, different types of insects, and several other factors. Family controls were required to be over 18 to reduce the chance that they would become cases.

One disadvantage of matching by age is that it may be difficult or impossible to find controls close enough in age to participate in the study. Sometimes a group-matching strategy is used to approximately balance age without matching one-to-one.

Another disadvantage of matching is that it is possible to *overmatch* by choosing a matching variable that is part of the causal pathway of the disease or condition. Overmatching tends to mask the relationship between the risk factor and the disease. For example, if obesity causes hypertension, which causes strokes, then matching on hypertension status would be overmatching because it would be removing part of the effect of obesity on strokes.

3.5. Biases in Case-Control Studies

In a case-control study, *selection bias* refers to the problem that people who agree to participate in a study may be different from people who do not agree to participate. Sometimes the nonparticipants can be compared with the participants in terms of gender and age to test comparability of the participants and the nonparticipants. An alternative is to use a *capture-recapture* strategy by acquiring data from a separate registry to characterize cases with the condition ***(16)***.

Unlike cohort studies, case-control studies are subject to *recall bias*. Recall bias refers to differential recall between the cases and the controls about exposure to the risk factor(s). For example, a questionnaire survey of mothers of babies with birth defects will likely recover much more detail on exposure compared with mothers of normal children. In some cases, access to the medical records will allow equal ascertainment of the exposure if the medical records are complete enough to have the information.

3.6. Cross-Sectional Studies

The *cross-sectional* study design is a unique kind of case-control study. This type of design is used if cases cannot be identified a priori or if the prevalence of the disease or condition needs to be determined. Subjects are sampled randomly and then classified according to whether or not they have the condition. From this point on, everything proceeds as if the study were a typical case-control study. Even the odds ratio can be determined from prevalence data, called the prevalence odds ratio. An example of a cross-sectional study is drawing blood samples from a population of interest and then cross-classifying them by biochemical or genetic markers after assays have been performed on the blood.

4. Outcomes

Outcomes in an observational study depend on the type of study. In a case-control study, the "disease" outcome is binary (present or absent) or ordinal (healthy, preclinical, clinical, and advanced). If the outcome is ordinal, the relationship is usually examined by comparing two states at a time.

Several types of outcomes are possible in a cohort study. As with a case-control study, the outcome may be binary. For example, the grouping factors may be smokers and nonsmokers and the outcome is the development of cardiovascular disease. If the key outcome is rare, like lung cancer, the study may need to be much larger to have sufficient disease events to allow comparisons of the two groups (see **Chapter 14 and Chapter 19**). The outcome variable(s) also usually includes the binary disease event and the time to occurrence of the disease. In this case, survival analysis statistical methods are used (see **Chapter 15**).

Outcomes in a cohort study may also be continuous. For example, a cohort study may look at the level of PSA (prostate-specific antigen) or a lung function measure. The advantage of this type of study is that changes may be detected before they are irreversible.

The outcome of a cohort study may also be a counting variable, such as the number of genetic abnormalities (breaks in the chromosomes). For example, in a study of the effect of human growth hormone (HGH) in children of very small stature compared with normal-height children, a retrospective cohort study was used with the outcome being a count of the number of chromosomal defects *(17)*.

Each type of outcome requires a different type of statistical analysis: logistic analysis or survival analysis for binary data (see **Chapter 14** and **Chapter 15**, respectively), Poisson regression for counts of the number of events *(3)*, and mixed model regression and analysis of variance (see **Chapter 11**) for continuous observations over time. When the outcomes are continuous, the effect of a discrete risk factor may be expressed as a difference in means. If the risk factor is continuous, it may be expressed as a correlation.

5. More on Odds Ratios and Relative Risks

5.1. Relative Risks

If the outcome is binary, then the probability of the event occurring is based on some risk factor being present. This is more often presented in terms of a *relative risk*: the ratio of the probability of the event with the factor present compared with (divided by) the probability of the event occurring with the factor absent. In general, the relative risk requires a time frame for the event to occur (e.g., a month, a year, 10 years). A secondary infection from someone who has a cold may only take a few days to develop, whereas the development of emphysema from cigarette smoking may take decades. Another example might be the probability of an anticancer drug achieving a 95% in vitro effective reduction of cancer cell activity. Clearly, in this case the probability of the drug being effective depends on the individual cell response.

Usually, relative risk is determined for two different levels of the risk factor. If the risk factor is continuous, the two levels must be chosen. For example, use the level of exposure to cotton dust in a cotton processing plant; the levels might be chosen to be $10\,\mu g \cdot ms/m^3$ and $200\,\mu g \cdot ms/m^3$ (a level that equals the National Institute of Occupational Safety and Health level of permissible exposure). The relative risk of a $200\,\mu g \cdot ms/m^3$ compared with a $10\,\mu g \cdot ms/m^3$ exposure, if the coefficient of the odds ratio per $\mu g \cdot ms/m^3$ is 0.00346 from a logistic regression (see **Chapter 14**), is

$$\text{RR} = \exp[0.00346 \times (200 - 10)] = 1.93.$$

If the variable were age, the choice of the two levels is often a decade apart. If the risk factor were discrete, for example, such as managers, foremen, and weavers in the cotton processing plant, the relative risk is determined pairwise. For example, if the risk of byssinosis (disease of the lungs caused by inhalation of cotton dust or dusts from other vegetable fibers) in a 5-year period is 3% for managers, 15% for foremen, and 30% for weavers, the relative risk of byssinosis of weavers to managers is 30% ÷ 3% = 10.0, for weavers to foremen is 30% ÷ 15% = 2.0, and for foremen to managers is 15% ÷ 3% = 5.0.

5.2. Odds Ratios

As discussed earlier, an alternative way of measuring risk is in terms of the odds ratio. To compare relative risk and the odds ratio, suppose the incidence of lung cancer in smokers is 1/1000 in a 5-year period and 1/10,000 in nonsmokers in the same time period. Then the relative risk is

$$\text{RR} = \frac{1/1000}{1/10{,}000} = 10.0.$$

The odds for smokers is 1/999 = (1/1000) / (999/1000) and the odds for the nonsmokers is 1/9999; thus the odds ratio is

$$\text{OR} = \frac{1/999}{1/9999} = 10.01.$$

If the disease is rare, the odds ratio is essentially the relative risk ***(12)***. If the disease is common, for example, 1/10 of children have colds compared with 1/100 adults, the relative risk is 1/10 divided by 1/100 = 10.00. However, the odds for children is 1/9 and for adults is 1/99, which gives an odds ratio of 99/9 = 11. Most diseases are rare in the population as a whole but may not be rare in a high-risk subgroup. For example, recurrence of breast cancer may be common in women who originally had breast cancer.

One advantage to using the odds ratio is the ability to calculate the odds ratio of *not* getting the disease given the risk factor. This is calculated as 1/{odds ratio of getting the disease given the risk factor}. For example, if the odds of heart disease given a good exercise program is 0.5, the odds of not getting heart disease with a good exercise program is 1/0.5 = 2.0. Using the odds ratio also gives us the ability to determine the odds ratio of getting the disease with the risk factor *not* present. This is calculated as 1/{odds ratio of getting the disease given the risk factor}. For example, if the odds ratio of heart disease in a nonsmoker is 0.4, then the OR in a smoker is 1/0.4 = 2.5. A final advantage is that the odds ratio can be computed from a case-control study even though the relative risk cannot ***(12)***. Using the odds ratio rather than the relative risk makes it easier to describe the relationship between the risk factor and the disease In addition, the coefficients of many of the regression models—logistic, Poisson, and proportional hazards regression analysis—can be directly interpreted in terms of the odds ratio (see **Chapter 14** and **Chapter 15**).

6. Conclusion

Observational studies are useful when randomization cannot be used to divide exposure into groups. Observational designs can also be used to compare factors when the groups are defined by the values of the outcome. Observational studies are not a replacement for randomized designs but allow formulation and testing of hypotheses in cases where experimental interventions are not possible. Experimental interventions are not possible when the characteristics of interest are innate parts of the experimental units or when using historical data. Each type of observational study—cohort and case control—can be used to characterize abnormal versus normal cells, mutant versus wild genes, or diseased versus nondiseased patients.

References

1. Becher, H. (2003) A case-control-family study for main effects and gene by environment interaction. *Int. J. Epidemiol.* **32**, 38–48.
2. Weinberg, C. R., and Umbach, D. M. (2000) Choosing a retrospective design to assess joint genetic and environmental contributions to risk. *Am. J. Epidemiol.* **152**, 197–203.
3. Rothman, K. J., and Greenland, S. (1998) *Modern Epidemiology*, 2nd ed. Philadelphia, Lippincott Williams & Wilkins.
4. Clayton, D., and Hillis, A. (1993) *Statistical Methods in Epidemiology*. Oxford, Oxford University Press.
5. Doll, R., and Hill, A. B. (1964) Mortality in relation to smoking: ten years observation of British doctors. *Br. Med. J.* **248**, 1399–1410.
6. Doll, R., and Peto, R. (1976) Mortality in relation to smoking: 20 years observation on male British doctors. *Br. Med. J.* **273**, 1525–1586.
7. Doll, R., Peto, R., Boreham, J., and Sutherland, I. (2004) Mortality in relation to smoking: 50 years' observations on male British doctors. *Br. Med. J.* **328**, 1519–1528.
8. Elias, M. F., Sullivan, L. M., D'Agostino, R. B., Elias, P. K., Beiser, A., Au, R., Seshadri, S., DeCarli, C., and Wolf, P. A. (2004) Framingham stroke risk profile and lowered cognitive performance. *Stroke* **35**, 404–409.
9. Saccone, N. L., Goode, E. L., and Bergen, A. W. (2003) Genetic Analysis Workshop 13: summary of analyses of alcohol and cigarette use phenotypes in the Framingham Heart Study. *Genet. Epidemiol.* **25**, S90–S97.
10. Rubin, D. B. (1997) Estimating effects from large data sets using propensity scores. *Ann. Intern. Med.* **127**(8), 757–763.
11. Heckman, J. J., and Hotz, V. J. (1989) Choosing among alternative nonexperimental methods for estimating the impact of social programs. *J. Am. Stat. Assoc.* **84**(408), 862–880.

12. Kahn, H. A., and Sempos, C. T. (1990) *Statistical Methods in Epidemiology*. Oxford, Oxford University Press.
13. Rimm, A. A., Hoffmann, R. G., Anderson, A. J., Gruchow, H. W., and Barboriak, J. J. (1983) The relationship between vasectomy and angiographically determined atherosclerosis in men. *Prev. Med.* **12**, 262.
14. Horwitz, R. I., and Feinstein, A. R. (1978) Alternative analytic methods for case-control studies of estrogens and endometrial cancer. *N. Engl. J. Med.* **299**, 1089–1094.
15. Aoki, V., Millikan, R. C., Rivitti, E. A., Hans-Filho, G., Eaton, D. P., Warren, S. J., Li, N., Hilario-Vargas, J., Hoffmann, R. G., and Diaz, L. A. (2004) Cooperative Group for Fogo Selvagem Research. Environmental risk factors in endemic pemphigus foliaceus (fogo selvagem). *J. Invest. Dermatol.* **9**(1), 34–40.
16. Lee, A. J., Seber, G. A., Holden, J. K., and Huakau, J. T. (2001) Capture-recapture, epidemilogy and list mismatches: several lists. *Biometrics* **57**, 707–713.
17. Slyper, A. H., Shadley, J. D., van Tuinen, P., Richton, S. M., Hoffmann, R. G., and Wyatt, D. T. (2000) A study of chromosomal aberrations and chromosomal fragility after recombinant growth hormone treatment. *Pediatr. Res.* **47**, 634–639.

3

Descriptive Statistics

Todd G. Nick

Summary

Statistics is defined by the Medical Subject Headings (MeSH) thesaurus as the science and art of collecting, summarizing, and analyzing data that are subject to random variation. The two broad categories of summarizing and analyzing data are referred to as descriptive and inferential statistics. This chapter considers the science and art of summarizing data where descriptive statistics and graphics are used to display data. In this chapter, we discuss the fundamentals of descriptive statistics, including describing qualitative and quantitative variables. For describing quantitative variables, measures of location and spread, for example the standard deviation, are presented along with graphical presentations. We also discuss distributions of statistics, for example the variance, as well as the use of transformations. The concepts in this chapter are useful for uncovering patterns within the data and for effectively presenting the results of a project.

Key Words: Box plot; location; logarithms; multivariate data; spread.

1. Introduction

Statistics is defined by the Medical Subject Headings (MeSH) thesaurus as the science and art of collecting, summarizing, and analyzing data that are subject to random variation. *(1)* The two broad categories of summarizing and analyzing data are referred to as *descriptive* and *inferential* statistics. This chapter deals with descriptive statistics. Inferential statistics involve the use of statistical tests such as Student's *t*-test, analysis of variance (ANOVA), and so forth. These will be covered in later chapters. When summarizing data, descriptive statistics and graphics are used to display data in a succinct manner. Displaying data is useful for uncovering patterns within the data and for effectively presenting the results of a project. In this chapter, we focus chiefly on describing one variable at a time, or univariate data. When describing relationships

From: *Methods in Molecular Biology, vol. 404: Topics in Biostatistics*
Edited by: W. T. Ambrosius © Humana Press Inc., Totowa, NJ

between two or more variables, bivariate or multivariate data displays should be considered. However, as will be shown here, the first step in describing bivariate data is to use simple descriptive statistics.

Descriptive statistics, or simply statistics, are often used on a sample to estimate characteristics of a population. Characteristics, or traits, that we measure on an individual or other source are often called *variables*, because they vary from individual to individual. Measurements obtained diverge for many reasons, including variability due to measurement error, environment, genotype, and the like. First, we describe the different types of variables.

2. Types of Variables

There are many types of variables that occur in molecular biology and medical fields. The two main categories of variables are *qualitative* and *quantitative* variables. Qualitative variables give rise to categorical data and are most often referred to as simply *categorical variables*. They are merely classifications, such as membership in one of a few groups; for example, race (black/African American, white, Asian, American Indian/Alaskan native, Native Hawaiian/other Pacific Islander) or cervical tissues (cancerous, normal). Labels, or names, are given to different diagnoses, but a magnitude cannot be given. If there is no natural ordering of the categories, then the categorical variable is *nominal*. In the case of exactly two categories (yes/no), we say the nominal variable is a *dichotomous* or a *binary* variable; for example, sex (male, female) and remission status (partial, complete). If there is a natural ordering, such as pain classified on a 4-point scale (none, mild, moderate, severe), the categorical variable is said to be *ordinal*. With ordinal variables, the magnitude is not important, but there is an order to the data. Occasionally, a variable can be classified as either nominal or ordinal. For example, genotype can be classified on a nominal scale with the three genotype categories of **AA**, **AB**, **BB**. More commonly, however, the number of **A** alleles are counted and the trait is treated as an ordinal scale, such as 0, 1, or 2 **A** alleles.

Quantitative variables can be measured according to an amount or quantity (e.g., expression levels) and are also called numeric, scaled, or metric variables. When the values only take integers or a small number of values, we say it is a *discrete,* or *discontinuous*, numeric scale. Both order and magnitude are important for discrete variables, but the values are usually restricted to integers (e.g., the count of the number of mutant alleles). When the values are not restricted to a set of specified values (e.g., weight of 175.25 lb), we say the variable is a *continuous* numeric variable. In practice, there is overlap between discrete and continuous quantitative variables, but this overlap is usually insignificant because typically the data type can be described by the same statistic (e.g., median). It is important to note that ratios can be taken only if the quantitative variable has a nonarbitrary zero point. For example, the Celsius temperature scale is a relative scale and not a ratio-scaled measure. For example, 50°C is not twice as much as 25°C. However, the Kelvin scale is an absolute scale so it would be appropriate to say 50 K is twice the heat as 25 K. Ratio-scaled variables can be either discrete or continuous. When the value of a variable is only known to occur in a certain period, the data are *censored* (e.g., timed to some event in months).

A *response* variable is a measure that is thought to be affected by the different conditions and is of primary interest. The response variable is also called a dependent or outcome variable. A response variable that leads to the completion of follow-up of an individual in a trial is an *end point* (e.g., death). An *explanatory* variable is one that is collected or actively controlled by the investigator to better understand the variation observed in the response variable *(2)*. An explanatory variable is also called a predictor, independent, or factor variable. Response and explanatory variables can either be quantitative or qualitative. Along with the scientific question of interest, the data type and number of response and explanatory variables decide which statistics and statistical tests will be appropriate *(3)*.

3. Describing Qualitative Data

Categorical variables, including ordinal variables, describe qualities or attributes. To describe these variables, counts or frequencies of individuals in each category are often displayed visually. Occasionally, the *mode*, or the most frequent observation, may be reported. The mode is useful in describing nominal data because there is no magnitude or order. However, it is usually sufficient to report the frequency and proportion for each category. If frequency is denoted by n and the total number of observations is denoted by N, then the proportion (or relative frequency) of each category is computed by $(n/N) \times 100\%$. Typically, a frequency table or distribution is given to show the values and frequency of the values of a variable. For example, to describe the race of subjects in a sample, a frequency table may be given (**Table 1**). Alternatively, a bar chart may be used to display the relative frequencies in a graph (**Fig. 1**).

Table 1
Frequency Chart of Race (N = 500)

	n	%
White	399	78
Black	78	17
Asian	15	3
American Indian	5	1
Pacific Islander	3	<1

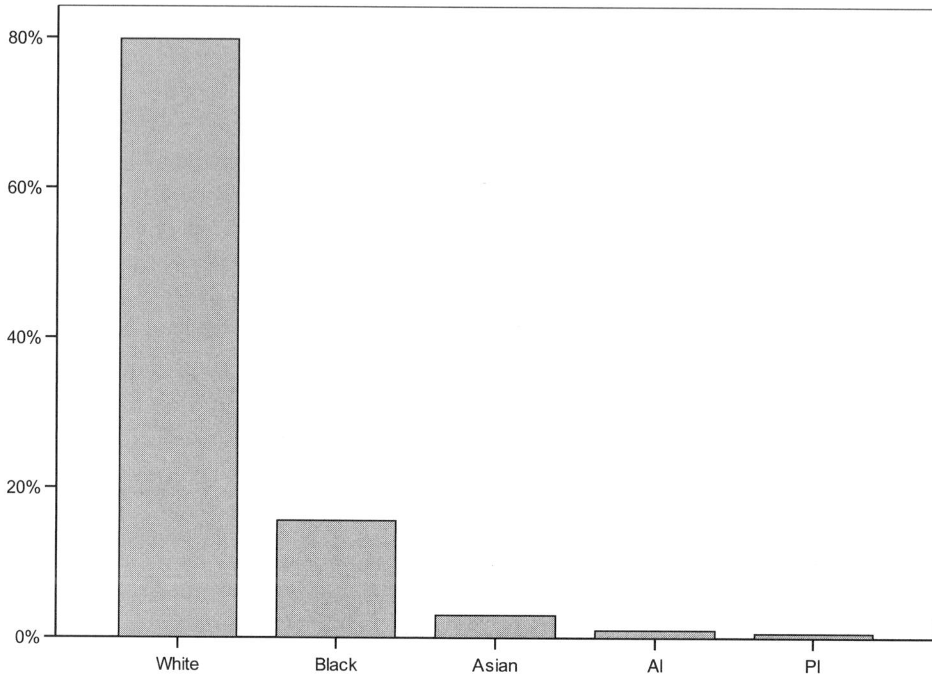

Fig. 1. Bar chart of race distribution (AI, American Indian; PI, Pacific Islander).

It is important to inspect the frequencies of the categories to determine whether a reduction of the categories is warranted for statistical analyses. For example, from the descriptive information of race in the example above, there are only three Pacific Islanders and five American Indians. Describing further comparisons by these five race categories would be wasteful and not provide any insight into trends by race. It would be more useful to collapse the Asian, American Indian, and Pacific Islander categories into a new category (Other), thereby reducing the number of categories to three. Another viable strategy would combine all non-white categories into a new category (minority), again reducing the number of categories to two (78% white and 22% non-white).

4. Describing Quantitative Variables

For quantitative variables, it is necessary to report two statistics. Regardless of the specific statistics used, the two measures that should be reported for quantitative variables are measures of the center and variability. These are referred to as measures of *location* and *spread*. The number of observations (the sample size, n) is also important, but it is generally not counted as one of the summary statistics. There are many statistics that measure the location, or typical values, as well as the spread, or the typical variability, of quantitative variables. For measuring location, two of the most common measures are the mean and the median, though there is a wide assortment of other measures of location to choose from. In contrast with qualitative data, the mode is little used when reporting quantitative data. For measuring spread, the two most useful statistics are the standard deviation and outer quartiles.

4.1. Measures of Location

The *mean* is the most typical measure of location on a sample. The mean, commonly called the arithmetic average, is typically denoted as $\bar{x}$. If $x_1, x_2, \ldots,$ x_n are the n observations in a sample, the sample mean is

$$\bar{x} = \frac{\sum_{i=1}^{n} x_i}{n},$$

where Σ is the uppercase Greek letter *sigma* denoting summation. If the ages of subjects in a study were 9, 11, 12, 14, and 18 years, then the mean age would be the sum of the ages divided by 5, so 64/5 = 12.8 years. The mean is sensitive to extreme values but is very useful for comparison methods because it has desirable mathematical properties ***(4)***.

The *median* is the middle-most value in a set of ranked data. It is more representative of "typical" subjects in the data than is the mean. When there is an odd number of subjects, the median is the middle value of the (ordered) sample values. In a sample of 5 subjects in the previous example, the third highest value is age 12 and represents the median age. When there is an even number of scores, the mean of the two middle scores is used. For the observations (–4, –2, 1, 5, 7, 8, 8, 10), the median is 6. The median is useful when there are extreme values because it is not as influenced by outliers as the mean.

The median is also called the 50th percentile, which is the second quartile. A more general term for the median is the 0.50 quantile. Quantiles split a sample of observations into equal and ordered parts. For example, the 0.50 quantile splits the data into two equal parts and is called the median. The 0.25, 0.50, and 0.75 quantiles are called quartiles because they split the data into four parts. Quintiles, deciles, and percentiles are other terms used to describe the quantiles that split the data into fifths, tenths, and hundredths. To compute any quantile, the data should first be ranked in ascending order from 1 to n. The qth quantile is then obtained by taking rank = $q \times (n + 1)$ and then interpolating between the two values with ranks on either side of the rank ***(5)***. For the observations (–4, –2, 1, 5, 7, 8, 8, 10), the 0.80 quantile corresponds with rank = 0.80 × (8 + 1) = 7.2. The rank of 7.2 falls in between the seventh and eighth rank, or the observations with the values 8 and 10. By using linear interpolation, we find the rank of 7.2 corresponds with an observation with the value of 8.4, which was derived using [8 + 0.2 (10 − 8)] = 8.4. Based on the rank of 7.2, after interpolation, the 0.80 quantile is 8.4.

There are other measures of location besides the mean, median, and quantiles. For example, the trimmed mean, sometimes called the Winsorized mean, is another measure of location that does not depend on extreme values. The data is trimmed on both ends, or *tails,* of the distribution, and the mean is computed using the remaining data. For example, 10% trimming refers to using the middle 90% of the data by removing the upper and lower 5% of the data. Although trimming and Winsorization avoids the overinfluence of outliers on statistics, one needs to use caution in computing the standard error of the trimmed mean because the usual formulas are not appropriate ***(6)***.

4.2. Measures of Spread

One is often interested in the amount of variability, or spread, in a set of observations. It should be reported along with the measure of location. The simplest measure of spread is the *range*. The range is the difference between the largest and smallest values. The range is most appropriate when there are less than 10 observations and there are no extreme values. When there are at least 10 observations, the outer quantiles are preferred as a measure of spread ***(2)***. The most common quantiles to use are the 0.25 and 0.75 quantiles, and the difference between these two quantiles is called the *interquartile range* (IQR). The 0.10 and 0.90 are alternative quantiles when reporting the spread. The use of quantiles as a measure of spread is less sensitive to extreme values than range and other measures of spread. Generally, the IQR, along with the median, is a useful method to summarize data with especially asymmetrical distributions ***(7)***.

Perhaps the most common measure of spread is the standard deviation because of its association with the mean and its use in common statistical tests such as the *t*-test. Although the standard deviation is sensitive to extreme values, it measures the spread around the mean. The *standard deviation*, commonly denoted as s or SD, depends on the extent to which individual observations differ from the mean of the observations. Its calculation involves an intermediate step, the variance. The *variance*, s^2, is the sum of the squared deviations of each individual value from the mean and is defined as

$$s^2 = \frac{\sum_{i=1}^{n} (x_i - \overline{x})^2}{n-1}.$$

The divisor, $n - 1$, is known as the number of *degrees of freedom* (d.f.) of the variance estimate. The d.f. represents the number of independent pieces of information on the variability inherent in the observations *(8)*. However, variance is expressed in squared units, and it is more intuitive to express the variability of a set of observations in original units. The standard deviation is obtained by taking the square root of the variance and is defined as

$$s = \sqrt{\frac{\sum_{i=1}^{n} (x_i - \overline{x})^2}{n-1}}.$$

For the observations (10, 20, 30), the mean is 20 and

$$s = \sqrt{\frac{(10-20)^2 + (20-20)^2 + (30-20)^2}{3-1}} = \sqrt{\frac{100+0+100}{2}} = \sqrt{100} = 10.$$

It is easy to conclude that the standard deviation of (1, 2, 3) is 1 and s of (100, 200, 300) is 100.

A practical implication of the standard deviation can be expressed in terms of coverage of the data. Regardless of the distribution of measurements, even if the data are not reasonably symmetric about the mean, at least 75% of the values fall within two standard deviations of the mean and at least 89% fall within three standard deviations of the mean. If the shape of the data is symmetric (see discussion below) with only one mode, approximately 68% of the observations are within one standard deviation of the mean, 95% are within two standard deviations, and almost all of the values fall within three standard deviations of the mean *(9)*.

4.3. Displaying Quantitative Variables

Plots are valuable in showing the shape of the distribution of a set of observations. Two plots that are useful in displaying univariate data are a histogram and a box plot. A histogram is a frequency distribution that shows how often certain values of a variable occur. The horizontal axis shows the range of the variable and the vertical axis shows either the frequency or the relative frequency of observations within each interval that is plotted. **Figure 2** shows histograms for a bell-shaped distribution (left) and a positively skewed distribution, or a distribution with a tail to the right (right). The variable on the left was generated with a normal distribution (see below) and the variable on the right with a log normal distribution (if X has a normal distribution then $Y = e^X$ is said to have a log normal distribution). If, instead, the tail was to the left, the

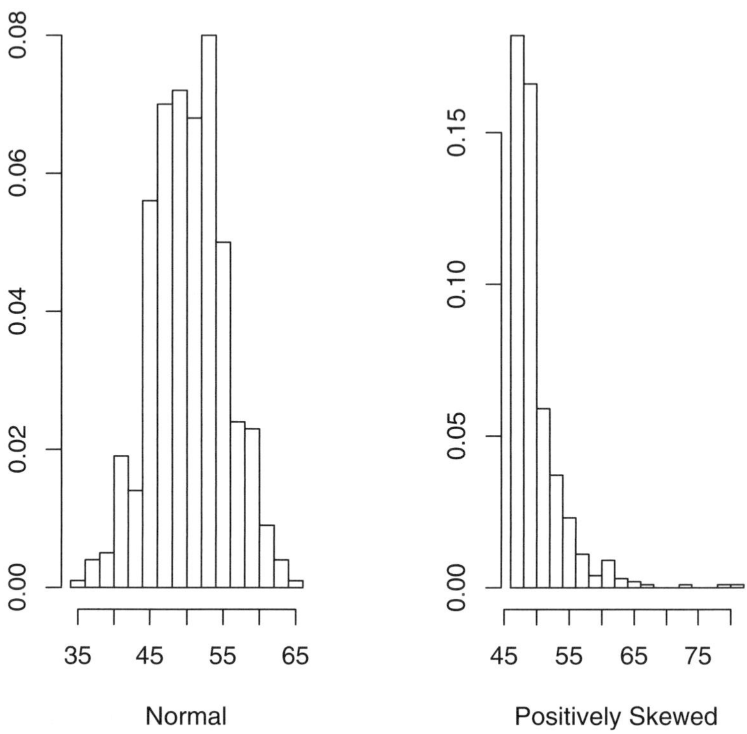

Fig. 2. Histograms of bell-shaped and positively skewed distributions with mean 50 and standard deviation 5 ($N = 500$).

distribution would be called a negatively skewed distribution. Both distributions have a mean of 50 and a standard deviation of 5. For the skewed distribution, an extreme value of 123 was set to 80 so it would not dominate the plot.

A bell-shaped distribution often follows a *Gaussian* or *normal distribution.* Unlike skewed distributions, the normal distribution is symmetrical, and the highest point at the center is the mean of the distribution. The normal distribution also has two points of inflection, which are points where the curve changes from convex to concave and vice versa. The distance between the mean and the first inflection point (where the curve changes from concave up to concave down) corresponds with one standard deviation and distinguishes the normal distribution from other bell-shaped distributions. Distributions that are bell-shaped and not normally distributed may be heavy-tailed or light-tailed distributions. A heavy-tailed distribution refers to a curve with more distinctive tails compared with the tails of a normal distribution. A light-tailed distribution refers to a curve with less distinctive tails than a normal distribution. For example, the uniform distribution, where the values of the variable are uniformly or equally distributed over an interval, is a light-tailed distribution. Kurtosis is a statistical measure of heavy tails and is positive when the tails are heavier than the normal distribution and negative when the tails are lighter. Skewness refers to one tail of the curve being heavier (or lighter) than the other.

To determine if a variable has a distribution similar to another (test) distribution, a quantile-quantile plot (q-q plot) may be used. The q-q plot graphs the quantiles of one distribution against the quantiles of another distribution. If the data follow closely to the diagonal line, then the distributions are considered similar. Specifically, to determine whether a variable has a normal distribution, one would compare the distribution of the variable to a normal distribution. A q-q plot that uses a normal distribution as the test distribution is called a normal quantile plot or sometimes a normal probability plot. For example, **Figure 3** shows normal quantile plots for two distributions that were generated and compared with a test distribution that is normal. The distribution on the left was based on a normal distribution with mean 50, standard deviation 5, and sample size 50. The distribution on the right was based on a uniform distribution with minimum and maximum of 35 and 65. As shown in **Figure 3**, the graph on the left shows most of the points falling on the diagonal line because

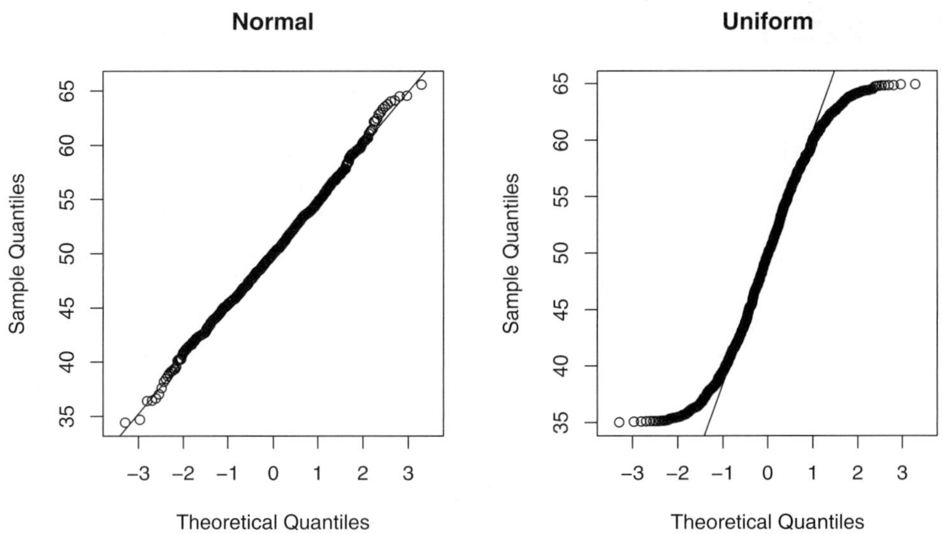

Fig. 3. Normal quantile plots of a normally distributed variable (left) and a uniformly distributed variable (right).

the data were based on a normal distribution. However, the graph on the right shows the middle-most 60% of the data fall on the diagonal line, but the lower and upper 20% of the data show departure from the line, indicating a non-normal distribution.

An often utilized property of the normal distribution is the probability that a normally distributed variable lies within two standard deviations of the mean is 95.46% and within one standard deviation of the mean is 68.26%.

For example, 10,000 observations were generated based on a normal distribution with a mean of 100 and a standard deviation of 10. **Figure 4** shows a histogram of the normally distributed variable. Because the distribution is normal, we would expect approximately 68% of the values to fall between 90 and 100 and approximately 95% of the values to fall between 80 and 120. The approximation is accurate because 96% of the observations fall between two standard deviations and 68% fall within one standard deviation. However, when

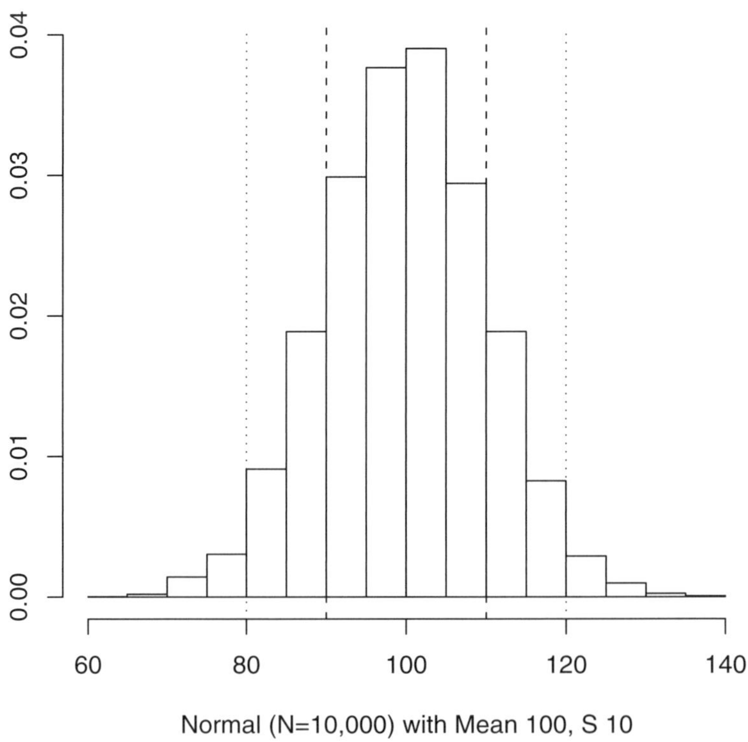

Fig. 4. Histogram of a normal distribution with mean 100 and standard deviation 10. The outer vertical lines (dots) show the values of mean $\pm 2 \times s$, and the inner vertical lines (dashes) show the values of mean $\pm s$.

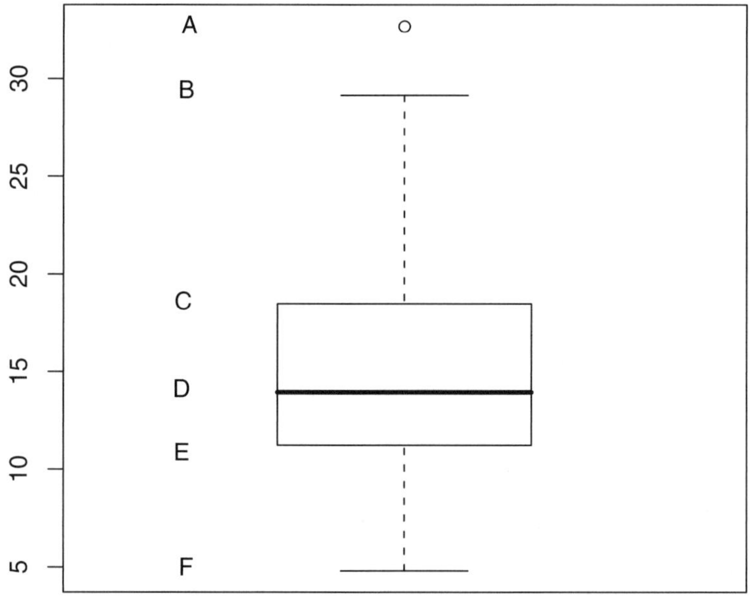

Fig. 5. Illustration of a box plot. **A**, outlier; **B** and **F**, whiskers; **C** and **E**, upper and lower quartiles; **D**, median.

the shape of the distribution is not a normal distribution, this approximation is not adequate.

A box plot is a graph to examine the overall shape of a variable and is especially useful for comparing distributions of different groups of data (e.g., treatment and control). **Figure 5** shows an example of a box plot. The box shows the range of the middle 50% of the data, or the values that fall between the lower and upper quartiles, labeled **C** and **E**, respectively. The horizontal line, labeled **D** in the figure, drawn inside the box is the median. Whiskers, labeled **B** and **F**, are the small horizontal lines above and below the box and go to the extremes of the data. Whiskers are defined based on the interquartile range, which is **C-E** in the figure. The upper whisker is defined as **C** + 1.5 (**C** – **E**) and the lower whisker is **E** – 1.5 (**C** – **E**). If the upper whisker is greater than the maximum, then the whisker is set to the maximum, and if the lower whisker is less than the minimum, then the whisker is set to the minimum. Points falling outside of the whiskers are shown by themselves and are considered extreme or outlining observations and are labeled **A** in **Figure 5**.

Figure 6 shows the same two distributions presented with histograms in **Figure 2**, both with a mean of 50 and a standard deviation of 5. The box plot

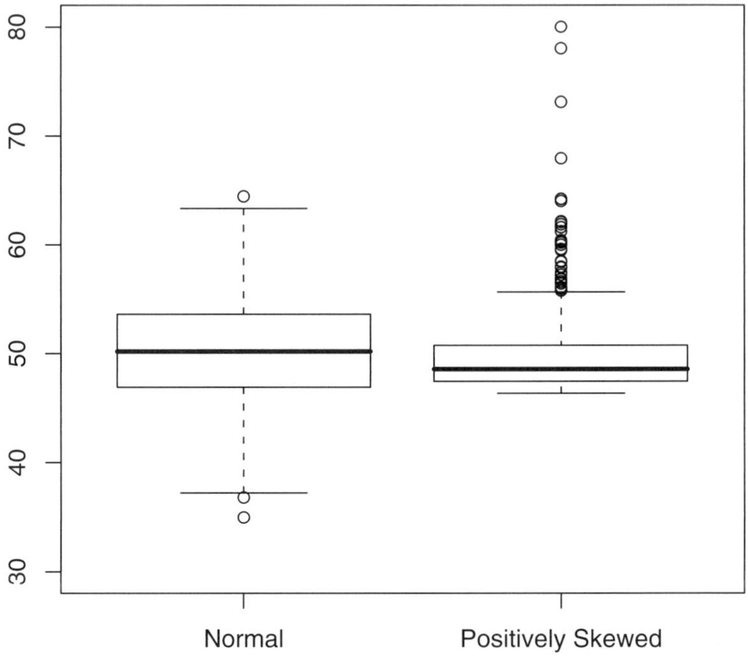

Fig. 6. Box plot of normal and positively skewed distributions with mean 50 and standard deviation 5 ($N = 500$).

on the left is symmetric and the one on the right is positively skewed. The normally distributed variable shows a symmetric distribution with the whiskers and the ends of the box falling equally away from the median and, as expected, has only a few outliers. The positively skewed variable has a box (the 25th to 75th percentiles) that is asymmetric, its quartiles and whiskers are not equally away from the median, and it has many outlier values.

5. Illustration

To illustrate the calculations above, data on time since liver transplantation, in years, for 24 female subjects is given in **Table 2**. The mean for time from liver transplant among females is 9.71 years and the median is 9.50 years. The range is 4 to 17 years and the IQR is $12.75 - 7.25 = 5.5$. The variance is 13.18 and the standard deviation is 3.63. The histogram and box plot are shown in **Figure 7**. Because the distribution is approximately normal, about 95% of the values would fall between 2.5 to 17.0 years $9.71 \pm 2 \times 3.63 = 9.71 \pm 7.26 \Rightarrow$ (2.5 to 17.0 years). Note that all of the data, or 100%, fall within two standard deviations.

Table 2
Data on Time from Liver Transplantation (in Years) for 24 Female Subjects

13	9	8	14	8	4
9	10	6	17	16	12
6	8	4	8	11	5
10	7	10	13	15	10

6. Presenting Variables Together

When reporting descriptive statistics, the precision used should be consistent both in the tables and in text. When reporting measures on categorical data, both percentages and frequencies should be given. Reporting both will avoid confusion in the calculation and also allows for percentages to be reported as integers. For example, a percentage of 82.45% can be reported as 82% if the

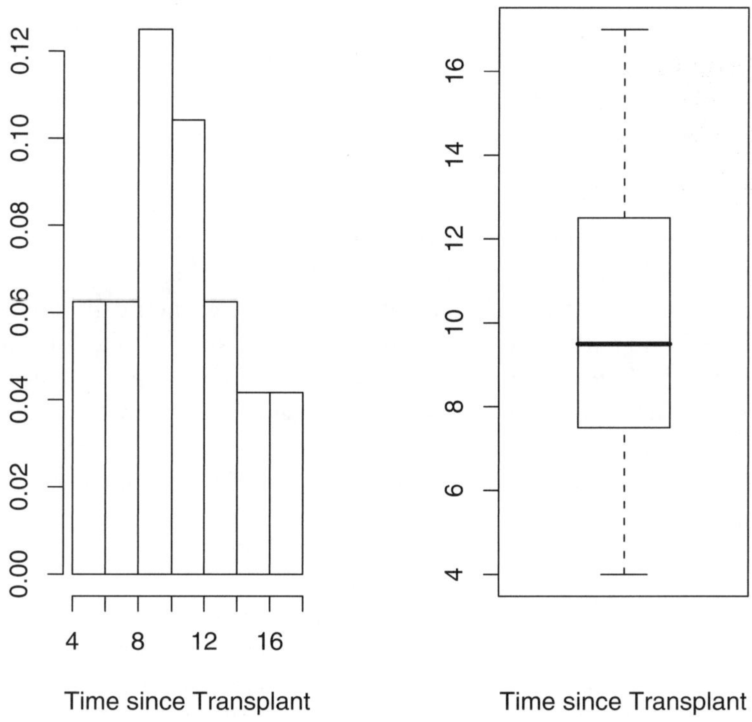

Fig. 7. Histogram (left) and box plot (right) of time since transplant.

Table 3
Frequency Table of Variables (N = 500)

Age (years), quartiles[a]	50 (46, 52)
Female sex, *n* (%)	245 (49%)
Race, *n* (%)	
Black	78 (17%)
White	399 (78%)
Other	23 (1%)
Genotype of marker, *n* (%)	
AA	294 (59%)
AG	177 (35%)
GG	29 (6%)
Weight (lb), mean (*s*)	170 (14)
Side effects, *n* (%)	
0	320 (64%)
1 or more	280 (36%)

[a]Quartiles are reported as 0.50 (0.25, 0.75) quantiles.

frequency is also reported. When reporting measures on quantitative measures, summary measures should not be reported to more than one extra decimal place over the raw data ***(9)***. For example, mean age in years may be 13.245, but this should be reported as 13 years or 13.2 years if age was recorded as an integer.

Typically, a descriptive table contains information on more than one variable and is an efficient method of reporting descriptive statistics. Tables are usually presented to describe variables because multiple bar charts would be wasteful of space. Both qualitative and quantitative variables should be combined in a table to describe particular sets or groups of data, such as baseline characteristics. **Table 3** shows an example of how both types of variables can be described together. Quartiles are frequently used for asymmetric distributions and means and standard deviations for symmetric distributions ***(10)***. Categories within a qualitative variable with only a few individuals should be combined if possible.

7. Logarithms and Other Transformations

The transformation and statistic selected is based on the distribution of a variable. The most important distribution for continuous variables is the normal distribution. If a variable is not normally distributed, one usually seeks a transformation to normalize the distribution.

Often, a transformation, not the raw variable itself, is used to describe and analyze data in molecular biology. For example, instead of using spot intensity data, the logarithm of spot intensity data is analyzed.

A logarithm, or log, transformation is most commonly used and is applied when the data are highly skewed to make it more symmetric. It is widely accepted that log-transformed expression data follows a bell-shaped curve ***(10)***. Square root and cube root transformations are also used when analyzing spot intensity data. Logs have some useful properties for handling data. Two useful properties are $\log(a \times b) = \log(a) + \log(b)$ and $\log(a/b) = \log(a) - \log(b)$. Therefore, if the logs are used, multiplicative relationships become simpler additive relationships ***(12)***.

Log to the base 10, also called common logs, is commonly used in medical research, and log to the base 10 of x is written as $\log_{10}(x)$. The three numbers 1000, 100, and 10 are transformed on the log to the base 10 scale to $\log_{10}(1000) = 3$, $\log_{10}(100) = 2$, and $\log_{10}(10) = 1$. If there are zero values where $\log(0)$ would be undefined, then $\log_{10}(x + 1)$ is often used.

The natural log, abbreviated ln, is another useful log. The natural log is log to the base e where e is 2.71828 . . . , a constant. Natural logs are very comparable to using log to the base 2, which are typically used in microarray analysis. Using a log to the base 2 will transform the four numbers 16, 8, 4, and 2 to $\log_2(16) = 4$, $\log_2(8) = 3$, $\log_2(4) = 2$, and $\log_2(2) = 1$. Log to the base 2 should be used instead of log to the base 10 when the data range through just a few powers of 10 to avoid fractional powers of 10 ***(13)***. Because some statistical software packages do not have a log to the base 2 transformation (e.g., SPSS 13.0; SPSS Inc., Chicago, Ill.), it may be necessary to use a formula for change of base. By using log to the base 10, $\log_2(x) = \log_{10}(x)/\log_{10}(2)$. For example, $\log_2(32) = \log_{10}(32)/\log_{10}(2) = 1.505/0.301 = 5$.

Fold change is a statistic used in differentiating gene expression and is the ratio of two observations. Typically, two- or threefold differences are considered important, but these are often arbitrary thresholds ***(14)***. Caution should be used when interpreting fold changes. For example, for the raw values of 16 and 4, the fold change is $16/4 = 4$. However, on the log base 2 scale, the fold change is only 2.

8. Describing Multivariate Data

When describing multivariate data, simple descriptive statistics are used to summarize each variable individually. Then it is important to describe the relationship between the variables. For this step, the relationships between pairs of quantitative variables are described using Pearson or Spearman correlation coefficients and are displayed with scatter plots (see **Chapter 8**). The relationships between pairs of qualitative variables are described using tests of association for categorical data and displayed using contingency tables (2×2 or $R \times C$ tables) (see **Chapter 5**). The relationships between pairs of variables conditional on the values of other variables should be displayed with a conditional plot or coplot ***(15)***.

9. Distribution of Statistics

If a random sampling process is involved when selecting a sample, then the variables are called random variables (see **Chapter 4**). In fact, the term *statistic*, or estimator, is used to refer to a rule that provides an estimate of a value that is characteristic of the population. The quantity computed from a sample is called an estimate, and the value from the population that is being estimated is called the parameter. The estimator is a random variable that varies from sample to sample and has its own distribution. The distribution of an estimator is called the sampling distribution of a statistic. The uncertainty or imprecision of an estimate is quantified based on the variability of the sampling distribution. This variability is not natural variability but variability due to error and is called the standard error of an estimator. The standard error is important for inferential statistics when estimating statistics and comparing groups of data because it is used to construct margin of errors and confidence intervals. Given the standard error, approximate 95% confidence intervals are constructed by taking $\pm\, 2 \times$ (standard error of the statistic) (see **Chapter 4**).

9.1. Distribution of the Mean

The sampling distribution of the mean has a sample mean, $\bar{x}$, and its sample standard deviation is often called the standard error of the mean (abbreviated SEM). Recall that the standard deviation measures natural variability. Therefore, for describing data, it is important to use the standard deviation as a measure of natural variability and not report the SEM. However, to quantify uncertainty of a statistic, such as the mean, the standard error of the statistic, such as the SEM, should be used. From a sample,

$$\text{SEM} = \frac{s}{\sqrt{n}}.$$

That is, the variability due to error, the SEM, will be smaller for samples that are larger and for samples whose variability is lower (less dispersed). It is important to note that the sampling distribution of the mean will be normally distributed if the observations are normally distributed; or, based on the central limit theorem, the sampling distribution will follow a normal distribution even if the observations do not follow a normal distribution as long as the sample size is large *(3)*. Because it follows a normal distribution, twice the SEM represents the margin of error of the estimate.

9.2. Distribution of a Proportion

For categorical data, proportions are usually the statistic of interest, and the sample distribution of a proportion approximately follows a normal distribution for large samples (see **Chapter 5**). The approximation holds for proportions between 10% and 90% for sample sizes of 50 or greater *(3)*. For dichotomous variables, the estimator is the sample proportion, $\hat{p}$, or the number of events divided by the sample size. The standard error of this estimator is

$$\sqrt{\frac{\hat{p}\times(1-\hat{p})}{n}}.$$

Twice the standard error will represent the margin of error of the estimate if the normal approximation is adequate. For example, a study of antiseizure medication in 100 epileptic patients resulted in 20 patients with continuing seizures, giving an estimate of a proportion of $\hat{p} = 20/100 = 0.20$. The standard error of this sample statistic is

$$\sqrt{\frac{0.20\times(1-0.20)}{100}} = 0.04.$$

Because the margin of error is defined as twice the standard error, the error of estimating the population proportion is $2 \times 0.04 = 0.08$. That is, we summarize the result as the proportion of patients without seizure freedom is estimated to be 0.20 with a margin of error of 0.08.

9.3. Distribution of the Variance

The sampling distributions for the mean and proportion follow a normal distribution for large sample sizes. However, the sampling distribution of the variance is asymmetrical. The estimator of the population variance is denoted as s^2, as noted above. For the simple one-sample study, s^2 has associated with it $n - 1$ d.f. If σ^2 is the variance in the population, then

$$\frac{s^2\times(n-1)}{\sigma^2}$$

follows a chi-square distribution, denoted χ^2, with $n - 1$ d.f. That is, the shape of the chi-square distribution depends on the number of d.f. For example, **Figure 8** shows the shape of three distributions with d.f. equal to 1, 9, and 19. The higher the d.f., the more symmetric the distribution becomes.

Solving for the population parameter σ^2, we can quantify the imprecision of the estimate of the variance, which is

$$\frac{s^2 \times (n-1)}{\chi^2_{0.975}} \leq \sigma^2 \leq \frac{s^2 \times (n-1)}{\chi^2_{0.025}},$$

where $\chi^2_{0.025}$ and $\chi^2_{0.975}$ represent the 2.5th and 97.5th percentiles of the chi-square distribution with $n-1$ d.f.

For example, for the three distributions below in **Figure 8**, the corresponding chi-square values are shown in **Table 4**. Tables are usually provided in statistical textbooks and give the chi-square value for a given d.f.

If a study had 20 subjects and $s^2 = 10$, then the uncertainty in estimating the variance of 10 is

$$\frac{10 \times (19)}{32.852} \leq \sigma^2 \leq \frac{10 \times (19)}{8.907} = 5.8 \text{ to } 21.3.$$

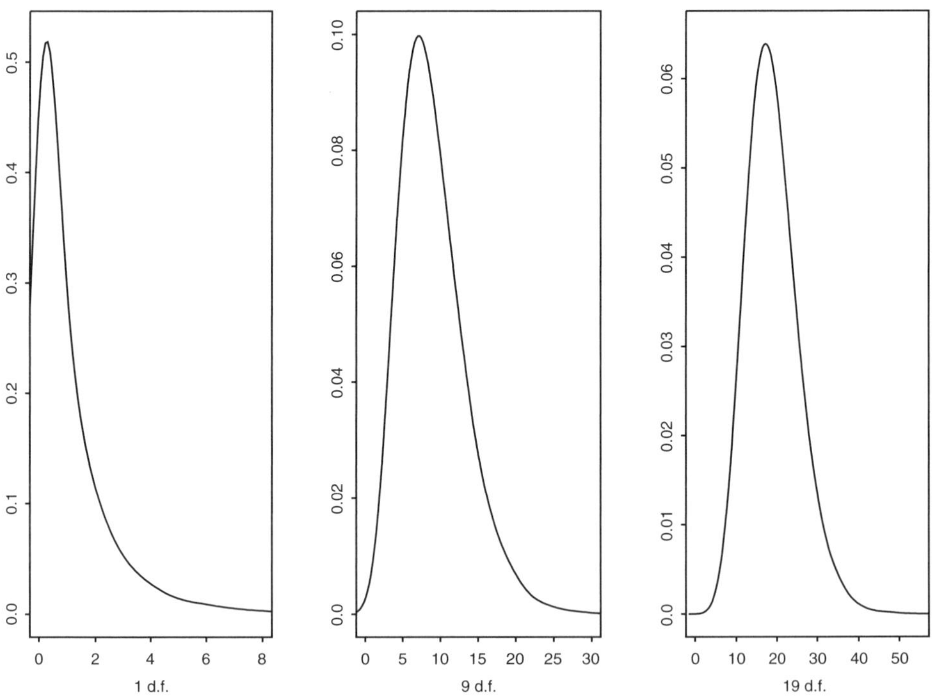

Fig. 8. Density plots of chi-square distribution with 1, 9, and 19 degrees of freedom (d.f.).

Table 4
Chi-Square Distribution Values for the 0.025 and 0.975 Quantiles for 1, 9, and 19 Degrees of Freedom

d.f.	$\chi^2_{0.025}$	$\chi^2_{0.975}$
1	0.001	5.024
9	2.700	19.023
19	8.907	32.852

It appears more subjects are needed to estimate the variance with a greater precision. Confidence intervals will be discussed in more detail in **Chapter 4**.

10. Conclusion

Describing the data and statistics of results is a vital part of any medical study. For any project, it is important to describe the characteristics of the study units so that readers can evaluate the relevance of the results to a particular population or setting ***(2)***. Additionally, the report of the standard error of a statistic is necessary to quantify the precision of the estimates, such as treatment effects. Whether the study results can be generalized to other populations and settings will be based on the use of adequate descriptive statistics and measures of uncertainty of the statistics. If little information is provided on the characteristics of the study units and the outcomes, then it is doubtful that the results will be applied elsewhere. On the other hand, if sound descriptive statistics and measures of imprecision of the inferential statistics are given, then the study can be assessed for generalizability and may eventually be applied to other situations.

Acknowledgments

The author wishes to thank Katarzyna Bryc for her review of this chapter and excellent suggestions.

References

1. Medical Subjects Headtions (MeSH). Bethesda: National Library Of Medicine. http://www.nlm.nih.gov/mesh/MBrowser.html. Accessed April 2, 2007.
2. McPherson, G. (1990) *Statistics in Scientific Investigation: Its Basis, Application and Interpretation*. New York, Springer-Verlag.

3. Altman, D. (1991) *Practical Statistics for Medical Research.* London, Chapman & Hall.
4. Bland, M. (2000) *An Introduction to Medical Statistics*, 3rd ed. New York, Oxford University Press.
5. Altman, D. G., and Bland J. M. (1994) Quartiles, quintiles, centiles, and other quantiles. *Br. Med. J.* **309**, 396.
6. Wilcox, R. (1998) Trimming and Winsorization. In: Armitage, P., and Colton, T. *Encyclopedia of Biostatistics.* New York, John Wiley & Sons, pp. 4588–4590.
7. Altman, D. G., Schulz, K. F., Moher, D., Egger, M., Davidoff, F., Elbourne, D., Gøtzsche, P. C., and Lang, T. for the CONSORT Group. (2001) The Revised CONSORT Statement for Reporting Randomized Trials: explanation and elaboration. *Ann. Intern. Med.* **134**, 663–694.
8. Finney, D. J. (1955) *Experimental Design and Its Statistical Basis.* Chicago, The University of Chicago Press.
9. Pagano, M., and Gauvreau, K. (2000) *Principles of Biostatistics*, 2nd ed. Pacific Grove, Duxbury.
10. Altman, D., and Bland, J. M. (1996) Presentation of numerical data. *Br. Med. J.* **312**, 572.
11. Aoki, K., and Petrov, A. (2003) A comprehensive microarray data management approach. In: Hardiman, G. *Microarrays Methods and Applications.* Eagleville, DNA Press LLC, pp. 171–192.
12. Bland, J. M., and Altman, D. G. (1996) Logarithms. *Br. Med. J.* **312**, 700.
13. Cleveland, W. S. (1993) *Visualizing Data.* Summit, N.J.: Hobart Press.
14. Draghici, S. (2003) *Data Analysis Tools for DNA Microarrys.* Boca Raton, Chapman & Hall/CRC.
15. Everitt, B. (2005) *An R and S-PLUS® Companion to Multivariate Analysis.* London, Springer-Verlag.

4

Basic Principles of Statistical Inference

Wanzhu Tu

Summary

In this chapter, we discuss the fundamental principles behind two of the most frequently used statistical inference procedures: confidence interval estimation and hypothesis testing. both procedures are constructed on the sampling distributions that we have learned in previous chapters. To better understand these inference procedures, we focus on the logic of statistical decision making and the role that experimental data play in the decision process. Numerical examples are used to illustrate the implementation of the discussed procedures. This chapter also introduces some of the most important concepts associated with confidence interval estimation and hypothesis testing, including P values, significance level, power, sample size, and two types of errors. We conclude the chapter with a brief discussion on statistical and practical significance of test results.

Key words: Hypothesis testing; P value; point and confidence interval estimation; power; sample size; significance level; simultaneous inference; student t distribution; type I and type II errors.

1. Introduction

Statistical inference is a decision-making process. Different decision-making processes follow different decision rules. Medical decisions, for example, are usually based on the physician's assessment of the patient, the physician's clinical judgment, and the physician's interpretation of treatment guidelines. A statistical decision process, or statistical inference, attempts to isolate the decision maker from his personal opinion and preference to achieve an *objective* conclusion that is supported by the data. Two commonly encountered forms of statistical inference are parameter estimation and hypothesis testing. Although each is designed to address a different type of research question, both rely upon the sample data to justify their conclusion.

From: *Methods in Molecular Biology, vol. 404: Topics in Biostatistics*
Edited by: W. T. Ambrosius © Humana Press Inc., Totowa, NJ

In an estimation procedure, one uses sample information to estimate the value of an unknown parameter associated with a certain population. Herein, the term *parameter* is used to describe a particular numerical characteristic of a population. For example, the mean of a population is a parameter. Denoted by μ, it depicts the central location of the population. The variance of the population, denoted by σ^2, is another parameter. As a population parameter, σ^2 quantifies the magnitude of dispersion or the variability of the population.

The other form of statistical inference is hypothesis testing, which is primarily used to adjudicate the truthfulness of certain preconceived statements concerning the value of a population parameter. Hypothesis testing is a particularly popular form of statistical inference in biomedical research because it directly assesses the strength of data evidence either for or against a scientific proposition.

There are numerous established inference procedures. Researchers choose appropriate procedures based on the parameters of interest and experimental conditions under which the data are collected. Underlying the varying forms of statistical inference, however, are a set of principles that are common to all inference procedures. The purpose of the current chapter is to introduce these common principles and illustrate their application through several frequently used procedures, including confidence interval estimation and hypothesis testing of population means and variances. The development of these procedures is mostly based on the distributions of the sample mean and sample variance introduced in the previous chapter. Also discussed are the interpretations of inference results, errors, and power. Finally, the chapter concludes with a brief discussion on the distinction between statistical and practical significance.

2. Parameter Estimation

2.1. Point Estimation

Parameter estimation is a useful technique when the primary goal of the analysis is to estimate a certain numerical characteristic of a population. For example, a geneticist is often interested in estimating the allele frequencies of certain genes in a target population. A clinician may want to estimate the amount of viral shedding in patients with a certain infection. A biochemist may be interested in estimating the average concentration of a certain protein in a patient population. In each of these situations, the interest is to estimate a certain numerical quantity associated with a particular *population*.

With few exceptions, it is difficult for an investigator to directly assess each and every single member of the population. Therefore, directly calculating the population parameter of interest becomes logistically difficult, if not impossible. A practical way to achieve the goal of estimation is to rely on a sample of subjects, drawn from the intended population, and then use the sample data to estimate the unknown population parameter.

Although it is easy to see the appeal of such a sample-based estimation approach, there are a few important factors that might influence the quality of the resulting estimate. First among them is the representativeness of the sample. The preferred scheme is to take a *random sample* from the target population. In this chapter, all of the procedures are presented under the assumption of random sampling. Another important consideration is the optimality of the estimate, which is the main topic of the remainder of the section.

A generic description for an estimation problem is as follows: Suppose that subjects, or elements, of a certain *population* are selected through a randomized experiment, and the observations obtained from these subjects form a *sample*. Assuming that these observations follow a particular distribution with an unknown parameter, denoted as θ, we attempt to use the sample data to estimate θ. To achieve this goal, we need to have a rule for the estimation, usually expressed as a formula, that tells us how to calculate a numerical estimate based on the information provided by the sample. Such a rule, or formula, is referred to as an *estimator*. A numerical value of the estimator, usually computed from the sample, is called an *estimate*. If the true parameter is θ, we usually write an estimator (or estimate) of θ as $\hat{\theta}$. For example, an estimator for a population mean can be written as $\hat{\mu}$.

The simplest form of estimation is to calculate a single-valued point from the sample and use it as an estimator of the unknown parameter. Such an estimator is called a *point estimator*. Let's consider the following example.

Example 1

Prostate specific antigen (PSA) is a protein produced by the cells from the prostate. The blood concentration of PSA is often used as a biomarker of prostate cancer. Results under 4 ng/mL are usually considered normal. The higher the PSA level, the more likely a patient has prostate cancer. Because of this relationship, postproctectomy PSA has also been used to measure the success of the operation. **Table 1** contains the PSA levels of 30 patients who had radical proctectomy, measured 6 months after the operation.

Table 1
Prostate Specific Antigen Levels (ng/mL) in 30 Patients Who Underwent Proctectomy Measured 6 Months after Operation

0.2	0.1	0.0	0.0	0.1	0.1
0.1	0.1	0.1	0.0	0.4	0.0
0.0	0.2	0.2	0.1	2.7	0.1
0.0	0.2	1.3	0.0	0.2	0.0
0.1	0.3	0.1	0.0	0.0	0.1

If we believe that this sample of 30 patients is representative of the population of prostate cancer patients who underwent radical proctectomy, we could use the data obtained from these patients to estimate the mean postoperation PSA value.

In a more general notation, we write $X_1, X_2, \ldots, X_n$ as a random sample from a population, where X_i represents the observation from the ith experimental unit, $i = 1, 2, \ldots, n$. In this example, the experimental unit is the individual patient, the sample size is $n = 30$, and sample observations are $X_1 = 0.2$, $X_2 = 0.1, \ldots, X_{30} = 0.1$. We now use this sample to estimate the mean postoperation PSA level μ of the population of prostate cancer patients who underwent proctectomy.

To estimate the population mean PSA level μ, we could use any of the 30 observed sample points as our estimator, that is $\hat{\mu} = X_i$ for any $i = 1, 2, \ldots, 30$. Depending on which sample observation we use, the value of the estimator takes a wide range, from 0 ng/mL to 2.7 ng/mL. Alternatively, we could use the sample mean as our point estimator, $\hat{\mu} = \bar{X} = \frac{1}{n}\sum_{i=1}^{n} X_i$, because, for a given sample, $\bar{X}$ takes only a single value. For the PSA example, sample mean is $\bar{X} = 0.2267$ ng/mL. As this example demonstrates, for a given parameter, there may exist many point estimators. The question is which one of these is the "best" estimator.

In order answer this question, we must first clarify what kind of properties we desire in an estimator. Intuitively, one of the desirable characteristics that we seek in an estimator is *unbiasedness*. In other words, a good estimator should neither consistently overestimate nor underestimate the parameter. Another property that we seek in an estimator is small variability. In other words, an ideal estimator should not vary too much from sample to sample. Combining these two characteristics, we want an estimator that has small variation while maintaining its unbiasedness. This leads to the concept of minimum variance unbiased estimator (MVUE). As its name indicates, an MVUE is an unbiased estimator, and it has the smallest variance among all unbiased estimators.

In the case of population mean, it can be shown that $\hat{\mu} = \bar{X}$ is indeed an MVUE, thus is superior to the estimator based on a single sample point, $\hat{\mu} = X_i$. To understand this, let us compare $\hat{\mu} = X_i$ and $\hat{\mu} = \bar{X}$. Formally, we assume $X_1, X_2, \ldots, X_n$ to be a random sample from a population with mean μ and variance σ^2. Because X_i is a sampling point from the population, we know that the expected value and variance of X_i are μ and σ^2, respectively; that is, $E(X_i) = \mu$ and $Var(X_i) = \sigma^2$. From the sampling distribution of the population mean, we know that $E(\bar{X}) = \mu$ and $Var(\bar{X}) = \sigma^2/n$. This demonstrates that both single sample point (X_i) and sample mean ($\bar{X}$) are unbiased estimators of μ, but between the two, the sample mean $\bar{X}$ has a smaller variance thus should be considered as the preferred estimator. The result of this comparison is hardly surprising: the sample mean depicts the central location of the sample. If the sample truly represents the subjects in the population, the sample mean will provide a reasonable estimate of the central location of the population. Indeed, it can be shown that $\bar{X}$ is an MVUE of μ *(1)*. Therefore, as a general principle, we usually use the sample mean $\bar{X}$ to estimate the population mean μ. Extending the same principle to the case of variance, we often use sample variance $s^2 = \frac{1}{n-1}\sum_{i=1}^{n}(X_i - \bar{X})^2$ to estimate an unknown population variance σ^2.

2.2. Confidence Interval Estimation

Although a good point estimator may possess desirable properties such as those just discussed, many researchers feel that without any assurance of its "correctness," a single-valued estimate is not particularly useful in a decision-making process. A simple way to overcome this deficiency is to specify a range, rather than a single value, for the estimation of a parameter. In other words, we hope to guarantee with a high probability that the true parameter is covered by this range. Such a range is called a *confidence interval* of a parameter. The probability that guarantees the coverage is called the *confidence level.*

To define an interval mathematically, we need two values: a lower limit and an upper limit. Formally, if θ is the true parameter and $\hat{\theta}_1$ and $\hat{\theta}_2$ are the two limit values that satisfy the following criterion,

$$P(\hat{\theta}_1 \le \theta \le \hat{\theta}_2) = 1 - \alpha,$$

then the interval $(\hat{\theta}_1, \hat{\theta}_2)$ is said to be a $100(1 - \alpha)\%$ confidence interval estimate of θ, where P is a probability measure, $\hat{\theta}_1$ and $\hat{\theta}_2$ are called lower and upper confidence limits, respectively, and $(1 - \alpha)$ is called the confidence level. In plain English, the above equation simply says that the probability that the interval $(\hat{\theta}_1, \hat{\theta}_2)$ contains the true parameter θ is $1 - \alpha$. In practice, we often specify α as a rather small number, say $\alpha = 0.05$, thus the resulting confidence level would be high, such as 95%. Later we shall see that α represents an error rate of the inference.

For confidence intervals, we seek two properties simultaneously: The first is a high confidence level. Confidence level can be viewed as an assurance of the correctness of the estimation. Without a high confidence level, the estimation loses its credibility. The second is a narrow interval range. When the range is too wide, the estimation becomes less informative. For example, when we estimate a person to be between 5 and 95 years of age, the estimate essentially fails to convey any information on how old the person actually is. Unfortunately, these two properties are inherently contradictory: With wider interval range, we become more confident that our estimate is correct. But when the range becomes wider, our estimation becomes less useful. Therefore, a careful balance must be maintained when we derive a confidence interval.

2.2.1. Large Sample Confidence Interval for the Mean

Formally deriving confidence intervals for the population parameters exceeds the scope of the current chapter. Instead, we shall present a heuristic argument for the construction of large sample confidence interval for the population means.

Suppose we have a population with an unknown mean parameter μ. For simplicity, we assume the population variance, σ^2, is known. To estimate the mean, we take a large random sample $X_1, \ldots, Xn$, where $n \geq 30$. As described in the previous section, we use the sample mean $\bar{X} = \Sigma_{i=1}^{n} X_i / n$ as a point estimate for μ.

To find a $(1 - \alpha)100\%$ confidence interval for μ, denoted by $(\hat{\mu}_1, \hat{\mu}_2)$, we start from its definition. Suppose that $\hat{\mu}_1$ and $\hat{\mu}_2$ are two quantities that satisfy $P(\hat{\mu}_1 \leq \mu \leq \hat{\mu}_2) = 1 - \alpha$. Simultaneously subtracting $\bar{X}$ from $\hat{\mu}_1$, μ, and $\hat{\mu}_2$, and then dividing the differences by $\sigma/\sqrt{n}$, we have

$$P\left(\frac{\hat{\mu}_1 - \bar{X}}{\sigma/\sqrt{n}} \leq \frac{\mu - \bar{X}}{\sigma/\sqrt{n}} \leq \frac{\hat{\mu}_2 - \bar{X}}{\sigma/\sqrt{n}}\right) = 1 - \alpha,$$

or equivalently,

$$P\left(\frac{\bar{X} - \hat{\mu}_1}{\sigma/\sqrt{n}} \geq \frac{\bar{X} - \mu}{\sigma/\sqrt{n}} \geq \frac{\bar{X} - \hat{\mu}_2}{\sigma/\sqrt{n}}\right) = 1 - \alpha.$$

According to the sampling distribution of $\bar{X}$, we know that when the sample size is large, $(\bar{X} - \mu)/(\sigma/\sqrt{n})$ can be approximated by the standard normal distribution N(0,1). Thus we have the following equations,

$$\frac{\bar{X} - \hat{\mu}_1}{\sigma/\sqrt{n}} = z_{\alpha/2}, \quad \text{and} \quad \frac{\bar{X} - \hat{\mu}_2}{\sigma/\sqrt{n}} = -z_{\alpha/2},$$

where $z_{\alpha/2}$ is the cutoff point corresponding with a right tail area of $\frac{\alpha}{2}$ under the standard normal distribution. Because of the symmetry of the standard normal distribution, $-z_{\alpha/2}$ is the cutoff point in the left tail.

Solving these equations, we obtain

$$\hat{\mu}_1 = \bar{X} - z_{\alpha/2}\frac{\sigma}{\sqrt{n}},$$

$$\hat{\mu}_2 = \bar{X} + z_{\alpha/2}\frac{\sigma}{\sqrt{n}}.$$

Confidence limits calculated from the above formulas will give us the desired $100(1 - \alpha)\%$ confidence interval estimate of μ.

A few things need to be clarified before we use this confidence interval estimation procedure. The first is the assumption that σ is known. This assumption, however, is unlikely to be true because the calculation of σ depends on

μ. Because μ is unknown and is to be estimated, how could we know σ? But for all practical purposes, this will not prevent us from using the procedure, because in large sample situations, unknown σ can be reliably approximated by the sample standard deviation s. Therefore, when the true value of σ is unknown, we calculate the large sample 100(1 – α)% confidence interval for the mean as

$$\bar{X} \pm z_{\alpha/2} \frac{s}{\sqrt{n}}.$$

The second issue is the interpretation of the above confidence interval. If we use $\alpha = 0.05$, the confidence level (1 – α) will be 95%. In a specific experiment (or the sample generated from that experiment), the calculated confidence interval either contains μ or it does not. Although we do not know whether the confidence interval that we calculated for this particular sample contains μ or not, we do know that if we repeat the experiment many times and use the above procedure to construct confidence intervals based on the resulting samples, about 95% of the confidence intervals that we obtain will contain the true value of the population mean ***(2)***.

Example 1 (Continued)

Finally, to illustrate the use of the confidence interval procedure introduced above, we revisit the PSA example: Using the sample of 30 prostate patients, we find the point estimate of the mean postoperative PSA value to be $\bar{X} = 0.2267$. The 95% confidence interval estimate is calculated as follows:

$$\bar{X} \pm z_{\alpha/2} \frac{s}{\sqrt{n}},$$

$$0.2267 \pm 1.96 \times \frac{0.5252}{\sqrt{30}},$$

$$(0.0388, 0.4146).$$

Therefore, we are 95% confident that the mean postprostatectomy PSA level is between 0.0388 and 0.4146 ng/mL.

2.2.2. Student t*-Distribution*

As we see in the previous section, the construction of a large sample confidence interval for μ is based on the sampling distribution of $\bar{X}$. We know that when the sample size is large, the standard normal distribution N(0,1) can be used to approximate the behavior of the quantity $\sqrt{n}(\mu - \bar{X})/s$. But when the sample size is not large, such approximation may not work as well. In this section, we discuss the sampling distribution of $\sqrt{n}(\mu - \bar{X})/s$ when the sample size is small or moderate. The behavior of this distribution, called t-distribution, is very similar to that of the standard normal distribution when the sample size is large (often defined as $n \geq 30$). But it works equally well for smaller samples.

We first give a verbal description of the t-distribution. The density curve of a t-distribution is quite similar to that of the standard normal distribution. It is bell-shaped, symmetric around zero, but with tails heavier than those of the standard normal curve. The t-distribution has only one parameter, its degrees of freedom (d.f.). The larger the degrees of freedom, the smaller the variance. The mean of a t-distribution is always zero; the variance is d.f./(d.f. – 2). It can also be shown that when the degrees of freedom of the distribution approaches infinity (d.f. $\to \infty$), a t-distribution approaches the standard normal distribution.

With the t-distribution, we are now able to describe the behavior of $\sqrt{n}(\mu - \bar{X})/s$. Suppose we select a random sample $X_1, \ldots, X_n$ from a normally distributed population, $N(\mu, \sigma^2)$. Let $\bar{X}$ and s^2 be the sample mean and variance, respectively. Then $T = (\bar{X} - \mu)/(s/\sqrt{n})$ follows a Student's t-distribution with $n - 1$ degrees of freedom, where n is the sample size.

In other words, *if the population is normally distributed*, statistic $T = (\bar{X} - \mu)/(s/\sqrt{n})$ has a t-distribution with $n - 1$ degrees of freedom. This is true regardless of the sample size. With this distribution, we will be able to make inferences concerning the population means, even when sample size is not large.

2.2.3. Small Sample Confidence Interval for the Mean

Mimicking the arguments in the large sample case, we construct the $100(1 - \alpha)\%$ confidence interval for μ using the t-distribution,

$$\bar{X} \pm t_{\alpha/2} \frac{s}{\sqrt{n}},$$

where $t_{\alpha/2}$ is the $\alpha/2$th quantile of the t-distribution with $n - 1$ degrees of freedom.

The t confidence interval has a structure that is parallel to the large sample confidence interval based on the standard normal distribution. Although it is designed for the estimation of population means in small sample cases, it can also be used for any sample size. In fact, as the sample size increases to $n \geq 30$, the numerical value of $t_{\alpha/2}$ approaches $z_{\alpha/2}$, thus the confidence limits calculated from the small sample procedure are essentially identical to those calculated from the large sample procedure. For this reason, many standard statistical software packages only provide confidence intervals based on the t-distribution.

Example 2

B-type natriuretic peptide (BNP) is released from the cardiac ventricles in response to increased wall tension and thus can be used as a marker for congestive heart failure *(3)*. In a local clinic, BNP levels of 20 heart failure patients were obtained from bedside assay. Fourteen of the patients were further classified as the New York Heart Association (NYHA) class II and six as class III. In the sample of NYHA class II patients, the mean BNP level was 412 pg/mL and the sample standard deviation was 231 pg/mL. For NYHA class III patients, the mean BNP level was 731 pg/mL and the sample standard deviation was 402 pg/mL. Using the sample information, we construct a 95% confidence interval for the mean BNP level of NYHA class II patients:

$$\bar{X} \pm t_{\alpha/2} \frac{s}{\sqrt{n}},$$

$$412 \pm 2.1604 \frac{231}{\sqrt{14}},$$

$$(278.6, 545.4),$$

where $t_{\alpha/2} = t_{0.025} = 2.1604$ was obtained from the *t*-distribution with d.f. = 14 – 1 = 13. Therefore, we conclude that based on the evidence provided by this small random sample, we are 95% confident that the mean BNP level of NYHA class II patients is between 278.6 pg/mL and 545.4 pg/mL.

2.2.4. Simultaneous Inference: Bonferroni's Multiplicity Adjustment

Similarly, one may proceed to use the information from the six NYHA class III patients to obtain a confidence interval estimate of the mean BNP level of NYHA class III patients. Although this is very easy to do computationally, one must be clear about the methodological consequence of making multiple inferences in one experiment. For one thing, the 95% confidence level that we are citing is only valid for each individual inference, not for the entire experiment. In other words, when we construct a 95% confidence interval for the mean BNP level of NYHA class II patients, we allow 5% error in our inference. When we repeat the same procedure for the NYHA class III patients, we allow another 5% error. Thus, the overall error rate for the entire experiment will be larger than the professed 5%. Intuitively, this is not difficult to understand: If there is only one inference, we have 5% chance to make a mistake. If we conduct 100 inferences, each with 5% error rate, we will make 5 mistakes in 100 inferences on average just by chance. Therefore, as the number of the inferences increases, the confidence level we have for the entire experiment decreases. This can be a very serious problem when a large number of inferences are involved. For example, one microarray experiment may involve several thousand genes. If we set the inference error rate at 5%, we could expect to see hundreds of invalid inferences in one experiment. To guard against such inflated error rate, we often need to control the experiment-wise error (i.e., the overall chance of making an incorrect inference) rather than the error associated with a specific inference. A general class of methods for such adjustments are often referred to as simultaneous inference.

One of the popular simultaneous inference procedures is called the Bonferroni method. The basic idea of Bonferroni's approach is to reduce the individual inference error rate to a lower level so that the experiment-wise error rate can be controlled at the nominal level of $\alpha 100\%$. Specifically, if there are k inferences, we construct the $(1 - \alpha)100\%$ confidence interval using $t_{\alpha/(2k)}$ instead of $t_{\alpha/2}$.

Bonferroni's adjustment is based on the well-known Bonferroni inequality, a probability *inequality*. So the method is not exact, (i.e., the resulting experiment-wise error rate would be *no more than* α). At times, the Bonferroni adjustment may result in an experiment-wise error rate much less than the stated level. Therefore, it is often regarded as one of the more conservative multiplicity adjustment procedures.

Using the Bonferroni procedure, we can calculate the mean BNP levels for NYHA classes II and III patients as (255.6, 568.4) and (315.4, 1146.6), respectively. Note that these intervals are considerably wider than the unadjusted ones, suggesting the loss of precision as a result of the increased error rate. It should be pointed out that the Bonferroni adjustment may be too conservative (i.e., resulting intervals are too wide) for many applications. To alleviate, many alternative adjustment methods have been proposed (see **Chapter 7** for a more in-depth discussion on the subject).

2.2.5. Confidence Interval for the Variance

Occasionally, our research requires inference on population variances. In **Section 2.1**, we have stated that the point estimate of the population variance σ^2 is its sample counterpart s^2 (i.e., $\hat{\sigma}^2 = s^2$). This section describes the construction of a confidence interval estimate of σ^2 using the sampling distribution of s^2. Specifically, let $X_1, \ldots, X_n$ be a random sample from a normally distributed population with mean μ and variance σ^2. The quantity $(n - 1)s^2/\sigma^2$ follows a chi-square distribution with $n - 1$ degrees of freedom, that is,

$$\frac{(n-1)s^2}{\sigma^2} \sim \chi^2_{(n-1)}.$$

We therefore have

$$P\left(\chi^2_{1-\alpha/2} \le \frac{(n-1)s^2}{\sigma^2} \le \chi^2_{\alpha/2}\right) = 1-\alpha,$$

where $\chi_{1-\alpha/2}$, and $\chi_{\alpha/2}$ respectively correspond with the $(1 - \alpha/2)$th and $(\alpha/2)$th quantiles in a chi-square distribution with $n - 1$ degrees of freedom. From this, we obtain the $100(1 - \alpha)\%$ confidence interval estimate of variance σ^2 as

$$\frac{(n-1)s^2}{\chi^2_{\alpha/2}} \le \sigma^2 \le \frac{(n-1)s^2}{\chi^2_{1-\alpha/2}}.$$

Example 3

To estimate the variation of a particular model of bone density scanner, the manufacturer of the scanners randomly selected four of its machines for testing. In measuring a standard of known density, the four selected scanners produced the following readings: 4.1, 4.0, 3.9, 3.9. Give a 95% confidence estimate of the variance of this model of bone scanner when the scanners are used to measure the standard. We first calculated the sample variance $s^2 = 0.0092$. The 95% confidence estimate of the variance of this model of bone scanner is then

$$\frac{(n-1)s^2}{\chi^2_{\alpha/2}} \le \sigma^2 \le \frac{(n-1)s^2}{\chi^2_{1-\alpha/2}}$$

$$\frac{(4-1)\times 0.0092}{9.3484} \le \sigma^2 \le \frac{(4-1)\times 0.0092}{0.2058}$$

$$0.0030 \le \sigma^2 \le 0.1341,$$

where 0.2058 and 9.3484 are the 2.5th and 97.5th percentiles, respectively, of the chi-square distribution with 3 degrees of freedom. Based on the above calculation, we can claim with 95% confidence that the variance of the measuring instruments is between 0.003 and 0.1341.

2.2.6. One-Sided Confidence Intervals

The discussion of confidence intervals so far has been of two-sided intervals. That is, we provide lower and upper limits on the parameter of interest. We can also define one-sided confidence intervals that consist only of a lower confidence limit or only an upper confidence limit. A lower one-sided $100(1 - \alpha)\%$ confidence interval estimate of θ satisfies the following criterion

$$P(\hat{\theta}_1 \le \theta) = 1 - \alpha.$$

An upper one-sided $100(1 - \alpha)\%$ confidence interval estimate of θ satisfies the following criterion

$$P(\theta \le \hat{\theta}_2) = 1 - \alpha.$$

One-sided confidence intervals are calculated using the similar formulae as for two-sided intervals. Instead of using the $\alpha/2$ and $1 - \alpha/2$ quantiles of the reference distributions, we use either the α or the $1 - \alpha$ quantile.

Example 2 (Continued)

The lower confidence limit of a one-sided 95% confidence interval for the mean BNP level of NYHA class II patients is

$$\bar{X} - t_\alpha \frac{s}{\sqrt{n}},$$

$$412 - 1.7709 \frac{231}{\sqrt{14}},$$

$$(302.7, \infty),$$

where $t_\alpha = t_{0.05} = 1.7709$ was obtained from the t-distribution with d.f. $= 14 - 1 = 13$. We would conclude that we are 95% confident that the mean BNP level of NYHA class II patients is greater than 302.7 pg/mL. The upper confidence limit of a one-sided 95% confidence interval for the mean BNP level of NYHA class II patients is

$$\bar{X} + t_\alpha \frac{s}{\sqrt{n}},$$

$$412 + 1.7709 \frac{231}{\sqrt{14}},$$

$$(-\infty, 521.3).$$

We would conclude with 95% confidence that the mean BNP level of NYHA class II patients is less than 521.3 pg/mL. In practice, we would use one of the one-sided confidence intervals or a two-sided interval.

3. Hypothesis Testing

3.1. Understanding Hypothesis Testing

Another form of statistical inference is hypothesis testing. Hypothesis testing uses sample information to decide the truthfulness of a prespecified statement concerning a certain population parameter. The procedure leads to a decision of either rejecting or not rejecting the statement. The statement being tested is often referred to as a *hypothesis*.

A typical hypothesis testing procedure involves 5 steps:

1. Formulating the null hypothesis (denoted as H_0) and the alternative hypothesis (H_a).
2. Specifying the significance level (α).
3. Computing the value of the test statistic.
4. Determining the rejection region.
5. Stating a conclusion.

The essence of this 5-step procedure is to specify a proposition and then use the sample data to disprove it. The first step is to formulate the proposition of interest into a testable hypothesis. To be more exact, we formulate two contradicting statements: the *null hypothesis* (H_0) and the *alternative hypothesis* (H_a). By giving two contradicting hypotheses, we will force the decision maker to either reject or not reject the null hypothesis. Theoretically speaking, it should not matter which statement is specified as the "null" and which as the "alternative." However, for the convenience of defining the types of errors later in the chapter, we consider statements representing the *status quo*, *no change*, or *equal* as the null. The alternative is then reserved for statements implying a *change*, *inequality*, *greater than*, *less than*, and so forth. Under this convention, when we formulate the hypotheses into mathematical expressions, the equals sign "=" always appears in the null statement.

Using a generic notation, we denote the parameter of interest as θ. For tests concerning θ, there are three commonly encountered pairs of hypotheses:

1. two-sided test: $H_0: \theta = \theta_0$ versus $H_a: \theta \neq \theta_0$;
2. one-sided test: $H_0: \theta \geq \theta_0$ versus $H_a: \theta < \theta_0$;
3. one-sided test: $H_0: \theta \leq \theta_0$ versus $H_a: \theta > \theta_0$,

where θ_0 is the hypothesized value for the test.

In theory and in practice, hypotheses of interest come in various forms and are by no means restricted to the three pairs listed above. In this chapter, however, we shall use the hypotheses of such simple forms to illustrate a typical testing procedure. The same reasoning process is easily extended to hypotheses of difference forms.

The second step in testing a hypothesis is to specify a *significance level*. The significance level (denoted as α) is a *prespecified* maximum probability of incorrectly rejecting the null hypothesis when it is true. It represents the maximum amount of risk that one is willing to take when he or she rejects the null hypothesis. Naturally, we would like to set this level low. For example, the most commonly used significance level is $\alpha = 0.05$. As we will discuss later in the chapter, the significance level represents the maximum allowable type I error of the inference procedure.

The third step is to compute the value of the test statistic. Hypothesis testing is about making a decision using the information that comes from a sample. A *test statistic* is a quantity that measures the discrepancy between the null hypothesis and the sample data. Its value depends on the sample[1] and its sampling distribution has to be known when the null hypothesis is true. In interpreting the statistic, a "large" value of the test statistic implies a more pronounced disagreement between the data information and the null hypothesis, suggesting that the data do not support the null statement and the null statement should be rejected.

But how large is large? That is the key question we address in the fourth step. From an operational point of view, we need a rule to help us to decide whether the test statistic is indeed too large for the null hypothesis to be true. Such a rule is often expressed as a *rejection region*. Loosely speaking, a rejection region is simply a range of values such that we reject the null hypothesis if the test statistic falls in this region.

Finally in the fifth step, we state our conclusion of the test in the context of the problem.

The basic idea behind these 5 steps is very similar to that of a *proof by contradiction*. Let's compare the two:

1. Proof by contradiction: We first assume that **A** is true. From **A** we derive **B**. If **B** is known to be false, then we claim that **A** *cannot* be true. Thus the assumption that **A** is true must be rejected.
2. Hypothesis testing: We first assume that the null hypothesis is true. Under this assumption, the test statistic follows a known sampling distribution. If the test statistic takes a value that implies the data are *probably* contradicting the null hypothesis, then we reject the null hypothesis and conclude that H_0 is *unlikely* to be true.

To illustrate, we consider a classic example of proof by contradiction by Euclid (c.325–265 BC):

Example 4

Suppose that we want to prove that there does not exist a "largest prime number." We first assume that there exists a largest prime number, which we refer to as p. Then we let x be 1 plus the product of all prime numbers between 1 and p (i.e., $x = 1 \cdot 2 \cdot 3 \cdot 5 \cdot 7 \cdot 11 \cdots p + 1$). Clearly, x has no prime factors between 1 and p because dividing x by any of the prime factors would leave a remainder of 1. Hence by definition, x is a prime number. Because $x > p$, we have found a prime number that is larger than the "largest prime number" p. Therefore, p cannot be the largest prime number, and we must reject the assumption that there exists a largest prime number.

[1] So it is a *statistic* and a *random* variable!

The logic of a hypothesis testing is almost entirely parallel to that used in the proof of contradiction. The only difference is that in hypothesis testing, we use probability statements instead of the "absolute" judgments.

3.2. One-Sample t-Test

We now reconsider the PSA example as a testing problem. In the process, we will construct a test statistic for inference about the population mean.

As we discussed earlier, prostate specific antigen (PSA) is an important biomarker for prostate cancer. Ideally, the postoperation PSA value should drop to an undetectable level (i.e., 0 ng/mL) shortly after surgery. But in most patients, PSA values will gradually increase over time. Visually examining the PSA values in the sample, we see that 1 year after surgery, the PSA values in most patients are no longer zero. The question is whether the mean PSA level of this patient population is significantly higher than zero. To answer this question, we test the following hypotheses: $H_0: \mu = 0$ ng/mL versus $H_a: \mu > 0$ ng/mL, where μ is the mean PSA level measured 12 months after the surgery. The rejection of the null hypothesis implies that the PSA has become detectable.

In a more generic notation, the null hypothesis can be written as $H_0: \mu = \mu_0$ and the alternative as $H_a: \mu > \mu_0$. We assume that the random sample $X_1, \ldots,$ X_n comes from a normal population. *When the null hypothesis is true*, the population mean is $\mu = \mu_0$ and the quantity $T = \sqrt{n}(\bar{X} - \mu_0)/s$ follows a t-distribution with $n - 1$ degrees of freedom.

Let us assume that H_0 is true. Because the density of a t-distribution can be characterized as a bell-shaped curve that is symmetrical around zero under the null hypothesis, $T = \sqrt{n}(\bar{X} - \mu_0)/s$ is likely to take a value near the distribution center (zero). Therefore, in the unlikely event of an extremely large T value, we have to wonder whether we have made a mistake in assuming $H_0: \mu = 0$ to be true. In fact, the further away T is from zero, the more likely that H_0 is false. Following this logic, when T takes a very large value, we will have no other choice but to reject the null hypothesis H_0.

Examining the structure of T, we see that the magnitude of the T value reflects the level of discrepancy between the null hypothesis $H_0: \mu = \mu_0$ and the sample information. When the two agree, the T value tends to be small in the absolute value. When the two disagree, the magnitude of $|T|$ will be large (i.e., T takes on an unlikely value). When T takes an unlikely value, we reject H_0. It is now clear that T plays a role that is instrumental in deciding whether we should reject the null hypothesis or not. In statistics, quantities such as T are often referred to as *test statistics*. They are statistics because their values depend on the sample data. They can be used in hypothesis testing because they have known sampling distributions under the null hypotheses so we know which values are likely and which ones are unlikely.

Table 2
Reject Regions for the Test of Population Mean μ

Alternative hypothesis	Rejection region
$\mu < \mu_0$	$T < -T_\alpha$
$\mu > \mu_0$	$T > T_\alpha$
$\mu \neq \mu_0$	$T < -T_{\alpha/2}$ or $T > T_{\alpha/2}$

With the value of the test statistic T calculated, we are ready to decide on the hypotheses. The range of T values that leads to the *rejection of the null hypothesis is called the rejection region*. Intuitively, the reject regions have to be in the tails of the *t*-curve. **Table 2** summarizes the rules we use to define the rejection regions.

Now we complete the PSA example following the 5-step procedure. Let μ be the mean PSA level measured 1 year after proctectomy.

1. $H_0: \mu = 0\,\text{ng/mL}$ versus $H_a: \mu > 0\,\text{ng/mL}$.
2. We set the significance level at $\alpha = 0.05$.
3. Computing the value of the test statistic, we have

$$T = \frac{\bar{X} - \mu_0}{s/\sqrt{n}} = \frac{0.2267 - 0}{0.5252/\sqrt{30}} = 2.3642.$$

4. For the *t*-distribution with d.f. = 30 – 1 = 29 degrees of freedom, $T_{0.05} = 1.699$. So we should reject the null hypothesis if $T > 1.699$. Because $T = 2.3642 > 1.699$, we reject the null hypothesis.
5. The sample evidence indicates that 1 year after proctectomy, the mean PSA level in prostate cancer patients is significantly greater than zero.

3.3. An Alternative Decision Rule: P *Value*

The 5-step procedure introduced above involves the use of a *preselected* significance level α. So the maximum tolerable risk of incorrectly rejecting the null hypothesis is *subjectively* determined. The procedure is not particularly flexible, because in rejecting the null hypothesis, we only know that the error rate is not more than α, and have no idea how large the error rate actually is.

In hypothesis testing, the actual probability of incorrectly rejecting the null hypothesis given the sample observations is often called the *P* value or the *observed* significance level of the test. By definition, if a test yields a "large" *P* value, then there will be a "large" risk of committing an error in rejecting the true null hypothesis. A small *P* value, on the other hand, implies that there is little chance of making a mistake if the null hypothesis is rejected. Therefore, as a decision rule, *we reject the null hypothesis when the P value is small*. When this happens, we say the test result is statistically significant.

The computation of the P value in a hypothesis test is simple: Because we reject the null hypothesis only when the value of the test statistic is in the tail(s), the tail area defined by the test statistic represents the probability of incorrectly rejecting the null hypothesis. To be more exact, we only need to compute the corresponding tail area under the t density curve using the value of the test statistic T as a cutoff point in a one-tail situation. For the cases involving two-sided tests, we simply multiply the one-side tail area by 2 to take into account the areas in both tails.

Thus, we have an alternative procedure of hypothesis testing using the P value:

1. Formulating the null hypothesis (H_0) and the alternative hypothesis (H_a).
2. Computing the value of the test statistic.
3. Computing the P value.
4. Stating a conclusion.

We now illustrate this new procedure by reanalyzing the PSA data. Let μ be the mean postoperation PSA level of prostate cancer patients.

1. $H_0: \mu = 0\,\text{ng/mL}$ versus $H_a: \mu > 0\,\text{ng/mL}$.
2. Computing the value of the test statistic, we have

$$T = \frac{\bar{X} - \mu_0}{s/\sqrt{n}} = \frac{0.2267 - 0}{0.5252/\sqrt{30}} = 2.3642.$$

3. The tail area that corresponds with $T \geq 2.3642$ is $P(T \geq 2.3642) = 0.0125$. This P value of 0.0125 says if we reject the null hypothesis, there will be a 1.25% chance for us to make a mistake. Because the risk for an error is very small, we choose to reject the null hypothesis.
4. This testing procedure again confirms that the mean PSA level in prostate cancer patients is greater than zero 1 year after proctectomy.

In practice, data analysts usually have the aid of various statistical computing programs to carry out computational tasks required by inference procedures. Thus, the issue is often not so much about the computation of the P values but how to interpret them. For example, in the PSA data example, we have a P value of 0.0125 and we decide that 1.25% error rate is a small risk to take when we reject the null hypothesis. But there is no scientifically compelling reason for others to share our decision rule. In fact, a frequently asked question is when is a P value considered small and, therefore, when should a significant test result be declared? Unfortunately, there is no short answer to this question that is universally agreed upon by all scientists. Whereas some scientists favor the adoption of 0.05 as the threshold for the rejection of H_0 in all situations, others argue for more flexibility. From the perspective of statistical decision making, the P value is simply an indicator of the strength of data evidence against the null hypothesis. The stronger the evidence against H_0, the smaller the P value and the less the chance of incorrect rejection. Under this general principle, with a given P value, the investigator always retains the flexibility to reject or not to reject the null hypothesis. But in case she decides to reject, we will know the chance of a mistake. By reporting this P value, the investigator has given the readers a chance to assess the strength of the evidence. This is why many scientists consider the P value method more advantageous than the rejection region method.

3.4. Errors, Power, and Sample Size

No decision-making process is completely error-free. Although our discussion has so far focused on the error rate of incorrectly rejecting a null hypothesis, statistical testing is actually subject to more than one type of error. One makes a mistake in rejecting the null hypothesis if the null is true (called a type I error); and one makes a different kind of mistake by not rejecting the null when it is false (called a type II error).

Table 3 summarizes the situations leading to the occurrence of errors.

Here, α is the type I error rate (i.e., the conditional probability of rejecting the null hypothesis given that the null hypothesis is true); β is the type II error rate (i.e., the conditional probability of failure to reject the null hypothesis when the null is false). Ideally, one hopes to keep both α and β low. Journal editors routinely require the authors to specify the maximum value of the type I error rate (α) or the P values for statistical tests, which is not particularly difficult to do. But they rarely ask for a type II error rate β. Indeed, specifying a type II error rate is a significantly more difficult task to accomplish, because this error can happen under many different values of the alternative hypothesis. Once the data are collected, we are left with no real options to control this error. Therefore, we often try to control it in the design stage of the experiment. To be more exact, we can control the *power* $(1 - \beta)$ of the test by properly adjusting the sample size n. We usually require the power to be at least 80% (which implies that the type II error rate β is at most 0.2).

Table 3
Types of Errors in Hypothesis Tests

Decision	H_0 is true	H_a is true
Reject H_0	Type I error (α)	Correct decision $(1 - \beta)$
Do not reject H_0	Correct decision $(1 - \alpha)$	Type II error (β)

For a large sample test of a population mean μ, the sample size required to guarantee a $(1-\beta)$ power at significance level α is

$$n = \left[\frac{(z_{\alpha/2} + z_{\beta})s}{\mu_1 - \mu_0}\right]^2 = \left[\frac{(z_{\alpha/2} + z_{\beta})s}{\Delta}\right]^2,$$

where Δ is the difference to be detected by the test, and $z_{\alpha/2}$ and z_{β} are the z values corresponding with the right tail areas of $\alpha/2$ and β, respectively, in a standard normal distribution.

Again, we consider the PSA example.

Example 5

Assuming a standard deviation of 0.6 ng/mL, if we intend to detect a 0.20 pg/mL change in the mean PSA level with 0.05 significance level and 80% power in a two-sided test, how many patients do we need for the test?

Sample size calculation for the PSA example: Following the above formula, the required sample size can be calculated as

$$n = \left[\frac{(z_{\alpha/2} + z_{\beta})s}{\Delta}\right]^2 = \left[\frac{(1.96 + 0.84) \times 0.6}{0.2}\right]^2 \approx 71.$$

3.5. Statistical Significance and Practical Significance

When we reject a null hypothesis with a small P value, we say the result is *statistically significant*. In most practical situations, this simply means that a change (or difference) has been detected by our test. It does not imply that the change (or the difference) itself is of any practical importance. For example, it is possible that we detect a change of 2 mm Hg in systolic blood pressure between two patient groups. The result is statistically significant, but clinically the change is quite trivial.

On the other hand, when there is a change of real clinical significance, a statistical test may not be able to detect it if the sample size is not large enough (insufficient sample size leads to the lack of testing power)! Keep in mind that these two concepts originate from two different decision-making processes, and we should not equate one with the other.

Having studied principles of testing a statistical hypothesis, we now face the ultimate challenge: how to make our statistical tests practically relevant? This question should be addressed from two different aspects:

1. The hypotheses should reflect practically significant changes. In other words, the amount of change we intend to test should be of some practical importance.
2. When we design the experiments, we should always choose a sample size that gives us a reasonable chance to detect a practically meaningful change. In the

absence of a careful sample size estimate, a statistically nonsignificant test result should never be interpreted as evidence of "no difference."

Acknowledgments

The author thanks Regenstrief Institute for the kind support that it has provided during the writing of this manuscript.

References

1. Mood, A. M., Graybill, F. A., and Boes, D. C. (1974) *Introduction to the Theory of Statistics*, 3rd ed. New York, McGraw-Hill.
2. Arnold, S. F. (1990) *Mathematical Statistics*. Englewood Cliffs, Prentice-Hall.
3. Maisel, A. S., Krishnaswamy, P., Nowak, R. M., McCord, J., Hollander, J. E., Duc, P., Omland, T., Storrow, A. B., Abraham, W. T., and Wu, A. H. B. (2002). Rapid measurement of B-type natriuretic reptide in the emergency diagnosis of heart failure. *N. Engl. J. Med.* **347**(3), 161–167.

5

Statistical Inference on Categorical Variables

Susan M. Perkins

Summary

Categorical data are data that capture a characteristic of an experimental unit (such as a tissue specimen) rather than a numerical value. In this chapter, we first describe types of categorical data (nominal and ordinal) and how these types of data are distributed (binomial, multinomial, and independent multinomial). Next, methods for estimation and making statistical inferences for categorical data in commonly seen situations are presented. This includes approximation of the binomial distribution with a normal distribution, estimation and inference for one and two binomial samples, inference for *2×2* and *R×C* contingency tables, and estimation of sample size. Relevant data examples, along with discussions of which study designs generated the data, are presented throughout the chapter.

Key Words: Binomial distribution; chi-square test; Fisher's exact test; McNemar's test.

1. Introduction

1.1. What Is Categorical Data?

Categorical data is data that captures a characteristic of an experimental unit (such as a tissue specimen) rather than a numerical value. For example, a Western blot is a laboratory procedure that can be used to identify and quantify protein levels in a sample. When performing a Western blot, a researcher may be interested in the presence or absence of a protein (a characteristic) rather than a numerical value indicating the relative amount of protein expression.

There are several types of categorical data:

Nominal data: The levels of a variable do not have an inherent ordering. Examples include race (African American, Asian, Caucasian, or Other) and blood phenotype (A, B, AB, or O).

From: *Methods in Molecular Biology, vol. 404: Topics in Biostatistics*
Edited by: W. T. Ambrosius © Humana Press Inc., Totowa, NJ

Ordinal data: There is an inherent ordering of the levels of a variable, but it cannot be assumed that the differences between two adjacent levels are the same in magnitude. Examples include disease severity (mild, moderate, or severe) and weight (underweight, normal weight, overweight, obese). Note that in these examples, there is no reason to assume that the increase in disease severity from mild to moderate is the same as the increase from moderate to severe or that differences in weight would be the same when comparing underweight to normal-weight individuals as when comparing normal-weight to overweight individuals.

Categorical data is often described using either percentages or counts. For example, in the general U.S. population, blood phenotype can be described as 42% A, 10% B, 45% O, and 3% AB. When the interest is in describing the cross-classification of 2 categorical variables, the data is often placed into a table as shown in **Example 1**.

Example 1

In a study looking at the associations between variants in the *IGF2* gene and Beckwith-Wiedemann syndrome (a fetal overgrowth disorder), researchers collected the data shown in **Table 1** on a particular polymorphism (*T123C*) of the gene in subjects with and without the syndrome ***(1)***. Such tables are referred to as frequency or contingency tables and are further characterized by the number of levels of the row variable (*R*) and column variable (*C*). **Table 1** is a 2×2 table because there are 2 levels of presence of disease (absent or present) and 2 levels of variant type (1 or 2).

1.2. Categorical Data Distributions

When there are only two levels of a categorical variable, the variable is often called a *binary* or *dichotomous* variable. Binary variables can be thought of and treated statistically as either nominal or ordinal. When there are more than 2 levels of a categorical variable, the variable is called a *multinomial* or *polytomous* variable. These variables can be classified as either nominal or ordinal

Table 1
Beckwith-Wiedemann Syndrome and T123C Genetic Variant

Beckwith-Wiedemann syndrome	Variant		
	Type 1	Type 2	Total
Absent	79	157	236
Present	50	96	146
Total	129	253	382

but not both. In order to develop an understanding of categorical data distributions, let us consider the following 2 examples, one a case-control study and the other a cross-sectional study.

Example 1 (Continued)

Using the Beckwith-Wiedemann example shown in **Table 1**, the researchers selected 2 patient samples, one sample having Beckwith-Wiedemann syndrome (cases) and the other not (controls). In each sample, the researchers then looked for the presence of the 2 variant types. In this case, within each sample, the numbers of subjects with variant type 1 and variant type 2 follow binomial distributions.

Example 2

Researchers were interested in developing a prediction rule to predict the development of diabetes in older adults **(2)**. In **Table 2**, all 1549 participants in the study were cross-classified on whether they had lower or higher triglyceride levels and whether or not they had a normal or abnormal glucose tolerance test (GTT). In this example, a single sample of subjects is observed at one point in time, so the 2×2 table is from a single multinomial sample.

1.3. General Notation

Consider a sample of n objects (people, rats, etc.) where the sample is divided into several levels (e.g., survived or died). Let n_j be the number of objects in level j, $j = 1$ to J. The counts $\{n_1, \ldots, n_J\}$ follow what is called a *multinomial distribution*. As a special case, if the number of levels is 2 ($J = 2$), the counts are said to follow a *binomial distribution*. Another representation of a multinomial distribution arises when the sample of n objects is cross-classified by 2 factors (such as triglyceride level and GTT levels in **Example 2**). In this case, the resulting counts are denoted as $\{n_{ij}, i = 1 \text{ to } I, j = 1 \text{ to } J\}$. The general convention is that i indexes the row variable and j the column variable.

Table 2
Glucose Tolerance Test and Triglyceride Level

	Triglycerides		
GTT level	<150 mg/dL	≥150 mg/dL	Total
Normal	858	177	1035
Abnormal	364	150	514
Total	1222	327	1549

When a researcher obtains I independent samples, instead of obtaining a sample of n objects, and classifies each sample into J levels (as in **Example 1**), the resulting counts have an *independent* multinomial distribution, still denoted by $\{n_{ij}, i = 1 \text{ to } I, j = 1 \text{ to } J\}$. In the notation above, we have specified the independent sample an object is taken from as the row variable and the levels each sample is classified into as the column variable; however, in the methods described in this chapter, it does not matter in practice which is the row or column variable.

The only distinction between the multinomial and independent multinomial distributions is how many independent samples are obtained (1 or more than 1, respectively). Fortunately, as will be noted further in **Section 4.1** and **Section 4.4**, the statistical tests have the same form, regardless of whether the counts have a multinomial or independent multinomial distribution.

1.4. Statistical Analysis Using Categorical Data

Many estimation and inferential methods used for performing statistical analyses for categorical data utilize either normal or chi-square distribution approximations. These approximations are valid for large sample sizes. When the sample size is small enough so that the statistics cannot be assumed to have these distributions, estimation and inference can sometimes be based on the exact distribution of the data.

In the following sections, we will first discuss how to approximate the binomial distribution with the normal distribution. Second, we will examine estimation and inference for a proportion from a single population and compare 2 proportions from 2 independent populations. Third, we will discuss how the chi-square distribution is used to test the association between two variables and how to test for association when the assumptions of the chi-square test do not hold. Finally, we will discuss inference for 2 dependent proportions.

2. The Binomial Distribution and the Normal Approximation to the Binomial Distribution

2.1. The Binomial Experiment

A binomial experiment is an experiment where an independent sample of n experiments is performed, with each experiment having the same binary outcome. A simple example would be 100 tosses of a coin. In this case, the experiment is tossing the coin, the sample size n is 100, and the binary outcome is whether the coin landed heads up or tails up. In general, assume the values of a binary outcome are either success (e.g., heads) or failure (e.g., tails). In the binomial experiment, it is assumed that the probability of success, p, is the same for each experiment. It is also assumed that the probability of failure is equal to $1 - p$. For example, when tossing a fair coin, p would be equal to 0.5.

Natural questions that arise in this experimental setting include: What is the probability of observing exactly X successes in n trials? What is the probability of observing at least X successes in n trials? The binomial distribution can be used to answer these questions.

2.2. The Binomial Distribution

In the binomial distribution, the random variable X is defined as the number of successes in n trials, where the probability of success for each trial is p. $P(X = x)$ can be calculated using the formula:

$$P(X = x) = \binom{n}{x} p^x (1-p)^{n-x}$$

where

$$\binom{n}{x} = \frac{n!}{x!(n-x)!}.$$

The symbol ! stands for *factorial*, and for any positive whole number a, $a!$ (spoken as "a factorial") = $a(a-1)\ldots 2\cdot 1$. For example, $4! = 4\cdot 3\cdot 2\cdot 1 = 24$. For the special case of 0, $0! = 1$. To calculate $P(X \le x)$, the above formula can be used to calculate the $P(X = x)$ for each value of X that is less than or equal to x, and then these probabilities are simply summed. This is written in mathematical notation as:

$$P(X \le x) = \sum_{i=0}^{x} \binom{n}{x} p^i (1-p)^{n-i}. \quad (1)$$

For any given n and p, the mean of X is np, and the variance of X is $np(1-p)$.

2.3. The Normal Approximation to the Binomial

When n is small, there are published tables that assist in the calculation of probabilities using the above formulas. However, this becomes cumbersome and unnecessary when n is large. According to the *central limit theorem*, for large n and when p is not too close to 0 or 1, X approximately follows a normal distribution with mean np and variance $np(1 - \mathrm{p})$. In this case, the calculations for the binomial distribution can be approximated using the normal distribution as follows:

$$P(X = x) = P\left(\frac{X - np}{\sqrt{np(1-p)}} = \frac{x - np}{\sqrt{np(1-p)}}\right) = P(Z = z)$$

where Z has a standard normal distribution. When is it appropriate to use this approximation? A rule of thumb is to use this approximation when both $n\hat{p} \ge 5$ and $n(1-\hat{p}) \ge 5$ where $\hat{p} = x/n$.

A correction called the *continuity correction* is usually applied in this setting. Because the normal distribution takes on continuous values and the binomial distribution only takes on integer values, ±0.5 should be added to the values of interest when approximating the integer values of the binomial distribution with the continuous values of the normal. That is, we approximate $P(X = x)$ by an interval $P(x - 0.5 \leq X \leq x + 0.5)$ and then use the normal approximation.

Example 3

Suppose the probability of having an abnormal fasting plasma glucose level is 0.3. In a sample of size 100, what is the approximate probability that between 20 and 40 subjects (inclusive) will have abnormal fasting plasma glucose levels?

Let X be the number of subjects with abnormal glucose levels. Note that the mean of this binomial is $np = 100(0.3) = 30$, and the standard deviation is $\sqrt{np(1-p)} = \sqrt{100(0.3)(0.7)} = 4.58$.

$$P(20 \leq X \leq 40) = P\left(\frac{20-0.5-30}{4.58} \leq \frac{X-30}{4.58} \leq \frac{40+0.5-30}{4.58}\right)$$
$$= P(-2.29 \leq z \leq 2.29)$$

Based on the standard normal table of probabilities, this probability is 0.9781. If we calculate this exactly as:

$$P(20 \leq X \leq 40) = \sum_{x=20}^{x=40} \binom{100}{x} (0.3)^x (1-0.3)^{100-x},$$

we get a probability of 0.9786.

3. Estimation and Testing of Single Proportions/Two Proportions

3.1. Estimation of a Single Proportion or the Difference Between Two Proportions

The following example will be used to demonstrate estimation of a proportion and differences in two proportions in the single- and two-population settings, respectively.

Example 4

In a prospective cohort study, troponin T levels were obtained for a sample of 801 subjects who had been hospitalized with acute myocardial ischemia *(3)*. Whether or not the subjects died within 30 days was then obtained and is summarized in **Table 3**. What is an overall estimate of dying within 30 days for the subjects with high troponin T levels? What is an estimate of the difference

Table 3
Survival versus Troponin T Levels

	Troponin T level		
Status	>0.1 ng/mL	≤0.1 ng/mL	Total
Alive	255	492	747
Dead	34	20	54
Total	289	512	801

in proportion of death between the subjects with high versus low troponin T levels?

In the single population setting, the point estimate of a single proportion p is $\hat{p} = x/n$ and the standard error of p is $\mathrm{SE} = \sqrt{p(1-p)/n}$. Thus, a $100 \times (1 - \alpha)\%$ interval estimate of p is

$$\hat{p} \pm z_{\alpha/2}\mathrm{SE} \quad or \quad \hat{p} \pm z_{\alpha/2}\sqrt{\frac{\hat{p}(1-\hat{p})}{n}}$$

where $z_{\alpha/2}$ corresponds with the upper $\alpha/2 \times 100$ percentile of the standard normal distribution. For **Example 4**, the point estimate of the proportion of subjects with high troponin T levels dying within 30 days is $34/289 = 0.12$, and the estimate of the standard error is $\sqrt{(0.12)(0.88)/289} = 0.02$. For a 90% interval estimate, the appropriate $z_{\alpha/2}$ is 1.64. Thus, a 90% interval estimate would be $0.12 \pm 1.64(0.02)$ or (0.09, 0.15).

For estimating the difference between the proportions from two independent populations, p_1 and p_2, note that the standard error of the difference in two independent proportions is

$$\sqrt{\frac{p_1(1-p_1)}{n_1}+\frac{p_2(1-p_2)}{n_2}}$$

so that a $100 \times (1 - \alpha)\%$ interval estimate of $p_1 - p_2$ is

$$\hat{p}_1 - \hat{p}_2 \pm z_{\alpha/2}\sqrt{\frac{\hat{p}_1(1-\hat{p}_1)}{n_1}+\frac{\hat{p}_2(1-\hat{p}_2)}{n_2}}.$$

For **Example 4**, the point estimate of the difference in the proportion of subjects dying between the patients with high versus low troponin T levels is $34/289 - 20/512 = 0.12 - 0.04 = 0.08$, and the standard error would be $\sqrt{\frac{(0.12)(0.88)}{289}+\frac{(0.08)(0.92)}{512}} = 0.02$. Again, if one were interested in a 90%

interval estimate of this difference, the estimate would be $0.08 \pm 1.64(0.02)$ or (0.05, 0.11).

3.2. Hypothesis Testing with a Single Proportion or the Difference Between Two Proportions

The data from **Example 2** will be used to demonstrate statistical inferences for a proportion and differences in two proportions in the single- and two-population settings, respectively. For example, among the participants with higher levels of triglycerides, is the proportion of those with abnormal glucose tolerance tests equal to 0.5? Is there a difference in the proportion of participants with abnormal glucose tolerance tests in those with lower and higher triglyceride levels?

Suppose you are interested in testing the hypothesis $H_0: p = p_0$ versus the alternative hypothesis $H_a: p \neq p_0$. This test is commonly referred to as the *binomial test.* For large *n*, this can be accomplished by calculating a *z*-statistic and comparing this statistic to the standard normal distribution. Note that the *z*-statistic is calculated under the assumption that the null hypothesis is true.

$$z = \frac{\hat{p} - p_0}{\text{SE}} = \frac{\hat{p} - p_0}{\sqrt{\dfrac{p_0(1 - p_0)}{n}}}, \quad \text{where} \quad \hat{p} = \frac{x}{n}.$$

For testing whether or not among those with high triglyceride levels the proportion with abnormal glucose tolerance tests is equal to 0.5 in **Example 2**, $\hat{p} = 150/327 = 0.46$, $p_0 = 0.5$ and $\text{SE} = \sqrt{(0.5)(0.5)/327} = 0.03$. In this case, $z = (0.46 - 0.5)/0.03 = -1.33$. Using a level of significance of $\alpha = 0.05$, the critical value for a 2-sided *z*-test is 1.96. Thus, the conclusion would be that there is no evidence that proportion of participants with higher levels of triglycerides with abnormal glucose tolerance tests is different than 0.5 at the 0.05 level of significance. Alternatively, a *P* value of 0.1836 could be obtained from a standard normal distribution table. The *P* value is the probability of observing a test statistic as or more extreme than the one observed given that the null hypothesis is true. A 1-sided test could be performed by simply finding the appropriate critical value for a 1-sided test. For example, if the alternative hypothesis above had been $H_a: p > p_0$, the appropriate critical value would be 1.64 (or the *P* value of 0.0918 could be calculated). See **Chapter 4** for additional discussion of 1-sided tests.

For testing the hypothesis $H_0: p_1 = p_2$ (which is equivalent to $p_1 - p_2 = 0$) versus $H_a: p_1 \neq p_2$, a similar strategy is used. Note that under the null hypothesis, $p_1 = p_2 = p$, and the best estimate of p ($\hat{p}$) is obtained by pooling the data from two samples.

$$z = \frac{(\hat{p}_1 - \hat{p}_2) - 0}{\text{SE}} = \frac{\hat{p}_1 - \hat{p}_2}{\sqrt{\frac{p_1(1-p_1)}{n_1} + \frac{p_2(1-p_2)}{n_2}}} \cong \frac{\hat{p}_1 - \hat{p}_2}{\sqrt{\frac{\hat{p}(1-\hat{p})}{n_1} + \frac{\hat{p}(1-\hat{p})}{n_2}}}$$

$$= \frac{\hat{p}_1 - \hat{p}_2}{\sqrt{\hat{p}(1-\hat{p})\left(\frac{1}{n_1} + \frac{1}{n_2}\right)}} \quad \text{where} \quad \hat{p}_1 = \frac{x_1}{n_1}, \hat{p}_2 = \frac{x_2}{n_2} \quad \text{and} \quad \hat{p} = \frac{x_1 + x_2}{n_1 + n_2}.$$

For the data in **Example 2**, $\hat{p}_1 = 364/1222 = 0.30$, $\hat{p}_2 = 150/327 = 0.46$ and $\hat{p} = (364+150)/(1222+327) = 0.33$. Thus, $\text{SE} = \sqrt{0.33(0.67)\left(\frac{1}{1222} + \frac{1}{327}\right)} = 0.03$ and $z = (0.30 - 0.46)/0.03 = -5.33$. Again, using a level of significance of $\alpha = 0.05$, the critical value for a 2-sided z-test would be 1.96, and the null hypothesis would be rejected. The conclusion would be that the proportion of participants with abnormal glucose tolerance tests are different for those with triglyceride levels lower than 150 mg/dL compared with those with triglyceride levels greater than or equal to 150 mg/dL at the 0.05 level of significance. Again, P values could also be obtained if desired, and 1-sided tests can be performed by using the appropriate critical value.

3.3. Assumptions

In order for the normal approximation to hold for the formulas in this section, n must be large enough, and p and $1 - p$ must be far enough away from zero. How is this determined? A simple rule of thumb is that $n\hat{p}$ and $n(1 - \hat{p})$ (in the one-population setting) or $n\hat{p}_1$, $n(1 - \hat{p}_1)$, $n\hat{p}_2$ and $n(1 - \hat{p}_2)$ (in the two-population setting) should all be greater than 5.

4. Tests of Association

Often, we are interested in whether 2 variables are associated with each other. For example, is the presence or absence of a particular disease associated with a particular gene? Is recurrence of a tumor associated with receiving a certain type of chemotherapy? In the following subsections, the null hypothesis (H_0) is that there is no association between the 2 variables and the alternative hypothesis (H_a) is that there is an association between the 2 variables. We will discuss this in various settings.

4.1. Two-by-Two Tables

Consider again the Beckwith-Wiedemann syndrome data from **Example 1**. An intuitive approach to testing the null hypothesis in this setting would be to compare the observed counts in the 2×2 table (O_{ij}) to what we would expect these counts to be if the null hypothesis is true (E_{ij}). For multinomial distributed counts, when there is no association between the 2 variables (i.e., the rows and the columns are independent), the probability of being in the *i*th row and *j*th column, p_{ij}, is simply equal to the probability of being in the *i*th row times the probability of being in the *j*th column ($p_i \times p_j$), so that the $E_{ij} = np_ip_j$ and $\hat{E}_{ij} = n(R_i/n)(C_j/n) = R_iC_j/n$ where R_i is the row total for the *i*th row and C_j is the column total for the *j*th column. For independent multinomial sampling, although the derivation is different, it is still the case that $\hat{E}_{ij} = R_iC_j/n$. For example, in **Example 1** the expected count when Beckwith-Wiedemann syndrome is *absent* and the variant is *type 1* would be (129)(236)/382 = 79.70.

The following statistic:

$$X^2 = \sum_{i=1}^{I}\sum_{j=1}^{J}\frac{\left(O_{ij} - \hat{E}_{ij}\right)^2}{\hat{E}_{ij}}, \quad \text{where} \quad \hat{E}_{ij} = \frac{R_iC_j}{n}, \tag{2}$$

called *Pearson's chi-square test statistic*, has a chi-square distribution with 1 degree of freedom when the null hypothesis is true. Large differences between the observed and expected counts would indicate that the assumption of no association between the 2 variables is wrong, so the null hypothesis would be rejected for large values of the test statistic. This test is commonly called a *chi-square test of association.*

In **Example 1**, the calculation of the Pearson chi-square test statistic would be

$$X^2 = \frac{(79-79.70)^2}{79.70} + \frac{(157-156.30)^2}{156.30} + \frac{(50-49.30)^2}{49.30} + \frac{(96-96.70)^2}{96.70} = 0.024.$$

For a single degree of freedom, the 95th percentile for the chi-square distribution is 3.84. Thus, the null hypothesis would not be rejected in this case. The conclusion would be that there is no association between the type of variant and Beckwith-Wiedemann syndrome at the 0.05 level of significance. Note that the *P* value in this case is 0.8768 (which is easiest to find using statistical software).

4.2. R×C Tables

Pearson's chi-square test statistic is also used for a general *R*×*C* table. In this case, the degrees of freedom are (*R* – 1)(*C* – 1). Consider the following example.

Example 5

In a retrospective chart review of 124 patients with recurrent laryngeal squamous cell carcinoma ***(4)***, the number of subjects surviving at 2 years was cross-classified by the stage of their disease (classified using the Tumor, Node,

Table 4
Survival Status by Tumor Stage at Diagnosis

Tumor, Node, Metastasis stage	Survival status		Total
	Survived	Died	
Stage I	23 (14.11)	12 (20.89)	35
Stage II	13 (11.29)	15 (16.71)	28
Stage III	10 (12.50)	21 (18.50)	31
Stage IV	4 (12.10)	26 (17.90)	30
Total	50	74	124

Terms in parentheses are the expected counts.

Metastasis [TNM] system) at the time of initial treatment (or decision not to treat). Within each cell of **Table 4**, the observed and expected counts are given, with the expected counts in parentheses. For these data, to test the null hypothesis that there is no association between death and disease stage, the test statistic would be

$$X^2 = \frac{(23-14.11)^2}{14.11} + \frac{(12-20.89)^2}{20.89} + \ldots + \frac{(4-12.10)^2}{12.10} + \frac{(26-17.90)^2}{17.90} = 19.73.$$

The degrees of freedom would be (2 – 1)(4 – 1) = 3. The 95th percentile of a chi-square distribution with 3 degrees of freedom is 7.81, so the null hypothesis would be rejected. Again, using statistical software, the P value is calculated as 0.0002. The conclusion would be that there is an association between death within 2 years of recurrence of laryngeal squamous cell carcinoma and disease stage at the 0.05 level of significance.

4.3. Relationship Between Tests of Independence and Homogeneity

Note that chi-square tests of association are also commonly called *chi-square tests of independence*. When they are conducted from samples drawn from independent multinomial populations, they are also referred to as *tests of homogeneity* because the tests essentially assess whether the multinomial distributions are homogenous across the independent samples. However, **Equation 2** and its associated degrees of freedom are identical regardless of whether one is testing independence or homogeneity.

4.4. Fisher's Exact Test

Using Pearson's chi-square test statistic to test association requires that the sample sizes be large enough so that the test statistics do indeed follow a chi-square distribution. How large is large enough? A rule of thumb is that all cells must have expected counts greater than 5. When the expected counts are smaller than this, an alternative approach is to use *Fisher's exact test.*

First consider the 2×2 table. If the 2 variables are indeed independent (i.e., the null hypothesis is true), and we assume both the row and column totals were prespecified, the resulting cell counts follow what is called a hypergeometric distribution. The idea underlying Fisher's exact test is to generate all 2×2 tables that have the same row and column totals as the observed table. Each of these tables has a probability of occurring (which can be calculated using the hypergeometric distribution formula):

$$P(table) = \frac{R_1!R_2!C_1!C_2!}{n!n_{11}!n_{12}!n_{21}!n_{22}!}.$$

Any table with a probability less than or equal to the table of observed data would support the alternative hypothesis, so the sum of the probabilities of all of these tables would represent the probability of observing a table as or more supportive of the alternative hypothesis as the observed table under the assumption that the null hypothesis is true. This is, in fact, the definition of a *P* value for testing the null hypothesis of no association.

Example 6

As part of a larger cross-sectional study ***(5)***, researchers assessed whether the alteration of a particular gene (*p53*) was associated with elevated expressions of another gene (*p73*). They collected data on 17 samples as presented in **Table 5**. The probability of observing this table under the null hypothesis of

Table 5
Alteration of *p53* and Expression of *p73*

	Elevated expression of p73		
Alteration of p53	Yes	No	Total
Yes	11	3	14
No	0	3	3
Total	11	6	17

no association is $\frac{14!3!11!6!}{17!11!3!0!3!} = 0.0294$. There are 3 other tables with the same row and column totals as the observed table, all of which have higher probabilities under the null hypothesis. The tables (only cell counts are shown for brevity) and their respective probabilities of occurrence are $P\begin{pmatrix} 10 & 4 \\ 1 & 2 \end{pmatrix} = 0.2426$; $P\begin{pmatrix} 9 & 5 \\ 2 & 1 \end{pmatrix} = 0.4853$; and $P\begin{pmatrix} 8 & 6 \\ 3 & 0 \end{pmatrix} = 0.2426$. Thus, the P value for testing no association in this data is equal to the probability for the observed table (P value $= 0.0294$). Note that even this simple example requires a fair amount of computation and would usually be performed with a statistical software package. Fisher's exact test can also be applied to $R \times C$ tables, although the computational effort involved greatly increases. As computers have become faster, this is less of an issue. In fact, larger tables can be routinely analyzed using current computers.

5. McNemar's Test

Occasionally, there is interest in comparing 2 proportions, but the proportions are not independent. Consider the following example:

Example 7

Researchers wanted to compare the usefulness of a new polymerase chain reaction (PCR) assay versus the culture method in the diagnosis of a particular bacterial infection. They obtained the data presented in **Table 6** from tissue specimens taken from 300 mice infected with the bacterium with regard to the presence (+) or absence (–) of the infection. The researchers wanted to know if the proportion of specimens identified as positive by PCR (113/300 = 0.377)

Table 6
PCR and Culture Methods of Diagnosis of Bacterial Infection

	PCR		
Culture	+	–	Total
+	66	7	73
–	47	180	227
Total	113	187	300

was significantly different than the proportion of specimens identified as positive by conventional methods (73/300 = 0.243). The method described in **Section 3.2** is not appropriate here because the 2 proportions are measured on the same sample and not 2 independent samples.

A statistical test that would be appropriate in this situation is *McNemar's test*. Note that in **Example 7**, if the number of specimens testing positive using the culture method and negative using the PCR method (n_{12}) were equal to the number of specimens testing negative using the culture method and positive using the PCR method (n_{21}), the 2 proportions of interest would be equal regardless of how many specimens tested positive using both methods or negative using both methods. Intuitively, the test statistic used to test equality should focus on the difference between these 2 numbers. The formula is in fact:

$$M = \frac{(n_{12} - n_{21})^2}{n_{12} + n_{21}},$$

which follows a chi-square distribution with 1 degree of freedom when the 2 proportions are equal. In **Example 7**, $M = (7 - 47)^2/(7 + 47) = 29.63$, and the critical value for a chi-square test with 1 degree of freedom and $\alpha = 0.05$ level of significance is 3.84. Thus, McNemar's test would indicate that the proportions are not equal. Note that for small samples sizes ($n_{12} + n_{21} \leq 25$), a continuity correction similar to the continuity correction described in **Section 2.3** is used.

6. Sample Size Estimation

Determining the appropriate sample size is a critical step in the development of any research protocol. Using too few subjects can result in the statistical tests having no ability (power) to show a statistically significant difference for the question of interest, and using too many subjects can be costly, both in terms of time and money. Formulas for sample size estimation for categorical variables, as in other settings, are based on the null and alternative hypothesis, the statistical test, α, β, and the difference you wish to detect. For example, the sample size needed *per group* to test the hypothesis that $H_0: p_1 = p_2$ versus $H_a: p_1 \neq p_2$ using the statistical test presented in **Section 3.2** is

$$n = \left[z_\alpha\sqrt{2\bar{p}(1-\bar{p})} + z_\beta\sqrt{p_1(1-p_1)+p_2(1-p_2)}\right]^2 \Big/ (p_1 - p_2)^2$$

where $\bar{p}$ is the average of p_1 and p_2. For example, suppose a researcher wished to design a study that could detect a difference between the proportion of individuals with high triglycerides of 0.3 in a sample of individuals with normal GTT and of 0.4 in a sample of individuals with abnormal GTT with $1 - \beta = 0.8$ (80% power) at a level of significance of $\alpha = 0.05$. In this case, $p_1 = 0.3$,

$p_2 = 0.4$, $\bar{p} = 0.35$, $z_\alpha = 1.96$, and $z_\beta = 0.8416$. Using the formula above, a sample size of $n = 356$ per GTT group would be required. Further discussion of sample size estimation in the categorical as well as in other settings can be found in **Chapter 19**.

7. Discussion

Aside from Fisher's exact test, there are a few other exact methods that are relevant to the settings in this chapter. For example, exact confidence intervals for both the single proportion and difference in 2 proportions are available, as is an exact test for a single proportion (the binomial test) and McNemar's test.

Note that in the equation for the Pearson chi-square statistic, the rows and columns in a table could be reordered without changing the value of the test statistic. In other words, any natural ordering of the levels of the variables if it exists is not taken into account and the variables are treated as if they were nominal. This may lead to a loss of information when either the rows or columns or both are ordinal. There is a more general class of statistical tests called *Cochran-Mantel-Haenszel tests* (or extended Mantel-Haenszel tests) that can more optimally utilize the ordinality of the data when either or both of the row and column variables are ordinally scaled and can also be used when one wants a single test of association between 2 variables and has collected the data on several samples (as is often the case in a multisite clinical trial) ***(6)***.

Finally, the careful reader will notice the similarity between the test of two independent proportions in **Section 3.2** and the chi-square test for association in the 2×2 table in **Section 4.1**. In fact, it can be shown algebraically that the square of the z-statistic in **Section 3.2** is identical to the chi-square statistic in **Section 4.1**. Thus, whether you conduct the test in **Section 3.2** or display the data as a 2×2 table and conduct the test in **Section 4.1**, both tests will produce the same P value. This is relevant particularly when using computer software because the chi-square test for association is commonly available in statistical software, but the z-test for 2 independent proportions is not.

Acknowledgments

Partial support for the development of this chapter was provided by the General Clinical Research Center at Indiana University School of Medicine (M01 RR00750).

References

1. Murrell, A., Heeson, S., Cooper, W. N., Douglas, E., Apostolidou, S., Moore, G. E., Maher, E. R., and Reik, W. (2004) An association between variants in the *IGF2* gene and Beckwith-Wiedemann syndrome: interaction between genotype and epigenotype. *Hum. Mol. Genet.* **13**(2), 247–255.

2. Kanaya, A. M., Wassel Fyr, C. L., de Rekeneire, N., Shorr, R. I., Schwartz, A. V., Goodpaster, B. H., Newman, A. B., Harris, T., and Barrett-Connor, E. (2005) Predicting the development of diabetes in older adults. *Diabetes Care* **28**(2), 404–408.
3. Ohman, E. M., Armstrong, P. W., Christenson, R. H., Granger, C. B., Katus, H. A., Hamm, C. W., O'Hanesian, M. A., Wagner, G. S., Kleiman, N. S., Harrell, F. E., Jr., and others. (1996) Cardiac troponin T levels for risk stratification in acute myocardial ischemia. *N. Engl. J. Med.* **335**(18), 1333–1341.
4. Lacy, P. D., and Piccirillo, J. F. (1998) Development of a new staging system for patients with recurrent laryngeal squamous cell carcinoma. *Cancer* **83**(5), 910–917.
5. Cai, Y. C., Yang, G., Nie, Y., Wang, L., Zhao, X., Song, Y., Seril, D. N., Liao, J., Xing, E. P., and Yang, C. S. (2000) Molecular alterations of *p73* in human esophageal squamous cell carcinomas. *Carcinogenesis* **21**(4), 683–689.
6. Stokes, M. E., Davis, C. S., and Koch, G. G. (1995) *Categorical Data Analysis using the SAS® System*. Cary, SAS Institute.

6

Development and Evaluation of Classifiers

Todd A. Alonzo and Margaret Sullivan Pepe

Summary

Diagnostic tests, medical tests, screening tests, biomarkers, and prediction rules are all types of classifiers. This chapter introduces methods for classifier development and evaluation. We first introduce measures of classification performance including sensitivity, specificity, and receiver operating characteristic (ROC) curves. We then review some issues in the design of studies to assess and compare the performance of classifiers. Approaches for using the data to estimate and compare classifier accuracy are then introduced. Next, methods for combining multiple classifiers into a single classifier are presented. Lastly, we discuss other important aspects of classifier development and evaluation. The methods presented are illustrated with real data.

Key Words: Accuracy; predictive value; receiver operating characteristic (ROC) curve; sensitivity; specificity; study design.

1. Introduction

Recent technologic and scientific advances have led to an explosion in the number of new screening tests, diagnostic tests, and biomarkers that are being developed for the early detection and diagnosis of medical conditions. Often, the earlier a medical condition can be detected or diagnosed, the better the outcome. Diagnostic tests, screening tests, and biomarkers are all types of classifiers that differ in the context in which they are applied. Classifiers can be used in other contexts also, including prognosis and predicting response to treatment.

Diagnostic tests are used to diagnose a particular medical condition. Examples of diagnostic tests include ultrasound and computed tomography for detection of appendicitis, optical immunoassay tests for detection of influenza, and bacterial culture for the detection of infectious diseases. Screening for disease

From: *Methods in Molecular Biology, vol. 404: Topics in Biostatistics*
Edited by: W. T. Ambrosius © Humana Press Inc., Totowa, NJ

is differentiated from diagnosis in that screening often occurs in nonsymptomatic populations, whereas diagnosis typically takes place in symptomatic populations. Examples of screening tests include prostate specific antigen (PSA) for prostate cancer, mammography for breast cancer, Papanicolaou (PAP) for cervical cancer, fecal occult-blood test for colorectal cancer, and serum cholesterol and blood pressure for cardiovascular disease. Prognosis can be considered a special type of diagnosis where the condition to be detected is a clinical outcome of interest. For example, the Framingham risk score uses the age, gender, total cholesterol, high-density lipoprotein (HDL) cholesterol, systolic blood pressure, and tobacco use of an individual to estimate the 10-year risk of a heart attack and coronary death.

Classifiers may have binary, ordinal, or continuous results. Examples of tests with binary results include home pregnancy tests and bacterial cultures, which are either positive or negative. A radiologist's interpretations of images to quantify the suspicion of cancer are usually based on the following 5-point ordinal scale: 1 = normal, 2 = benign, 3 = probably benign, 4 = suspicious for cancer, and 5 = highly suspicious for cancer. Examples of continuous tests include tumor-marker concentrations such as PSA for detecting prostate cancer and otoacoustic emissions tests for detecting hearing impairment.

To help make the ideas more concrete, screening and diagnostic tests are primarily used to illustrate the concepts throughout the chapter. However, the material presented applies more generally to any classification task. Disease status (D) as determined by a *gold standard* is assumed to be measured without error ($D = 1$ indicates disease and $D = 0$ indicates no disease). **Section 6.2** describes the impact of errors in measuring disease status on evaluating the performance of classifiers. Let Y be the test result. Without loss of generality, it is presumed that larger values of Y are more indicative of disease.

Before classifiers are applied in practice, it is imperative that the performance of the classifiers be evaluated. In **Section 2**, we define and motivate measures of classification performance. Study design issues to answer relevant questions, to avoid bias, and to ensure efficient use of resources are discussed in **Section 3**. For a detailed discussion on sample size calculations, see **Chapter 8** of Pepe *(1)* for classifiers in particular and **Chapter 19** of this text for a general treatment of sample size considerations. Approaches for using the data to estimate and compare test accuracy are provided in **Section 4**. In settings where the accuracy of a single test is not satisfactory, combining the results of multiple tests is often considered with the hope of the combined test having better performance (**Section 5**). Other important aspects of classifier development and evaluation are presented in **Section 6**.

2. Measures of Classification Accuracy

The diagnostic accuracy of a test is the test's ability to discriminate among alternative states of health; for example, cancer versus cancer-free or more generally diseased versus not diseased. In this section, we define several measures of accuracy.

2.1. True- and False-Positive Fractions

Consider binary tests that are either positive for disease ($Y = 1$) or negative for disease ($Y = 0$). The accuracy of these tests is often summarized with the *true-positive fraction* (TPF) and the *false-positive fraction* (FPF). The TPF is the proportion of diseased subjects detected by the screening test. That is, TPF is the conditional probability that a diseased subject screens positive, $P(Y = 1|D = 1)$. On the other hand, the FPF is the proportion of nondiseased subjects erroneously deemed positive by the screening test, $P(Y = 1|D = 0)$. TPF is also referred to as the *sensitivity* and FPF is equal to 1 – *specificity*. The TPF quantifies the key benefit of screening, (i.e., disease detection), whereas the FPF quantifies a key disadvantage of screening because subjects that have false-positive results are sent for workup procedures or treatments that are often costly in both human and monetary aspects. Therefore, when comparing tests, it is important to consider how tests compare in regard to both TPF and FPF. A perfect test has TPF = 1 and FPF = 0. Conversely, a noninformative test (i.e., no better than flipping a coin) is such that TPF = FPF.

Example 1

A study was conducted to determine the ability of ultrasound (US) to diagnose childhood appendicitis *(2)*. **Table 1** summarizes the results of US for 283 children compared with their true appendicitis status. Histopathology and follow-up questionnaire were the gold standard used to determine the true appendicitis status for the children. Note that US is a binary test that is positive for appendicitis (US = 1) or negative for appendicitis (US = 0). **Table 1**

Table 1
Results of Ultrasound and Appendicitis Status

	Appendicits		
	$D = 1$	$D = 0$	
US = 1	94	9	103
US = 0	15	165	180
	109	174	283

indicates that 109 of the children were determined to have appendicitis by the gold standard. Of these 109 children, 94 tested positive with US. Therefore, the TPF for US is 94/109 = 0.862. Of the 174 children determined to not have appendicitis, 9 falsely had positive US results. Thus, the FPF for US is 9/174 = 0.0517, or equivalently the specificity is 1 – 0.0517 = 0.9483.

2.2. Predictive Values

The true- and false-positive fractions quantify how well the test reflects true disease status. Clinicians and patients may be more interested in the predictive value of a test (i.e., the probability a subject has the disease given the results of a test) rather than the TPF and FPF. *Positive predictive value* (PPV) is the probability of disease in those with a positive test result, $P(D = 1|Y = 1)$. *Negative predictive value* (NPV) is the probability of not having the disease when the test result is negative, $P(D = 0|Y = 0)$. A perfect test has PPV and NPV both equal to 1. Conversely, a noninformative test has PPV equal to the population prevalence of disease and NPV equal to one minus the prevalence of disease.

Example 1 (Continued)

Of the 103 children who tested positive with US, 94 had appendicitis (**Table 1**). Therefore, the PPV is 94/103 = 0.91. Similarly, of the 180 children who tested negative with US, 165 did not have appendicitis. Thus, NPV is 165/180 = 0.92.

Predictive values depend not only on the TPF and FPF of the test but also on the *prevalence of disease* in the population in which the test is performed. Specifically,

$$\text{PPV} = \frac{\text{TPF} \times \text{prevalence}}{\text{TPF} \times \text{prevalence} + \text{FPF} \times (1 - \text{prevalence})} \quad \text{and}$$

$$\text{NPV} = \frac{(1 - \text{FPF}) \times (1 - \text{prevalence})}{(1 - \text{FPF}) \times (1 - \text{prevalence}) + (1 - \text{TPF}) \times \text{prevalence}}.$$

Note that this is just Bayes' rule (see **Chapter 16** for more on this). It is evident from the first expression that settings with low disease prevalence can yield low PPV even for tests with good TPF and FPF. The second expression suggests that settings with high disease prevalence can yield low NPV for tests with good TPF and FPF.

2.3. Diagnostic Likelihood Ratios

Diagnostic likelihood ratios (DLRs) are another way to quantify the performance of a binary test. The positive DLR (DLR^+) is defined as the probability of a positive test result in diseased subjects divided by the probability of a positive test result in nondiseased subjects. Similarly, the negative DLR (DLR^-) is defined as the probability of a negative test result in diseased subjects divided by the probability of a negative test result in nondiseased subjects. That is,

$$\text{DLR}^+ = \frac{P(Y=1|D=1)}{P(Y=1|D=0)} = \frac{\text{TPF}}{\text{FPF}} \quad \text{and}$$

$$\text{DLR}^- = \frac{P(Y=0|D=1)}{P(Y=0|D=0)} = \frac{(1-\text{TPF})}{(1-\text{FPF})}.$$

Example 1 (Continued)

In **Section 2.1**, it was estimated that the TPF and FPF of US were 0.862 and 0.0517, respectively. Inserting these estimates into the above equations yields that DLR^+ is 0.862/0.0517 = 16.7 and DLR^- is (1 – 0.862)/(1 – 0.0517) = 0.15. The DLR^+ estimate indicates that for every 16.7 children with appendicitis correctly classified, one child without appendicitis is incorrectly classified. Similarly, the estimate of DLR^- indicates that for every 1/0.15 = 6.7 children without appendicitis correctly classified, one child with appendicitis is incorrectly classified.

Before a diagnostic test is performed, the *odds of disease* is $P(D = 1)/P(D = 0)$, where $P(D = 1)$ is the prevalence of disease in the population. This is also referred to as the pretest odds. The odds of disease after the test is performed (i.e., posttest odds) is $P(D = 1|Y)/P(D = 0|Y)$. DLRs relate the pretest and posttest odds as follows:

$$\text{posttest odds with positive test result} = \text{DLR}^+ \times \text{pretest odds}$$
$$\text{posttest odds with negative test result} = \text{DLR}^- \times \text{pretest odds}.$$

Therefore, the DLRs quantify the change in the odds of disease obtained by knowledge of the result of the diagnostic test. A perfect test has DLR^+ and DLR^- of ∞ and 0, respectively. Conversely, a noninformative test has DLR^+ and DLR^- both equal to 1.

Example 1 (Continued)

The prevalence of appendicitis in this example is 109/283 = 0.39. Therefore, the pretest odds of appendicitis is 0.39/(1 – 0.39) = 0.63. Inserting the estimate of pretest odds as well as DLR^+ = 16.7 and DLR^- = 0.15 (calculated above) into the equations above, we obtain that the pretest odds of appendicitis are increased to 16.7 × 0.63 = 10.5 by a positive US. Conversely, the pretest odds are decreased to 0.15 × 0.63 = 0.09 by a negative US.

2.4. ROC Curves

Receiver operating characteristic (ROC) *curves* are a well-accepted measure of accuracy for tests with continuous or ordinal results. Our presentation here focuses on continuous tests. ROC curves display the trade-offs between the TPF and FPF of the test as the definition of a positive result is varied. By choosing a cutpoint c on the continuous scale, a binary test may be defined such that a test result with $Y \geq c$ is considered positive and if $Y < c$, the test is considered negative. An ROC curve is a plot of the TPF versus FPF associated with such binary tests as the cutpoint c is varied from $-\infty$ to $+\infty$. That is, an ROC curve is a plot of TPF(c) versus FPF(c) for all c, where

$$\text{TPF}(c) = P(Y \geq c|D = 1) \text{ and}$$
$$\text{FPF}(c) = P(Y \geq c|D = 0).$$

ROC curves measure the amount of separation between the distribution of test results in the diseased population from the distribution of test results in the nondiseased population (Fig. 1). When the distributions of test results for the diseased and nondiseased populations completely overlap, then the ROC curve is the 45-degree line from (0, 0) to (1, 1), with FPF(c) = TPF(c) for all c indicating a noninformative test. The more separated the distributions, the closer

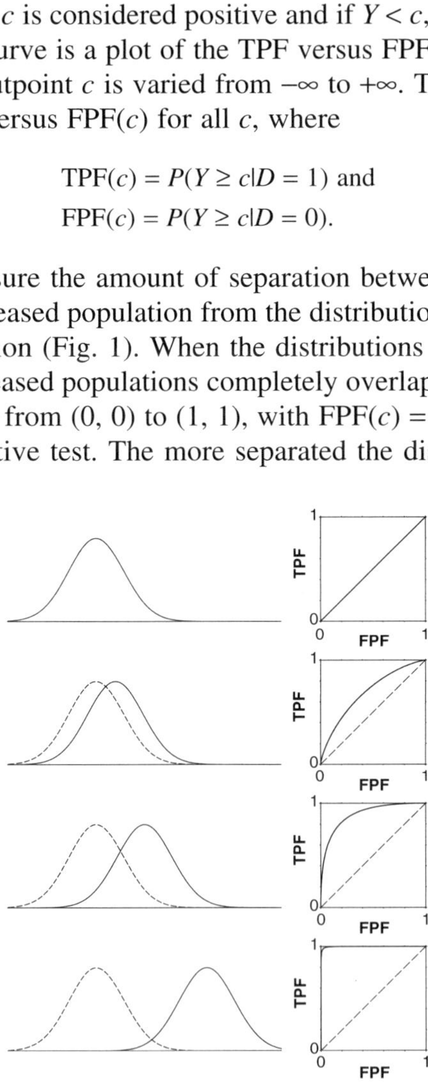

Fig. 1. Left panels: Distributions of test results for diseased population (solid curve) and nondiseased populations (broken curve). Right panels: Corresponding ROC curves (solid curve). Broken line corresponds with a noninformative test.

the ROC curve is to the upper left-hand corner. A curve that reaches the upper left corner, with FPF(c) = 0 and TPF(c) = 1 for some c, corresponds with a perfect test. A key feature of ROC curves is that they can be used to visually compare the accuracy of different tests even when tests are measured either in different units or on completely different scales.

A standard way to summarize the accuracy of a continuous test is to calculate the *area under the* ROC *curve* (AUC). This is equal to 1 for a perfect test and is equal to 0.5 for an uninformative test. Interestingly, the AUC corresponds with the probability that the test result for a randomly chosen diseased subject exceeds that for a randomly chosen nondiseased subject. That is, AUC is $P(Y_1 > Y_2 | D_1 = 1, D_2 = 0)$, where the subscripts 1 and 2 correspond with a diseased and nondiseased subject, respectively. AUC can also be interpreted as the average TPF over the whole range of possible FPF. Restricting attention to a certain region of the ROC curve may be appropriate in certain settings. For screening tests applied to apparently healthy populations, one may only be interested in thresholds that yield low FPF, say. Restricting attention to the ROC curve where FPF is adequately low and calculating the area under that region yields a summary known as the *partial* AUC (pAUC).

2.5. Selecting a Measure of Accuracy

The measure of accuracy appropriate for a particular study will depend on the objective of the study. The study objectives can be used to categorize a study in one of five phases for the development of a medical test *(3)* (**Table 2**). The first phase (phase 1) consists of exploratory investigations to identify promising tests and determine how best to use tests. Typically, phase 1 studies employ a case-control design (see **Section 3.1**) where conveniently available cases (diseased subjects) and controls (nondiseased subjects) are used. Phase 2 rigorously estimates the accuracy (TPF and FPF) of the test using a case-control design where the cases and controls are carefully selected from the population of interest. Phase 3 studies involve defining screen positivity, determining factors that affect test accuracy, comparing promising tests, and developing algorithms for combining tests. They are usually large population-based case-control studies. Prospective application of tests using a cohort study is referred to as phase 4. These studies determine predictive values as well as TPF and FPF when the tests are applied in practice. Phase 5 usually involves randomized prospective trials that compare a new test with the standard of practice. These studies evaluate the costs and benefits and take treatment effects into account. Often, the primary outcomes of interest in phase 5 are mortality and long-term morbidities associated with the disease in the presence and in the absence of the testing program.

Table 2
Phases of Research for the Development of a Medical Test

Phase	Description	Typical objectives	Typical design
1	Exploratory investigations	Identify promising tests and settings for application	Case-control study with convenience sampling
2	Retrospective validation	Rigorously estimate accuracy (TPF and FPF)	Population-based case-control sampling
3	Retrospective refinement	Define screen positivity. Determine factors that affect test accuracy. Compare promising tests. Develop algorithms for combining tests	Large-scale population-based case-control study
4	Prospective application	Determine predictive values, TPF, and FPF when test is applied in practice	Cohort study
5	Disease impact	Determine effects of testing on cost and mortality associated with disease	Randomized prospective trial comparing new test with standard of practice

3. Basics of Study Design

Many of the well-accepted principles for the design of therapeutic studies discussed in **Chapter 1** also apply to the design of studies to compare the accuracies of tests. The Standards for Reporting of Diagnostic Accuracy (STARD) group provided a checklist for reporting results of studies of diagnostic accuracy and hence a list of aspects of design and analysis that should be considered in planning a study *(4,5)*. In this section, we review some basic issues in the design of comparative accuracy studies.

3.1. Case-Control versus Cohort Designs

Studies to compare the accuracy of tests can be performed prospectively or retrospectively. Retrospective studies involve selecting subjects on the basis of their true disease status as determined by the gold standard and performing the tests on them. These retrospective studies are often called case-control studies, where cases are those with disease and controls are those without disease. Prospective studies involve applying the tests to a random sample from the population of interest and determining true disease status for all study subjects. True disease status can be determined concurrently for a cross-sectional cohort study or over a follow-up period for a truly prospective cohort study.

The main advantage of a case-control study is that the overall study size is generally much smaller than it is for a cohort study. The case-control design, however, has the following disadvantages: (i) the spectrum of cases might not be representative of those in the population because the cases were identified in the absence of the screening tests under study. This is known as spectrum bias (**Table 3**); (ii) because PPV and NPV depend on disease prevalence in the population and this cannot be estimated from the data, neither can predictive values; (iii) the impact of screening on mortality and morbidity and on costs associated with screening cannot be assessed. It is for these reasons that case-control studies in phases 1, 2, and 3 often precede the conduct of a cohort screening study in phase 4, the latter being more definitive but also more costly.

Table 3
Common Sources of Bias

Type of bias	Description
Verification bias	Nonrandom selection for definitive assessment of disease with the gold standard reference test
Errors in the reference	True disease status is subject to misclassification because the gold standard is imperfect
Spectrum bias	Types of cases and controls included are not representative of the population
Test interpretation bias	Information is available that can distort the diagnostic test result
Unsatisfactory tests	Tests that are uninterpretable or incomplete do not yield a test result
Extrapolation bias	The conditions or characteristics of populations in the study are different from those in which the test will be applied
Lead time bias[a]	Earlier detection by screening may erroneously appear to indicate beneficial effects on the outcome of a progressive disease
Length bias[a]	Slowly progressing disease is overrepresented in screened subjects relative to all cases of disease that arise in the population
Overdiagnosis bias[a]	Subclinical disease may regress and never become a clinical problem in the absence of screening, but is detected by screening

[a]Applies only to screening for preclinical states of a progressive disease.

3.2. Paired versus Unpaired Designs

Studies designed to compare multiple screening tests can have a *paired design*, where all tests are performed on each individual. Alternatively, an *unpaired design* can be employed, where each individual is allocated (ideally randomly) to one of the tests. Pairing is often desirable because it can reduce variability in making comparisons among tests by eliminating between-subject variance. Therefore, pairing is usually a more efficient design requiring smaller sample sizes. However, if the adminstration of one test interferes with the results of another test, an unpaired design may be necessary.

3.3. Blinding

Blinding is just as important in the context of screening test studies as it is in therapeutic studies. In particular, the screening tests and gold standard tests should be performed without knowledge or interference of other test results. Furthermore, in cohort studies the determination of true disease status with the gold standard test should be performed without knowledge of the screening test results. On the other hand, in case-control studies, the determination of the screening test results should be obtained in the absence of knowledge of true disease status so the performance of the screening test in the study will more closely reflect its performance when applied in practice.

3.4. Avoiding Bias

It is important that studies of diagnostic tests are carefully designed and analyzed to avoid bias. **Table 3** describes biases that are frequently encountered in diagnostic test studies. (See **Chapter 1** of Pepe *(1)* for a discussion of each bias.)

3.5. Factors Affecting Test Performance

There are many factors that can affect the performance of a test. Such factors could include demographic attributes of the subjects tested (e.g., age, gender, race), characteristics or severity of their disease (e.g., histology and stage in cancer), characteristics of controls (e.g., benign disease or nondiseased), characteristics of testers (e.g., experience, institution), and conditions under which the tests are performed. It is important to identify and understand the influence of these factors because: (i) populations and settings where a test is more or less accurate can be identified, which can be useful in determining how best to use a test (this can be accomplished using regression analysis, which is discussed briefly in **Section 6.3**); (ii) study results may not be relevant to populations with different conditions or characteristics. This is referred to as extrapolation bias (**Table 3**). In unpaired designs where subjects receive only one of the screening tests, subjects should be randomized to the test that they receive, and care should be taken to balance the randomization in regard to factors that might influence the performance of each test.

4. Estimating Performance from Data

4.1. Single Binary Test

Consider a cohort study with n study subjects who receive a binary test. Using the notation in **Table 4**, the performance of the test can be estimated as follows:

$$\widehat{\text{TPF}} = \frac{n_D^+}{n_D}, \quad \widehat{\text{PPV}} = \frac{n_D^+}{n^+}, \quad \widehat{\text{DLR}}^+ = \widehat{\text{TPF}}/\widehat{\text{FPF}},$$

$$\widehat{\text{FPF}} = \frac{n_{\bar{D}}^+}{n_{\bar{D}}}, \quad \widehat{\text{NPV}} = \frac{n_{\bar{D}}^-}{n^-}, \quad \widehat{\text{DLR}}^- = (1-\widehat{\text{TPF}})/(\widehat{\text{FPF}}),$$

where the hat symbol (ˆ) denotes an estimate of the quantity.

The estimators of TPF, FPF, and predictive values are proportions, so confidence intervals can be constructed using standard approaches appropriate for binomial proportions. When the parameters are near 0 or 1, normal approximation confidence intervals may extend beyond 0 and 1. In these settings, logistic transformed confidence intervals may be preferred.

Confidence intervals for $\log \text{DLR}^+$ and $\log \text{DLR}^-$ can be constructed using the following variance expressions:

$$\text{var}(\log \widehat{\text{DLR}}^+) = \frac{1-\text{TPF}}{n_D \text{TPF}} + \frac{1-\text{FPF}}{n_{\bar{D}} \text{FPF}} \quad \text{and}$$

$$\text{var}(\log \widehat{\text{DLR}}^-) = \frac{\text{TPF}}{n_D(1-\text{TPF})} + \frac{\text{FPF}}{n_{\bar{D}}(1-\text{FPF})}.$$

Table 4
Data from a Study Investigating the Performance of a Test *Y* Relative to the True Disease Status *D*

	$D = 1$	$D = 0$	
$Y = 1$	n_D^+	$n_{\bar{D}}^{\pm}$	n^+
$Y = 0$	n_D^-	$n_{\bar{D}}^-$	n^-
	n_D	$n_{\bar{D}}$	n

The estimators of TPF, FPF, DLR^+, and DLR^- given above can also be applied to data from a case-control study. The above estimators of predictive values are not appropriate for a case-control study. However, the formulas in **Section 2.2** can be used to estimate the predictive values if population prevalence is known or can be estimated.

Example 1 (Continued)

Using the ultrasound data in **Table 1**, we obtain the following estimates and corresponding 95% logistic confidence intervals: $\widehat{TPF} = 0.86$ (0.78, 0.92), $\widehat{FPF} = 0.05$ (0.03, 0.10), $\widehat{PPV} = 0.91$ (0.84, 0.95), and $\widehat{NPV} = 0.92$ (0.87, 0.95). Furthermore, DLR^+ is estimated to be 16.7 ($\log \widehat{DLR}^+ = 2.81$), and DLR^- is estimated to be 0.15 ($\log \widehat{DLR}^+ = -1.93$). Note that these estimates of accuracy are identical to those obtained in **Section 2**. The 95% confidence intervals for $\log DLR^+$ and $\log DLR^-$ are (2.17, 3.45) and (–2.40, –1.46), respectively. Taking the exponential function of the confidence limits yields confidence intervals of (8.8, 31.6) for DLR^+ and (0.09, 0.23) for DLR^-.

Because it is important to consider both TPF and FPF when assessing the performance of a test, joint confidence regions rather than univariate confidence intervals should be used. A joint $(1 - \alpha)$ confidence region can be formed by the rectangle $(TPF_L, TPF_U) \times (FPF_L, FPF_U)$, where (TPF_L, TPF_U) and (FPF_L, FPF_U) are $(1 - \alpha^*)$ level confidence intervals for TPF and FPF, respectively, and $1 - \alpha^* = (1 - \alpha)^{½}$.

Example 1 (Continued)

To calculate a joint 95% confidence region for TPF and FPF, we first note that $\alpha = 5\%$. Therefore, $1 - \alpha^* = (1 - 0.05)^{½} = 0.975$. So we calculate 97.5% confidence intervals for TPF and FPF of ultrasound. These are (0.77, 0.92) and (0.02, 0.10), respectively. These results indicate that with 95% confidence the TPF is between 0.77 and 0.92 *and* the FPF is between 0.02 and 0.10.

4.2. Comparison of TPF and FPF for Two Binary Tests

We next consider comparing TPFs and FPFs for two binary screening tests (test **A** and test **B**). Subscripts are added to the parameters to indicate the test. There are different metrics that can be used for comparisons. Consider, for example, the TPFs of two tests: one can calculate the absolute difference $TPF_A - TPF_B$, the odds ratio $\frac{TPF_A(1-TPR_B)}{TPF_B(1-TPF_A)}$, or the ratio TPF_A/TPF_B. The ratio has a straightforward interpretation and also has advantages in that statistical inference on the relative scale is less difficult than inference on the absolute scale, and the interpretation on the relative scale is less awkward than that for odds ratios ***(1)***. For these reasons, the relative true-positive fraction rTPF(A, B) = $\text{TPF}_A/\text{TPF}_B$ and relative false-positive fraction rFPF(A, B) = $\text{FPF}_A/\text{FPF}_B$ are considered in this chapter (where the lowercase r denotes ratio).

By noting that the null hypothesis $H_0: \text{TPF}_A = \text{TPF}_B$ is equivalent to $H_0: \text{rTPF(A, B)} = 1$, we focus on constructing confidence intervals for rTPF and rFPF. It can then be concluded that tests have different classification probabilities if the confidence intervals do not contain the value 1.

4.2.1. Unpaired Design

Using data collected from an unpaired design where $n(A)$ subjects receive test **A** and $n(B)$ subjects receive test **B** (**Table 5**), rTPF and rFPF can be estimated as the ratios of the estimated true- and false-positive fractions

$$r\widehat{\text{TPF}}(A, B) = \frac{\widehat{\text{TPF}}_A}{\widehat{\text{TPF}}_B} = \frac{n_D^+(A)/n_D(A)}{n_D^+(B)/n_D(B)} \quad \text{and}$$

$$r\widehat{\text{FPF}}(A, B) = \frac{\widehat{\text{FPF}}_A}{\widehat{\text{FPF}}_B} = \frac{n_{\bar{D}}^+(A)/n_{\bar{D}}(A)}{n_{\bar{D}}^+(B)/n_{\bar{D}}(B)}.$$

To construct confidence intervals, we first do so for the log rTPF and log rFPF: $\log r\widehat{\text{TPF}} \pm Z_{1-\alpha/2}\text{var}(\log r\widehat{\text{TPF}})^{1/2}$ and $\log r\widehat{\text{FPF}} \pm Z_{1-\alpha/2}\text{var}(\log r\widehat{\text{FPF}})^{1/2}$, where $Z_{1-\alpha/2}$ is the $100(1-\alpha/2)$ percentile from the standard normal distribution. The expressions for the variances are

$$\text{var}(\log r\widehat{\text{TPF}}) = \frac{1-\text{TPF}_A}{n_D(A)\text{TPF}_A} + \frac{1-\text{TPF}_B}{n_D(B)\text{TPF}_B} \quad \text{and}$$

$$\text{var}(\log r\widehat{\text{FPF}}) = \frac{1-\text{FPF}_A}{n_{\bar{D}}(A)\text{FPF}_A} + \frac{1-\text{FPF}_B}{n_{\bar{D}}(B)\text{FPF}_B}.$$

Table 5
Data from an Unpaired Study

	$D = 1$			$D = 0$		
	$Y = 1$	$Y = 0$		$Y = 1$	$Y = 0$	
Test **A**	$n_D^+(A)$	$n_D^-(A)$	$n_D(A)$	$n_{\bar{D}}^+(A)$	$n_{\bar{D}}^-(A)$	$n_{\bar{D}}(A)$
Test **B**	$n_D^+(B)$	$n_D^-(B)$	$n_D(B)$	$n_{\bar{D}}^+(B)$	$n_{\bar{D}}^-(B)$	$n_{\bar{D}}(B)$

Taking the exponential of the confidence limits yields confidence intervals for rTPF and rFPF.

Example 1 (Continued)

We have already assessed the ability of US to detect pediatric appendicitis. The aim of this study was to compare the abilities of US alone and computed tomography (CT) performed in addition to US to diagnose childhood appendicitis ***(2)***. The study was a randomized prospective cohort study where 600 children with suspected appendicitis were randomized to receive US alone or US with abdominal CT (US + CT). **Table 6** summarizes the results of US and US + CT compared with the true appendicitis status. Using these data, we estimate that TPF(US + CT) = 133/135 = 0.985 and TPF(US) = 94/109 = 0.862 so rTPF(US + CT, US) = 0.985/0.862 = 1.14. Similarly, we estimate that FPF(US + CT) = 20/182 = 0.110 and FPF(US) = 9/174 = 0.052 so rFPF(US + CT, US) = 0.110/0.052 = 2.12. Thus, the combination of US and CT appears to detect 14% more cases of appendicitis than US alone. However, 2.12 times more subjects without appendicitis test positive with US + CT than with US alone. The 95% confidence intervals for $\log r\widehat{TPF} = 0.133$ and $\log r\widehat{FPF} = 0.754$ are (0.055, 0.211) and (–0.005, 1.512), respectively. Exponentiating the confidence limits yields confidence intervals of (1.06, 1.23) for rTPF and (0.99, 4.54) for rFPF. Similar to the approach in **Section 4.1**, a 95% joint confidence region for rTPF and rFPF can be constructed using $\alpha^* = 1 - (1 - 0.05)^{1/2} = 0.025$ instead of $\alpha = 0.05$. Estimation of this joint confidence region suggests that with 95% confidence, the rTPF lies in (1.05, 1.25) *and* rFPF lies in (0.89, 5.05). Because the confidence interval for rTPF excludes 1 but the interval for rFPF does not, we conclude that US + CT has superior TPF, but there is no evidence of a difference in FPF between US + CT and US alone.

Table 6
Results of Ultrasound and Ultrasound with CT (US + CT) in Subjects with Appendicitis (*D* = 1) and without Appendicitis (*D* = 0)

	$D = 1$			$D = 0$		
	$Y = 1$	$Y = 0$		$Y = 1$	$Y = 0$	
US	94	15	109	9	165	174
US + CT	133	2	135	20	162	182

4.2.2. Paired Design

Using data collected from a paired design in which all n study subjects receive binary tests **A** and **B** (**Table 7**), we estimate $\widehat{\text{TPF}}_A = n_D^+(A)/n_D$, $\widehat{\text{TPF}}_B = n_D^+(B)/n_D$, $\widehat{\text{FPF}}_A = n_{\bar{D}}^+(A)/n_{\bar{D}}$, and $\widehat{\text{FPF}}_B = n_{\bar{D}}^+(B)/n_{\bar{D}}$. In the ratios, the numbers of diseased (n_D) and nondiseased ($n_{\bar{D}}$) cancel out. Therefore, rTPF(A, B) and rFPF(A, B) can be estimated as $r\widehat{\text{TPF}} = n_D^+(A)/n_D^+(B)$ and $r\widehat{\text{FPF}} = n_{\bar{D}}^+(A)/n_{\bar{D}}^+(B)$, respectively. The following variance estimates can be used to construct confidence intervals for log rTPF and log rFPF *(6)*:

$$\widehat{\text{var}}(\log r\widehat{\text{TPF}}) = (b + c)/[(a + b)(a + c)] \text{ and}$$
$$\widehat{\text{var}}(\log r\widehat{\text{FPF}}) = (f + g)/[(e + f)(e + g)].$$

The resulting confidence intervals can be exponentiated to obtain confidence intervals for rTPF and rFPF.

Example 2

Christenson and others *(7)* described the results of a cardiac troponin T rapid assay (RA) test and an enzyme-linked immunosorbent assay (ELISA) test in 717 hospital patients: 510 with cardiac disease and 207 without cardiac disease. This is a paired design because all patients received both tests. Using the data in **Table 8**, we can compare the abilities of RA and ELISA to diagnose cardiac disease. We calculate $r\widehat{\text{TPF}}$(ELISA, RA) = 222/187 = 1.19 with 95% confidence, interval (1.11, 1.26). Thus, ELISA detects 19% more subjects with cardiac disease than RA. However, $r\widehat{\text{FPF}}$(ELISA, RA) = 68/53 = 1.28 indicating that ELISA also detects 1.28 times more subjects without cardiac disease than the RA test. The corresponding 95% confidence interval for rFPF is (1.07, 1.54). Estimation of the 95% joint confidence region suggests that with 95% confidence rTPF(ELISA, RA) lies in (1.10, 1.28) *and* rFPF(ELISA, RA) lies

Table 7
Paired Study Design Data with Test Results Y_A for Test A and Y_B for Test B

	Diseased ($D = 1$)			Nondiseased ($D = 0$)		
	$Y_B = 1$	$Y_B = 0$		$Y_B = 1$	$Y_B = 0$	
$Y_A = 1$	a	b	$n_D^+(A)$	e	f	$n_{\bar{D}}^+(A)$
$Y_A = 0$	c	d	$n_D^-(A)$	g	h	$n_{\bar{D}}^-(A)$
	$n_D^+(B)$	$n_D^-(B)$	n_D	$n_{\bar{D}}^+(B)$	$n_{\bar{D}}^-(B)$	$n_{\bar{D}}$

Table 8
Results of Rapid Assay and ELISA Tests for Diagnosing Cardiac Disease

	Cardiac ($D = 1$)			Noncardiac ($D = 0$)		
	RA = 1	RA = 0		RA = 1	RA = 0	
ELISA = 1	183	39	222	45	23	68
ELISA = 0	4	284	288	8	131	139
	187	323	510	53	154	207

in (1.04, 1.58). Because both do not contain 1, we conclude that the ELISA test has superior TPF but inferior FPF compared with the RA test.

McNemar's test can also be used to test the null hypothesis $H_0: r\text{TPF}(A, B) = 1$ and similarly to test $H_0: r\text{FPF}(A, B) = 1$ for two binary tests in a paired data setting *(8)*. Specifically, to test the null hypothesis that $r\text{TPF}(A, B) = 1$, the McNemar's statistic $M_D = (b - c)^2/(b + c)$ is compared with a chi-square distribution with 1 degree of freedom. Similarly, the null hypothesis that $r\text{FPF}(A, B) = 1$ is tested by comparing the statistic $M_{\bar{D}} = (f - g)^2/(f + g)$ with a chi-square distribution with 1 degree of freedom.

Example 2 (Continued)

To test the hypotheses that the TPFs and FPFs for the rapid assay and ELISA tests are the same or equivalently that the relative accuracies equal 1, we calculate $M_D = (39 - 4)^2/(39 + 4) = 28.5$ and $M_{\bar{D}} = (23 - 8)^2/(23 + 8) = 7.3$. Comparing the values of M_D and $M_{\bar{D}}$ to a chi-square distribution with 1 degree of freedom yields P values of <0.0001 and 0.007, respectively. These results indicate that the TPFs and FPFs for the two tests are different. These conclusions are consistent with the confidence intervals constructed above.

4.3. Estimating ROC Curves and Summary Indices

Next we discuss two approaches for estimating ROC curves and corresponding AUC summary statistics. The approaches differ in the assumptions they make. Let $Y_{Di}, i = 1, \ldots, n_D$ and $Y_{\bar{D}j}, j = 1, \ldots, n_{\bar{D}}$ be the continuous test results for the diseased and nondiseased subjects, respectively.

4.3.1. Empirical ROC Curve

The first approach we consider is the empirical approach, which makes no assumptions about the distribution of the data. Specifically, this approach estimates the ROC curve using the simple observed estimates of TPF(c) and FPF(c) for all cutpoints. Let $\widehat{\text{TPF}}(c)$ $\widehat{\text{FPF}}(c)$ be the proportion of diseased (nondiseased) subjects with test results at or exceeding c. The empirical ROC curve $\widehat{\text{ROC}}_e$ is then formed by plotting $\widehat{\text{TPF}}(c)$ versus $\widehat{\text{FPF}}(c)$ for each cutpoint. Extrapolation of the empirical ROC curve to all possible cutpoints can be made by connecting observed data points linearly. For data with no ties, adjacent points are connected with horizontal and vertical lines resulting in a step function. As the threshold changes, inclusion of a true-positive result produces a vertical jump of size $1/n_D$, and inclusion of a false-positive result produces a horizontal jump of size $1/n_{\bar{D}}$. When there are ties in the data between diseased and nondiseased test results, both the true-positive and false-positive fractions change simultaneously, resulting in a point displaced both horizontally and vertically from the last point.

Example 3

A case-control study with 90 cases of pancreatic cancer and 51 controls without pancreatic cancer was conducted where all subjects had two serum biomarkers measured: a cancer antigen CA125 and a carbohydrate antigen CA19-9 ***(9)***. The empirical ROC curves for CA125 and CA19-9 are provided in **Figure 2**. The curves indicate that CA19-9 does a better job of discriminating those with pancreatic cancer from those without than does CA125.

In some settings, there is a particular threshold c for which there is interest to determine the variability of the corresponding point on the ROC curve [FPF(c), TPF(c)]. For example, PSA is routinely used to screen for prostate cancer. PSA is measured on a continuous scale, but the conventional criterion

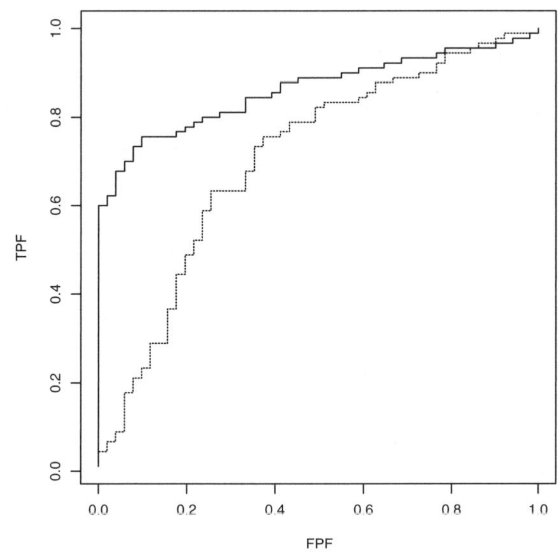

Fig. 2. Empirical ROC curves for CA19-9 (solid curve) and CA125 (broken curve).

for a positive PSA test is PSA > 4.0 ng/mL. Therefore, when evaluating the performance of PSA, there may be interest in estimating the FPF and TPF and corresponding variability when the threshold is fixed at 4. The joint confidence region can be calculated using the methods discussed in **Section 4.1**.

Alternatively, when a predefined threshold does not exist, one may be interested in estimating the TPF that corresponds with the threshold that yields a particular FPF. For example, one might be interested in estimating the TPF for PSA that corresponds with an FPF of 0.2. A variance expression for the estimated TPF is provided on page 101 of Pepe ***(1)***. A method for estimating a confidence band for the ROC curve is also available ***(10)***.

The AUC can be estimated as the area under the empirical ROC curve. It can be shown that this is equal to

$$\widehat{\mathrm{AUC}}_e = \sum_{j=1}^{n_{\bar{D}}} \sum_{i=1}^{n_{\bar{D}}} \left\{ I[Y_{Di} > Y_{\bar{D}j}] + \frac{1}{2} I[Y_{Di} = Y_{\bar{D}j}] \right\} \Big/ n_D n_{\bar{D}}, \tag{1}$$

where $I[\ ]$ is equal to 1 or 0 according to whether or not the expression in square brackets is true. This is equal to the Wilcoxon–Mann–Whitney two-sample statistic for comparing the distributions of test results in the diseased and non-diseased populations ***(11)***. Observe that when there are no tied data points, then **Equation 1** is calculated by comparing each disease test result with each non-disease result. The proportion of pairs where the ordering is "correct" (i.e., $Y_{Di} > Y_{\bar{D}j}$) is the empirical AUC. Variance expressions for $\widehat{\mathrm{AUC}}_e$ are available ***(11–13)***.

Example 3 (Continued)

Applying **Equation 1** to the pancreatic cancer biomarker data yields estimates of the empirical AUC of 0.71 for CA125 and 0.86 for CA19-9. As expected from **Figure 2**, the AUC is larger for CA19-9. Using the variance expression provided by Hanley and McNeil ***(12)***, we estimate that the variance of the empirical AUC is 0.0022 for CA125 and 0.00094 for CA19-9. To construct a 95% confidence interval for the AUC for CA125, we first calculate a 95% confidence interval for logit AUC = log(AUC/(1 − AUC))

$$log\left(\frac{\widehat{\mathrm{AUC}}_e}{1-\widehat{\mathrm{AUC}}_e}\right) \pm Z_{1-\alpha/2} \frac{var(\widehat{\mathrm{AUC}}_e)^{1/2}}{\widehat{\mathrm{AUC}}_e(1-\widehat{\mathrm{AUC}}_e)} = (CI_L, CI_U)$$

which is (0.431, 1.317). Then a 95% confidence interval for the AUC can be calculated as

$$\left(\frac{\exp(CI_L)}{1+\exp(CI_L)}, \frac{\exp(CI_U)}{1+\exp(CI_U)}\right)$$

which equals (0.61, 0.79) for CA125. Similarly, we calculate that the 95% confidence interval for logit AUC for CA19-9 is (1.322, 2.331), which yields a confidence interval of (0.79, 0.91) for the AUC.

The pAUC for a restricted range of FPF can be estimated directly from the estimated ROC curve as the area under the portion of the $\widehat{ROC}_e$ of interest. Bootstrap methods ***(14)*** are recommended for estimating the variance of p$\widehat{AUC}$ ***(15)***.

4.3.2. Binormal ROC Curve

Other approaches can be used for ROC curve estimation. If the test results from the diseased population and the test results from the nondiseased population have normal distributions with means μ_D and $\mu_{\bar{D}}$ and standard deviations σ_D and $\sigma_{\bar{D}}$, then the corresponding ROC curve has the classic *binormal* function form:

$$\mathrm{ROC}(t) = \Phi(a + b\Phi^{-1}(t)), \tag{2}$$

where t is the FPF on the x-axis, ROC(t) is the TPF on the y-axis, Φ is the standard normal cumulative distribution function, a is the intercept, and b is the slope. This binormal model only requires that there exists a monotone transformation of the data that will make the diseased and nondiseased test results normally distributed.

There are different approaches for estimating a and b. One approach assumes that the test results from the diseased population are normally distributed with mean μ_D and standard deviation σ_D, and the test results from the nondiseased population are normally distributed with mean $\mu_{\bar{D}}$ and standard deviation $\sigma_{\bar{D}}$. Then a and b can be estimated using sample means and standard deviations

$$\hat{a} = (\hat{\mu}_D - \hat{\mu}_{\bar{D}})/\hat{\sigma}_D \quad \text{and} \quad \hat{b} = \hat{\sigma}_{\bar{D}}/\hat{\sigma}_D.$$

Other approaches assume that there exists some unknown transformation that makes the test results normally distributed. For example, Pepe ***(16)*** uses regression models to estimate a and b. On the other hand, Metz and others ***(17)*** categorize the continuous test results and then estimate a and b using maximum likelihood by applying the Dorfman-Alf algorithm ***(18)***.

Using the binormal model, ROC(t), the TPF at a fixed FPF = t can be estimated as $\Phi(\hat{a} + \hat{b}\Phi^{-1}(t))$ and the AUC can be estimated using

$$\widehat{AUC} = \Phi\left(\frac{\hat{a}}{(1+\hat{b}^2)^{1/2}}\right). \tag{3}$$

Wieand and others ***(9)*** provide an expression for the variance of this estimate. An approach for constructing confidence bands for the binormal ROC curve is also available ***(19)***.

Example 3 (Continued)

For the pancreatic cancer data, we transformed the marker values to a natural logarithmic scale. The means and standard deviations of the transformed marker values are $(\hat{\mu}_D, \hat{\sigma}_D, \hat{\mu}_{\bar{D}}, \hat{\sigma}_{\bar{D}}) = (3.26, 0.99, 2.67, 0.78)$ for CA125 and $(\hat{\mu}_D, \hat{\sigma}_D, \hat{\mu}_{\bar{D}}, \hat{\sigma}_{\bar{D}}) = (5.42, 2.34, 2.47, 0.86)$ for CA19-9. The slope and intercept for the binormal ROC curve for CA125 are estimated to be $(3.26 - 2.67)/0.99 = 0.60$ and $0.78/0.99 = 0.79$, respectively. Similarly, the slope and intercept for CA19-9 are estimated to be $(5.42 - 2.47)/2.34 = 1.26$ and $0.86/2.34 = 0.37$, respectively. These estimates yield the binormal ROC curves provided in **Figure 3**. Using **Equation 3**, the AUC estimates are 0.68 and 0.88 for CA125 and CA19-9, respectively. These estimates are similar to the empirical estimates 0.71 and 0.86.

4.4. Comparing ROC Curves

In this section, approaches are discussed for comparing two ROC curves. We focus on comparing empirical ROC curves and binormal curves.

4.4.1. Empirical ROC Curves

Two ROC curves can be compared by comparing the corresponding AUC estimates. Let $\Delta\widehat{AUC}_e = \widehat{AUC}_{Ae} - \widehat{AUC}_{Be}$ be the difference in the empirical

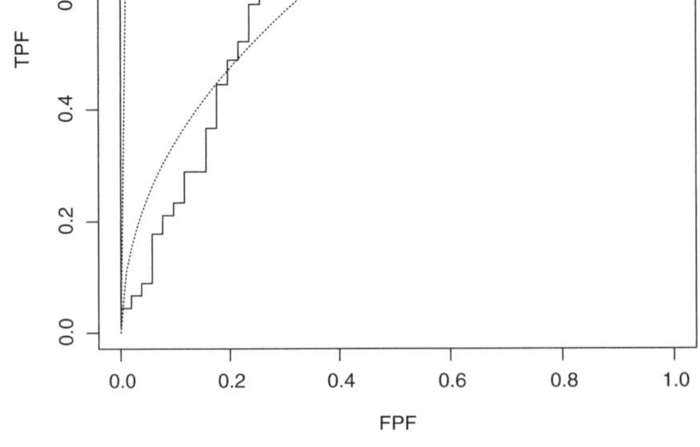

Fig. 3. ROC curves for the pancreatic cancer biomarkers estimated using binormal ROC model (solid curves). Empirical ROC curves are included (dotted curves). The top two curves correspond with CA19-9. The bottom two curves correspond with CA125.

estimates of AUC for test **A** and test **B**. Then the null hypothesis of equal ROC curves for test **A** and test **B** can be tested by comparing $\Delta\widehat{AUC}_e/\text{var}(\Delta\widehat{AUC}_e)^{1/2}$ with a standard normal distribution. If the two ROC curves are estimated using different sets of samples (i.e., an unpaired design), then the variance of $\Delta\widehat{AUC}_e$ can be estimated as the sum of the variances for the AUC estimates for tests **A** and **B**. If the two ROC curves are estimated using the same set of study subjects (i.e., a paired design), then the correlation between the AUC estimates must be taken into account ***(1)***.

Similar to comparing differences in empirical estimates of AUC, if there is particular interest in a restricted portion of the ROC curve, then comparisons of empirical estimates of pAUC can be used ***(9)***.

Example 3 (Continued)

The difference in the AUCs for the empirical ROC curves of the pancreatic cancer biomarkers is $0.86 - 0.71 = 0.15$. Because this study has a paired design, the test statistic must take into account the correlation between the AUC estimates for the two biomarkers. Using the variance expression provided on page 108 of Pepe ***(1)***, we find that the difference in AUCs is statistically significantly different from 0 ($P = 0.007$).

4.4.2. Binormal ROC Curve

There are two approaches for comparing two binormal ROC curves. The first approach is to compare the estimated intercept and slope parameters for the two curves ***(20)***. The second approach is to compare the estimated AUCs for the two tests ***(9)***. Specifically, when the AUCs are equal, $\hat{\Delta}/\widehat{var}(\hat{\Delta})^{1/2}$ has a standard normal distribution where

$$\hat{\Delta} = \left(\frac{\hat{a}_1}{(1+\hat{b}_1^{\,2})^{1/2}}\right) - \left(\frac{\hat{a}_2}{(1+\hat{b}_2^{\,2})^{1/2}}\right)$$

and $(\hat{a}_1, \hat{b}_1)$ and $(\hat{a}_2, \hat{b}_2)$ are the estimated intercept and slope parameters for test **A** and test **B**, respectively.

5. Combining Tests

Most diagnostic tests are not perfect. Sometimes tests yield too many false positives or false negatives to be used in clinical practice. When there are multiple imperfect tests available for detecting a particular disease (as is often the case), there is interest in determining whether combining multiple tests could yield a composite test that more accurately detects presence of disease. This is true for cancer biomarkers. For example, there has been research to determine whether combining the ratio of free to total PSA and total PSA could improve the ability to discriminate those with and without prostate cancer. Combining tests is also popular in clinical practice when a clinician has to make a diagnosis based on numerous sources of information. This information can include test results, signs, symptoms, and medical history. In this section, we review methods available for combining multiple tests with the goal of developing a more accurate composite test.

5.1. Binary Tests

First consider the setting where there are two binary tests to be combined. Two binary tests can be combined by classifying a subject as diseased if both tests are positive and nondiseased otherwise. This is referred to as the *believe the negative (BN) rule* **(21)**, or the *and rule*. The BN rule is more stringent than either test alone. It decreases FPF and decreases TPF relative to the individual tests but maintains the TPF of the composite test above $TPF_A + TPF_B - 1$. Therefore, this combination strategy is used when both tests have high TPF but also have FPF that is too high because the rule decreases FPF while hopefully not reducing TPF very much.

Another approach to combining two binary tests is to consider a subject diseased if either test is positive. This is referred to as the *believe the positive (BP) rule*, or the *or rule*. The BP rule increases TPF relative to the component tests. It also increases the FPF, but by no more than $FPF_A + FPF_B$. This combination strategy is used when the tests have low FPFs but inadequate TPFs.

Example 2 (Continued)

Consider combining the results of the rapid assay (RA) test and ELISA test (**Table 8**). The BP combination has TPF = (183 + 39 + 4)/510 = 0.443 and FPF = (45 + 23 + 8)/207 = 0.37, which are not much better than ELISA, rTPF(BP combination, ELISA) = 0.443/0.435 = 1.02 and rFPF(BP combination, ELISA) = 0.37/0.33 = 1.12. The BN combination is not much better than the RA test, rTPF(BN combination, RA) = 0.36/0.37 = 0.97 and rFPF(BN combination, RA) = 0.22/0.26 = 0.85. Therefore, we conclude that neither combination provides a particularly useful improvement.

5.2. Continuous Tests

Combining multiple continuous tests requires an algorithm to classify subjects as either diseased or not. For example, protein mass spectrometry profiles and gene expression arrays (see **Chapter 20**) yield high-dimensional data. The data are often combined to discriminate between those with and without disease. There are many classification algorithms available. Some of these algorithms estimate the risk scores or predicted probabilities of disease given the test values. We denote the set of test values **Y**. Let RS(**Y**) = $P(D = 1|\mathbf{Y})$ denote the risk score. The risk score has the important property that it or any monotone increasing function of it has the best ROC curve among all possible functions of **Y** *(22)*. It is the best combination for discriminating diseased from nondiseased subjects.

Simple logistic regression (see **Chapter 14**) is the most commonly used method for estimating risk scores. Logistic regression has the appealing property that it can handle data from case-control studies, which are common study designs when evaluating biomarkers. Modifications to logistic regression have been proposed ***(22)*** as well as nonparametric approaches ***(23)***. Other classification algorithms that are used include Bayesian methods, *logic regression* ***(24)***, *classification trees* ***(25)***, *artificial neural networks*, *support vector machines* ***(26)***, and *boosting* ***(27,28)***. The best approach for estimating the risk score depends on the particular data available and goal of the analysis.

Example 3 (Continued)

Using the pancreatic cancer biomarker data, we combine the CA125 and CA19-9 marker values using the following logistic regression model:

$$\operatorname{logit} P(D = 1|\mathbf{Y}) = \beta_0 + \beta_1 \log(\text{CA125}) + \beta_2 \log(\text{CA19-9}).$$

The estimated parameters are $\hat{\beta}_0 = -5.78$, $\hat{\beta}_1 = 0.931$, and $\hat{\beta}_2 = 1.029$. The empirical ROC curve for the combined test is provided in **Figure 4**. The

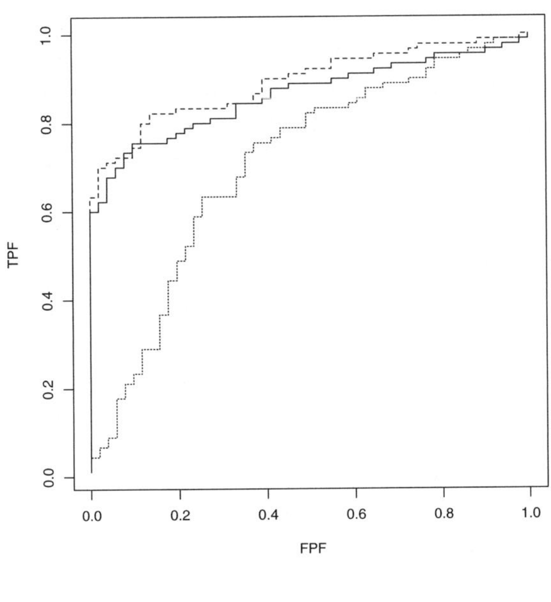

Fig. 4. Empirical ROC curves for CA19-9 (solid line), CA125 (dotted line), and combined test (dashed line).

empirical AUC for the combined test is 0.89, which is only slightly better than the AUC value of 0.86 for CA19-9 alone.

It is well-known that risk scores or prediction rules tend to have worse performance when applied to a different population than the one in which the rule was constructed. There are two reasons for why this occurs. First, the new population may differ from that studied in regard to factors affecting disease or elements in the risk score. Second, there is the statistical phenomenon of *shrinkage* ***(29,30)***. Therefore, it is important to carefully validate the risk score before it is applied in practice. This can be achieved using *external* or *internal validation* methods. Ideally, the performance of the prediction rule is constructed in one data set and validated in another. This is known as external validation. Alternatively, the data set can be split into a training set in which the risk score is developed and a test set where the risk score is evaluated. This is referred to as cross-validation. Approaches for adjusting for shrinkage are available ***(31)***.

6. Additional Topics

6.1. Verification Bias

Sometimes the gold standard test to definitively assess presence or absence of disease is too costly or invasive to be applied to all study subjects. When this is the case, subjects who appear to be at high risk may be more likely to have disease status assessed via the gold standard test than those who appear to be at lower risk. For example, subjects testing positive with the new test may be more likely to be verified for disease than those that screen negative. This can result in biased estimates of accuracy if the estimation methods do not properly account for nonrandom disease ascertainment. This bias is known as *verification bias* (**Table 3**). Methods for estimating accuracy that properly account for the nonrandom disease ascertainment are available ***(32,33)***.

6.2. Errors in the Reference Test

It was assumed in this chapter that disease status, *D*, is defined and measured by a perfect gold standard. Sometimes only an imperfect reference is measured. Imperfect reference tests are sometimes referred to as *bronze* or *alloy standards*. For example, diagnostic tests for *Chlamydia trachomatis*, among other infections, must be evaluated using specimens from persons whose true infection status cannot be known with certainty. Lacking a perfect gold standard, cell culture has been used as an imperfect reference. It is generally accepted that FPF is close to 0 for culture, but TPF is less than 1. It is important to be aware that small errors in imperfect reference tests can lead to a large bias in the estimated TPF and FPF ***(34)***. In addition, the bias can lead to overoptimistic TPF and FPF values or to underestimation of test performance.

6.3. Regression

As noted in **Section 3.5**, it is important to determine factors that affect test performance. This can be achieved using regression methodology. Regression analysis can also be used to compare the performance of multiple tests while controlling for covariates. Specifically, binary regression methods can be used to assess and compare effects on the TPF and FPF ***(35)*** and on predictive values ***(36)*** for one or more binary tests. For continuous and ordinal tests, there are several approaches. One approach is to model covariate effects on the test results using standard regression methods and then calculate the induced effects of covariates on the ROC curves ***(37,38)***. Another approach is to model covariate effects on the ROC curves ***(16,39)*** or on their AUCs ***(15,40,41)***.

6.4. Evaluating Usefulness

After a test is shown to adequately identify presence or absence of disease, the next step is to determine the practical usefulness of the test in managing patients. It is important to perform large prospective phase 5 population-based studies that evaluate the impact that testing has on disease mortality and morbidity, subject willingness to be screened, costs of screening, costs of erroneous screening test results, and benefits relating to detection of subclinical disease. Even an accurate test may not be useful if subjects are not willing to be screened, to undergo further workup and treatment after screening positive, or if treatment is ineffective for disease detected by the test.

7. Conclusion

This chapter reviewed methods for developing and evaluating the performance of classifiers. Diagnostic tests, screening tests, medical tests, biomarkers, and prediction rules can all be considered classifiers. Different measures of classification accuracy along with corresponding estimation methods were presented. In addition, methods for combining multiple tests were introduced. The topics discussed in this chapter are described in greater detail in books by Pepe ***(1)*** and Zhou and others ***(42)***. Software to implement many of the methods discussed in this chapter is available on-line at http://www.fhcrc.org/labs/pepe/book and ftp://ftp.wiley.com/public/sci_tech_med/statistical_methods/.

References

1. Pepe, M. S. (2003) *The Statistical Evaluation of Medical Tests for Classification and Prediction*. Oxford, Oxford University Press.

2. Kaiser, S., Frenckner, B., and Jorulf, H. K. (2002) Suspected appendicitis in children: US and CT—a prospective randomized study. *Radiology* **223**, 633–638.
3. Pepe, M. S., Etzioni, R., Feng, Z., Potter, J. D., Thompson, M., Thornquist, M., Winget, M., and Yasui, Y. (2001) Phases of biomarker development for early detection of cancer. *J. Natl. Cancer Inst.* **93**, 1054–1061.
4. Bossuyt, P. M., Reitsma, J. B., Bruns, D. E., Gatsonis, C. A., Glasziou, P. P., Irwig, L. M., Lijmer, J. G., Moher, D., Rennie, D., and Vet, H. C. W. D. (2003) Towards complete and accurate reporting of studies of diagnostic accuracy: the STARD initiative. *Clin. Chem.* **49**, 1–6.
5. Bossuyt, P. M., Reitsma, J. B., Bruns, D. E., Gatsonis, C. A., Glasziou, P. P., Irwig, L. M., Lijmer, J. G., Moher, D., Rennie, D., and Vet, H. C. W. D. (2003) The STARD statement for reporting studies of diagnostic accuracy: explanation and elaboration. *Clin. Chem.* **49**, 7–18.
6. Cheng, H., and Macaluso, M. (1997) Comparison of the accuracy of two tests with a confirmatory procedure limited to positive results. *Epidemiology* **8**, 104–106.
7. Christenson, R. H., Fitzgerald, R. L., Ochs, L., Rozenberg, M., Frankel, W. L., Herold, D. A., Duh, S. H., Alonsozana, G. L., and Jacobs, E. (1997) Characteristics of a 20-minute whole blood rapid assay for cardiac troponin T. *Clin. Biochem.* **30**, 27–33.
8. Schatzkin, A., Connor, R. J., and Taylor, P. R. (1987) Comparing new and old screening tests when a reference procedure cannot be performed on all screenees. *Am. J. Epidemiol.* **125**, 672–678.
9. Wieand, S., Gail, M. H., James, B. R., and James, K. L. (1989) A family of nonparametric statistics for comparing diagnostic markers with paired or unpaired data. *Biometrika* **76**, 585–592.
10. Hsieh, F., and Turnbull, B. W. (1996) Nonparametric and semiparametric estimation of the receiver operating chacterisitic ROC curve. *Ann. Stat.* **24**, 25–40.
11. Bamber, D. (1975) The area above the ordinal dominance graph and the area below the receiver operating characteristic graph. *J. Math. Psychol.* **12**, 387–415.
12. Hanley, J. A., and McNeil, B. J. (1982) The meaning and use of the area under a receiver operating characteristic (ROC) curve. *Radiology* **142**, 29–36.
13. DeLong, E. R., DeLong, D. M., and Clarke-Pearson, D. L. (1988) Comparing the areas under two or more correlated receiver operating characteristic curves: a nonparametric approach. *Biometrics* **44**, 837–845.
14. Efron, B., and Tibshirani, R. J. (1993) *An Introduction to the Bootstrap*. New York, Chapman & Hall.
15. Dodd, L. E., and Pepe, M. S. (2003) Semiparametric regression for the area under the receiver operating characteristic curve. *J. Am. Stat. Assoc.* **98**, 409–417.
16. Pepe, M. S. (2000) An interpretation for the ROC curve and inference using GLM procedures. *Biometrics* **56**, 352–359.
17. Metz, C. E., Herman, B. A., and Shen, J. H. (1998) Maximum likelihood estimation of receiver operating characteristic (ROC) curves from continuously-distributed data. *Stat. Medi.* **17**, 1033–1053.

18. Dorfman, D. D., and Alf, E. (1969) Maximum likelihood estimation of parameters of signal detection theory and determination of confidence intervals: rating data. *J. Math. Psychol.* **6**, 487–496.
19. Ma, G., and Hall, W. J. (1993) Confidence bands for receiver operating characteristic curves. *Medical Decision Making* **13**, 191–197.
20. Metz, C. E., and Kronman, H. B. (1980) Statistical significance tests for binormal ROC curves. *J. Math. Psychol.* **22**, 218–243.
21. Marshall, R. J. (1989) The predictive value of simple rules for combining two diagnostic tests. *Biometrics* **45**, 1213–1222.
22. McIntosh, M., and Pepe, M. S. (2002) Combining several screening tests: optimality of the risk score. *Biometrics* **58**, 657–664.
23. Baker, S. G. (2000) Identifying combinations of cancer markers for further study as triggers of early intervention. *Biometrics* **56**, 1082–1087.
24. Ruczinski, I., Kooperberg, C., and LeBlanc, M. L. (2003) Logic regression. *J. Computati. Graphical Stat.* **12**, 475–511.
25. Breiman, L., Freidman, J. H., Olshen, R. A., and Stone, C. J. (1984) *Classification and Regression Trees*. Belmont, Wadsworth.
26. Cristianini, N., and Shawe-Taylor, J. (2000) *An Introduction to Support Vector Machines: And Other Kernel-Based Learning Methods*. Cambridge, Cambridge University Press.
27. Schapire, R., Freund, Y., Bartlett, P., and Lee, W. (1998) Boosting the margin: a new explanation for the effectiveness of voting methods. *Ann. Stat.* **26**, 1651–1686.
28. Friedman, L. M., Hastie, T., and Tibshirani, R. (2000) Additive logistic regression: a statistical view of boosting. *Ann. Stat.* **28**, 400–407.
29. Efron, B., and Morris, C. (1977) Stein's paradox in statistics. *Sci. Am.* **236**, 119–127.
30. Copas, J. B. (1997) Using regression models for prediction: shrinkage and regression to the mean. *Stat. Methods Med. Res.* **6**, 167–183.
31. Moons, K. G. M., Donders, A. R. T., Steyerberg, E. W., and Harrell, F. E. (2004) Penalized maximum likelihood estimation to directly adjust diagnostic and prognostic prediction models for overoptimism: a clinical example. *J. Clin. Epidemiol.* **57**, 1262–1270.
32. Begg, C. B., and Greenes, R. A. (1983) Assessment of diagnostic tests when disease is subject to selection bias. *Biometrics* **39**, 207–216.
33. Alonzo, T. A., and Pepe, M. S. (2005) Assessing accuracy of a continuous screening test in the presence of verification bias. *Appl. Stat.* **54**, 173–190.
34. Gart, J. J., and Buck, A. A. (1966) Comparison of a screening test and a reference test in epidemilogic studies. II. A probabilitic model for the comparison of diagnostic tests. *Am. J. Epidemiol.* **83**, 593–602.
35. Leisenring, W., Pepe, M. S., and Longton, G. (1997) A marginal regression modelling framework for evaluating medical diagnostic tests. *Stat. Med.* **16**, 1263–1281.
36. Leisenring, W., Alonzo, T., and Pepe, M. S. (2000) Comparisons of predictive values of binary medical diagnostic tests for paired designs. *Biometrics* **56**, 345–351.

37. Tosteson, A., and Begg, C. B. (1985) A general regression methodology for ROC curve estimation. *Medical Decision Making* **8**, 204–215.
38. Toledano, A. Y., and Gastonis, C. A. (1996) Ordinal regression methodology for ROC curves derived from correlated datta. *Stat. Med.* **15**, 1807–1826.
39. Pepe, M. S. (1997) A regression modelling framework for receiver operating characteristic curves in medical diagnostic testing. *Biometrika* **84**, 595–608.
40. Dorfman, D. D., Berbaum, K. S., and Metz, C. E. (1992) Receiver operating characteristic rating analysis: generalization to the population of readers and patients with the jack-knife method. *Invest. Radiol.* **27**, 723–731.
41. Obuchowski, N. A. (1995) Multireader, multimodality receiver operating characteristic curve studies: hypothesis testing and sample size estimation using an analysis of variance approach with dependent observations. *Acad. Radiol.* **2**, S22–S29.
42. Zhou, X. H., Obuchowski, N. A., and McClish, D. K. (2002) *Statistical Methods in Diagnostic Medicine*. New York, John Wiley & Sons.

7

Comparison of Means

Nancy Berman

Summary

This chapter describes statistical methods to test for differences between means or other measures of central tendency of 2 or more populations. Parametric tests and nonparametric tests are included. Methods for pairwise comparisons when more than 2 groups are being compared are included.

Key Words: Analysis of variance; contrasts; Kruskal-Wallace test; *t*-test; Wilcoxon-Mann-Whitney test; Wilcoxon signed rank test.

1. Introduction

In this chapter, we will present methods to test hypotheses about means. We will use Student's *t*-test to test hypotheses for paired samples or to compare 2 independent groups and we will use analysis of variance (ANOVA) to compare means in more than 2 independent groups. We will also describe tests that may be used when the *t*-test or ANOVA would not be appropriate. These are called nonparametric tests, which compare measures of central tendency that are different from the mean. In addition, we will introduce the F distribution, which is used in ANOVA and also in comparing variances of independent groups.

We will use the following notation:

Parameter	Population value	Sample value
Mean	μ	$\bar{x}$
Standard deviation	σ	s
Variance	σ^2	s^2
Sample size		n
Sample value		x_i

From: *Methods in Molecular Biology, vol. 404: Topics in Biostatistics*
Edited by: W. T. Ambrosius © Humana Press Inc., Totowa, NJ

When there is more than one sample to discuss, we will use numeric subscripts to indicate the groups (e.g., μ_1, $\bar{x}_1$, s_1, n_1 for group 1; μ_2, $\bar{x}_2$, m_2, s_2, and n_2 for group 2, etc.). We will use double subscripts for values in each group (e.g., x_{ij}, where i is the group and j is the sample number). Other notations will be introduced as needed.

2. Test Statistics

2.1. The t-Test

In **Chapter 4**, you were introduced to the t-test for testing the hypothesis that the mean of a sample was equal to a given value. In this chapter, we will show the use of the t-test for two other comparisons of means, first when the samples are paired and second for independent groups. We recall that the t-statistic has the general form

$$t = \frac{\text{difference of means}}{\text{standard error}},$$

where the standard error is the standard error of the numerator. The t-distribution is a "sampling" distribution, that is, it does not assume any knowledge about the true parameters. It is important to remember that the t-distribution is actually a family of distributions, each one identified by a parameter called the *degrees of freedom* (d.f.), which is related to the sample size. It is similar in shape to the standard normal distribution, and as the degrees of freedom increase, it becomes more similar (**Fig. 1**).

If you have computed a t-statistic and are looking it up in a table to calculate the P value, you will need to use the degrees of freedom to get the correct P value. If you are using a computer program, it will probably compute this for you from the sample size.

2.2. The F Distribution

The F statistic is another sampling distribution that will be used in this chapter. The F statistic is the ratio of two independent random variables that have chi-square distributions (**Chapter 5**). In this chapter and in many other applications, the F statistic is the ratio of 2 independent sample variances:

$$F = \frac{s_1^2}{s_2^2}. \tag{1}$$

Note that there are two parameters that go with an F distribution, the degrees of freedom for the numerator and the degrees of freedom for the denominator.

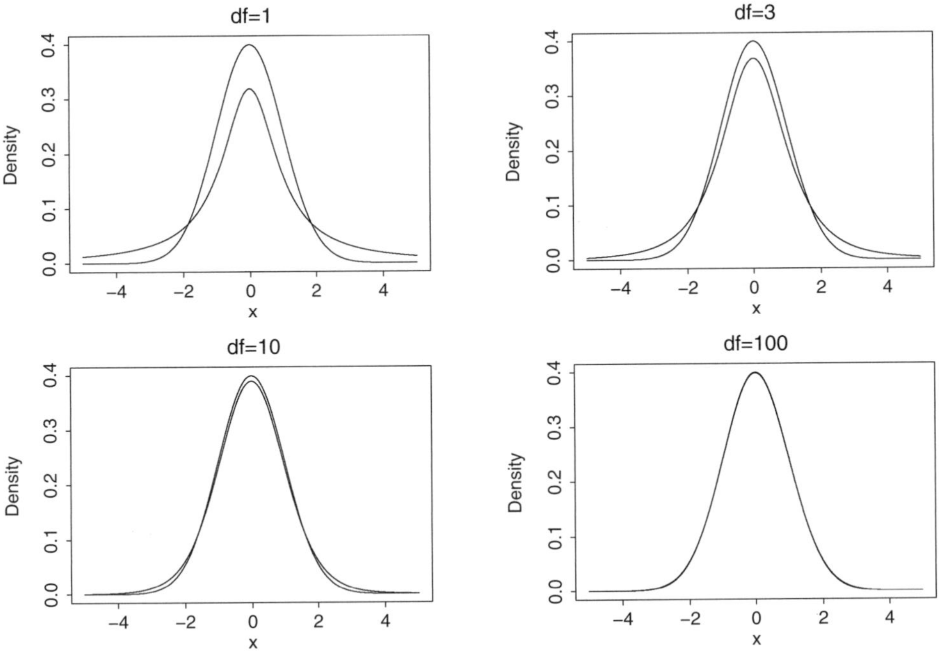

Fig. 1. Values of the t-distribution for different degrees of freedom.

3. *t*-Tests Comparing Two Means

3.1. Paired Samples

Paired samples means that there are two groups of samples, that each sample in the first group is related to one and only one sample in the second group, and that there are no unpaired samples in either group. Both groups have the same sample size, n. The most common types of paired samples are measurements on the same subject before and after an intervention, measurements on a group of experimental animals and their pair-fed matches, or measurements on subjects from an affected group and matched subjects from a control group. In order to compute the t-statistic, you must first compute the difference between each of the paired values,

$$d_i = x_{i1} - x_{i2}, \tag{2}$$

and then compute the mean and standard deviation, $\bar{x}_d$ and s_d, of the differences. It is assumed that the d_i have a normal distribution. This is guaranteed if both groups are assumed to come from underlying normal distributions.

Usually the difference is compared with zero, so the null and alternate hypotheses are

$$H_0: \delta = 0$$

$$H_1: \delta \neq 0,$$

where δ is the (unknown) true value of the mean difference.

The *t*-statistic is simply

$$t = \frac{\bar{x}_d - c}{s_d / \sqrt{n}}. \tag{3}$$

If you have detected that this looks a lot like the *t*-test on a single sample, you are right! There are a few points to remember. One is that although the mean of the difference, $\bar{x}_d$, is equal to the difference of the means, $\bar{x}_1 - \bar{x}_2$, you can't compute the standard deviation for the difference from the standard deviations of the 2 groups; you must compute the differences and then their standard deviation. Another is that you can call either group 1 or 2, but you usually do this based on your hypothesis. If you want to compare the difference to a constant, *c*, other than zero, and your software doesn't support it, then you can subtract *c* from the values or mean in the first group, but not in both.

Example 1

Table 1 shows pre- and posttreatment data from a sample of subjects who participated in a study of the psychobiology of depression ***(1)***. The sample value is cerebrospinal fluid: 3-methoxy-4-hydroxyphenylglycol (CSF MHPG). The groups are composed of unipolar depressed subjects and bipolar depressed subjects. We include the difference between values before and after treatment. We will test the null hypothesis that the difference is not zero in the unipolar group.

Using **Equation 3**, the *t*-statistic for the unipolar group is

$$t = \frac{31.61}{8.20/\sqrt{12}} = 13.35 \quad \text{with 11 degrees of freedom.}$$

The *P* value for this statistic is <0.0001, so we conclude that there is a significant change in CSH MHPG in the unipolar group.

We can test the same hypothesis for the bipolar group. The *t*-statistic is

$$t = \frac{16.33}{6.89/\sqrt{12}} = 8.21 \quad \text{with 11 degrees of freedom.}$$

The *P* value for this statistic is also <0.0001. Therefore, we conclude that there is also a significant change in CSH MHPG in the bipolar group.

In the next section, we will use a *t*-test to see if there are differences between the groups in the amount of change.

Table 1
Pre- and Posttreatment Levels and Change in CSF MHPG for 2 Diagnostic Groups (expressed in pmol/ml)

	Unipolar			Bipolar		
	Pretreatment	Posttreatment	Change	Pretreatment	Posttreatment	Change
	63.7	36.7	27	26.5	18.3	8.2
	61	37.8	23.2	38.6	25.8	12.8
	59	32.8	26.2	59	24.6	34.4
	65.2	24.6	40.6	42.8	28.4	14.4
	59.6	24.4	35.2	28.7	19.2	9.5
	59.7	22.7	37	47.7	32.6	15.1
	79.3	38.5	40.8	44.7	27.9	16.8
	60.6	28.9	31.7	47.7	28.6	19.1
	69.5	32.7	36.8	52.7	35.3	17.4
	54.9	37.7	17.2	54.6	37.8	16.8
	54.2	31.8	22.4	40.8	19.8	21
	67.5	26.3	41.2	40.8	30.3	10.5
Mean	62.85	31.24	31.61	43.72	27.38	16.33
Var	47.68	33.26	67.26	93.76	38.81	47.46
SD	6.91	5.77	8.20	9.68	6.23	6.89
Median			33.45			15.95
n	12	12	12	12	12	12

3.2. The Two-Sample t-Test in Independent Groups

The t-test is also used to compare means in 2 independent groups. In this case, the groups are independent, there is no matching, and the sample sizes may be different. There are 2 assumptions. The first is that each sample has an underlying normal distribution. This may be relaxed somewhat if the variables are continuous and the distribution is symmetric. The other assumption is that the standard deviations in the 2 samples are the same. This is more restrictive, and we will address what to do if this is violated later in this section. For now, we will assume that both assumptions are met. The null and alternate hypotheses are

$$H_0: \mu_1 = \mu_2$$

$$H_1: \mu_1 \neq \mu_2.$$

The quantity we are interested in is the difference between the population means, $\mu_1 - \mu_2$. We use both sample standard deviations to compute the "pooled" standard deviation as

$$s_p = \sqrt{\frac{(n_1 - 1)s_1^2 + (n_2 - 1)s_2^2}{n_1 + n_2 - 2}}. \quad (4)$$

The *t*-statistic comparing the means is simply

$$t = \frac{\bar{x}_1 - \bar{x}_2}{s_p\sqrt{\frac{1}{n_1} + \frac{1}{n_2}}}, \tag{5}$$

which has $n_1 + n_2 - 2$ degrees of freedom.

You will recall that one of the assumptions for the *t*-test on independent samples is that they have the same variance. If the data in 2 groups are normally distributed, this assumption may be tested using an *F* statistic. The hypotheses are

$$H_0: \sigma_1^2 = \sigma_2^2$$

$$H_1: \sigma_1^2 \neq \sigma_2^2.$$

The *F* statistic to test this hypothesis is the ratio of the sample variances (see **Section 2.1**), with the larger variance in the numerator:

$$F = \frac{s_1^2}{s_2^2}. \tag{6}$$

The degrees of freedom for the numerator are $n_1 - 1$ and the degrees of freedom for the denominator are $n_2 - 1$. If *F* statistic is very large, so that the *P* value is small, then there is evidence that the variances are not equal.

The *F* statistic is most widely used for testing equality of variances for 2 independent samples, however it is not always accurate when the distribution of the two samples is not precisely normal. An alternative is the Brown-Forsythe (BF) test *(2)*, which looks at the spread of the data in absolute terms. It is usually used with comparison of multiple groups in ANOVA (see below) but may be used with 2 groups. To compute the BF statistic, we first calculate the median of each group, md_i, and then compute the absolute value of the difference between each sample and its group median:

$$bf_{ij} = |x_{ij} - md_i|. \tag{7}$$

The BF test is simply a 2-sample *t*-test comparing the bf_{ij} in the two groups.

If the variances of the 2 groups are not equal, a modified estimate of the standard error of the difference is used and the degrees of freedom for the *t*-statistic are adjusted. The standard error of the difference is estimated as

$$s_{np} = \sqrt{\frac{s_1^2 + s_2^2}{n_1 + n_2}} \tag{8}$$

and the t-statistic is

$$t' = \frac{\overline{x}_1 - \overline{x}_2}{s_{np}}. \tag{9}$$

The distribution of this statistic can be approximated by a t-distribution with a different number of degrees of freedom. Several statisticians have developed methods for estimating the best matching distribution; the one most commonly used at this time is due to Satterthwaite *(3)*. The formula for computing these degrees of freedom is complicated and will not be given here.

Example 2

In **Example 1**, we computed the change after treatment in CSF MHPG in both the unipolar and bipolar groups (**Table 1**). We would like to test whether the change is the same in both groups. The null hypothesis is that the mean change in unipolar and bipolar groups are equal. To test whether variances are equal in the 2 groups, using **Equation 6**, the F statistic is

$$F = \frac{67.26}{47.46} = 1.42.$$

The n for the unipolar group is 12, the n for the bipolar group is also 12, so that the F statistic has 11 and 11 degrees of freedom. From a table, the P value for an F statistic with value 1.42 and 11 and 11 degrees of freedom is 0.57, so we can assume that the variances are equal.

To use the BF test, the absolute differences from the medians for the changes in CSH MHPG in **Table 1** are shown in **Table 2**. The t-statistic for these data is 1.28 with 22 degrees of freedom, which has a P value of 0.2131. This test also does not reject the null hypothesis that the variances are equal.

Therefore we compute the test statistic using **Equation 4** and **Equation 5**:

$$s_p = \sqrt{\frac{11 \cdot 67.26 + 11 \cdot 7.46}{22}} = 7.57$$

and

$$t = \frac{31.61 - 16.33}{7.57 \cdot \sqrt{\frac{1}{12} + \frac{1}{12}}} = 4.94 \quad \text{with 22 degrees of freedom.}$$

The P value for this statistic is <0.0001 so we can conclude that there is a difference between the 2 groups in the amount of change.

A good software package may automatically perform the F test and give you the results of the F test along with the t-value, degrees of freedom, and P value

Table 2
Absolute Deviation from the Median

	Unipolar	Bipolar
	6.45	7.75
	10.25	3.15
	7.25	18.45
	7.15	1.55
	1.75	6.45
	3.55	0.85
	7.35	0.85
	1.75	3.15
	3.35	1.45
	16.25	0.85
	11.05	5.05
	7.75	5.45
Mean	6.99	4.58
Var	17.63	24.70
SD	4.20	4.97
n	12	12

for both equal and unequal variances. Other packages may allow you to do the *F* test separately and then choose the *t*-test that goes with the results. Most packages do not do the BF test as a *t*-test option, but you may be able to do it by computing the deviations and using a *t*-test. Alternatively, the *t*-test on independent groups is equivalent to an ANOVA on 2 groups (**Section 5.1**), so you may use ANOVA software to test this. If your software assumes equal variances and does not allow for adjusted *t*-values, then, unless you can be sure that the variances are equal, you should try to run it on a different package or use a nonparametric approach (**Section 4**).

4. Tests of Central Tendency When the Distribution Is Not Normal

The basic assumption of a *t*-test is that the underlying distribution of the group(s) is normal. It is a fairly robust test, so that it may be appropriate if the sample size is large and the distribution is symmetric about the central value. Sometimes the data can be transformed so that it meets the assumption of a normal distribution. For example, many biological variables have a distinct positive skew. Often, the logarithm of the variables has a normal distribution, so that you can compare the samples using the log-transformed values. Just remember that you are comparing the geometric mean, not the arithmetic mean when you compare log-transformed variables.

Table 3
Twelve Numbers with Ranks

Value	2	19	23	31	35	47	56	56	59	61	98	98
Rank	1	2	3	4	5	6	7.5	7.5	9	10	11	12

If the data cannot be transformed to have a normal distribution, then the *t*-test will not give the correct *P* value. There are alternatives to the *t*-test for non-normally distributed variables. These tests are referred to as nonparametric tests. They are usually based on ranks rather than actual values. To compute the ranks for a sample, the values are arranged in increasing order and then numbered sequentially from 1 to the sample size, *n*. Sometimes 2 or more samples will have the same value. In this case, each one gets the average of the 2 or more ranks that are tied. **Table 3** shows the ranking for 12 numbers with 1 tie. These tests also use the median as the measure of central tendency, rather than the mean.

Most software packages can compute nonparametric tests and give the *P* values for test statistic. Alternatively, the *P* values may be given in tables. Many tables for nonparametric tests just give the critical values for certain sample sizes and other statistics. If the sample size is large enough, >20, then there is a large sample statistic that can be used that has a distribution close to a standard distribution, such as the normal or chi-square distribution. Most software packages compute this large-sample statistic and *P* value, as well as or instead of the exact statistic.

4.1. The Sign Test for a Single Sample

The null hypothesis for this test is that the median of the sample is a certain value, *C*. Using θ to represent the true value of the median this is stated as

$$H_0 : \theta = C$$

$$H_1 : \theta \neq C.$$

The sign test is based on the binomial test (**Chapter 5**). The data is arranged in order and the number of samples that are bigger than the median is computed. If the median is *C*, then you would expect half the samples to be larger than *C* and half to be smaller, so that $P(>C) = P(<C) = 0.5$. Suppose you have a sample size of *n* and *r* samples are larger than *C*; then using the binomial distribution the probability of getting *r* samples larger than *C* is

$$p(r) = \frac{n!}{r!(n-r)!} 0.5^r (1-0.5)^{n-r} = \frac{n!}{r!(n-r)!} \left(\frac{1}{2}\right)^r. \quad (10)$$

Table 4
Mood Scores in 12 Unipolar Subjects

3.45	4.56	4.68	5.06	5.41	6.63	6.99	7.81	8.22	8.81	9.14	9.86

Many statistical books have tables of the binomial probability for different combinations of n and r. Because hypothesis testing is based on the probability of getting the observed result or something rarer for a 2-sided test, we must compute the binomial probability for $r + 1$, $r + 2$, and so on, up to n. Then these are added together to give the answer (i.e., the probability of getting this number of samples larger than C if the median value is C).

Example 3

Table 4 shows the behavioral scores for the same group of unipolar depressed subjects that were in **Table 1**. We assume from the literature that the true median score is 5. There are 9 samples larger than C.

Using the binomial distribution the probability of getting 9 or more samples is computed as in **Table 5**.

For a 2-sided test, we calculate $2 \times 0.0730 = 0.1460$ and we do not reject the null hypothesis that the median is C.

For large samples, $n > 20$, the following has approximately a standard normal distribution:

$$z = \frac{X - 0.5n}{0.5\sqrt{n}}, \tag{11}$$

where $X = r - 0.5$ if $r \geq n/2$ or $X = r + 0.5$ if $r < n/2$.

Example 3 (Continued)

We illustrate using the large sample approximation (although the exact distribution is more appropriate for this example). We let $X = 9 - 0.5 = 8.5$, then

Table 5
Binomial Probabilities for $n = 21$ and $r \geq 9$

	r	$P(r)$
	9	0.0537
	10	0.0161
	11	0.0029
	12	0.0002
Sum		0.0730

$z = 1.44$ and the 2-sided P value is 0.150, which agrees closely with the exact distribution.

4.2. The Wilcoxon Signed Rank Test for Paired Samples

The most commonly used nonparametric test to compare paired samples is the Wilcoxon signed rank test. This is similar to the sign test but has the advantage that the magnitudes of the ranks are included in the calculations. It assumes that the differences are distributed symmetrically about the median difference, which would be expected if there were no real underlying difference between the paired values. As with the paired t-test, the analysis is based on the difference between the paired values. Using θ to represent the median of the differences, the hypotheses are

$$H_0: \theta = 0$$

$$H_1: \theta \neq 0.$$

The process is as follows. The differences are ranked on the basis of their magnitude, ignoring the sign. Then the signs are restored to the rankings. The sum of the positive ranks, S_p, and of the negative ranks, S_n, are computed. The test criterion, T, is the smaller of these 2 sums. If the sample is small (<16), then the critical value for T must be looked up in a table *(4)*. If the sample is large, then you may use the statistic in **Equation 12**, which has an approximately standard normal distribution:

$$z = \frac{T + 0.5 - 0.25n(n+1)}{\sqrt{n(n+1)(2n+1)/24}}. \tag{12}$$

Example 4

Table 6 shows the behavioral scales for the unipolar depressed subjects before and after treatment. The third column shows the differences, the fourth column shows the ranks on unsigned values, and the fifth column shows the signed ranks.

For this example, the sum of the positive ranks is 57 and the sum of the negative ranks is 21, so $T = 21$. Tables *(4)* show that this is above the critical value for 0.05 for the signed rank test on 12 subjects, so we do not reject the null hypothesis and assume that there was no change.

An n of 12 is too small for the normal approximation to be correct, but if we substitute $T = 21$ and $n = 12$ in **Equation 12**, then we get z = 1.37 and a P value of 0.17.

If the set of differences between pairs in a paired sample are used, then this is logically the same as the test of a single sample, so we could use the sign

Table 6
Baseline and Treatment Scales and Change Ranks and Signed Ranks

Baseline	Treatment	Difference	Absolute value	Rank	Signed rank
7.81	7.88	–0.07	0.07	1	–1
4.56	0.40	4.16	4.16	12	12
8.22	9.20	–0.98	0.98	5	–5
9.14	6.29	2.85	2.85	9	9
5.41	3.34	2.07	2.07	8	8
8.81	8.12	0.69	0.69	2	2
6.99	6.14	0.85	0.85	3	3
9.86	6.99	2.86	2.86	10	10
4.68	3.04	1.65	1.65	7	7
3.45	4.40	–0.95	0.95	4	–4
5.06	8.91	–3.85	3.85	11	–11
6.63	5.23	1.40	1.40	6	6

test to test the null hypothesis that the median is zero. However, the Wilcoxon signed rank test assumes that the values are distributed symmetrically about the median, uses more information, and, in general, is more powerful than the sign test (see **Chapter 4**). We could also use the Wilcoxon signed rank test with a single sample if the assumption of symmetry about the median was reasonable.

Example 4 (Continued)

If we use the sign test to test the hypothesis of a median equal to zero, we have 8 of 12 samples above the median. This gives us a 2-sided P value of 0.3877. The large sample approximation is $z = 0.86$, which gives us a 2-sided P value of 0.3897. This result is approximately double the P value obtained from the Wilcoxon signed rank test and therefore, although neither value is in the critical region, agrees with the notion that the Wilcoxon signed rank test is more powerful when the assumptions are met.

4.3. The Wilcoxon-Mann-Whitney Test to Compare Two Groups

We next look at the test known as the *Wilcoxon rank sum test* and sometimes as the *Mann-Whitney U test*. The underlying concept is that if two populations have the same distribution, then if you mix them together you would expect to find equal numbers from both groups above and below the overall median. Using θ to represent the true shift between the 2 distributions, the null hypothesis is

$$H_0: \theta = 0$$

$$H_1: \theta \neq 0.$$

We begin by computing the ranks for the combined group of subjects. To use the standard notation, we assume that one group has m subjects and the other has n subjects, and m and n need not be equal.

We then compute S_m and S_n, which are the sums of the ranks of the subjects in the groups with m and n subjects, respectively. The test statistic can be calculated for either m or n:

$$\begin{aligned} U_m &= S_m - 0.5m(m+1) \\ U_n &= S_n - 0.5n(n+1). \end{aligned} \tag{13}$$

Because $U_m = mn - U_n$, only one need be calculated. For small samples, the critical region must be obtained from tables ***(4)*** or from software, and is usually based on the smallest sample. The large sample approximation (for either m or $n > 20$ or both) is

$$z = \frac{U + 0.5 - 0.5mn}{\sqrt{mn(m+n+1)/12}}, \tag{14}$$

where U is the smaller of U_m and U_n. This statistic has approximately a standard normal distribution. (Note the 12 in the denominator is fixed. It is not related to the sample size, although it may seem so in the next example.)

Example 5

Table 7 shows the depression scales for unipolar and bipolar subjects and the rank for each value using the total group, where $m = n = 12$. The U statistics from **Equations 13** are

$$U_m = 176 - 0.5 \times 12 \times 13 = 98 \quad \text{and}$$

$$U_n = 12 \times 12 - 98 = 46.$$

From tables ***(4)***, the 5% critical region for a 2-tailed test is $U \leq 37$, therefore we do not reject the null hypothesis.

The large sample statistic from **Equation 14** is

$$z = \frac{46 + 0.5 - 0.05 \times 12 \times 12}{\sqrt{12 \times 12 \times 25/12}} = -1.47.$$

The 2-sided P value for -1.47 is 0.1410, so we do not reject the null hypothesis.

Table 7
Depression Scores and Overall Ranks for 2 Groups of Subjects

	Unipolar		Bipolar	
	Score	Rank	Score	Rank
	7.81	19	8.52	21
	4.56	7	7.07	17
	8.22	20	6.34	13
	9.14	23	7.32	18
	5.41	11	5.78	12
	8.81	22	6.77	15
	6.99	16	3.68	3
	9.86	24	4.93	9
	4.68	8	3.27	1
	3.45	2	4.27	6
	5.06	10	4.19	5
	6.63	14	4.09	4
Sum of ranks		176		124

5. Comparisons of Means in More Than Two Groups: ANOVA

5.1. ANOVA

In many situations, you want to compare the means in more than 2 groups. The null hypothesis is that the means in all the groups are equal. One way to do this would be to run *t*-tests as in **Section 3.2** on all possible pairs of groups. There are several problems with this approach. First, you are ignoring the structure of the design but not looking first to see if there is any difference between any of the groups. Second, you are also ignoring the design because each *t*-test will use a measure of the variation in the data based on only the 2 groups being tested, which will differ for each test, instead of using all the data to estimate the variation in the total sample. Third, you are increasing the probability of finding a difference where none exists (type I error; **Chapter 4**). To understand it, suppose you have 3 groups so you do 3 *t*-tests. You set the critical value of α to 0.05 for each test. Then the probability of 1 or more significant results if the null hypothesis is true is $1 - (1 - 0.05)^3 \approx 0.15$. This is called the problem of multiple comparisons, and you will see it addressed further on in this chapter.

Analysis of variance was developed to analyze this type of data without creating these problems. The basic analysis is a global test for any differences in means between groups. These can be followed by secondary tests to locate these differences, using the information from the total sample and adjusting for multiple comparisons. The groups are assumed to represent some characteristic referred to as the *factor* or independent variable, and the analysis tests the *effect* of this factor on the outcome. (Sometimes the terms *factor* and *effect* are used interchangeably.) The different groups are referred to as *levels* of the factor, whether or not that is a literal description. Levels may refer to groups that represent different doses of a drug or groups which represent different diagnoses.

The logic behind ANOVA is as follows. Assume you would like to compare the means between g groups, where $g \geq 3$. Each group has an underlying normal distribution, and, very importantly, all g groups have the same underlying variance. We denote the mean of each group as $\bar{x}_i$, the standard deviation as S_i, the sample size as n_i, and the jth sample in group i as x_{ij}. Treating all samples as a single group, the total sample size is N, the grand mean is denoted $\bar{x}$, and the total variation is S_t^2. Recall that the variance is calculated as the sum of squares around the mean divided by the sample size −1. Each component of the sum of squares around the grand mean can be rewritten as follows:

$$x_{ij} - \bar{x} = (x_{ij} - \bar{x}_i) + (\bar{x}_i - \bar{x}). \tag{15}$$

Thus the total variation can be divided into 2 parts, the variation due to the difference of each sample from its group mean, and the variation due to the difference between the group means and the grand mean. If the group means are equal, then the variation due to the difference between them and the grand mean should be small; if at least 2 of the group means are different, then this component should be larger. This is illustrated in **Figures 2A** and **2B**. Both have 3 groups of measurements, each group has the same variance. The three groups are shown on the left-hand side of the figure. The right-hand side of each figure shows the 3 means compared to the overall mean. In **Figure 2A**, the means of the 3 groups are approximately equal and the dispersion of the group means is also small. In **Figure 2B**, the means are very different, and the variation of the group means from the overall mean is large.

Calculations for the ANOVA are based on the sum of squares of deviations from the mean as shown in **Table 8**. Each sum of squares has a degrees of freedom associated with it.

SSB is the *between groups sum of squares* of deviations of the group means from the overall mean. The degrees of freedom are equal to the number of groups minus 1.

SSE is the sum of squares of the deviation of each sample from its group mean, often called *the error sum of squares*. It has degrees of freedom equal to the total sample size minus the number of groups.

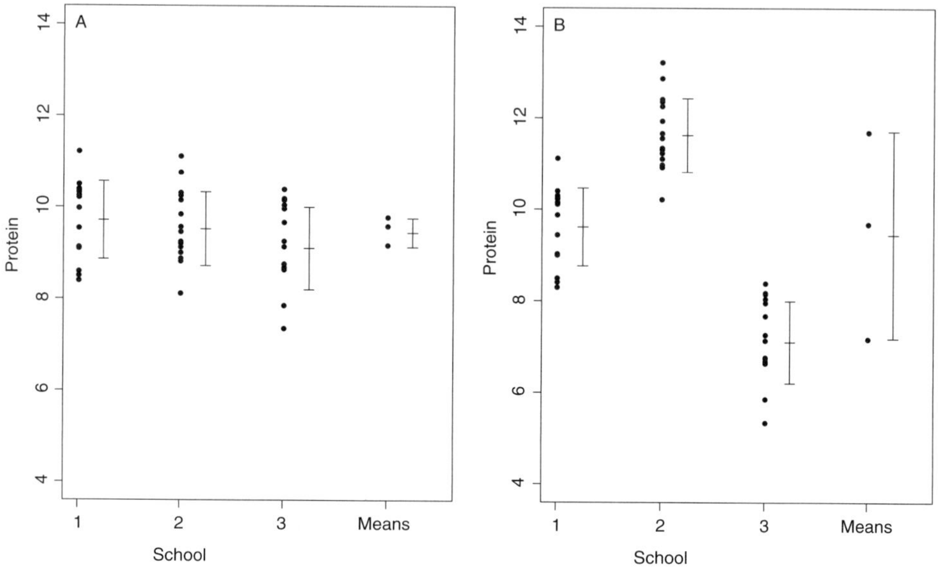

Fig. 2. Three groups with means and variances shown separately and for the combined group. **(A)** Groups with equal means and variances; **(B)** groups with equal variances and unequal means.

SST is the *total sum of squares* of each observation about the group mean and is equal to SSB + SSE. It has $N - 1$ degrees of freedom.

The values in the fourth column of **Table 8** are called the *mean squares* for SSB and SSE. MSB and MSE are computed by dividing the sums of squares by their degrees of freedom. MSB is the estimates of the variation due to differences between group means, and MSE is the estimates of the variation due to the variation within groups. The test statistic is

$$F = \text{MSB}/\text{MSE}. \tag{16}$$

Table 8
Calculations for Analysis of Variance

Source	Sum of squares	d.f.	Mean square
Between groups	$\text{SSB} = \Sigma_i(\bar{x}_i - \bar{x})^2$	$g - 1$	$\text{MSB} = \text{SSB}/(g - 1)$
Error	$\text{SSE} = \Sigma_i\Sigma_j n_j(x_{ij} - \bar{x}_i)^2$	$N - g$	$\text{MSE} = \text{SSE}/(N - g)$
Total	$\text{SST} = \Sigma_i\, \Sigma_j(x_{ij} - \bar{x})^2$	$N - 1$	

This statistic has an *F* distribution with the degrees of freedom as given for the numerator and denominator. If the means are very different, the MSB will be large compared with MSE, the *F* ratio will be very large, and the *P* value for the test will be small.

To test the assumption of equal variances, we can use the BF test defined in **Section 3.2**. As before, we compute the dispersion of the samples about their median as the absolute value of the difference between the sample value and its group median. We use ANOVA to test for differences between these dispersions. If the ANOVA is significant, then it means that the assumption of equal variances is not satisfied, and a modification of the ANOVA should be used. The simplest modification is by Box *(5)*, which, similar to the modifications of the *t*-test, computes different degrees of freedom for the *F* statistic. Box's test requires equal sample sizes; a test by Welch *(6)*, which computes a different *F* statistic and degrees of freedom, does not require this. These tests should be available in most software packages.

Example 6

Table 9 shows the log of testosterone levels in a sample of hypogonadal men who participated in a study of testosterone replacement after 30 days of treatment *(7)*. There were 3 groups, each treated with either a different dosage or method of replacement of the hormone. Log transformed values were used because the raw value was not normally distributed. We wish to determine if there are any differences between groups.

Table 10 shows the computation of mean absolute deviance for the BF test of equality of variances. The *F* statistic for this test is 2.16 with 2 and 54 degrees of freedom. The *P* value is 0.1248, thus we can assume the variances are equal.

The results of the ANOVA are presented in **Table 11**, which shows the sums of squares, the degrees of freedom, the means squares, and *F* statistic.

The degrees of freedom for the *F* statistic are 2 and 54. The *P* value for the *F* statistic is <0.0001, so we reject the null hypothesis.

Table 11 is the format used by most software packages for presenting the results of an ANOVA, although the labeling of the mean squares may vary.

The above describes the simplest form of ANOVA. The result is called the *effect* of different groups. In this discussion, we are only testing *fixed effects*, that is we assume that the groups represent all levels of the group effect we are interested in. For example, if we were testing the effect of different doses of a medication on an outcome measure, if only 3 doses would be used and we had a group using each, the dosage would be a fixed effect. The opposite of a fixed effect is a random effect, where a random set of some levels of the parameter

Table 9
Log Testosterone Levels in 3 Groups of Subjects

	Group 1	Group 2	Group 3
	6.44	6.17	6.46
	6.42	6.09	6.03
	6.30	6.16	5.25
	5.72	6.83	5.90
	6.20	6.00	3.84
	5.89	7.17	4.40
	6.59	6.41	6.35
	6.01	5.57	5.06
	5.87	5.66	5.91
	5.67	5.93	4.74
	5.88	6.27	5.31
	5.77	6.14	5.66
	5.34	6.84	4.93
	5.37	6.08	5.20
	6.82	6.83	5.78
	6.23	6.37	4.74
	6.22	7.05	6.42
	5.88	6.93	5.18
		5.64	5.25
		5.80	
N	18	20	19
Mean	6.03	6.30	5.39
Std	0.40	0.49	0.71
SS	2.74	4.60	8.96
Median	5.95	6.17	5.25

of interest is represented by the groups. If several doses were possible, but we just studied the lowest, middle, and highest, then this would be a random effect. Random effects will be discussed in **Chapter 11**.

We note also that this analysis can be used for only two groups in lieu of the *t*-test. That is because the *F* statistic with 1 degree of freedom in the numerator is the square of a *t*-statistic with the degrees of freedom equal to the degrees of freedom of the denominator of the *F* statistic. In this case, the *F* statistic would be the square of the *t*-statistic, and the *P* value will be the same. This is why we recommend using ANOVA software to compute the BF test for 2 groups.

Table 10
Absolute Deviation from the Median

	Group 1	Group 2	Group 3
	0.49	0.00	1.20
	0.46	0.07	0.78
	0.35	0.00	0.00
	0.23	0.66	0.65
	0.25	0.17	1.42
	0.06	1.01	0.86
	0.64	0.25	1.09
	0.06	0.59	0.19
	0.08	0.50	0.65
	0.28	0.24	0.51
	0.07	0.10	0.06
	0.18	0.03	0.41
	0.61	0.68	0.33
	0.58	0.09	0.05
	0.87	0.66	0.53
	0.28	0.20	0.52
	0.27	0.89	1.16
	0.07	0.77	0.07
		0.53	0.01
		0.36	
N	18	20	19
Mean	0.32	0.39	0.55
Std	0.24	0.33	0.44
SS	9.96	10.96	11.96

Example 2 (Continued)

In this example, the *t*-statistic comparing the amount of change in the unipolar and bipolar groups was 4.94, with 22 degrees of freedom. For ANOVA, the *F* statistic is 24.41 with 1 and 22 degrees for freedom, and the square root of 24.41 is 4.94.

Table 11
Analysis of Variance Table for the Data in Table 9

Source	Sum of squares	d.f.	Mean square	*F* statistic
Between groups	8.40262	2	4.23131	13.92
Error	16.29622	54	0.30178	
Total	24.59884	56		

Similarly, ANOVA can be expanded to test the effects of multiple factors and their interaction. It can also be used to test the effect of multiple related samples, such as measurements over time in the same individual, known as repeated measures. These topics will be discussed in **Chapter 10** and **Chapter 11**.

5.2. Contrasts

The F statistic from the ANOVA can tell us whether or not there are differences between the means of the groups, but it cannot tell where those differences are. In order to do that, we must make additional comparisons of group means, or *post hoc* tests, keeping in mind the issues described in the beginning of this section. These comparisons may be simple pairwise comparisons of pairs of means or may be more complex. For example, in **Example 6**, we could compute the pairwise differences between group 1 and group 2, between group 1 and group 3, and between group 2 and group 3 and test whether each is different from zero. We could also ask if the mean of group 1 is equal to the average of the means of group 2 and group 3. This would be stated as:

$$\mathrm{H}_0: \mu_1 = \frac{1}{2}(\mu_2 + \mu_3) \quad \text{or} \quad \mu_1 - \frac{1}{2}\mu_2 - \frac{1}{2}\mu_3 = 0.$$

$$\mathrm{H}_1: \mu_1 - \frac{1}{2}\mu_2 - \frac{1}{2}\mu_3 \neq 0.$$

For testing, the null hypothesis is rewritten as a linear combination of the sample means

$$L = \bar{x}_1 - \frac{1}{2}\bar{x}_2 - \frac{1}{2}\bar{x}_3 = 0. \tag{17}$$

An equation in this format is called a contrast. It has 2 characteristics:

1. It is a linear combination of the group means with the right-hand side of the equation = 0; and
2. The coefficients of the means always add up to 0.

In **Equation 17**, the coefficients of the means are, in order, 1, $-\frac{1}{2}$ and $-\frac{1}{2}$, so it meets these requirements. Similarly a simple comparison of the first 2 groups would be written as

$$\mathrm{L} = \bar{x}_1 - \bar{x}_2 - 0 \cdot \bar{x}_3 = 0. \tag{18}$$

The coefficients for **Equation 18** are, in order, 1, −1, and 0, which sum to zero. Note that zero is the coefficient for the term not in the comparison.

Post hoc tests may be *a priori* (comparisons that were planned before the ANOVA was performed) or *a posteriori* (run after the ANOVA is performed and run only if the F statistic from the ANOVA is significant). These types of contrasts are tested using different methods.

5.3. A Priori *Comparisons*

A priori contrasts may be tested using a *t*-test for a single group (**Chapter 4**, the comparison value is 0), with the following modification to the estimate of standard error. The general form of the contrast for *g* groups with a sum of *N* total samples is

$$L = \sum c_i \bar{x}_i = 0, \quad \text{where} \quad \sum c_i = 0, \tag{19}$$

where the summation is over all *g* groups, the c_i are the coefficients of the means, and some may be equal to 0.

The standard error for the contrast (SEC) is based on the MSE from the ANOVA

$$\text{SEC} = \sqrt{\text{MSE}\left(\sum \frac{c_i^2}{n_i}\right)}. \tag{20}$$

The test statistic is simply

$$t = \frac{L}{\text{MSE}} \quad \text{with } N - g \text{ degrees of freedom} \tag{21}$$

where *L* is the value of the contrast at the sample means (**Equation 19**).

If more than one contrast is being computed, then the probability of a type I error is increased as described above, so that it is necessary to correct for this. One method, known as Dunnett's multiple comparison procedure or the Bonferroni *t* method, computes a multiplier for the SEC, which is based on the overall type I error, the number of comparisons being made, and the degrees of freedom of MSE. The new critical level of the *F* statistic for each comparison is equal to this multiplier times SEC. The calculation of the multiplier is complicated and is best done by computer software.

Example 6 (Continued)

Suppose we wanted to test the *a priori* hypothesis that the mean in the first group was equal to the average of the means in the two other groups. The contrast is

$$L = \bar{x}_1 - \frac{1}{2}(\bar{x}_2 + \bar{x}_3) = 0.$$

The coefficients are 1, –, and –. The standard error (**Equation 20**) is

$$\text{SEC} = \sqrt{0.3018\left(\frac{1^2}{18} + \frac{-0.5^2}{20} + \frac{-0.5^2}{19}\right)} = 0.1566.$$

The *t*-statistic (**Equation 21**) is

$$t = \frac{6.0348 - 0.5(6.2165 + 5.3902)}{0.1566} = 1.2168 \quad \text{with 57 degrees of freedom.}$$

The *P* value for this *t* is 0.2265, so we can assume that the mean of the first group is not different from the combined mean of the second and third groups.

5.4. A Posteriori *Contrasts*

A posteriori, or unplanned contrasts, are more common in research. As stated above, they are only computed if the *F* statistic for the overall ANOVA is significant. The most common *a posteriori* comparisons are pairwise comparisons of means. Although a *t*-test with a denominator based on the MSE may seem to be appropriate, there is still the problem of multiple comparisons, lack of independence between tests (e.g., sharing of information), and the fact that they were unplanned. One approach is to simply divide the overall type I error by the number of tests performed. This is known as a Bonferroni correction and is the simplest way to correct for multiple tests. Unfortunately, the Bonferroni correction results in required significance levels that are very small, which may drastically reduce the power of the study (see **Chapter 19**), and which does not take the other problems into account. Several better methods for multiple testing have been developed. The most common approach is to take advantage of the fact that there is overlap between tests (e.g., if $\bar{x}_1 < \bar{x}_2$ and $\bar{x}_2 < \bar{x}_3$, then $\bar{x}_1 < \bar{x}_3$) and use this fact to develop new critical values, based on the sample size, number of tests, and SE. There are 4 tests that are most commonly used. The computation of the new critical values for significance is complicated and will not be presented here.

1. Tukey's HSD (honest significant difference) is used to test all pairwise comparisons. It is not used for other contrasts. It is also very useful because it gives confidence intervals for the pairwise differences and is thus very commonly used.
2. The Newman-Keuls procedure, also known as the Student-Newman-Keuls (SNK) procedure, uses a stepwise approach to comparing means by first ordering them by magnitude then using a critical value that is also based on the number of steps between means. It is more powerful than Tukey's method but does not give confidence intervals, so it is not always useful.
3. Scheffe's procedure may be used to test contrasts other than pairwise comparisons. It is not as powerful as the others, so that it should not be used if only pairwise comparisons are required.
4. Dunnett's procedure is used when you want to compare one mean to all the others. Usually, the one mean is a control and the others are different levels of treatments, but this is not necessary. Dunnett's test is not used for all pairwise comparisons, but it is more powerful than either of the others in this special situation.

We are not going into more detail for these methods, because computing the critical values can be complicated, and we expect you will be doing this using computer software. Make sure your software will use the procedure you choose. Be aware that any of the first three will automatically do all pairwise comparisons and that not all packages allow you to use Scheffe's method to do other types of contrasts. Also, if you want to use Dunnett's procedure, make sure you know how the package selects the control mean. Some packages automatically select a value such as the first numeric or alphabetic level, others allow you to specify it.

Example 6 (Continued)

The F statistic for the overall ANOVA was significant, so the SNK method was used to test for pairwise differences. The means (in order of magnitude) were 6.2965 for the second group, 6.0348 for the first group, and 5.3902 for the third group. The SNK results showed that means in the first and second groups were both significantly larger than the mean in the third group.

6. Kruskal-Wallis Test

When the assumptions of normality and/or equal variance cannot be met even by transforming the data, then a nonparametric procedure known as the *Kruskal-Wallis test* may be used to compare central tendencies. This is an extension of the Wilcoxon-Mann-Whitney test of **Section 4.3**, and the process is similar. Assume we have g groups ($g \geq 3$) with n_i samples in each. We begin by computing the ranks for each value in the total sample, which we label r_{ij}. Then we compute the sum of the ranks in each group, S_i.

We then compute the sum of squares

$$ST^2 = \sum_j (S_i^2 / n_i) \tag{22}$$

and the Kruskal-Wallis statistic

$$K = \frac{12ST^2}{N(N-1)} - 3(N+1). \tag{23}$$

For small values of the n_i, the critical values of K may be obtained from tables. For reasonably large n_i, the distribution of K is approximately chi-square with $g - 1$ degrees of freedom. If there are more than a few ties, then the calculations become more complicated. As with the nonparametric tests in **Section 4**, most software packages include this test and the appropriate P values.

Example 7

Table 12 shows the mood scales after 30 days for the same men that were in **Example 6** *(8)* and the ranks within the entire group for each sample. There are no ties. We wish to determine if there are differences in the distribution of scores between the 3 groups. The sum of ranks are 580, 481, and 592, respectively.

We compute

$$ST^2 = \frac{580}{18} + \frac{481}{20} + \frac{592}{19} = 48{,}702.41$$

Table 12
Mood Scores in 3 Groups of Subjects and Overall Ranks

	Group A		Group B		Group C	
	Score	Rank	Score	Rank	Score	Rank
	11.43	3	1.43	1	3.57	2
	21.19	6	16.96	4	17.14	5
	25.36	11	22.50	7	29.64	14
	28.21	13	23.93	8	31.85	16
	32.50	17	24.29	9	33.33	19
	33.93	21	25.00	10	33.93	20
	34.29	22	26.43	12	34.64	23
	34.64	24	30.36	15	35.13	26
	37.50	30	33.21	18	36.79	29
	40.36	35	35.00	25	37.92	31
	41.79	39	35.36	27	40.56	36
	42.86	43	35.71	28	40.96	37
	45.00	49	38.57	32	41.07	38
	45.30	50	38.75	33	41.94	41
	47.14	52	38.93	34	44.59	46
	47.50	54	41.89	40	44.84	48
	47.56	55	42.50	42	46.43	51
	47.86	56	43.57	44	47.14	53
			44.29	45	49.00	57
			44.64	47		
n_i	18		20		19	
S_i	580		481		592	

and

$$K = \frac{12 \times 48{,}702.41}{57 \times 58} - 3 \times 58 = 2.7783.$$

The *P* value for *K*, based on a chi-square distribution with 2 degrees of freedom, is 0.2493, therefore we do not reject the hypothesis that the distributions of mood scores in the 3 groups are the same.

7. Sample Size Considerations

The concept of power and sample size is discussed in **Chapter 19**. For the *t*-test, the power increases as the assumed difference between the means increases and decreases as the standard error in the denominator increases. In the paired *t*-test, the standard deviation of the change is not the same as the standard deviation of either group and is usually smaller, so that using paired samples can result in a more powerful test. Similarly for the ANOVA, power increases as the difference between some or all of the means increases and decreases as the variance increases. Note that pre- or posttest comparisons usually have a smaller *P* value as the critical level, so that the power for these comparisons should be estimated independently of the power for the test, if they are critical.

There are some methods for estimating the power in the nonparametric tests, however they are more complex. In general, the larger the difference in medians that you can assume, the more powerful your test will be.

8. Conclusion

We have described and illustrated the use of the *t*-test and ANOVA for testing differences between means. Both of these tests are based on the assumption that there is an underlying normal distribution in the groups. Further, the *t*-test for independent samples and the ANOVA require that the groups have equal variances. Methods to test this requirement and to obtain results when it is violated are described. In the case where the assumptions cannot be met, we have described tests from the class of tests known as nonparametric tests, which do not depend on any distribution assumptions. These tests are based on ranks and look for differences in the median rather than the mean. These tests are useful for the relatively simple problems addressed in the chapter but have not been fully developed for more complex studies.

References

1. Maas, J. W., Koslow, S., Davis, J. M., Katz, M. M., Mendels, J., Robins, E., Stokes, P. E., and Bowden, C. L. (1980) Biological component of the NIMH Clinical Research Branch Collaborative Program on the psychobiology of depression: I. Background and theoretical considerations. *Psychol. Med.* **10**, 759–776.
2. Brown, M. B., and Forsythe, A. B. (1974) Robust tests for equality of variances. *J. Am. Stat. Assoc.* **69**, 364–367.
3. Satterthwaite, F. W. (1946) An approximate distribution of estimates of variance components. *Biometrics Bulletin* **2**, 110–114.
4. Sprent, P. (1989) Tables of critical values for nonparametric methods. In: *Applied Nonparametric Statistical Methods*. New York, Chapman & Hall, pp. 231–241.
5. Box, G. P. E. (1954) Some theorems on quadratic forms applied in the study of analysis of variance problems. *Ann. Math. Stat.* **25**, 290–302.
6. Welch, B. L. (1951) On the comparison of several mean values: An alternative approach. *Biometrika* **38**, 330–336.
7. Swerdloff, R. S., Wang, C., Cunningham, G., Dobs, A., Iranmanesh, A., Matsumoto, A. M., Snyder, P. J., Weber, T., Longstreth, J., Berman, N., and the Testosterone Gel Study Group (2000) Long term pharmacokinetics of transdermal testosterone gel versus testosterone patch in hypogonadal men. *JCEM* **85**, 4500–4510.
8. Wang, C., Swerdloff, R. S., Iranmanesh, A., Dobs, A., Snyder, P. J., Cunningham, G., Matsumoto, A. M., Weber, T., Berman, N., and the Testosterone Gel Study Group (2000) Transdermal testosterone gel improves sexual function, mood, muscle strength and body composition parameters in hypogonadal men. *JCEM* **85**, 2839–2853.

8

Correlation and Simple Linear Regression

Lynn E. Eberly

Summary

This chapter highlights important steps in using correlation and simple linear regression to address scientific questions about the association of two continuous variables with each other. These steps include estimation and inference, assessing model fit, the connection between regression and ANOVA, and study design. Examples in microbiology are used throughout. This chapter provides a framework that is helpful in understanding more complex statistical techniques. such as multiple linear regression, linear mixed effects models, logistic regression, and proportional hazards regression.

Key words: Coefficient of determination; diagnostics; extrapolating; homoscedastic; least squares; mean square error; outlier; Pearson correlation; power and sample size; Spearman correlation; studentized residuals; Working-Hotelling confidence band.

1. Introduction

A *correlation* is a numerical summary that describes the degree to which two continuous variables, *X* and *Y*, are linearly related to each other. A *simple linear regression* of *Y* on *X* takes this one step further and formalizes a statistical model between the two variables. Statistical tests of the linear relation of *Y* with *X* can then be carried out, and predictions of *Y* are based on the estimated linear relation. Regressions can be carried out using either experimental or nonexperimental (observational) data. In this chapter, we introduce the fundamentals of correlation and regression, including estimation and inference, assessing model fit, the ANOVA-regression connection, and study design.

Example 1

A research team is interested in understanding transcription levels and subsequent translation to a protein for a particular gene. mRNA expression and

From: *Methods in Molecular Biology, vol. 404: Topics in Biostatistics*
Edited by: W. T. Ambrosius © Humana Press Inc., Totowa, NJ

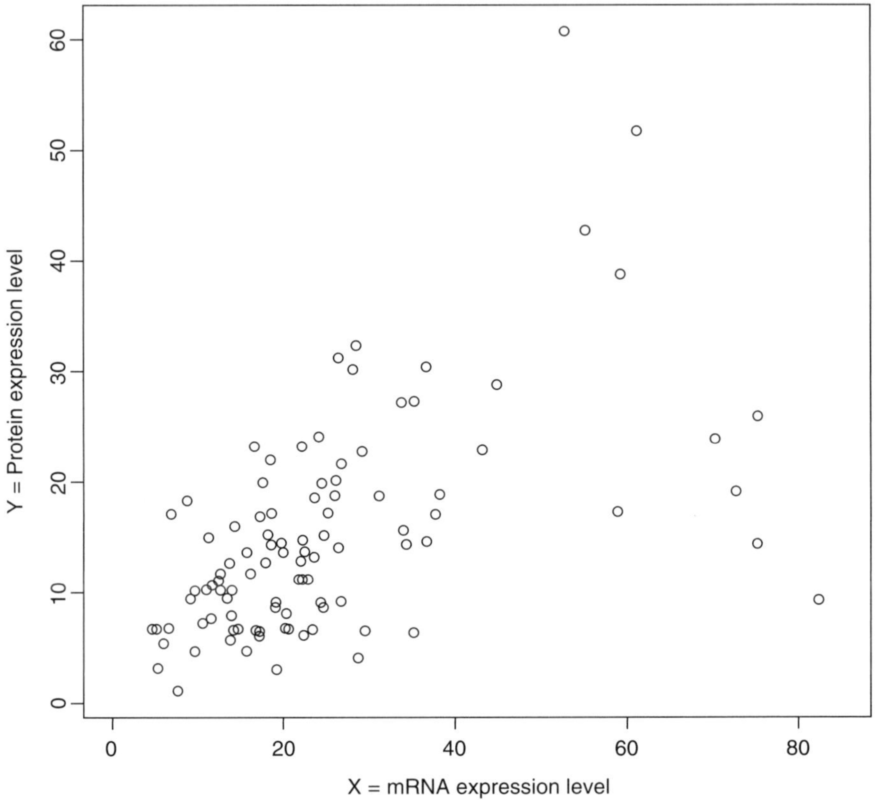

Fig. 1. Protein (Y) versus mRNA (X) expression levels in melanoma samples from 100 patients.

protein expression are each measured relative to a control using melanoma samples from 100 patients. The team will explore how the two expression levels are related; see **Figure 1**.

2. Correlation

2.1. Pearson Product-Moment Correlation Coefficient

2.1.1. Estimation and Interpretation

A correlation ρ is a summary of the degree to which two continuous random variables, X and Y, are linearly related to each other. ρ can take on any value from −1 through 1 with 1 indicating a perfect positive correlation and 0 indicating no correlation between X and Y (**Figs. 2a–2d**). From a sample (x_1, y_1), (x_2, y_2), . . . , (x_n, y_n) of size n, this correlation can be estimated with the *Pearson product-moment correlation coefficient*,

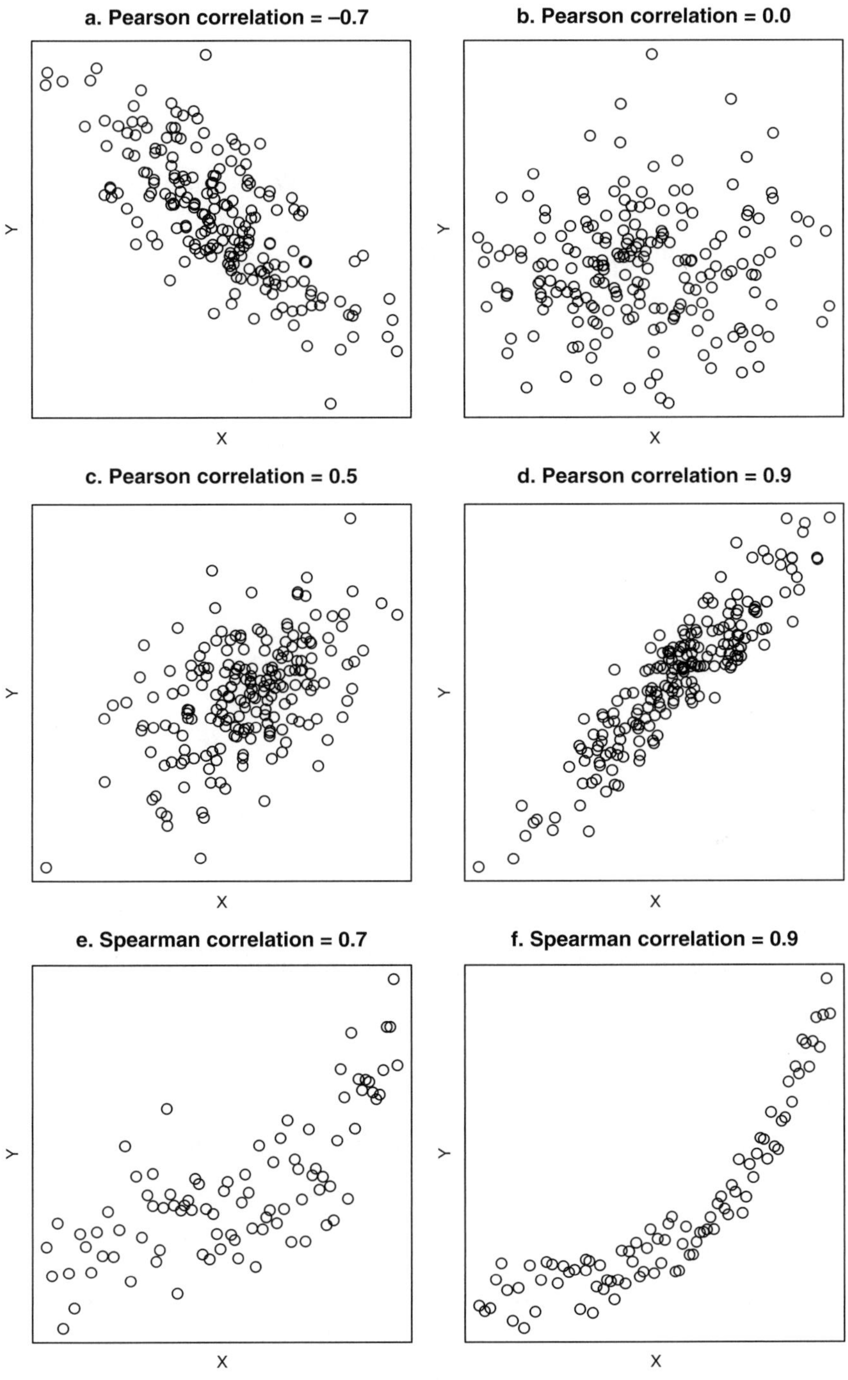

Fig. 2. Examples of data with various strengths of correlation.

$$\hat{\rho} = \frac{\sum_{i=1}^{n}(x_i - \bar{x})(y_i - \bar{y})}{\sqrt{\sum_{i=1}^{n}(x_i - \bar{x})^2 \sum_{i=1}^{n}(y_i - \bar{y})^2}}. \quad (1)$$

Note that the correlation of X with Y is identical to that of Y with X. Here i indexes the units of observation in the sample. When the (x_i, y_i) are jointly normally distributed, $\hat{\rho}$ is a biased estimate of the population correlation between X and Y, but the bias decreases as n increases.

2.1.2. Inference

Scientific conclusions about the correlation are made with inferential techniques such as hypothesis testing and confidence intervals; see **Chapter 4** for a review of these concepts. When the (x_i, y_i) are jointly normally distributed, or approximately so, a t-test for $H_0: \rho = 0$ is carried out using the test statistic

$$t^* = \frac{\hat{\rho}\sqrt{n-2}}{\sqrt{1-\hat{\rho}^2}}.$$

We reject H_0 in favor of $H_1: \rho \neq 0$ when $|t^*| > t_{1-\alpha/2,n-2}$. An approximate $(1 - \alpha)100\%$ confidence interval is computed using a transformation

$$r = \frac{1}{2}\ln\left(\frac{1+\hat{\rho}}{1-\hat{\rho}}\right);$$

because $\hat{\rho}$ takes values in $[-1, 1]$, r takes values in $(-\infty, \infty)$, thus allowing a better approximation to normality. A $(1 - \alpha)100\%$ confidence interval for r has lower and upper bounds

$$(r_L, r_U) = \left(r - z_{1-\alpha/2}/\sqrt{n-3}, r + z_{1-\alpha/2}/\sqrt{n-3}\right).$$

From these we compute the lower and upper bounds of a confidence interval for ρ,

$$\left(\frac{e^{2r_L} - 1}{e^{2r_L} + 1}, \frac{e^{2r_U} - 1}{e^{2r_U} + 1}\right).$$

Both the test and confidence interval are large sample approximations; a rule of thumb is to use these only when $n > 30$.

Example 1 (Continued)

The Pearson correlation between X = mRNA expression and Y = protein expression is 0.55. To test $H_0: \rho = 0$, $t^* = 0.55\sqrt{100-2}/\sqrt{1-0.55^2} = 6.52$. Because $|t^*| = |6.52| > t_{1-0.05/2,100-2} = 1.98$, we reject the null in favor of $H_1: \rho \neq 0$ at level $\alpha = 0.05$ and conclude that the correlation is significantly different from zero. To get a confidence interval for $\hat{\rho}$, we first need $r = \frac{1}{2}\ln((1+0.55)/(1-0.55)) =$

0.62 and its confidence interval $0.62 \pm 1.96/\sqrt{100-3} = (0.42, 0.82)$. Thus the 95% confidence interval for $\hat{\rho}$ is

$$\left(\frac{e^{2(0.42)}-1}{e^{2(0.42)}+1}, \frac{e^{2(0.82)}-1}{e^{2(0.82)}+1}\right) = (0.40, 0.68).$$

2.2. Spearman Rank Correlation Coefficient

If the relation between X and Y is not quite linear (e.g., quadratic, exponential), or if the (x_i, y_i) are not jointly normally distributed, then the *Spearman rank correlation coefficient* may be a more appropriate summary of the strength of the relation (**Figs. 2d–2e**). This is computed with the same formula as above (**Equation 1**) but replacing each x_i with its rank among $x_1, \ldots, x_n$, and replacing each y_i with its rank among $y_1, \ldots, y_n$. (That is, if the data are 10, 23, 13, and 9, they would be replaced by 2, 4, 3, and 1. The computation is more difficult if there are ties, but this is the general idea.) The interpretation is thus slightly different than the Pearson correlation coefficient: The Spearman correlation coefficient measures the tendency of Y to increase or decrease with X, where that tendency is not constrained to a linear relation. The t-test and confidence intervals shown above can also be constructed similarly using the ranks instead of the original data.

3. Simple Linear Regression

3.1. The Linear Relation

A *simple linear regression* model expands upon the idea of correlation and formalizes a statistical relation between the two variables such that Y is linearly related to X. X is variously known as the covariate, or the predictor, explanatory, or independent variable. Correspondingly, Y is known as the outcome, or the predicted, response, or dependent variable. This is in contrast with a correlation, which does not make this distinction between which variable is explanatory and which is outcome.

The first step when considering a regression is to plot the data (x_1, y_1), (x_2, y_2), . . . , (x_n, y_n) as in **Figure 1**, with the x_i on the horizontal axis and the y_i on the vertical axis, to verify that there is an approximate linear relation between the two. This linearity is formalized in the mathematical relation $Y = \beta_0 + \beta_1 X$, where β_0 represents the *intercept* (the value of Y when $X = 0$) and β_1 represents the *slope* (the magnitude of the change in Y when X is larger by one unit).

The purposes of a regression analysis are generally to estimate and test β_0 and/or β_1 and to form predictions for Y based on X. The general procedure for carrying out a regression analysis is as follows:

(i) verify through a data plot that a linear relation is likely to be appropriate;
(ii) estimate the linear relation with a regression model;

(iii) assess through diagnostics whether the model provides an appropriate fit to the data; and
(iv) if so, use the model to draw inferences about the linear relation.

Step (ii) through **step (iv)** will be outlined here and in **Section 4** and **Section 5**.

3.2. Estimation of the Linear Relation

In a regression of Y on X using the sample (x_i, y_i) $(i = 1, 2, \ldots, n)$, the x_i are assumed to be fixed and known without error, whereas the y_i are assumed to be random. These assumptions inform our procedure for estimating the linear relation (i.e., estimating the intercept β_0 and slope β_1). Examine **Figure 3**; a well-chosen estimated line should lie "close" to as many of the (x_i, y_i) as possible. Because the x_i are considered fixed, then for each x_i "close" is taken to be the distance from each observed y_i to the estimated line (i.e., the vertical

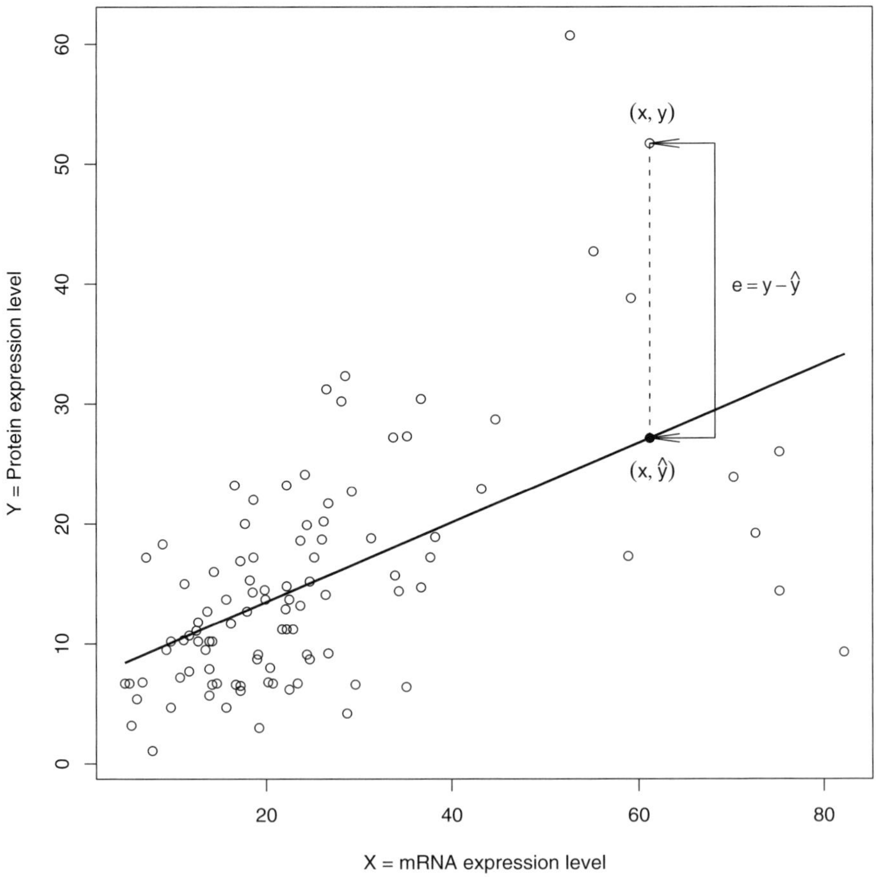

Fig. 3. Estimated line from the regression of Y on X.

distance). We treat distance below the line equivalently to distance above the line by considering the squared distances. This is the *least squares criterion*: we choose β_0 and β_1 to minimize the sum of squared vertical distances

$$\sum_{i=1}^{n} (y_i - (\beta_0 + \beta_1 x_i))^2.$$

Mathematical solution of this criterion gives us the estimators

$$\hat{\beta}_1 = \frac{\sum_{i=1}^{n} (y_i - \bar{y})(x_i - \bar{x})}{\sum_{i=1}^{n} (x_i - \bar{x})^2} \qquad (2)$$

$$\hat{\beta}_0 = \bar{y} - \hat{\beta}_1 \bar{x}$$

assuming a specific distribution, such as normality, for the y_i is not required to do this estimation.

Example 1 (Continued)

From a regression of Y = protein expression on X = mRNA expression, our data result in $\hat{\beta}_0 = 6.86$ and $\hat{\beta}_1 = 0.33$. Thus, each 1 unit higher mRNA expression level is associated with a 0.33 unit higher mean protein expression level. The intercept is interpreted as mean protein expression when mRNA expression is 0, which is not a biologically useful value in this context.

3.3. The Simple Linear Regression Model

In order to proceed beyond estimation of intercept and slope, additional assumptions are required to turn our mathematical linear relation into a statistical linear model:

(i) the y_i are independent of each other;
(ii) the y_i each follow a normal distribution;
(iii) the mean of that distribution is a linear function of x_i; and
(iv) the variance of that distribution is the same for all y_i (constant variance, or *homoscedasticity*).

We write this in one expression as

$$y_i \overset{indep}{\sim} N(\beta_0 + \beta_1 x_i, \sigma^2)$$

or equivalently as (3)

$$y_i = \beta_0 + \beta_1 x_i + \varepsilon_i$$

$$\varepsilon_i \overset{indep}{\sim} N(0, \sigma^2).$$

Techniques for verifying whether or not these model assumptions are met are discussed in **Section 4**. Under these assumptions, least squares estimation is equivalent to *maximum likelihood estimation* (see **Chapter 11**), and $\hat{\beta}_0$ and $\hat{\beta}_1$ are unbiased estimates of β_0 and β_1, respectively. *Fitted values* (or predicted values) for Y, denoted by $\hat{y}$, are estimates of the mean of Y for a given X and are computed using the estimated regression model

$$\hat{y}_i = \hat{\beta}_0 + \hat{\beta}_1 x_i;$$

these are shown as the solid line in **Figure 3**, drawn only for the available range of X values. *Extrapolating* beyond the range of the data is risky; the linear relation may no longer be valid. σ^2 is estimated with the *mean square error*, also known as the residual mean square

$$\hat{\sigma}^2 \equiv s^2 = \frac{1}{n-2} \sum_{i=1}^{n} (y_i - \hat{y}_i)^2 \quad (4)$$

with corresponding degrees of freedom $n - 2$: from a sample of size n, two regression parameters β_0 and β_1 must be estimated.

3.4. Regression Through the Origin

Consider our study where X is a measure of transcription and Y is a measure of translation. This seems like a natural application for which *regression through the origin* could be used. In such a model, we force $\beta_0 = 0$, which forces the mean translation level to be 0 when the transcription level is 0. However, regression through the origin is rarely applicable and should only be used when *all of* the following conditions are met: (i) it is biologically plausible for the mean of Y to be 0 when X is 0; (ii) there is strong evidence that the relation of Y with X is linear at the origin; and (iii) the sample of X values used in the study includes 0 ***(1)***. In our gene expression example, condition (i) is met but (ii) and (iii) are likely not.

4. Diagnostics: Assessing the Regression Model Fit

4.1. What to Assess

An important part of any statistical analysis is assessment of how well the chosen model fits the data. In regression, estimation of the linear slope ($\hat{\beta}_1$) is not sufficient to understand whether a linear model is appropriate. Six aspects of the model should be assessed:

(i) independence;
(ii) normality;
(iii) linearity;

(iv) constant variance;
(v) presence of outliers; and
(vi) need for additional predictor variables.

Because the errors ε_i represent the difference between the observed data and the assumed model, we use *residuals*,

$$e_i = y_i - \hat{y}_i,$$

the difference between the observed data and the estimated model, to assess model fit. **Figure 3** shows the residual $e = 51.7 - 27.1 = 24.6$ computed from the observed data point $(x, y) = (61.2, 51.7)$ and the estimated model fit $(x, \hat{y}) = (61.2, 27.1)$ for that point. However, for diagnostics it is more useful to use a standardized version of the residuals, the *studentized residuals*

$$r_i = \frac{y_i - \hat{y}_i}{\sqrt{s^2(1-h_i)}} \quad \text{where} \quad h_i = \frac{1}{n} + \frac{(x_i - \bar{x})^2}{\sum_{j=1}^{n}(x_j - \bar{x})^2}. \tag{5}$$

The h_i represent scaling factors, dependent on X, that are derived so that the r_i approximately follow $N(0, 1)$. Your statistical package should compute them for you.

4.2. Tools Used to Assess Model Fit

4.2.1. Plot of Residuals versus X

A plot of r_i versus x_i (or r_i vs. $\hat{y}_i$) is used to assess constant variance, linearity, and presence of outliers. If the assumptions of constant variance and linearity are satisfied, then this plot will show a random-like scatter of points approximately equally spread above and below the horizontal at 0; an example is shown in **Figure 4a**.

A common sign of nonconstant variance is a plot that shows a cone-shaped scatter of points, typically exhibiting increasing variance with the mouth of the cone at the larger values of X, as shown in **Figure 4b**. When the sample size is small, plotting $|r_i|$ versus x_i instead of r_i versus x_i can better highlight such a pattern.

A common sign of nonlinearity is a curvilinear or other systematic pattern in the points, indicating that there is a pattern in the residuals that is still related to X, even after the linear trend has been accounted for by the model, as shown in **Figure 4c**. Thus, the relation of Y to X is not appropriately represented by a line.

Outliers can result from an outlying X value, an outlying Y value, or both; see **Figure 4d**, where two Y-outliers are marked with arrows. Identifying outliers is important because outlying observations can "pull" the regression

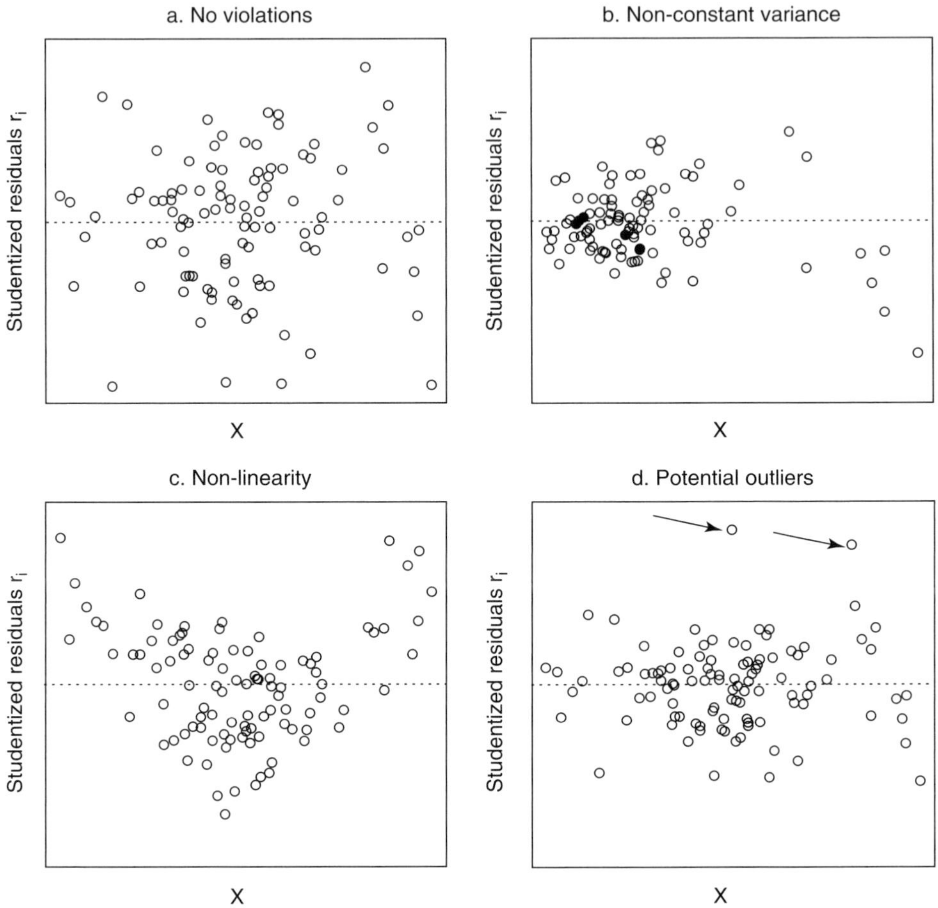

Fig. 4. Examples of residuals exhibiting various violations of model assumptions.

line toward them, thus influencing the estimation. Potential outliers should always first be checked for data entry or calculation errors or for equipment, technician, or other experimental condition errors. Data found to be erroneous should be corrected (if possible) or omitted from the analysis. Any remaining outliers may be informative in that they could represent an important way in which the data violate the model. Only approximately 5% of the r_i should be outside of ±2 and only approximately 0.1% should be outside ±3. Other measures for outliers are *leverage* and *influence*; see **Chapter 9**. If an outlier is found, a type of sensitivity analysis could be done: the outlier could be deleted, the analysis redone, and the results compared with the results obtained when including the outlier. If the two sets of results are not substantially different, the outlier is not of concern and does not need to be deleted. If the results *are* substantially different, the research team should consider possible scientific explanations for having obtained such an observation; both sets of results could then be reported.

Example 1 (Continued)

For the gene expression data, **Figure 5a** shows a typical cone-shaped pattern indicative of a variance that increases with *X*. Linearity cannot be assessed from

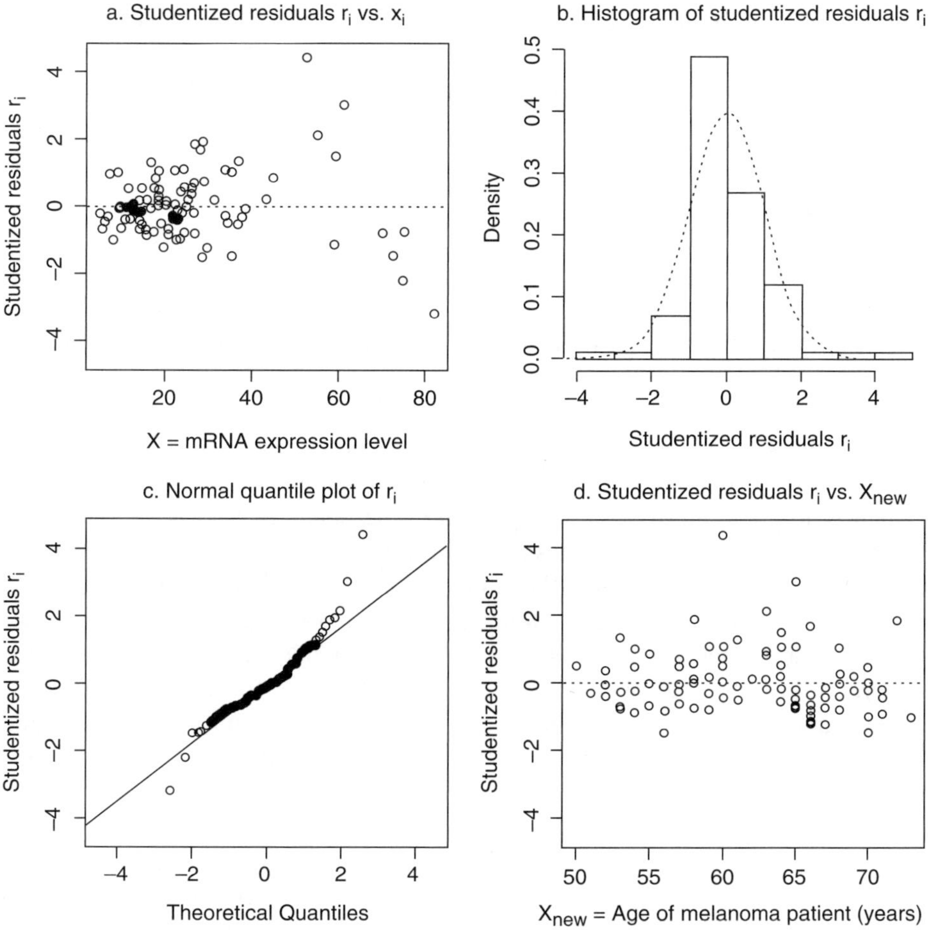

Fig. 5. Diagnostic plots from the regression of *Y* on *X*.

this plot until the constant variance assumption is satisfied. There is one potential outlier in these data with studentized residual value 4.44 from data point $(x, y) = (52.7, 60.7)$. However, there is no reason to believe this is an incorrect observation, and it may become less of an outlier if we are able to correct the nonconstant variance; we will not delete it.

4.2.2. Summary Plots of Residuals

Summary plots of the r_i are used for assessing normality and independence. Stem-and-leaf plots, box-and-whisker plots, and histograms are all useful tools for summarizing the distribution of the r_i. Are the r_i symmetrically distributed around 0? Is their distribution approximately bell-shaped, like the normal distribution? A normal quantile (or probability) plot of the r_i also indicates whether the residuals fall close to normality, and, if not, whether they are left- or right-skewed or heavy- or light-tailed (see **Chapter 3**). Normality is difficult to assess and as a rule of thumb requires $n > 30$.

Example 1 (Continued)

Figure 5b and **Figure 5c** show two summary plots of the r_i. The histogram indicates that the studentized residuals are approximately symmetric around 0, but the normal quantile plot indicates that they are slightly heavy-tailed in both the left and right tails.

Independence can be assessed if the sequence across time (or space) in which the data were collected is known. If so, then a sequence plot of the r_i versus sequence number can indicate any pattern in the residuals across the sequence. Presence of a pattern (e.g., increasing, oscillating, or curvilinear trend) indicates that the conditions under which the data were collected changed systematically, leading to nonindependent observations. No pattern (a random scatter of points) indicates that independence is likely. Lack of independence can also be inferred under certain sampling schemes. For example, if 5 mice from each of 10 litters are used in a study, then likely the measurements taken on mice within the same litter are correlated. Special models are needed for such data; see **Chapter 11**.

4.2.3. Plot of Residuals versus Additional Predictor Variables

Oftentimes, additional information on each unit of observation is collected during the course of a study. A pattern in the plot of r_i versus such an additional variable (denote it by X_{new}) indicates that the variable has an important relation with the response, above and beyond the relation of Y with X. In this case, *multiple linear regression* should be considered to regress Y on both X and X_{new} simultaneously; see **Chapter 9**.

Example 1 (Continued)

Figure 5d shows a plot of the r_i versus an additional available predictor, the ages of the 100 melanoma patients. Because there is no pattern in this plot, age does not have a strong relation with Y above and beyond the relation of X with Y.

4.3. When Assessments Show a Problem

Nonconstant variance and nonnormality in Y often appear together. In such situations, first assess for nonlinearity and nonconstant variance, and then, once these are satisfied, reassess whether the normality assumption is violated. Checks for outliers and the need for additional predictor variables should be left for last, after any other violations have been corrected.

A nonlinear relation between Y and X can sometimes be transformed to linearity with either a transformation of Y or X or both. For example, if Y increases approximately exponentially with X, then regress Y on e^X or regress $\ln(Y)$ on X. If Y increases approximately with $\ln(X)$, then regress Y on $\ln(X)$ or regress e^Y on X. In some fields, it is preferable to try transforming X first, whereas in others it is preferable to try transforming Y first. If Y increases in a quadratic (or other smooth nonlinear) relation with X, then *polynomial regression*, a type of multiple linear regression, is needed; see **Chapter 9**.

Nonconstant variance and nonnormality can also often be corrected with transformations of Y. For example, in exposure or other laboratory studies, it is common to take $\ln(Y)$ as the outcome to correct for skewness in the distribution of Y. Another common transformation is $\sqrt{Y}$. Whenever any transformation of Y is used, the transformed Y and its estimated line are used for all diagnostics and inference (covered in **Section 5**).

Example 1 (Continued)

Figure 1 shows that both X and Y have skewed distributions, with observations tending to fall more toward the lower ends of their ranges. This can often be corrected with a log transformation (log2, log10, or ln). We next fit a regression of $\ln(Y)$ on $\ln(X)$. The data with fitted line are shown in **Figure 6**, and diagnostic plots are shown in **Figure 7**. The plot of r_i versus x_i now shows a random-like scatter of points around the horizontal at 0, indicating no violations of linearity or constant variance. There is only one outlying observation, with a studentized residual of –3.53; this is not unusually large, and there is no reason to eliminate it from the analysis. The histogram and normal quantile plot indicate the studentized residuals now approximately follow a normal distribution. There is again no strong relation of the studentized residuals with patient age. Because all model assumptions appear to be satisfied, we can proceed with model inference.

Any type of logarithmic transformation, for example, cannot be used directly on data with negative or zero values. In such cases, it is common to add a small number to *every* observation before transforming. After model estimation, it is good practice to verify that one's results are not sensitive to one's choice of that small number by repeating the model estimation with two or three other choices of small number.

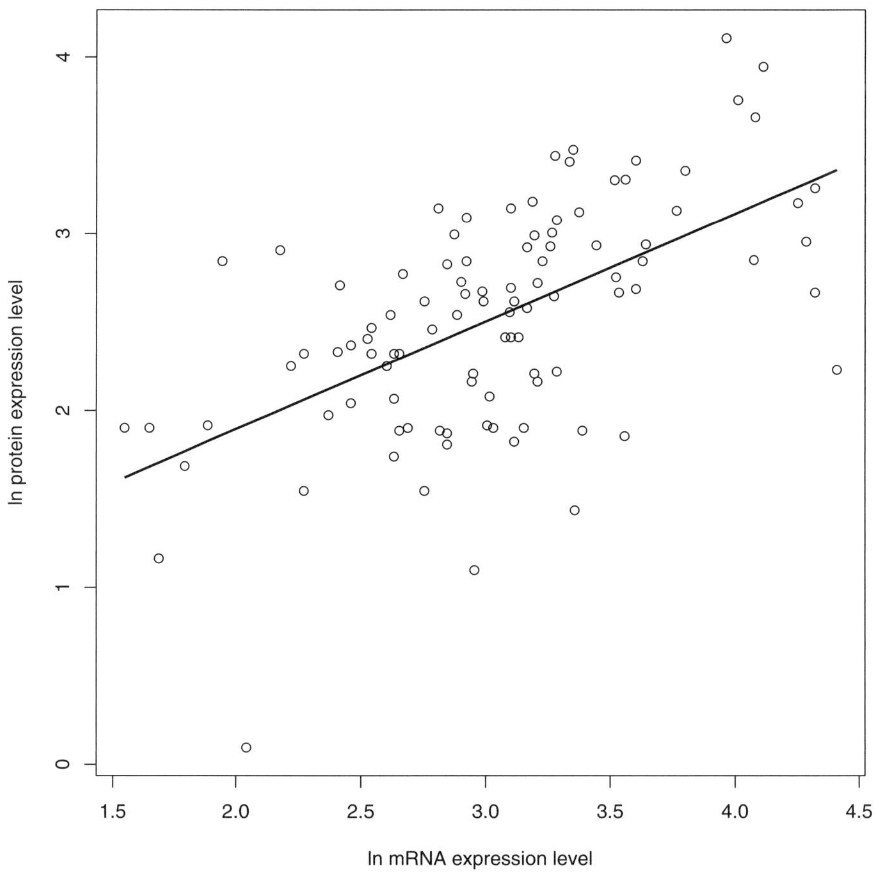

Fig. 6. Estimated line from the regression of ln(Y) on ln(X).

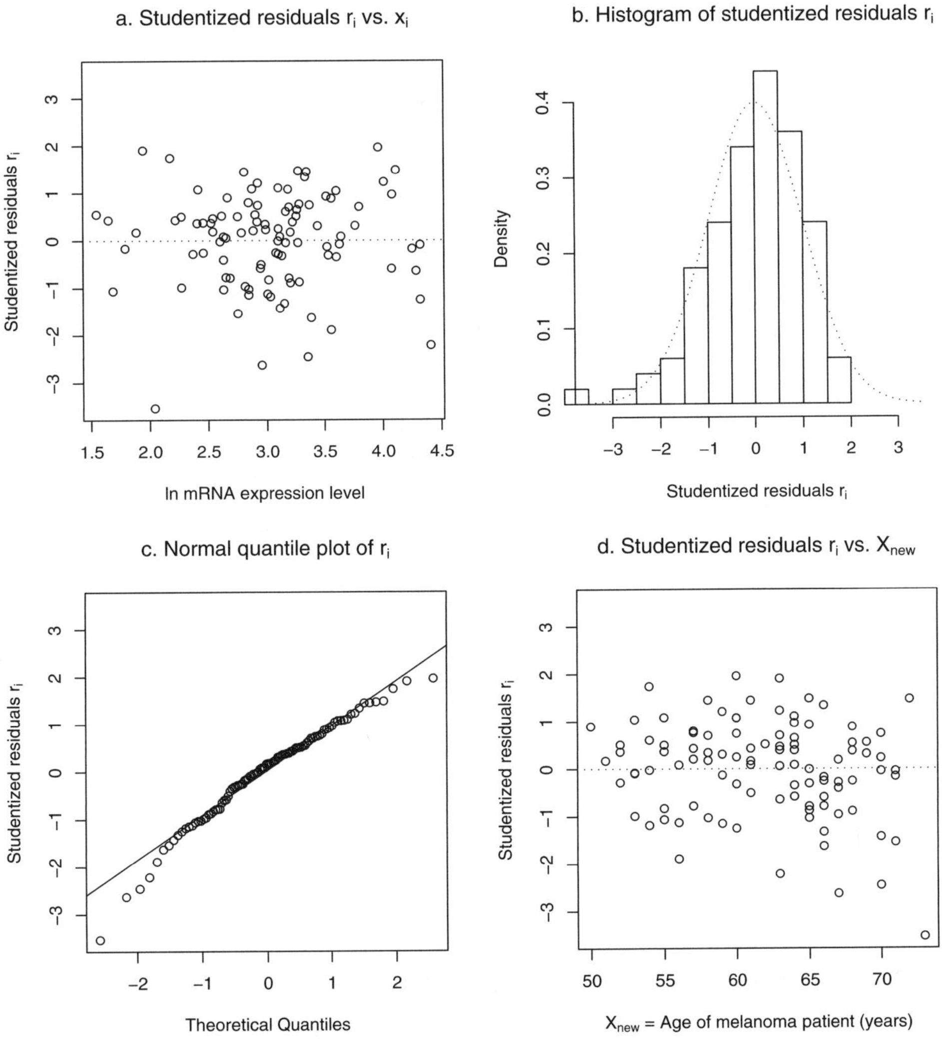

Fig. 7. Diagnostic plots from the regression of ln(Y) on ln(X)

5. Inferences from the Regression Model

5.1. Inferences About the Estimated Linear Relation

Once a model has been determined to provide an appropriate fit to the data, we can draw scientific conclusions from the analysis using hypothesis tests and confidence intervals. Both of these require standard errors for our estimates $\hat{\beta}_0$ and $\hat{\beta}_1$,

$$\text{s.e.}[\hat{\beta}_0] = \sqrt{s^2\left(\frac{1}{n} + \frac{\bar{x}^2}{\sum_{i=1}^{n}(x_i - \bar{x})^2}\right)}$$

$$\text{s.e.}[\hat{\beta}_1] = \sqrt{s^2\left(\frac{1}{\sum_{i=1}^{n}(x_i - \bar{x})^2}\right)}, \tag{6}$$

using s^2 as defined in **Equation 4**. The (1 – α)100% confidence intervals for the intercept and slope are computed as

$$\hat{\beta}_0 \pm t_{1-\alpha/2,n-2}\ \text{s.e.}[\hat{\beta}_0]$$

$$\hat{\beta}_1 \pm t_{1-\alpha/2,n-2}\ \text{s.e.}[\hat{\beta}_1],$$

again with $n - 2$ degrees of freedom. A hypothesis test for the intercept is constructed from the test statistic

$$t^* = \frac{\hat{\beta}_0}{\text{s.e.}[\hat{\beta}_0]}.$$

The null hypothesis $H_0: \beta_0 = 0$ is rejected at level α in favor of $H_1: \beta_0 \neq 0$ when

$$|t^*| > t_{1-\alpha/2,n-2}.$$

A test is similarly constructed for the slope from the test statistic

$$t^* = \frac{\hat{\beta}_1}{\text{s.e.}[\hat{\beta}_1]}. \tag{7}$$

The null hypothesis $H_0: \beta_1 = 0$ is rejected at level α in favor of $H_1: \beta_1 \neq 0$ when

$$|t^*| > t_{1-\alpha/2,n-2}.$$

One-sided tests and intervals can also be constructed; see **Chapter 4**. Rejection of $H_0: \beta_1 = 0$ indicates that there is a significant linear slope of Y on X, but it does not indicate that a linear relation is the most appropriate relation; diagnostics are needed to assess the adequacy of the linear assumption. In addition, in studies where the values of X are not randomly assigned to the units of observation, rejection of $H_0: \beta_1 = 0$ does not indicate that a change in X *causes* a change in Y; we can only infer association, not causation. Kleinbaum and others ***(2)*** have a summary of the type and weight of evidence required to reasonably infer causation from observational studies.

Example 1 (Continued)

From our regression of ln(Y) on ln(X), we get $\hat{\beta}_0 = 0.68$ (s.e.[$\hat{\beta}_0$] = 0.27), $\hat{\beta}_1 = 0.61$ (s.e.[$\hat{\beta}_1$] = 0.09), and $s^2 = 0.28$. $\hat{\beta}_0$ now estimates the mean of ln(Y) when ln(X) = 0 (i.e., when $X = 1$). $\hat{\beta}_1$ represents the mean change in ln(Y) when ln(X) is higher by 1 unit; for every 1 unit higher ln mRNA expression level, the model estimates on average a 0.61 unit higher ln protein expression level. Because $|t^*| = |0.61/0.09| = 6.78 > t_{0.975,98} = 1.98$, we reject $H_0: \beta_1 = 0$ in favor of $H_1: \beta_1 \neq 0$ at level $\alpha = 0.05$ and conclude that there is a significant linear relation between ln(Y) and ln(X). A 95% confidence interval for β_1 is $0.61 \pm (1.98)(0.09) = (0.43, 0.79)$. This interval does not span 0, another indication that H_0 is rejected at level $\alpha = 0.05$.

5.2. *Inferences About* Y

The mean of the distribution of Y, for any value of $X = x^*$, is estimated with the fitted value computed at that x^*,

$$\hat{y} = \hat{\beta}_0 + \hat{\beta}_1 x^*,$$

which has standard error

$$\text{s.e.}[\hat{y}] = \sqrt{s^2\left(\frac{1}{n} + \frac{(x^* - \bar{x})^2}{\sum_{i=1}^{n}(x_i - \bar{x})^2}\right)}. \tag{8}$$

The standard error increases the farther x^* is from $\bar{x}$. A$(1 - \alpha)100\%$ confidence interval is computed as $\hat{y} \pm t_{1-\alpha/2,n-2}$ s.e.[$\hat{y}$].
This interval has appropriate type I error rate only when it is constructed for a single x^* value. When confidence intervals are constructed simultaneously for all X values within the range of the data, we need a *confidence band* for the entire regression line

$$\hat{y} \pm (\sqrt{2F_{1-\alpha,2,n-2}})\,\text{s.e.}[\hat{y}].$$

This is the Working-Hotelling $1 - \alpha$ confidence band, and at each x^* it is wider than the corresponding confidence interval at x^*.

Interest may lie also in predicting a single value for Y for the given x^*, rather than estimating the mean of Y for the given x^*. The fitted value $\hat{y}$ above is also used for such a prediction, but we will denote it by $\hat{y}_{new} = \hat{\beta}_0 + \hat{\beta}_1 x^*$ to distinguish our intentions. It then has a larger standard error

$$\text{s.e.}[\hat{y}_{new}] = \sqrt{s^2\left(1 + \frac{1}{n} + \frac{(x^* - \bar{x})^2}{\sum_{i=1}^{n}(x_i - \bar{x})^2}\right)} \tag{9}$$

because it must account for both the variability within the distribution for Y and the variability in estimating the mean of that distribution. A confidence interval is constructed as above. Estimation of the mean and prediction of a single value should be restricted to x^* values within the range of the original sample of x_i.

When the regression is carried out on a transformed outcome (such as ln protein expression in our example), then fitted values, and their confidence intervals or prediction intervals, can be back-transformed to the original scale as long as the transformation is a monotone function. Standard errors should *not* be back-transformed; for example, exponentiating the standard error of a fitted ln protein expression level does not accurately represent the standard error of the untransformed protein expression level.

Example 1 (Continued)

For an mRNA expression level of 7.0, $\ln(x^*) = \ln(7.0) = 1.95$, and the corresponding estimated mean ln protein expression level is $\hat{\beta}_0 + \hat{\beta}_1 1.95 = 0.68 + (0.61)(1.95) = 1.87$ and has standard error 0.11 (**Equation 8**). A 95% confidence interval for the mean ln protein expression level is thus $1.87 \pm (1.98)(0.11) = (1.65,2.09)$. With a prediction standard error of 0.54 (**Equation 9**), a 95% prediction interval for a new single ln protein expression level is $1.87 \pm (1.98)(0.54) = (0.80,2.94)$. Back-transforming, a 95% confidence interval for the mean protein expression level is $(e^{1.65},e^{2.09}) = (5.21,8.08)$ and a 95% prediction interval for a new single protein expression level is $(e^{0.80},e^{2.94}) = (2.23,18.92)$.

5.3. Effect of Departures from Normality

When the assumption of normality for Y is violated, the tests and confidence intervals shown above for $\hat{\beta}_0$, $\hat{\beta}_1$, and $\hat{y}$ will still be approximately correct, because $\hat{\beta}_0$ and $\hat{\beta}_1$ have asymptotic normal distributions even when Y does not. The approximation will improve as the sample size n increases. In contrast, confidence and prediction intervals for $\hat{y}_{new}$ are sensitive to the violation of normality.

6. ANOVA Tables for Regression

An *analysis of variance* (ANOVA; see **Chapter 7**) table is constructed by partitioning the total variation in Y into variation due to two pieces: model variation (based on the available predictors) and error (residual) variation. This partitioning is represented as $y_i - \bar{y} = (\hat{y}_i - \bar{y}) + (y_i - \hat{y}_i)$ (total = model + error). This table can be constructed for any regression model (but is still called an

Table 1
ANOVA Table for Simple Linear Regression

Source	Sum of squares (SS)	Degrees of freedom	Mean squares (MS)	F^*
Model	$\sum_i (\hat{y}_i - \bar{y})^2$	1	$\sum_i (\hat{y}_i - \bar{y})^2$	$\frac{\text{MS(Model)}}{\text{MS(Error)}}$
Error	$\sum_i (y_i - \hat{y}_i)^2$	$n - 2$	$\frac{\sum_i (y_i - \hat{y}_i)^2}{n-2}$	
Total	$\sum_i (y_i - \bar{y})^2$	$n - 1$		

ANOVA table) with the $\hat{y}_i$ estimated by the regression model and the degrees of freedom computed from the number of regression parameters in the model (Table 1).

The F-test is a test of the slope, so we reject $H_0: \beta_1 = 0$ at level α in favor of $H_1: \beta_1 \neq 0$ when $F^* > F_{1-\alpha,1,n-2}$; this test statistic is the square of the t-test statistic seen in **Equation 7**, so the two tests are equivalent. The ANOVA table does not provide an F-test of $H_0: \beta_0 = 0$.

The quantity R^2 = SS(Model)/SS(Total) is called the *coefficient of determination* and takes values from 0 through 1. R^2 gives the proportion of SS(Total) due to the regression on X. A large R^2 thus indicates that the estimated linear relation will provide good predictions for Y, but R^2 is not sufficient to assess model fit.

Example 1 (Continued)

For our regression of ln protein expression on ln mRNA expression, the ANOVA table is shown in Table 2. The R^2 of 13.52/40.72 = 0.33 indicates that only about a third of the total variability in ln protein expression levels is explained by its relation with ln mRNA expression levels. Relatively low R^2 such as this are common in human biology studies.

Table 2
ANOVA Table for the Gene Expression Data

Source	Sum of squares (SS)	Degrees of freedom	Mean squares (MS)	F^*
Model	13.52	1	13.52	13.52/0.28 = 48.29
Error	27.20	98	0.28	
Total	40.72	99		

7. Study Design for Simple Linear Regression

The following decisions need to be made in order to design a study that will use simple linear regression as the primary analysis:

(i) which X values to use and how to space them; and
(ii) how many observations to collect.

When a linear relation between Y and X is appropriate for all X values being considered, a wider range in X values leads to a larger value for $\sum_i (x_i - \bar{x})^2$ and hence smaller standard errors for $\hat{\beta}_0$ and $\hat{\beta}_1$ (**Equation 6**). Thus, power to detect a significant intercept or a significant slope can be increased with widely spaced X values. Note that in some studies, it may not be possible to choose either the X values or their spacing.

In most cases, the estimated slope is of primary interest. How many X values to use is then determined by the researcher's prior knowledge of the relation and by the purpose of the study. If the researcher is confident that the relation is linear, then a *minimum* of two X values are needed to estimate the slope. If the purpose is to determine whether a linear relation is appropriate, then a *minimum* of three X values are needed. See Ryan ***(1)*** for a more detailed discussion of designs for regression studies.

In general, observations collected should be equally divided among the chosen X values. How many observations to collect in total (n) can be computed for a specified slope estimate, standard error, type I error probability α, and power using the following equation:

$$\text{Power} = P\left(t_{n-2} > t_{1-\alpha/2,n-2} - \frac{\hat{\beta}_1}{\text{s.e.}[\hat{\beta}_1]} \right) + P\left(t_{n-2} < -+t_{1-\alpha/2,n-2} - \frac{\hat{\beta}_1}{\text{s.e.}[\hat{\beta}_1]} \right).$$

See **Chapter 19** for an in-depth treatment of power and sample size computation.

Example 2

Consider a study that will regress Y on X where the researchers would like to be able to detect a slope of $\beta_1 = 0.5$ or larger with power of at least 0.90. From a pilot study, they have estimated s.e.$[\hat{\beta}_1]$ to be approximately 0.15. When the new study is complete, they will use a simple linear regression t-test (**Equation 7**) with type I error rate of 0.05. How many observations do they need? First try $n = 37$, and we compute

$$\begin{aligned}\text{Power} &= P\left(t_{37-2} > t_{0.975,37-2} - \frac{0.5}{0.15}\right) + P\left(t_{37-2} < -t_{0.975,37-2} - \frac{0.5}{0.15}\right)\\ &= P(t_{35} > 2.0301 - 3.3333) + P(t_{35} < -2.0301 - 3.3333)\\ &= P(t_{35} > -1.3032) + P(t_{35} < -5.3634)\\ &= 0.8995 + 0,\end{aligned}$$

which is not quite 0.90; $n = 38$ results in a power of 0.8999, and $n = 39$ results in

$$\begin{aligned}\text{Power} &= P\left(t_{39-2} > t_{0.975,39-2} - \frac{0.5}{0.15}\right) + P\left(t_{39-2} < -t_{0.975,39-2} - \frac{0.5}{0.15}\right)\\ &= P(t_{37} > 2.0162 - 3.3333) + P(t_{37} < -2.0162 - 3.3333)\\ &= P(t_{37} > -1.3071) + P(t_{37} < -5.3595)\\ &= 0.9004 + 0;\end{aligned}$$

thus the researchers need $n = 39$ observations in their sample to achieve a power of at least 0.90.

8. Discussion

This chapter has provided an overview of simple linear regression concepts. A more thorough coverage is available in many books. In particular, see Neter and others *(3)* for more details on expected mean squares, tests for outliers, tests for randomness, tests for constant variance, an F-test for lack of fit to the linear relation, correlation models (models where it is not necessary to designate one variable as outcome and the other as predictor), and extensively worked examples and case studies.

Ryan *(1)* covers in addition more advanced topics on regression, including regression through the origin, weighted least squares, and alternative regression techniques for when the model assumptions are violated. In particular, Ryan *(1)* covers ridge, robust, and nonparametric regression techniques for when the normality assumption is violated (e.g., when many sample points appear to be outliers) and splines and nonlinear regression for when the linearity assumption is violated.

Vittinghoff and others *(4)* cover regression as well as several extensions, including approaches for censored outcomes, for probability-weighted outcomes, and for nonnormal outcomes such as binary (logistic regression) and count (Poisson regression). Harrell *(5)* also describes approaches for censored outcomes and for ordinal outcomes. Censoring is common in laboratory studies, where equipment can only determine a measure above a certain value (lower limit of detection) or below a certain value (upper limit of detection), and in survival and time-to-event studies, where some study participants survive beyond the study's end. Not accounting for censoring can lead to biased estimates of model parameters.

Mickey and others ***(6)*** cover the sampling of *X* values and random *X* models (correlation models) for when the *X* values are not assumed fixed and known. They also discuss the consequences of measurement error in *X*, which can also lead to biased estimates of model parameters.

Kleinbaum and others ***(2)*** discuss inferring causation in observational studies. Vittinghoff and others ***(4)*** discuss confounding, causal effects, and counterfactual experiments. All of these techniques assume the sample consists of independently collected observations. When this assumption is violated, other types of regression models are needed (see **Chapter 11**).

Acknowledgments

The author would like to thank Dr. Tracy Bergemann for helpful comments that improved the examples in this chapter.

References

1. Ryan, T. P. (1997) *Modern Regression Methods.* New York, John Wiley & Sons.
2. Kleinbaum, D. G., Kupper, L. L., Muller, K. E., and Nizam, A. (1997) *Applied Regression Analysis and Multivariable Methods*, 3rd ed. New York, Duxbury.
3. Neter, J., Kutner, M. H., Wasserman, W., and Nachtsheim, C. J. (1996) *Applied Linear Statistical Models*, 4th ed. New York, McGraw-Hill/Irwin.
4. Vittinghoff, E., Glidden, D. V., Shiboski, S. D., and McCulloch, C. E. (2005) *Regression Methods in Biostatistics: Linear, Logistic, Survival, and Repeated Measures Models.* New York, Springer.
5. Harrell, F. E. Jr. (2001) *Regression Modeling Strategies With Applications to Linear Models, Logistic Regression, and Survival Analysis.* New York, Springer.
6. Mickey, R. M., Dunn, O. J., and Clark, V. A. (2004) *Applied Statistics: Analysis of Variance and Regression*, 3rd ed. New York, John Wiley & Sons.

9

Multiple Linear Regression

Lynn E. Eberly

Summary

This chapter describes multiple linear regression, a statistical approach used to describe the simultaneous associations of several variables with one continuous outcome. Important steps in using this approach include estimation and inference, variable selection in model building, and assessing model fit. The special cases of regression with interactions among the variables, polynomial regression, regressions with categorical (grouping) variables, and separate slopes models are also covered. Examples in microbiology are used throughout.

Key Words: Adjusted R-square; Bonferroni correction; coefficient of multiple determination; collinearity; dummy variable; indicator variable; influence; interaction; leverage; parallel lines model; partial coefficients; partial sums of squares; polynomial; separate slopes model; sequential sums of squares; stepwise selection; variable selection.

1. Introduction

Oftentimes, several pieces of information on each unit of observation are collected during the course of a study, and interest lies in simultaneously examining the associations of these predictor variables with the outcome. In this chapter, we examine *multiple linear regression* estimation, inference, model building, and assessment. We will examine several special cases of this model, including polynomials, interactions, and categorical predictor variables. This chapter assumes the reader is familiar with the terminology and concepts in **Chapter 7** and **Chapter 8**.

Example 1

The level of natural killer cell production in a cancer patient may be related to the expression levels of certain genes. A research team measures the mRNA

From: *Methods in Molecular Biology, vol. 404: Topics in Biostatistics*
Edited by: W. T. Ambrosius © Humana Press Inc., Totowa, NJ

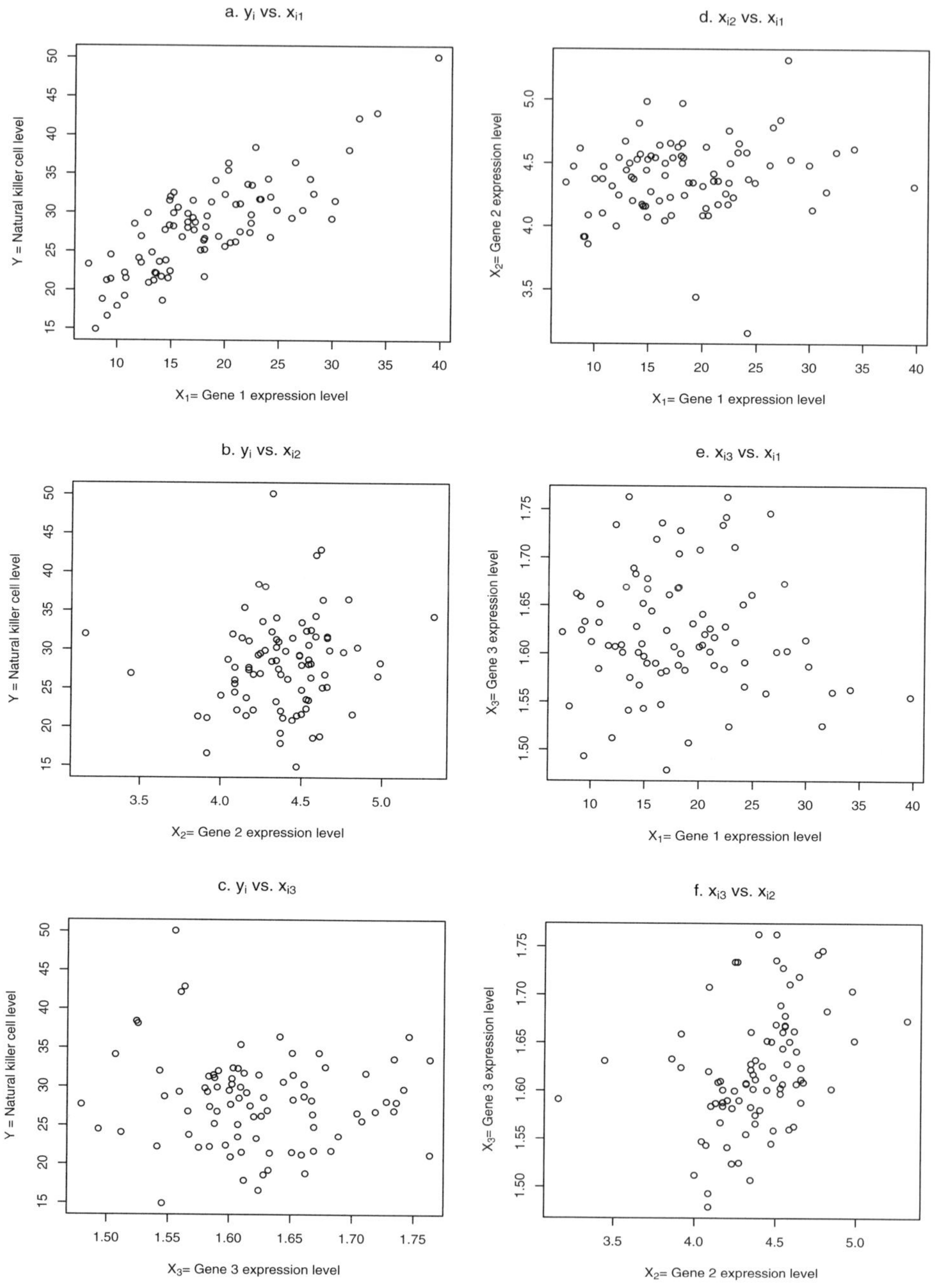

Fig. 1. Natural killer cell levels and gene expression levels for three genes.

expression levels relative to a control for three different genes in tumor samples and the natural killer cell levels in serum samples from 86 people with lung cancer; see **Figure 1**.

2. Regression with Multiple Explanatory Variables

2.1. Multiple Linear Regression Model

A *multiple linear regression* model is an extension of the simple linear regression model for data with multiple predictor variables and one outcome $(x_{i1}, x_{i2}, \ldots, x_{i,p-1}, y_i)$ for $i = 1, 2, \ldots, n$ units of observation. It formalizes a simultaneous statistical relation between the single continuous outcome Y and the predictor variables X_k $(k = 1, 2, \ldots, p - 1)$:

$$y_i = \beta_0 + \beta_1 x_{i1} + \beta_2 x_{i2} + \ldots + \beta_{p-1} x_{i,p-1} + \varepsilon_i$$

$$\varepsilon_i \overset{indep}{\sim} N(0, \sigma^2) \qquad (1)$$

where β_0 represents the intercept (the mean of Y when all $X_k = 0$), and each β_k represents a slope with respect to X_k (the magnitude of change in the mean of Y when X_k is larger by one unit and all other predictors are held constant). The β_k are thus sometimes called *partial regression coefficients*. As for simple linear regression, this model can be equivalently written as

$$y_i \overset{indep}{\sim} N(\beta_0 + \beta_1 x_{i1} + \beta_2 x_{i2} + \ldots + \beta_{p-1} x_{i,p-1}, \sigma^2).$$

The assumptions are thus the same as for simple linear regression:

(i) the y_i are independent of each other;
(ii) the y_i each follow a normal distribution;
(iii) the mean of that distribution is a linear function of each x_{ik}; and
(iv) the variance of that distribution is the same for all y_i (constant variance, or *homoscedasticity*).

The general procedure for carrying out a regression analysis is as follows:

(i) for each predictor, verify through a data plot that a linear relation is likely to be appropriate;
(ii) estimate the linear regression model;
(iii) assess through diagnostics whether the model provides an appropriate fit to the data;
(iv) if so, use the model to draw inferences about the regression coefficients;
(v) reduce the model by removing nonsignificant predictors, if appropriate for the study goals; and
(vi) reassess through diagnostics whether the model provides an appropriate fit to the data.

Step (ii) through **step (vi)** will be outlined here and in **Section 3**.

As for simple linear regression, estimation of this model is done with the least squares criterion: we choose the β_k to minimize the sum of squared vertical distances between the observed y_i and the fitted model:

$$\sum_{i=1}^{n}(y_i-(\beta_0+\beta_1 x_{i1}+\beta_2 x_{i2}+\ldots+\beta_{p-1}x_{i,p-1}))^2.$$

Assuming a specific distribution, such as normality, for the y_i is not required to do this estimation but is required for any statistical inference.

2.2. Inference

The formulas for the estimated regression parameters β_k and their standard errors are not easily expressed; your statistical software will print the estimates out for you. For details, see Neter and others ***(1)***. Standard errors depend on s^2, the *mean square error*,

$$s^2=\frac{1}{n-p}\sum_{i=1}^{n}(y_i-\hat{y}_i)^2,$$

with $n - p$ degrees of freedom because of the p regression coefficients being estimated. The fitted values $\hat{y}_i$ are computed from the estimated model:

$$\hat{y}_i=\hat{\beta}_0+\hat{\beta}_1 x_{i1}+\hat{\beta}_2 x_{i2}+\ldots+\hat{\beta}_{p-1}x_{i,p-1}.$$

A $(1-\alpha)100\%$ confidence interval for each coefficient β_k is

$$\hat{\beta}_k \pm t_{1-\alpha/2,n-p}\ \text{s.e.}[\hat{\beta}_k]$$

and a test of $H_0 : \beta_k = 0$ is constructed from the test statistic

$$t^*=\frac{\hat{\beta}_k}{\text{s.e.}[\hat{\beta}_k]}. \qquad (2)$$

We reject H_0 at level α in favor of $H_1 : \beta_k \neq 0$ when

$$|t^*| > t_{1-\alpha/2,n-p}.$$

One-sided tests and intervals can also be constructed; see **Chapter 4**. These tests and intervals do not have the appropriate type I error rate when they are constructed for each of several coefficients. For simultaneous intervals (or tests) of several coefficients, replace $t_{1-\alpha/2,n-p}$ with a Bonferroni corrected quantile, $t_{1-\alpha/(2s),n-p}$, for example where s is the number of coefficients for which intervals or tests are needed. See **Chapter 4** and **Chapter 7** for an introduction to and other uses for the Bonferroni correction.

As for simple linear regression, rejection of this hypothesis indicates that there is a significant linear trend present between the predictor being tested and the outcome; it does not indicate whether the linear assumption is appropriate or whether the model fits the data well. In addition, in observational studies, it does not indicate that changes in the predictor *cause* changes in Y; see also **Chapter 8, Section 5.1**.

Example 1 (Continued)

Gene 1 has the most significant association with natural killer cell levels, with regression coefficient $\hat{\beta}_1 = 0.73$, standard error 0.06, and test statistic $t^* = |0.73/0.06| = 12.17$. We reject $H_0 : \beta_1 = 0$ at level 0.05 because $|t^*| > t_{0.975,82} = 1.99$ and conclude that mean natural killer cells are 0.73 units higher for each 1 unit higher **gene 1** expression level. **Gene 2** and **gene 3** expressions do not show significant associations with natural killer cell levels at level 0.05. We fail to reject $H_0 : \beta_2 = 0$ because $t^* = |0.42/1.41| = 0.30 < 1.99$. We also fail to reject $H_0 : \beta_3 = 0$ at level 0.05 because $t^* = |-4.81/6.80| = 0.71 < 1.99$. A Bonferroni correction of these three tests would compare each t^* to $t_{1-0.05/(2*3),82} = t_{0.992,82} = 2.44$, and our conclusions would not change. The intercept is interpreted as the mean level of natural killer cells when all three gene expression levels are 0 and is not a biologically useful value in this context.

A $(1 - \alpha)100\%$ confidence interval for the estimated mean of Y at the specified values $x^* = (x_1^*, x_2^*, \ldots, x_{p-1}^*)$ using $\hat{y} = \hat{\beta}_0 + \hat{\beta}_1 x_1^* + \hat{\beta}_2 x_2^* + \ldots + \hat{\beta}_{p-1} x_{p-1}^*$ is

$$\hat{y} \pm t_{1-\alpha/2,n-p} \text{ s.e.}[\hat{y}].$$

A $(1 -\alpha)100\%$ confidence interval for a single predicted value of Y at x^* (denote it $\hat{y}_{\text{new}} = \hat{\beta}_0 + \hat{\beta}_1 x_1^* + \hat{\beta}_2 x_2^* + \ldots + \hat{\beta}_{p-1} x_{p-1}^*$) is $\hat{y}_{\text{new}} \pm t_{1-\alpha/2,n-p}$ s.e.$[\hat{y}_{\text{new}}]$; this is often called a *prediction interval*. Your statistical software will compute both these types of intervals for you. Only use x^* that fall within the range of the observed data.

Example 1 (Continued)

A 95% confidence interval for the mean natural killer cell level at gene expression levels $x^* = (13.2, 4.5, 1.7)$ (for **genes 1**, **2**, and **3**, respectively) is (23.08, 25.37), while a 95% prediction interval is (16.92,31.53). Note that neither a confidence interval nor a prediction interval for $x^* = (10, 10, 10)$ would be appropriate because this value is not within the range of the observed data. Less obviously, $x^* = (10, 5, 1.5)$ is also not within the range of the data; the data do have observed **gene 2** values of 5 and observed **gene 3** values of 1.5, but such small **gene 3** values do not occur with such large **gene 2** values; see **Figures 1d–1f**.

2.3. Overall ANOVA Table

An ANOVA table with sums of squares, degrees of freedom, and F-test can be constructed as for simple linear regression. It shows how the total sum of squares is partitioned into the model and the error sums of squares (**Table 1**). With $p - 1$ predictors, we have $p - 1$ degrees of freedom for the model and $n - p$ degrees of freedom for the error. The F-test is a test of the null hypothesis $H_0 : \beta_1 = \beta_2 = \ldots = \beta_{p-1} = 0$. We reject H_0 at level α when $F^* = \text{MS(Model)}/\text{MS(Error)} > F_{1-\alpha,p-1,n-p}$. This is a simultaneous test of all predictors, hence, unlike simple linear regression, there is no one t-test to which it is equivalent.

$R^2 = \text{SS(Model)/SS(Total)}$ is now called the *coefficient of multiple determination* and measures the proportion of total variation in Y that is associated with the $p - 1$ predictors. R^2 takes values from 0 through 1; $R^2 = 0$ when all $\hat{\beta}_1$, $\hat{\beta}_2, \ldots, \hat{\beta}_{p-1}$ are 0, and $R^2 = 1$ when the model fits the data perfectly (i.e., $y_i = \hat{y}_i$ for all i).

SS(Total) is the same no matter which model is fit; it depends on only the observed Y values. When a new model is fit, then, it is the partitioning of SS(Total) into SS(Model) and SS(Error) that changes. As an additional predictor is added to a model, more of SS(Total) will be "explained" by the model, leading to an increase in SS(Model) and a corresponding decrease in SS(Error). Because SS(Model) increases, R^2 increases, but only by small amounts for nonsignificant predictors. This means that a higher R^2 does not necessarily mean a better model. For this reason, we often instead consider the *adjusted* R^2, which

Table 1
Overall ANOVA Table for Multiple Linear Regression

Source	Sum of squares (SS)	Degrees of freedom	Mean squares (MS)	F^*
Model	$\sum_i (\hat{y}_i - \bar{y})^2$	$p - 1$	$\sum_i (\hat{y}_i - \bar{y})^2$	$\dfrac{\text{MS(Model)}}{\text{MS(Error)}}$
Error	$\sum_i (y_i - \hat{y}_i)^2$	$n - p$	$\dfrac{\sum_i (y_i - \hat{y}_i)^2}{n-p}$	
Total	$\sum_i (y_i - \bar{y})^2$	$n - 1$		

is computed using mean squares instead of sums of squares so that degrees of freedom (and hence number of predictors) are taken into account: $R_a^2 = 1 -$ MS(Error)/MS(Total). This in effect penalizes the R^2 value for models that include nonsignificant predictors.

2.4. Partitioning the ANOVA Table by Predictor

A sum of squares can be computed for each predictor separately as the increase in SS(Model) [or equivalently the decrease in SS(Error)] when that predictor is added to a model. This means that the order in which predictors are added to a model will change the computation. *Sequential sums of squares* are computed after adding each predictor to the model one at a time in a sequence; thus the sequential sum of squares for a predictor is "adjusted" for all predictors earlier in the sequence but not for those predictors later in the sequence. For example, with three predictors, we compute the sum of squares for X_1 [denote it by SS(X_1)], then the sum of squares for X_2 after adjusting for X_1 [SS($X_2|X_1$)], and finally the sum of squares for X_3 after adjusting for both X_1 and X_2 [SS($X_3|X_1, X_2$)]. For any model being considered, sequential sums of squares add up across the included predictors to equal the model sum of squares. In this example, SS(Model) = SS(X_1) + SS($X_2|X_1$) + SS($X_3|X_1, X_2$). Thus, the sequential sums of squares represent a partitioning of SS(Model) into $p - 1$ components, one for each predictor.

Partial sums of squares in contrast are computed by considering each predictor to be the last one in the sequence, thus they are "adjusted" for all other predictors in the model. For our example, we compute SS($X_1|X_2, X_3$), SS($X_2|X_1, X_3$), and SS($X_3|X_1, X_2$). If predictors are uncorrelated with each other, then for each predictor its partial and sequential sums of squares will be identical. Your statistical software will compute these sums of squares for you, and the documentation should make clear which types of sums of squares are printed in your output tables. **Chapter 10** covers these concepts in more detail.

The t-test of $H_0 : \beta_k = 0$ in **Equation 2** has an equivalent F-test using the partial sum of squares

$$F^* = \frac{\mathrm{SS}(X_k|X_1, \ldots, X_{k-1}, X_{k+1}, \ldots, X_{p-1})}{\mathrm{MS(Error)}}$$

where we reject H_0 at level α in favor of $H_1 : \beta_k \neq 0$ when $F^* > F_{1-\alpha,1,n-p}$. Several β_k can be tested simultaneously with a *general linear* F-test; see **Chapter 10**. Your statistical package will likely present both the sequential and partial F-tests in the output; your documentation should explain their presentation. Your specific scientific hypothesis can dictate whether the partial or sequential F-test is appropriate: should the test for the predictor of interest be adjusted for the

Table 2
Sequential ANOVA Table for the Natural Killer Cell Data

Source	Sequential sum of squares (SS)	Degrees of freedom	Mean squares (MS)	F^*	P value
Model	1,937.13	3	645.71	49.07	<0.0001
Gene 1	1,930.53	1	1,930.53	146.70	<0.0001
Gene 2	0.02	1	0.02	0.002	0.96
Gene 3	6.58	1	6.58	0.50	0.48
Error	1,079.00	82	13.16		
Total	3,016.13	85			

associations of the other predictors in the model (partial) or not (sequential)? In general, partial F-tests are used; some exceptions to this are described in many experimental design textbooks.

Example 1 (Continued)

The ANOVA table with sequential sums of squares for our model is shown in **Table 2**. Because $F^* = 49.07 > F_{0.95,3,82} = 2.72$, we reject $H_0 : \beta_1 = \beta_2 = \beta_3 = 0$ and conclude that at least one coefficient is non-zero (i.e., at least one of the three genes' expression level is significantly associated with natural killer cell levels). $F^* = 146.70$ tests whether **gene 1** is associated with the outcome, $F^* = 0.002$ tests whether **gene 2** is associated with the outcome after adjusting for the effect of **gene 1**, and $F^* = 0.50$ tests whether **gene 3** is associated with the outcome after adjusting for the effects of **gene 1** and **gene 2**.

If instead we compute partial sums of squares (**Table 3**), we see that the partial sums of squares do not add up to equal the model sum of squares. The

Table 3
Partial ANOVA Table for the Natural Killer Cell Data

Source	Partial sum of squares (SS)	Degrees of freedom	Mean squares (MS)	F^*	P value
Gene 1	1,819.01	1	1,819.01	138.22	<0.0001
Gene 2	1.19	1	1.19	0.09	0.76
Gene 3	6.58	1	6.58	0.50	0.48
Error	1,079.00	82	13.16		

square of each t-test statistic (**Equation 2**) is equal to its corresponding partial F-test statistic; these test the association of each gene with the outcome after adjusting for both other genes. The partial and sequential sums of squares for **gene 3** are identical, because both represent $SS(X_3|X_1, X_2)$.

3. Assessing Model Fit

3.1. What to Assess

An important part of any analysis is assessment of how well the chosen model fits the data. Seven aspects of the model should be assessed:

(i) independence;
(ii) normality;
(iii) linearity for each predictor variable;
(iv) constant variance (homoscedasticity);
(v) presence of outliers;
(vi) correlation of predictor variables with each other (*collinearity*); and
(vii) need for additional predictor variables.

As for simple linear regression, we use a standardized version of the residuals, the *studentized residuals* $r_i = (y_i - \hat{y}_i)/\sqrt{s^2(1-h_i)}$ (**Chapter 8, Equation 5**) where h_i is now computed using all $p - 1$ predictor variables; see Neter and others *(1)* for details. The quantity h_i is a measure of the distance between $(x_{i1}, x_{i2}, \ldots, x_{i,p-1})$ and $(\bar{x}_1, \bar{x}_2, \ldots, \bar{x}_{p-1})$ and is called the *leverage*. This standardization results in an approximate $N(0, 1)$ distribution for r_i.

3.2. Tools Used to Assess Model Fit

All diagnostics used for simple linear regression can be used for multiple linear regression as well. Summary plots of the r_i (stem-and-leaf plots, box-and-whisker plots, histograms, normal probability or quantile plots, and sequence plots) are used for assessing normality, independence, and the presence of outliers. Normality is difficult to assess and as a rule of thumb requires $n > 30$. A plot of r_i versus x_{ik}, for each of the $k = 1, 2, \ldots, p - 1$ predictors, can be used to assess constant variance and each predictor's linearity; see **Chapter 8** for details. An additional tool is a plot of r_i versus $\hat{y}_i$. When assumptions of normality, linearity, and constant variance are met, this plot should show a random-like scatter of points approximately equally spread above and below the horizontal at 0.

Example 1 (Continued)

Diagnostic plots are shown in **Figure 2**. **Figures 2a–2c** each show no indication of violations of linearity or constant variance; this is confirmed by

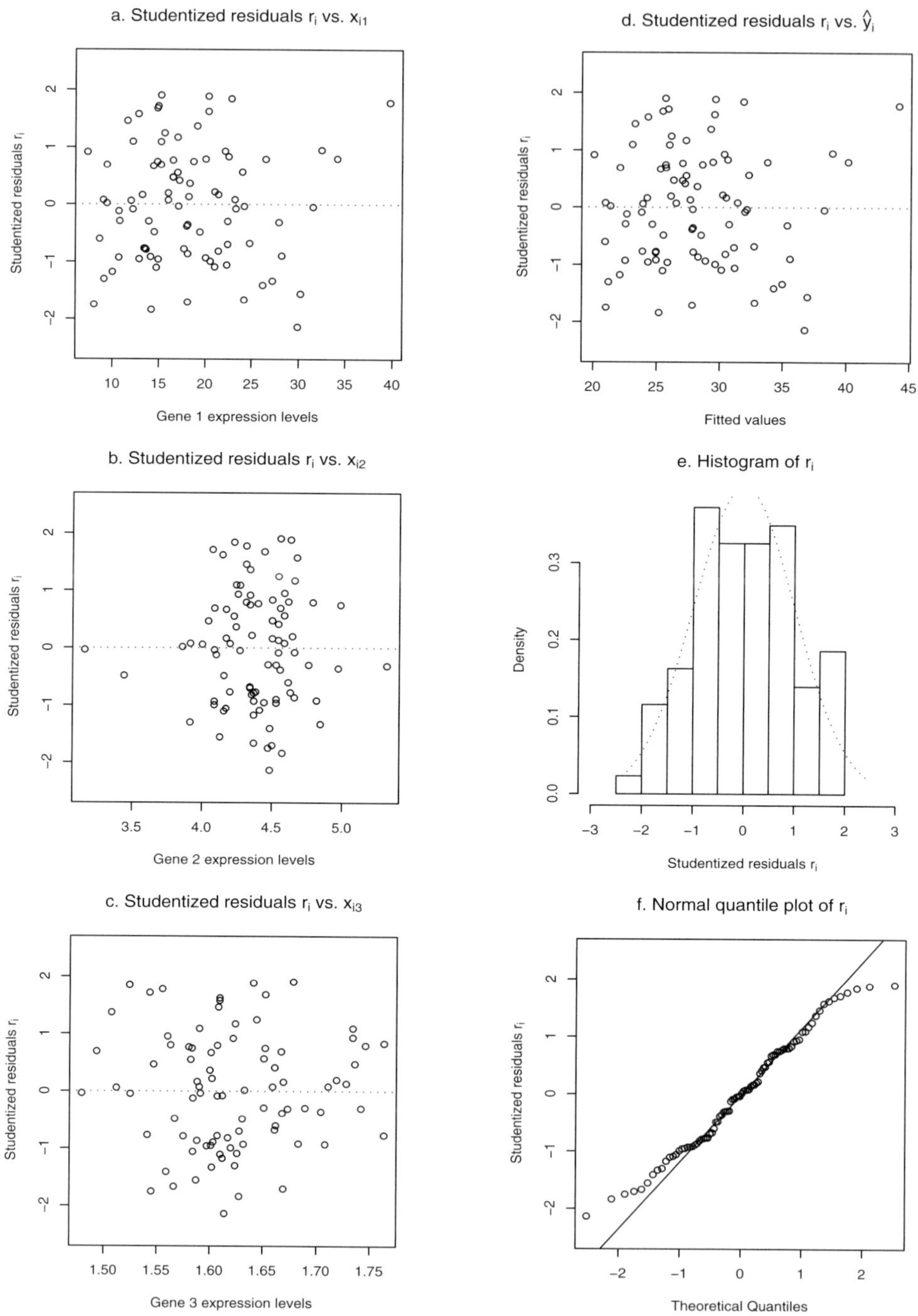

Fig. 2. Diagnostic plots from the regression of Y on X_1, X_2, and X_3.

Figure 2d. **Figures 2e–2f** show the residuals are approximately normally distributed, with slightly light tails. None of the plots indicate any outliers.

As mentioned in **Chapter 8**, outliers can be outlying in Y, in X, or both. Identifying outliers is important because outlying observations can "pull" the regression line toward them, thus influencing the estimation. Potential outliers should always first be checked for errors; see **Chapter 8**. Because of the multidimensional nature from having multiple predictors, outliers are more difficult to visually spot in any 2-dimensional plot. There are several ways to instead quantitatively measure the degree to which an observation may be outlying. Studentized residuals are a measure of how outlying an observation is in Y; only approximately 5% of the r_i should lie outside of ± 2 and only approximately 0.1% should lie outside of ± 3. Another is leverage, defined above, which is a measure of how outlying an observation is in the multidimensional X direction. A third type of measure is *influence*; an observation is highly influential if its exclusion results in substantial changes in the fitted line. DFFITS measures the influence of the ith observation on its own fitted value $\hat{y}_i$. DFBETAS measures the influence of the ith observation on each estimated regression coefficient. Cook's distance measures the influence of the ith observation on all n fitted values. Your software package should be able to compute these for you; see Ryan *(2)* for details and guidelines for use.

Collinearity occurs when predictors are correlated with each other. Pairwise collinearity can be assessed by estimating the correlation (see **Chapter 8**) between each pair of predictors. Collinearity has several consequences: (1) Variances of the $\hat{\beta}_k$ may be inflated, thus reducing the power of tests. This can be measured for each X_k by the *variance inflation factor* (VIF), which is a function of the R^2 from a regression of X_k on all other predictors. See Ryan *(2)* for details and guidelines for use. (2) Magnitudes and possibly directions (signs) of regression coefficients may change depending on which predictors are included in the model. For this reason, it is always good to verify whether the direction and magnitude of each regression coefficient is biologically plausible within the context of your study. (3) Lastly, a t-test of $H_0 : \beta_k = 0$ (equivalent to an F-test using that predictor's partial sum of squares) may give a different conclusion about H_0 from the F-test based on that predictor's sequential sum of squares. For example, suppose X_1 and X_2 are correlated and are entered into a model in that order. The t-test for β_1 represents a test of X_1 after adjusting for X_2, and the t-test for β_2 represents a test of X_2 after adjusting for X_1. In contrast, a sequential F-test of X_1 represents a test of the relation of X_1 with Y by itself (ignoring X_2), and a sequential F-test of X_2 represents a test of the relation of X_2 with Y after adjusting for X_1. As discussed above, you should use whichever test is appropriate for your scientific questions.

Example 1 (Continued)

The Pearson correlation among our predictors is 0.36 between **gene 2** and **gene 3** expression levels, 0.15 between **gene 1** and **gene 2** expression levels, and –0.07 between **gene 1** and **gene 3** expression levels. The correlation of 0.36 is not a particularly high correlation, and the directions and magnitudes of all our regression coefficients make sense when compared with the data plots in **Figure 1**. We thus do not suspect any problematic multicollinearity.

Unlike simple linear regression, a plot of r_i versus a predictor not used in the model (denote it as X_{new}) may not indicate whether that variable has an important relation with the response, above and beyond the relation of Y with all X_k in the model. If X_{new} is correlated with any of the X_k, then a plot of r_i versus X_{new} will not show the true relation and we must instead consider *partial regression plots*. For example, let $r_i(Y|X_1, X_2)$ denote the residuals from the regression of Y on the two predictors X_1 and X_2, and let $r_i(X_{new}|X_1, X_2)$ denote the residuals from the regression of X_{new} on X_1 and X_2. A plot of $r_i(Y|X_1, X_2)$ versus $r_i(X_{new}|X_1, X_2)$ will indicate the form and strength of the relation between Y and X_{new} after adjusting for the other predictors. If a pattern is seen in the plot, then X_{new} should be added to the model. The correlation between $r_i(Y|X_1, X_2)$ and $r_i(X_{new}|X_1, X_2)$ is called the *partial correlation* between Y and X_{new}. This is easily extended to more predictors.

3.3. When Assessments Show a Problem

As for simple linear regression, transformations of Y or of any X_k can be attempted to correct for nonlinearity, nonconstant variance, and/or nonnormality; see **Chapter 8** for details. Robust regression, ridge regression, weighted least squares regression, and nonparametric regression are other possibilities beyond the scope of this text; for example, see Ryan *(2)*.

4. Special Cases: Polynomials and Interactions

4.1. Polynomial Regression

When linearity is violated and transformations do not correct the nonlinearity, a model with a nonlinear trend of Y with X can be considered. The easiest such trend to construct is a polynomial (e.g., a quadratic trend or a cubic trend). *Polynomial regression* can be used either when the true relation is likely a polynomial or when the relation is complex and a polynomial is thought to be a good approximation. A polynomial model is built by including each term of the polynomial as a predictor. A common example is a quadratic regression:

$$y_i = \beta_0 + \beta_1 x_{i1} + \beta_2(x_{i1} * x_{i1}) + \varepsilon_i$$
$$= \beta_0 + \beta_1 x_{i1} + \beta_2 x_{i1}^2 + \varepsilon_i. \tag{3}$$

β_1 then represents the linear effect and β_2 represents the quadratic effect. In order to preserve this interpretation of the coefficients, a general rule is that when a higher-order polynomial term (such as a cubic) is included, all lower-order terms (linear and quadratic) must also be included, even if they are statistically nonsignificant. For example, not including the linear term in a quadratic regression forces the curve to be symmetric around $X_1 = 0$ (the vertical axis), which is rarely appropriate. It is common to center the predictor by subtracting its mean value and then use the centered predictor in the model: $y_i = \beta_0 + \beta_1 (x_{i1} - \bar{x}_1) + \beta_2(x_{i1} - \bar{x}_1)^2 + \varepsilon_i$; this can improve the computational stability of the estimation procedure. Inference and diagnostics are carried out as described in **Section 2** and **Section 3**.

4.2. Regression with Interactions

It is often the case that the relation of one predictor with Y depends on the level of another predictor. A simple way of introducing this into a model is through an *interaction* (or product) of the two predictors with each other. For example, suppose Y is natural killer cell level and X_1 is gene expression level for a particular gene of interest. The association of Y with X_1 may be affected by age of the patient (X_2). The third predictor in the following model is thus the interaction of the first two with each other:

$$y_i = \beta_0 + \beta_1 x_{i1} + \beta_2 x_{i2} + \beta_3(x_{i1} * x_{i2}) + \varepsilon_i. \tag{4}$$

How does this affect interpretation of our model? The intercept and slope of Y (natural killer cell level) with X_1 (expression) both depend on the value for X_2 (age), and likewise the intercept and slope of Y with X_2 (age) both depend on the value for X_1 (expression). Rewriting the model in order to focus on the role of X_2 will make this clear:

$$y_i = [\beta_0 + \beta_1 x_{i1}] + [\beta_2 + \beta_3 x_{i1}]x_{i2} + \varepsilon_i. \tag{5}$$

When $x_{i1} = 2$, then the intercept of Y with X_2 is $\beta_0 + 2\beta_1$, and the slope of Y with X_2 is $\beta_2 + 2\beta_3$. When $x_{i1} = 4$, then the intercept of Y with X_2 is $\beta_0 + 4\beta_1$, and the slope of Y with X_2 is $\beta_2 + 4\beta_3$. What is the interpretation of β_2 by itself? It represents the slope of Y with X_2 (age) when x_{i1} (expression) is 0, and similarly β_1 represents the slope of Y with X_1 (expression) when x_{i2} (age) is 0. If an interaction such as $X_1 * X_2$ (called a second-order interaction) is included in a model, then each of X_1 and X_2 should also be included, even if statistically nonsignificant. In general, when an interaction is present, all lower-order terms must also be included.

The choice of which interactions to include in a model, especially when many predictors are available, is difficult. Even with only five predictors, there are 10 second-order, 10 third-order, five fourth-order, and one fifth-order interaction. Omitting interactions that have important relations with the outcome can lead to biased estimates for the predictors that are in the model. On the other hand, including interactions that are not needed can inflate variances and thus reduce the power to test other predictors in the model.

In many situations, there will be interactions of particular interest that can be prespecified before model fitting begins. These interactions may be dictated, for example, by the study's goals, by previous research, by an understanding of the underlying science, or by policy- or decision-making needs. In other situations, such as when the study is meant to be exploratory, the analysis may contain all second-order interactions but no higher-order because they are more difficult to interpret. The approach to be taken during analysis should be specified in advance.

5. Parallelism: Comparing the Linear Trend Across Groups

5.1. The ANOVA-Regression Connection: Class Variables for Groups

One form of multiple linear regression occurs when one or more of the predictors X_k correspond with categorical (class or grouping) variables rather than with continuous variables. Consider, for example, gender, where we specify $X_1 = 1$ for an observation that is female and $X_1 = 0$ for an observation that is male; X_1 is thus an *indicator* (dummy) variable for female gender. Other common examples are treatment group (e.g., 1 for treatment, 0 for placebo) and race/ethnicity, with levels for Hispanic, non-Hispanic black, non-Hispanic Asian, and non-Hispanic white. A class variable with four levels requires three indicator variables. Here we might use an indicator for non-Hispanic black as X_1, an indicator for non-Hispanic white as X_2, and an indicator for non-Hispanic Asian as X_3; then Hispanics have $X_1 = X_2 = X_3 = 0$ and this group is known as the *reference group*. A class variable with q levels requires $q - 1$ indicator variables.

Suppose we regress Y on the three indicator variables for race/ethnicity:

$$y_i = \beta_0 + \beta_1 x_{i1} + \beta_2 x_{i2} + \beta_3 x_{i3} + \varepsilon_i.$$

These indicator variables do not have to be defined exactly as we did above; your statistical software will do it automatically, and the documentation should make it clear which definitions are being used. It is critical to understand your software's definitions, so that the interpretation of the corresponding regression coefficients is clear. To get fitted values, we compute for example

$\hat{y}_i = \hat{\beta}_0 + \hat{\beta}_1(1) + \hat{\beta}_2(0) + \hat{\beta}_3(0) = \hat{\beta}_0 + \hat{\beta}_1$ for a non-Hispanic black, and
$\hat{y}_i = \hat{\beta}_0 + \hat{\beta}_1(0) + \hat{\beta}_2(0) + \hat{\beta}_3(0) = \hat{\beta}_0$ for a Hispanic.

Thus, in a regression model that contains *only* indicator variable predictors, the interpretation of the intercept β_0 becomes the mean of Y for the reference group, and the interpretation of a "slope" such as β_1 becomes how much larger the mean of Y is for that group *relative* to the reference group. This regression model is equivalent to a one-way ANOVA of Y on race/ethnicity. The ANOVA tables (sums of squares, degrees of freedom, and F-test) produced by a regression procedure and an ANOVA procedure will be identical.

Example 2

The level of natural killer cell production in a cancer patient may be affected by a course of treatment for the cancer. A researcher measures the natural killer cell levels in serum samples from 172 people with lung cancer who have recently received one of two standard treatments (**Fig. 3**). Mean killer cell level was 32.39 (standard deviation 7.25) for the first treatment group and 28.10 (standard deviation 5.96) for the second treatment group. From a regression of killer cell levels on treatment group, the groups were found to be significantly different ($F^* = 18.01 > F_{0.95,1,170} = 3.90$, $P < 0.0001$). The fitted regression line was $\hat{y}_i = \hat{\beta}_0 + \hat{\beta}_1 I(treatment1) = 28.10 + 4.29 I(treatment1)$ where $I(treatment1)$ is the indicator variable for **treatment 1**: equal to 1 for **treatment 1** and equal to 0 for **treatment 2**. Thus $\hat{\beta}_0$ estimates the mean for the reference group (**treatment 2**) and $\hat{\beta}_0 + \hat{\beta}_1$ estimates the mean for **treatment 1**.

5.2. Regressions with Continuous and Class Variables

Often, a multiple linear regression contains both a continuous predictor and a categorical predictor. Suppose we consider the model

$$y_i = \beta_0 + \beta_1 x_{i1} + \beta_2 x_{i2} + \varepsilon_i$$

where X_1 is an indicator variable for two treatments and X_2 is gene expression level for a particular gene of interest. With $X_1 = 1$ for **treatment 1** and 0 for **treatment 2**, we compute fitted values as

$$\hat{y}_i = \hat{\beta}_0 + \hat{\beta}_1(1) + \hat{\beta}_2(expression) = [\hat{\beta}_0 + \hat{\beta}_1] + \hat{\beta}_2(expression)$$

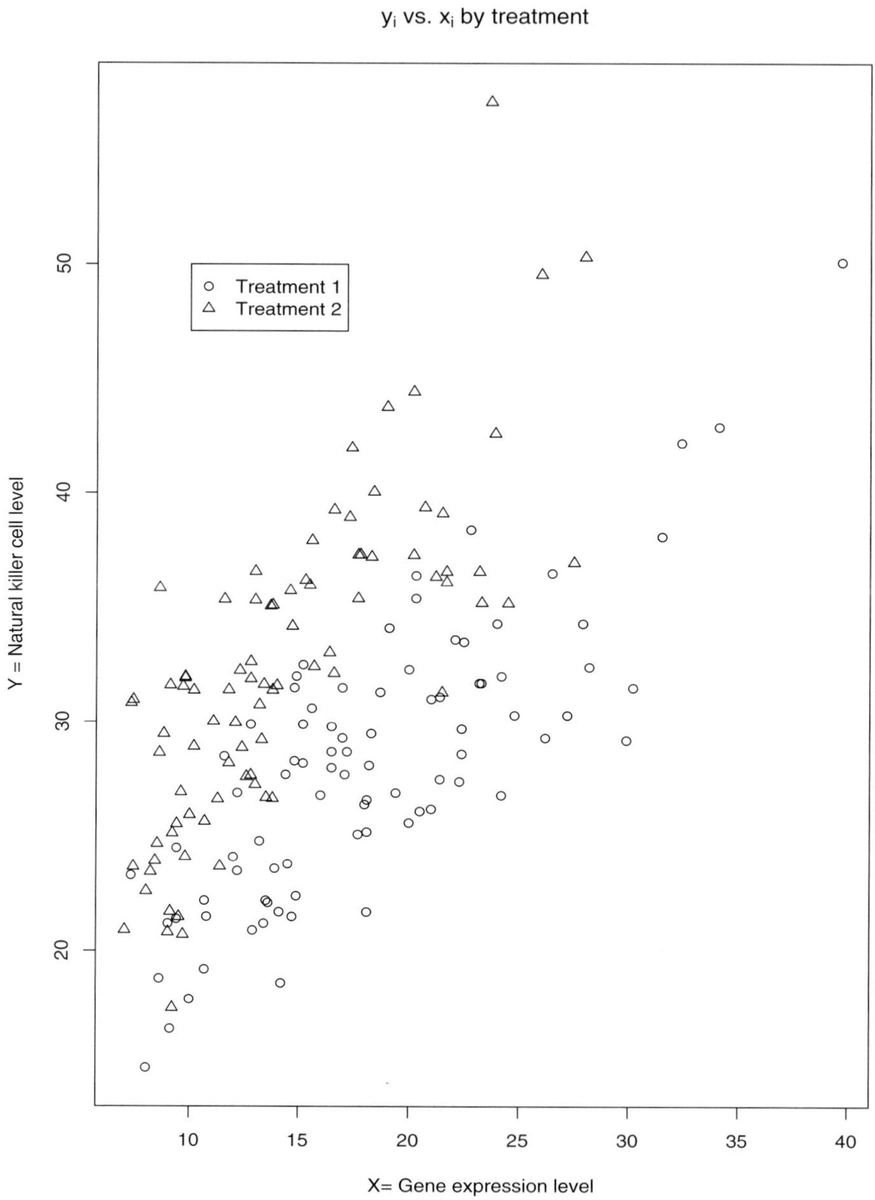

Fig. 3. Natural killer cell levels versus gene expression for two treatment groups.

for a patient with this level of gene expression receiving **treatment 1**, and

$$\hat{y}_i = \hat{\beta}_0 + \hat{\beta}_1(0) + \hat{\beta}_2(expression) = [\hat{\beta}_0] + \hat{\beta}_2(expression)$$

for a patient with this level of gene expression receiving **treatment 2**.

We now have two separate fitted lines that describe the relation of Y to the predictor expression level: one for **treatment 1** patients and one for **treatment 2** patients. For the **treatment 1** patients, the intercept is $\hat{\beta}_0 + \hat{\beta}_1$ and the slope is $\hat{\beta}_2$. For the **treatment 2** patients, the intercept is $\hat{\beta}_0$ and the slope is $\hat{\beta}_2$. This is called *parallelism* or a *parallel lines model*. Although the two groups share a common slope of Y on X_2, they have different intercepts.

Example 2 (Continued)

The researcher has collected gene expression relative to a control for one gene of interest and would also like to test the association of gene expression with natural killer cell levels. Our fitted model is now

$$\hat{y}_i = 12.10 + 0.88(1) + 7.81(expression) = [12.10 + 0.88] + 7.81(expression)$$

for a patient with this level of gene expression receiving **treatment 1**, and

$$\hat{y}_i = 12.10 + 0.88(0) + 7.81(expression) = [12.10] + 7.81(expression)$$

for a patient with this level of gene expression receiving **treatment 2**.

Partial F-tests show that both treatment ($F^* = 132.95 > F_{0.95,1,169} = 3.90$, $P < 0.0001$) and gene expression ($F^* = 255.46 > F_{0.95,1,169} = 3.90$, $P < 0.0001$) are significantly associated with natural killer cell levels. A fitted line plot is shown in **Figure 4**, where we can see that the slope (the mean increase in killer cell levels associated with each 1 unit higher gene expression level) is the same for the two treatment groups, but the line for **treatment 1** lies higher than the line for **treatment 2**.

5.3. Interactions with Class Variables

An extension of the multiple linear regression model above occurs when we consider the interaction of a continuous predictor with a categorical predictor. Suppose we consider the model

$$y_i = \beta_0 + \beta_1 x_{i1} + \beta_2 x_{i2} + \beta_3(x_{i1} * x_{i2}) + \varepsilon_i$$

where X_1 is an indicator variable for treatment group, X_2 is gene expression level, and the third predictor is equal to the interaction (product) of the two. Recalling that $X_1 = 1$ for **treatment 1** and 0 for **treatment 2**, we now compute fitted values as follows:

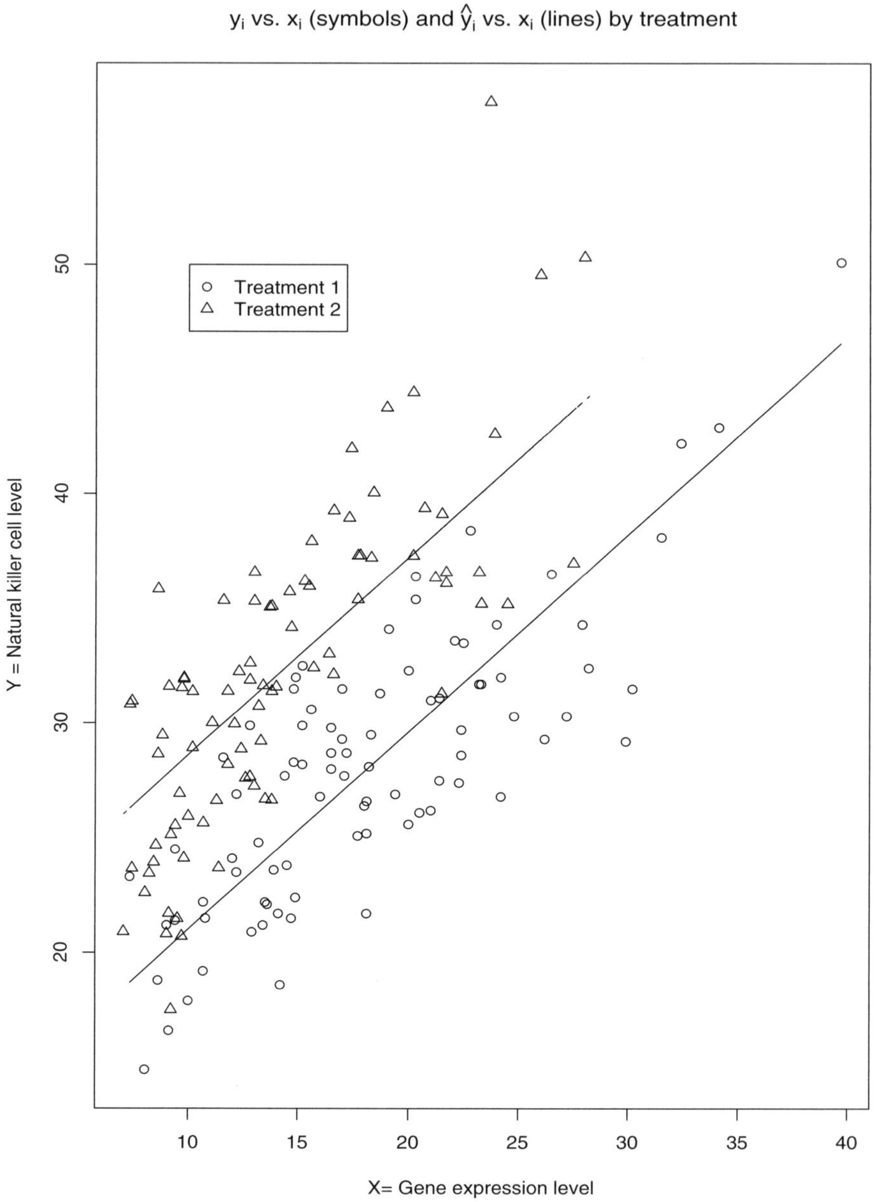

Fig. 4. Fitted line plot from the regression of Y on X_1 (categorical predictor) and X_2 (continuous predictor).

$$\begin{aligned}\hat{y}_i &= \hat{\beta}_0 + \hat{\beta}_1(1) + \hat{\beta}_2\,(expression) + \hat{\beta}_3\,(1 * expression)\\ &= [\hat{\beta}_0 + \hat{\beta}_1] + [\hat{\beta}_2 + \hat{\beta}_3]\,(expression)\end{aligned}$$

for a patient with this level of gene expression receiving **treatment 1**, and

$$\begin{aligned}\hat{y}_i &= \hat{\beta}_0 + \hat{\beta}_1(0) + \hat{\beta}_2\,(expression) + \hat{\beta}_3(0 * expression)\\ &= [\hat{\beta}_0] + [\hat{\beta}_2]\,(expression)\end{aligned}$$

for a patient with this level of gene expression receiving **treatment 2**.

We now have two separate fitted lines that describe the relation of Y to the predictor X_2, one for each treatment group. For the **treatment 1** patients, the intercept is $\hat{\beta}_0 + \hat{\beta}_1$ and the slope is $\hat{\beta}_2 + \hat{\beta}_3$. For the **treatment 2** patients, the intercept is $\hat{\beta}_0$ and the slope is $\hat{\beta}_2$. The lines are no longer parallel, because each group is allowed its own slope; this is sometimes called the *separate slopes model*.

Why not just fit two models that regress Y on X_2: one model with only the **treatment 1** patients and one model with only the **treatment 2** patients? The advantage of a model with all observations together is that the sample size is larger than when the sample is split into groups. A larger sample size often leads to a better estimate of σ^2 and to higher power for hypothesis tests. The disadvantage of a model with all observations together occurs when the different groups do not all share the same variance σ^2 (i.e., nonconstant variance across groups); using all observations is then a violation of the constant variance assumption.

Example 2 (Continued)

The researcher is concerned that the previous analysis may misrepresent the association of gene expression with natural killer cell levels; this would be true if the gene–killer cell association differed by treatment group. We thus fit a new model including a treatment by gene expression interaction $y_i = \beta_0 + \beta_1 I(treatment1) + \beta_2 expression + \beta_3(I(treatment1) * expression) + \varepsilon_i$:

$$\begin{aligned}\hat{y}_i &= 14.59 + 2.65(1) + 0.74\,(expression) + 0.32\,(1 * expression)\\ &= [14.59 + 2.65] + [0.74 + 0.32]\,(expression)\end{aligned}$$

for a patient with this level of gene expression receiving **treatment 1**, and

$$\begin{aligned}\hat{y}_i &= 14.59 + 2.65(0) + 0.74\,(expression) + 0.32\,(0 * expression)\\ &= [14.59] + [0.74]\,(expression)\end{aligned}$$

for a patient with this level of gene expression receiving **treatment 2**.

The first fitted line equation represents the relation of gene expression with natural killer cell levels for those in the **treatment 1** group, and the second fitted line is for those in the **treatment 2** group (**Fig. 5**).

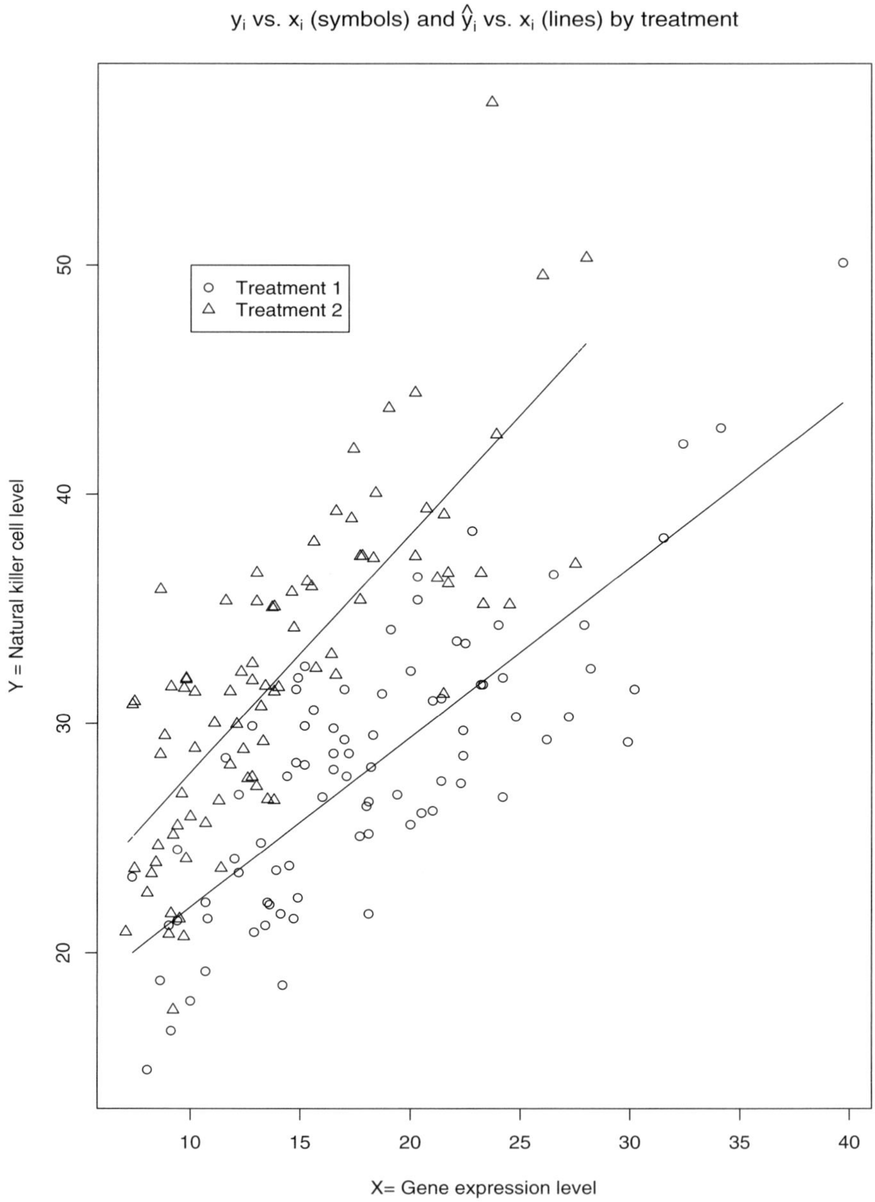

Fig. 5. Fitted line plot from the regression of Y on X_1 (categorical predictor) and X_2 (continuous predictor) with an interaction between X_1 and X_2.

A test of β_3 yields $t^* = 2.93$ (or its equivalent partial $F^* = 9.80$) with P value 0.002, thus we reject $H_0: \beta_3 = 0$ and conclude that there is a significant treatment by gene interaction. This confirms that the association of gene expression level with natural killer cell level differs by treatment group.

6. Variable Selection: Choosing Among Many Explanatory Variables

6.1. Overview of Automatic Selection Procedures

In some studies, there are many predictors available for consideration, any of which may have an important relation with the outcome. In general, a useful model has relatively few predictors, so that it is easy to interpret (such a model is called *parsimonious*), but if important predictors are omitted, regression coefficient estimates (and hence fitted values) can be biased. Automatic predictor *variable selection procedures* were developed to provide objective, systematic methods for selecting predictors to include in a final model. Selection procedures, however, are best used to identify several possible models that contain important predictors, so that then those few models can be examined in more detail by the analyst (e.g., taking into consideration collinearity, diagnostics, and study goals).

The *all possible regressions* procedure examines every possible combination of predictors and chooses the model with the "best" combination; there are several definitions of "best" that may be used, for example, highest R^2; see Ryan *(2)*. The *backward elimination* procedure starts with a model containing all possible predictors and then successively eliminates the least significant predictor, one at a time, until a final model is reached that contains only statistically significant predictors. The *forward selection* procedure does the same thing in reverse: it successively identifies the most significant predictor, one at a time, adding it to the model until all included predictors are statistically significant. For the backward (forward) procedure, the least (most) significant predictor is identified by the magnitude of its partial F-test statistic.

6.2. Stepwise Selection Procedures

Perhaps the most popular selection procedure is the *forward stepwise* procedure, which combines the backward and forward selection ideas. Beginning with the model with no predictors, at each step in the selection, a predictor can be added to the model if its partial F-test is statistically significant at a prespecified level, or a predictor can be deleted from the model if its partial F-test is no longer significant at a (possibly different) prespecified level. This identifies a single final model. The *backward stepwise* procedure operates similarly but begins with the model with all predictors.

6.3. Cautionary Notes

It must be emphasized that if automatic selection procedures are to be used, they are best used as a screening tool to identify several possible "good" models worthy of further consideration. Selection of variables with these procedures can result in a model with a downwardly biased estimate of σ^2, especially for small sample sizes; do not use them if $n - p \leq 10$ and avoid using them if $n - p \leq 40$.

Using these procedures when there are interactions thought to be important is difficult. These methods do not automatically include interactions; they need to be computed in advance and included in the list of potential predictors. However, the software running these procedures does not understand that each of the predictors that went into computing that interaction must also be included in any model containing that interaction.

7. Discussion

This chapter has provided an overview of multiple linear regression techniques. Neter and others *(1)* and Ryan *(2)* provide more complete treatment. For more detailed coverage in particular on leverage, influence, collinearity, and selection procedures, see Ryan *(2)*. Neter and others *(1)* discuss model validation. Vittinghoff and others *(3)* cover confounding, causal effects, counterfactual experiments, mediation, interactions, and variable selection. Gelman and others *(4)* cover Bayesian regression models. Harrell *(5)* discusses more advanced topics, such as regression trees, imputation/missing data issues, model validation, and resampling. All of these techniques assume the sample consists of independently collected observations. When this assumption is violated, other types of regression models are needed (see **Chapter 11**).

Acknowledgments

The author would like to thank Dr. Tracy Bergemann for helpful comments that improved the examples in this chapter.

References

1. Neter, J., Kutner, M. H., Nachtsheim, C. J., and Wasserman, W. (1996) *Applied Linear Statistical Models*, 4th ed. New York, McGraw-Hill/Irwin.
2. Ryan, T. P. (1997) *Modern Regression Methods*. New York, John Wiley & Sons.
3. Vittinghoff, E., Glidden, D. V., Shiboski, S. D., and McCulloch, C. E. (2005) *Regression Methods in Biostatistics: Linear, Logistic, Survival, and Repeated Measures Models*. New York, Springer.

4. Gelman, A., Carlin, J. B., Stern, H. S., and Rubin, D. B. (2003) *Bayesian Data Analysis*, 2nd ed. New York, Chapman & Hall/CRC.
5. Harrell, F. E. Jr. (2001) *Regression Modeling Strategies With Applications to Linear Models, Logistic Regression, and Survival Analysis*. New York, Springer.

10

General Linear Models

Edward H. Ip

Summary

This chapter presents the general linear model as an extension to the two-sample *t*-test, analysis of variance (ANOVA), and linear regression. We illustrate the general linear model using two-way ANOVA as a prime example. The underlying principle of ANOVA, which is based on the decomposition of the value of an observed variable into grand mean, group effect and random noise, is emphasized. Further into this chapter, the F test is introduced as a means to test for the strength of group effect. The procedure of F test for identifying a parsimonious set of factors in explaining an outcome of interest is also described.

Key Words: Common mean model; data decomposition; F test; signal-to-noise ratio; sum of squares; two-way ANOVA.

1. Introduction

The *general linear model* is a broad and encompassing class of statistical models that include 2-sample *t*-test (**Chapter 7**), analysis of variance (ANOVA; **Chapter 7**), and simple and multiple regression (**Chapters 8** and **9**). Viewed from such a vantage point, the general linear model is perhaps the most commonly used statistical procedure among clinical researchers.[1] Historically, the 2-sample *t*-test, ANOVA, and multiple regression were invented to address seemingly different data analytic issues. For example, the 2-sample *t*-test was designed for testing whether or not outcome measures from 2 groups, usually a treatment group and a control group, are different. Linear regression was first

[1] The general linear model should not be confused with the *generalized linear model*, which includes the general linear model as a special case. In a generalized linear model ***(see Ref. 4)***, the response variables need not be continuous.

From: *Methods in Molecular Biology, vol. 404: Topics in Biostatistics*
Edited by: W. T. Ambrosius © Humana Press Inc., Totowa, NJ

developed in the context of prediction in a heredity study—one variable, such as a father's height, was used to predict another, such as the child's height. It did not take long for statisticians to recognize that the 3 methods—the *t*-test, ANOVA, and multiple regression—all share some common characteristics ***(1–3)***. Stated in a general but somewhat loose form, the 3 methods are all based upon the following equation:

Response = Sum of effects due to different factors + noise

The 2-sample *t*-test is set to detect an effect that arises from being a member of one group versus another in the midst of sampling error (noise). ANOVA handles the case of more than 2 groups. Multiple regression, on the other hand, postulates models that dictate that the response (the dependent variable) arises from the combination of one or more *effects* masked by random noise. The whole point of deriving the mathematical machinery that underlies these methods is to identify signs of effects, or signals, that are sufficiently strong to be distinguished and separated from noise, which is inherent in the measurement process because of the random nature of selecting a representative sample from a population.

While the general linear model can be an indispensable tool for clinical researchers for analyzing a broad range of data situations, its operating characteristics and limitations need to be recognized. Briefly, they are

1. The response variable must be a continuous outcome measure.
2. The assumed effects due to different factors must be additive (i.e., linear).
3. The noise component must follow the same normal distribution regardless of a subject's characteristics, treatment status, and level of response.

Mathematically, the General Linear Model can be expressed as:

$$y_i = \beta_0 + \beta_1 x_{i1} + \ldots \beta_{p-1} x_{i,p-1} + \varepsilon_i, \tag{1}$$

where y_i denotes the data value of the dependent variable *Y* from the *i*th observation; $i = 1, \ldots, n$, $x_{i1}, \ldots, x_{i,p-1}$ denote the $p - 1$ data values of the independent variables (predictors) *X* from the *i*th observation; and ε_i denotes the error component (noise) from the *i*th observation. Furthermore,

$$\varepsilon_i \sim N(0, \sigma^2)$$

and is independent of y_i. **Equation 1** can also be written in matrix form: $\mathbf{y} = \mathbf{Xb} + \mathbf{e}$, where $\mathbf{X}$ is the design matrix of the independent variables and $\mathbf{y}$ and $\mathbf{e}$ are vectors of y_i and ε_i being stacked up.

Equation 1 indeed looks exactly the same as the equation for multiple regression (**Equation 1, Chapter 9**). However, there are subtle differences. First, the dependent variable *Y* in a general linear model may be a vector. For example, when measurements of both systolic and diastolic blood pressures are taken from the same individual as joint responses, then $Y = (Y_{systolic}, Y_{diastolic})$ forms the dependent variable. Second, the emphasis of the general linear model is that the independent variables X are categorical. Indeed, it can be shown that when there is only one categorical predictor X_1, which is coded $X_1 = 1$ if the subject is in the treatment group and $X_1 = 0$ if he or she is not (i.e., is in the control group), then the difference between the group means of treatment and control is β_1. Thus, a *t*-test is equivalent to testing whether or not β_1 is statistically significant. The control group is said to be a reference group. Analogously, ANOVA can be formulated in terms of tests of significance of coefficients β in **Equation 1**. **Table 1** lists various methods that can be considered special cases of the general linear model, and they are illustrated using an example of testing for the efficacy of a hormone-therapy drug for treating cancer in a clinical trial conducted on mice.

2. ANOVA Table

The ANOVA table is an essential tool for applying the general linear model to clinical data analysis. The fundamental idea underlying ANOVA is to decompose each data value into pieces that reveal how the factors, singly or in combination, contribute to the variation in the data. By examining the individual pieces and the overall structure of the decomposition, a clinical researcher can glean important information not available from traditional statistical tests.

The decomposition of data occurs at 2 levels. At the first level, the original data value can be split into various pieces, and the table corresponding with each piece is called an overlay. The values in the overlay tables are then squared and summed to form a sum of squares (SS). At a second level, the SS are decomposed, and the results are often depicted in the form of an ANOVA table. The sums of squared values provide information about both the strength of the effect and the *noise level* contained in the values of a data set. With such information available, it is then possible to test whether or not a specific effect is purely due to chance.

2.1. A Simple ANOVA Table for the Common Mean Model

A simple *common mean model* is perhaps a good way of starting to illustrate the 2 levels of decomposition of a data value. Consider an example (**Table 2**, panel 1) in which patients presenting to an emergency unit in a hospital reported pain intensity on a 10-point numerical scale. The sample mean value of pain intensity of 5 patients is 5.0. Each data point can be decomposed into a mean component and a residual component (the first level of decomposition). **Table 2** shows how the components—the overlays—can be used to recover the data value *(5)*. When the value in each of the data tables is squared, it results in a

Table 1
An Illustration of Several Statistical Methods under the Umbrella of the General Linear Model

Method	Response	Predictor 1	Predictor 2	Predictor 3
Two-sample *t*-test	Size of tumor Y	X_1, whether drug is used. Two levels[a]: treatment, control.		
One-way ANOVA	Size of tumor Y	X_1, which variant of drug is used. Three levels: variant 1, variant 2, placebo.		
Two-way ANOVA	Size of tumor Y	X_1, which variant of drug is used. Three levels: variant 1, variant 2, placebo.	X_2, radiation treatment. Two levels: absence, presence.	
ANCOVA[b]	Size of tumor Y	X_1, which variant of drug is used. Three levels: variant 1, variant 2, placebo.	X_2, radiation treatment. Two levels: absence, presence.	X_3, weight of mouse, continuous measure
Multiple regression (without categorical predictor)	Size of tumor Y	X_1, dosage of drug, continuous measure	X_2, dosage of radiation treatment, continuous measure.	X_3, weight of mouse, continuous measure
MANOVA[c]	Size of tumor, Y_1, and white blood cell number, Y_2, $Y = (Y_1, Y_2)$	X_1, which variant of drug is used. Three levels: variant 1, variant 2, placebo.	X_2, radiation treatment. Two levels: absence, presence.	
Multivariate multiple regression (without categorical predictor)[d]	Size of tumor, Y_1, and white blood cell number, Y_2, $Y = (Y_1, Y_2)$	X_1, dosage of drug at various levels, continuous measure	X_2, dosage of radiation treatment, continuous measure	X_3, weight of mouse, continuous measure

[a]For a factor, the level of the factor refers to a particular value or a state of the factor.

[b]ANCOVA stands for *analysis of covariance*. ANCOVA involves a continuous variable that may have a confounding effect that needs to be controlled for in an ANOVA experiment. See **Chapter 9** for a discussion.

[c]Multivariate analysis of variance.

[d]Multiple regression in its broadest sense includes categorical variables as predictors. Thus, in that context, it is equivalent to the general linear model, although they have different emphases.

Table 2
Decomposition of Data Value into Common Mean Value and Residual

3	2	6	4	10	Data value (y_i)
		=			
5	5	5	5	5	Common mean value overlay $\bar{y}$
		+			
−2	−3	1	−1	5	Residual overlay ($y_i - \bar{y}$)

Table 3
Decomposition of Sum of Squares

Squares					Sum of squares
9	4	36	16	100	**165 SS (total)**
					=
25	25	25	25	25	**125 SS (mean)**
					+
4	9	1	1	25	**40 SS (error)**

corresponding set of tables of squared values (**Table 3**). The most important relation that can be observed from **Table 3** is that SS(Total) = SS(mean) + SS(error). Squares of data values have highly tractable mathematical properties. We would not have obtained the neat decomposition of the SS had the data value in **Table 2** been raised to the fourth power. The mathematical tractability of the SS provides the foundation for all kinds of ANOVA analyses. **Table 4** summarizes information in **Table 3** into the ANOVA table for the common mean model.

Table 4
The ANOVA Table Summarizes the Values within the Overlays

Source of variation	Sum of squares (SS)	Degrees of freedom (d.f.)	Mean square (MS)
Mean	125	1	125
Residual	40	4	10
Total	165	5	

The degrees of freedom (d.f.) in **Table 4** refers to the number of independent pieces of information that contribute to the source of variation, and mean square (MS) is the SS from a specific source of variation divided by its corresponding degrees of freedom.

Formally, the common mean model can be presented with the following characterizations:

$$y_i = \mu + \varepsilon_i, \tag{2}$$

where we denote the *i*th data value by y_i, $i = 1, \ldots, n$; the common mean of the population by μ; and the residual by ε_i. Note that the population common mean μ is estimated by the sample mean $\bar{y}$ in the actual decomposition.

Assumptions of the Common Mean Model

1. The sample y_i follows a normal distribution.
2. The observations are independent.
3. The common population mean (unknown) is μ, and the common variance (unknown) is σ^2.

The *common mean model* is summarized in **Table 5**.

2.2. One-Way ANOVA Table

One-way ANOVA is often used is to compare an outcome variable among 3 or more groups that are independent but possibly have different means. Using the same principle of decomposing the sums of squares that we have seen in the naïve common mean model, which may not be very useful for analyzing complex data, one-way ANOVA is especially suitable for analyzing data from designed experiments (**Chapters 1** and **12**) in which data are collected over several levels of the same factor. In clinical trials, one of the most common applications of one-way ANOVA is the test for differences in efficacy of several drugs, among which a candidate drug and some existing drugs would be included. The membership of a subject to a drug group, or the way in which a subject is classified,

Table 5
The Common Mean ANOVA Table

Source of variation	Sum of squares (SS)	Degrees of freedom (d.f.)	Mean square (MS)
Mean	$n\bar{y}^2$	1	$n\bar{y}^2$
Residual	$\sum_{i=1}^{n}(y_i - \bar{y})^2$	$n - 1$	$\frac{1}{n-1}\sum_{i=1}^{n}(y_i - \bar{y})^2$
Total	$\sum_{i=1}^{n} y_i^2$	n	

is considered a *factor*. Hence, one-way ANOVA is also called ANOVA with one-way classification. It is a simple case of the general linear model.

The following example illustrates the procedure that leads to the formation of the one-way ANOVA table. In a study of Hodgkin's disease ***(6)***, plasma bradykininogen levels were measured in 3 populations: normal subjects, patients with active Hodgkin's disease, and patients with inactive Hodgkin's disease. The globulin bradykininogen is the precursor substance for bradykinin, which is thought to be a chemical mediator of inflammation. The original data (in micrograms of bradykininogen per milliliter of plasma) contained 23 observations for the first group, 17 for the second, and 28 for the third. For illustration purposes, we only selected the first 10 observations in each group. The data values are presented in **Table 6**.

The one-way ANOVA model states that a data value can be decomposed into a common mean, a group effect, and the rest—meaning a residual or an error term. The residual term should contain only noise and is assumed to be a random variate from a normal distribution with mean 0 and variance σ^2. Accordingly, the formal model can be expressed as $y_{ij} = \mu + \alpha_i + \varepsilon_{ij}$, $i = 1,2,3$, or in the matrix form $\mathbf{y} = \mathbf{Xb} + \mathbf{e}$. The vector $b = (\mu,\alpha_1,\alpha_2,\alpha_3)$, and the row of matrix **X**, consists of 1s, and 0s. For example, for members of the second group, the corresponding row in **X** is (1,0,1,0). To make sure that the model does not contain more moving parts (parameters) than it needs, we place constraints on α_i. For example, $\alpha_1 + \alpha_2 + \alpha_3 = 0$.

The collection of overlays of the data values consists of the common mean overlay, the group effect overlay, and the residual overlay. The group effects can be estimated from the differences between individual group means and the common mean, which are shown in **Table 7**. The decomposition of data values is shown in **Table 8**. Note that under this decomposition, there is a fitted value

Table 6
Data Values of Bradykininogen Level in 3 Groups of Patients

Normal control	Active Hodgkin's disease	Inactive Hodgkin's disease
5.37	3.96	5.37
5.80	3.04	10.60
4.70	5.28	5.02
5.70	3.40	14.30
3.40	4.10	9.90
8.60	3.61	4.27
7.48	6.61	5.75
5.77	3.22	5.03
7.15	7.48	5.74
6.49	3.87	7.85

Table 7
Values of the Estimated Common Mean and Group Effects for the Bradykininogen Example

	Common mean = 5.962		Inactive Hodgkin's disease
	Normal control	Active Hodgkin's disease	
Group mean	6.046	4.457	7.383
Group effect	0.084	−1.505	1.421

$\hat{y}$ for each observation y. For example, for the first value in the Normal control group $y = 5.37$, the fitted value $\hat{y} = \hat{\mu} + \hat{\alpha}_1 = 5.962 + 0.084 = 6.046$. It is common practice in statistics to use the hat notation to indicate an estimated value or fitted quantity. The sample mean of the entire sample is denoted by $\bar{y}_{..}$; the number of observations in **group 1** is denoted by n_i and $\bar{y}_{i.} = \frac{1}{n_i}\sum_{j=1}^{n_i} y_{ij}$ is the sample group mean. Each of the values in an overlay in **Table 8** is squared to form an overlay of squared values (**Table 9**). Analogous to the decomposition of the sum of squares for the common mean model, the sum of squared values from each squared overlay table adds up to that of the squared original value table: SS(Total) = SS(mean) + SS(group) + SS(error). In the Hodgkin's disease example, the values that correspond with these SS add up as follows: 1244.3 = 1066.4 + 42.9 + 135.0. The ANOVA table (overall mean included) for the bradykininogen example is presented in **Table 10**. Note that most software packages do not print out the mean row.

Generally, the one-way ANOVA procedure is characterized by:

$$y_{ij} = \mu + \alpha_i + \varepsilon_{ij}, \tag{3}$$

where we denote the jth data value in the ith group by y_{ij}, $i = 1, \ldots, I$, $j = 1, \ldots, n_i$; the common overall mean of the population by μ, the population group effect of **group 1** by α_i; and the residual by ε_{ij}.

Assumptions of One-Way ANOVA

1. The sample y_{ij} for each $i = 1, \ldots, I$ follows a normal distribution.
2. The observations are independent.
3. The population mean (unknown) of the ith group is α_i, and the groups share a common but unknown variance (unknown) σ^2 (the homoscedasticity assumption).

The ANOVA table is presented in **Table 11**, where n is the total number of observations; that is, $n = \sum_{i=1}^{I} n_i$.

Table 8
The Decomposition of the Original Data Value into Overlays for the Bradykininogen Example

Original data (y_{ij})

GI Normal control	GII Active Hodgkin's disease	GIII Inactive Hodgkin's disease
5.37	3.96	5.37
5.80	3.04	10.60
....	....	....
6.49	3.87	7.85

=*

Common mean ($\bar{y}_{..}$) overlay

GI Normal control	GII Active Hodgkin's disease	GIII Inactive Hodgkin's disease
5.96	5.96	5.96
5.96	5.96	5.96
....	....	....
5.96	5.96	5.96

+*

Group effect ($\bar{y}_{i.} - \bar{y}_{i..}$) overlay

GI Normal control	GII Active Hodgkin's disease	GIII Inactive Hodgkin's disease
0.08	−1.50	1.42
0.08	−1.50	1.42
....	....	....
0.08	−1.50	1.42

+*

Residual ($y_{ij} - \bar{y}_{i.}$) overlay

GI Normal control	GII Active Hodgkin's disease	GIII Inactive Hodgkin's disease
–0.68	–0.50	–2.01
–0.25	–1.42	3.22
....	....	....
0.44	–0.59	0.47

*The arithmetic operations apply to each value in the overlays.

It is customary in ANOVA procedures to subtract the overall mean from the observed value (known formally as sweeping the mean from the data table when producing overlays) and to use the resulting corrected value as the basis for decomposition. Popular programs such as SAS and SPSS all report

Table 9
The Squared Overlay Tables and the Decomposition of Sum of Squares for the Bradykininogen Example

Original data (y_{ij}^2)

GI Normal control	GII Active Hodgkin's disease	GIII Inactive Hodgkin's disease
28.84	15.68	28.84
33.64	9.24	112.36
....	....	....
42.12	14.98	61.62
	Total sum of squares	**1244.3 =**

Common mean ($\bar{y}_{..}^2$)

GI Normal control	GII Active Hodgkin's disease	GIII Inactive Hodgkin's disease
35.52	35.52	35.52
35.52	35.52	35.52
....	....	....
35.52	35.52	35.52
	Total sum of squares	**1066.4 +**

Group effect ($(\bar{y}_{i.} - \bar{y}_{..})^2$)

GI Normal control	GII Active Hodgkin's disease	GIII Inactive Hodgkin's disease
0.006	2.25	2.02
0.006	2.25	2.02
....	....	....
0.006	2.25	2.02
	Total sum of squares	**42.9 +**

Residual ($(y_{ij} - \bar{y}_{i.})^2$)

GI Normal control	GII Active Hodgkin's disease	GIII Inactive Hodgkin's disease
0.46	0.25	4.05
0.06	2.01	10.35
....	....	....
0.20	0.34	0.22
	Total sum of squares	**135.0**

corrected sums of squares (or corrected total sums of squares). Formally, one can think of the corrected procedure as applying decomposition of the following form:

$$y_{ij} - \mu = \alpha_i + \varepsilon_{ij}, \tag{4}$$

Table 10
The ANOVA Table Summarizing Results in the Bradykininogen Example

Source of variation	Sum of squares (SS)	Degrees of freedom (d.f.)	Mean square (MS)
Mean	1066.4	1	1066.4
Group effect	42.9	2	21.5
Residual	135.0	27	5.0
Total	1244.3	30	

with the corresponding ANOVA table, shown in **Table 12**. The SS due to the group effect is often referred to as the between-group SS, and the residual is referred to as the within-group SS. It can also be seen from the column for MS that the MS (group effect) is the sample variance estimate for variation in group means, whereas the MS (residual) is the variance estimate for the within-group variance, assuming that all of the groups share a common variance (the homoscedasticity assumption).

2.3. Two-Way ANOVA Table

One-way ANOVA is useful only when the effect of one factor on the outcome is considered and manipulated. In many designed experiments, several experimental factors are manipulated at the same time. Two-way or higher-way ANOVA are general linear models developed for situations in which there are 2 or more factors being varied. Two-way ANOVA will be described in this section. Although the principle of data value and variance decomposition

Table 11
One-Way ANOVA Table (Overall Mean Included)

Source of variation	Sum of squares (SS)	Degrees of freedom (d.f.)	Mean square (MS)
Mean	$n\bar{y}_{..}^2$	1	$n\bar{y}_{..}^2$
Group effect	$\sum_{i=1}^{I} n_i(\bar{y}_{i.} - \bar{y}_{..})^2$	$I - 1$	$\frac{1}{I-1}\sum_{i=1}^{I} n_i(\bar{y}_{i.} - \bar{y}_{..})^2$
Residual	$\sum_{i=1}^{I}\sum_{j=1}^{n_i} (y_{ij} - \bar{y}_{i.})^2$	$n - I$	$\frac{1}{n-1}\sum_{i=1}^{I}\sum_{j=1}^{n_i} (y_{ij} - \bar{y}_{i.})^2$
Total	$\sum_{i=1}^{I}\sum_{j=1}^{n_i} y_{ij}^2$	n	

Table 12
One-Way ANOVA Table (Mean Corrected)

Source of variation	Sum of squares (SS)	Degrees of freedom (d.f.)	Mean square (MS)
Group effect	$\sum_{i=1}^{I} n_i(\bar{y}_{i.} - \bar{y}_{..})^2$	$I - 1$	$\frac{1}{I-1}\sum_{i=1}^{I} n_i(\bar{y}_{i.} - \bar{y}_{..})^2$
Residual	$\sum_{i=1}^{I}\sum_{j=1}^{n_i}(y_{ij} - \bar{y}_{i.})^2$	$n - I$	$\frac{1}{I-1}\sum_{i=1}^{I}\sum_{j=1}^{n_i}(y_{ij} - \bar{y}_{i.})^2$
Total	$\sum_{i=1}^{I}\sum_{j=1}^{n_i} y_{ij}^2 - n\bar{y}_{..}^2$	$n - 1$	

remains the same, the two-way ANOVA involves an important feature that is not seen in one-way ANOVA—interaction between factors.

The following example, adapted from Northway and others *(7)*, illustrates two-way ANOVA and how interaction is handled. In a designed experiment for a drug, 16 male and 16 female dogs were assigned randomly to treatment with either a vehicle control (**group 0**) or one of three graded doses (8, 25, 75 mg/kg for **groups 1, 2,** and **3**, respectively) of an investigational compound. By design, each treatment group contained 4 dogs of each sex. Oral dosing by gavage feeding was performed once daily. Prior to treatment, blood samples were collected from the jugular vein of each animal after overnight starvation for alkaline phosphatase measurement. The 2 factors in this experiment are dosage and gender. **Table 13** shows the data set. A plot of the means of the alkaline phosphatase levels at various dosages shows that the mean response in female dogs looks rather different from that of the male dogs (**Fig. 1**), which suggests that there may exist both a main effect (due to gender) and an interaction effect (the female and male dogs react differently to different dosage levels). An interaction effect would translate into unequal slopes of the male and female profile lines in **Figure 1**. The two-way ANOVA table in **Table 14** indicates that indeed the mean square value of the gender effect is almost 9 times that of the residual (error), suggesting that there could be a gender effect. The interaction effect seems mild. Formal statistical tests of such an effect will be discussed in the subsequent section. If the interaction term is not included in the model, then its SS is included as part of the error term. Note that the sample sizes in each cell (e.g., **group 1 male** forms a cell) are equal. This kind of experiment is said to have a balanced design. In the above example, the number of replications per cell is $n = 4$. Unbalanced designs are more dif-

Table 13
Alkaline Phosphate Level in 32 Dogs for Different Treatment Effects

Alkaline phosphatase level	Sex	Group	Alkaline phosphatase level	Sex	Group
169	M	0	125	F	0
291	M	0	138	F	0
158	M	0	113	F	0
122	M	0	137	F	0
203	M	1	170	F	1
178	M	1	139	F	1
141	M	1	131	F	1
181	M	1	125	F	1
101	M	2	113	F	2
199	M	2	150	F	2
141	M	2	155	F	2
149	M	2	133	F	2
135	M	3	113	F	3
153	M	3	102	F	3
147	M	3	128	F	3
157	M	3	91	F	3

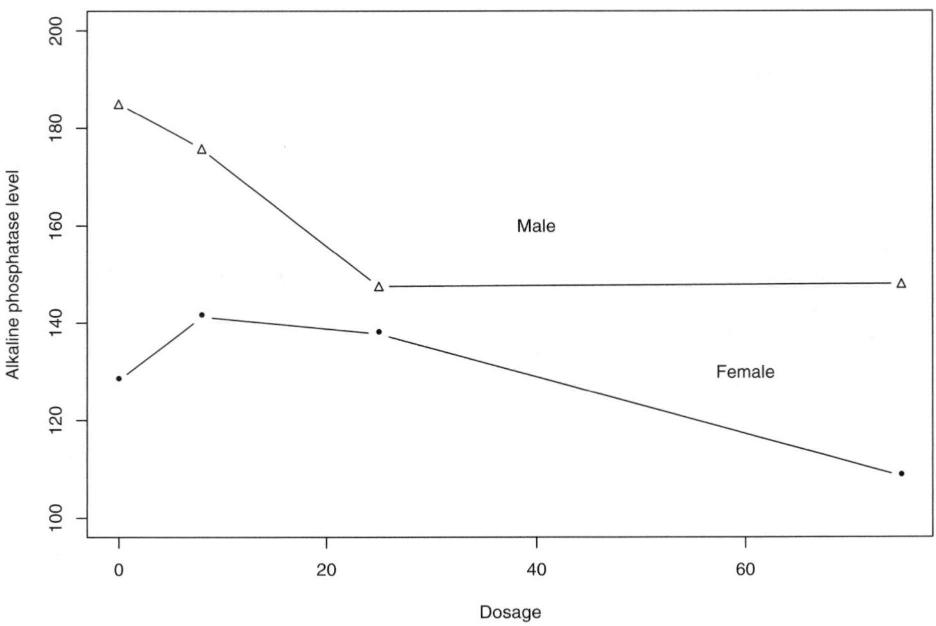

Fig. 1. Mean alkaline phosphatase levels by gender and dosage.

ficult to handle mathematically but are easily analyzed using statistical software. We focus on balanced design in this chapter. The two-way ANOVA model is

$$y_{ijk} = \mu + \alpha_i + \beta_j + \alpha\beta_{ij} + \varepsilon_{ijk}, \tag{5}$$

where we denote the kth data value in the (ij)th cell by y_{ijk}; for $i = 1, \ldots, I$, $j = 1, \ldots, J$, $k = 1, \ldots, n$. The common mean of the population is μ; the population group effect due to the first factor (**factor A**) is α_i; the population group effect due to the second factor (**factor B**) is β_j; the interaction effect between the 2 factors is $\alpha\beta_{ij}$, and the residual is ε_{ijk}.

Assumptions of Two-Way ANOVA

1. The sample Y_{ijk} for the (ij)th cell $i = 1, \ldots, I$, $j = 1, \ldots, J$ follows a normal distribution.
2. The observations are independent.
3. The population means (unknown) of classification due to the **factors A** and **B** are respectively α_i and β_j, and the groups share a common but unknown within-group variance σ^2 (the homoscedasticity assumption).

Table 15 summarizes the two-way ANOVA table. In **Table 15**, y_{ijk} is the kth observed data value in the (ij)th cell, **i**; $i = 1, \ldots, I$, $j = 1, \ldots, n_i$, $\bar{y}...$ is the overall sample mean of the entire sample; $\bar{y}_i.. = \frac{1}{nJ}\sum_{j=1}^{J}\sum_{k=1}^{n} y_{ijk}$ is the sample group mean of the first factor; $\bar{y}._j. = \frac{1}{nI}\sum_{i=1}^{I}\sum_{k=1}^{n} y_{ijk}$ is the sample group mean of the second factor; and n is the number of observations within a single cell.

Table 14
The ANOVA Table Summarizing Results in the Alkaline Phosphatase Level Experiment

Source of variation	Sum of squares (SS)	Degrees of freedom (d.f.)	Mean square (MS)
Gender	9870.1	1	9870.1
Group	4757.8	3	1585.6
Gender × group interaction[a]	2262.1	3	754.0
Error	26755.0	24	1114.8
Corrected total	43644.0	31	

[a]It is conventional notation to use the multiplication sign for interaction, but the multiplication should not be taken literally.

Table 15
Two-Way ANOVA Table (Mean Corrected)

Source of variation	Sum of squares (SS)	Degrees of freedom (d.f.)	Mean square (MS)
Factor (A)	$\sum_{i=1}^{I} nJ(\bar{y}_{i..} - \bar{y}_{...})^2$	$I - 1$	$\frac{1}{I-1}\sum_{i=1}^{I} n_i(\bar{y}_{i..} - \bar{y}_{...})^2$
Factor (B)	$\sum_{j=1}^{J} nI(\bar{y}_{.j.} - \bar{y}_{...})^2$	$J - 1$	$\frac{1}{J-1}\sum_{j=1}^{J} nI(\bar{y}_{.j.} - \bar{y}_{...})^2$
Interaction (AB)	$\sum_{i=1}^{I}\sum_{j=1}^{J} n(\bar{y}_{ij.} - \bar{y}_{i..} - \bar{y}_{.j.} + \bar{y}_{...})^2$	$(I - 1)(J - 1)$	$\frac{1}{(I-1)(J-1)}\sum_{i=1}^{I}\sum_{j=1}^{J} n(\bar{y}_{ij.} - \bar{y}_{i..} - \bar{y}_{.j.} + \bar{y}_{...})^2$
Residual	$\sum_{k=1}^{n}\sum_{i=1}^{I}\sum_{j=1}^{J}(y_{ijk} - \bar{y}_{ij.})^2$	$IJ(n - 1)$	$\frac{1}{IJ(n-1)}\sum_{k=1}^{n}\sum_{i=1}^{I}\sum_{j=1}^{J}(y_{ijk} - \bar{y}_{ij.})^2$
Corrected total	$\sum_{k=1}^{n}\sum_{i=1}^{I}\sum_{j=1}^{J} y_{ijk}^2 - nIJ\bar{y}_{...}^2$	$IJn - 1$	

2.4. Other Cases of One- and Two-Way ANOVA

1. In two-way ANOVA, when the number of observations per cell is 1, it is not possible to detect interaction. One can still test for main effects by assuming that there is no interaction (refer to **Sections 3** and **4**).
2. Sometimes a group can be considered as a member of a random sample drawn from a larger population. In such cases, the group effect parameters (α_i in one-way ANOVA, and either one or both of α_i and β_j in two-way ANOVA) are assumed to follow a normal distribution. This kind of so-called random-effect model will be discussed in **Chapter 11**. If some effects in a model are random and some are fixed (like the ones that are treated in this chapter), the model is called a mixed model (**Chapter 11**).
3. A subject in an ANOVA study may be measured repeatedly, such as in a longitudinal study. In the alkaline phosphatase level experiment, measurements are actually taken from each dog at several time points (week 4, week 8, and week 12). In such cases, the independence assumption between observations is no longer valid, and the within-subject variation should also be taken into account. This is also discussed in **Chapter 11**.

3. *F* Tests

It is not uncommon to see an F statistic and a P value reported as part of an ANOVA table. Being a workhorse that biostatisticians use for scientifically and rigorously testing whether there is a difference between groups, the F test plays a central role in the general linear model.

The underlying idea of the F test is to empirically compare the SS of the signal to that of the noise. If the variation in the signal level, as measured by the mean square (MS) of the effect, is small compared with the variation in the noise, then the F test would inform a researcher that the difference in group variation is likely to arise only from chance variation. On the other hand, if the magnitude of the MS in a signal is large compared with the magnitude of the level of noise, then the F test would suggest that the difference is likely to be genuine. However, we still need to answer the question "how large is large?" Knowing the properties of the distribution of the *signal-to-noise* ratio would allow us to quantify the likelihood of observing specific values of the ratio.

3.1. Distributions of Sums of Squares

The SSs in an ANOVA table follow specific distributions given the assumption that the error term in **Equations 2–5** is normal. Recall that if X is a random variable that is drawn from a standard normal distribution, then X^2 will be distributed as a central chi-square distribution with 1 degree of freedom (see **Chapter 4**). The sums of squared values of standard normal random variables also follow central chi-square (χ^2) distributions.[2] If we assume that a factor in an ANOVA analysis does not contribute to the dependent variable (i.e., in technical terms, the null hypothesis is true), then its associated SS will follow the central chi-square distributions. The ratio between 2 chi-square distributions, modified by their respective degrees of freedom, follows an F distribution. In symbols, we write

$$\frac{\chi_k^2/k}{\chi_m^2/m} \sim F_{k,m}, \tag{6}$$

where k and m denote the degrees of freedom of the respective chi-square distributions. The F distribution is ideal for testing the signal-to-noise ratio. An F distribution always has 2 degrees of freedoms (k, m), and the shape of its distribution varies with the values of k and m.

3.2. Example of F Test

We use the phosphatase level experiment as an example to illustrate the F test. The entire procedure can be described as follows:

1. State the model $Y_{ijk} = \mu + \alpha_i + \beta_j\ \alpha\beta_{ij} + \varepsilon_{ijk}$, where Y_{ijk} denotes the phosphate level, α_i; $i = 1,2$ denotes the gender effect, β_j; $j = 1, \ldots, 4$ denotes the group effect; and $\alpha\beta_{ij}$ denotes the interaction effect.

[2] However, if the individual normal random variables do not all have zero means, then the resulting chi-square is not *central* and is said to have noncentrality parameters.

2. State the null and alternative hypotheses for each effect. For group effect, they are

$$H_0 : \beta_1 = \beta_2 = \beta_3$$

$$H_a : H_0 \text{ is not true, or at least one group differs from the others}$$

3. Form the ANOVA table. The F statistic is computed using the following equation:

$$F = \frac{\text{MS}(\textit{effect})}{\text{MS}(\textit{error})}. \tag{7}$$

4. Find the P value of the F statistic by referring it to the F probability distribution with the 2 appropriate degrees of freedom, one from the effect and the other from the error.

The expression within each pair of parentheses in **Table 16** shows how a number is computed for the phosphatase example. $P(F_{k,m} > c)$ denotes the probability of observing a number higher than c in an F distribution with k and m degrees of freedom. **Figure 2** shows the graph for the $F_{3,24}$ distribution and the P value associated with testing the effect of group. If a significance level α is specified, then the P value allows for drawing a conclusion from the hypothesis test. For testing the group effect in the alkaline phosphatase level example, **Table 17** describes the decision rule and the conclusion when the significance level $\alpha = 0.05$.

Table 16
The ANOVA Table Summarizing Results in the Alkaline Phosphatase Level Experiment with *F* Statistic and *P* Value

Source of variation	Sum of squares (SS)	Degrees of freedom (d.f.)	Mean square (MS)	F	*P* value
Gender	9,870.1	1	9,870.1	**8.85** $\left(\frac{9870.1}{1114.8}\right)$	**0.007** $(P(F_{1,24} > 8.85))$
Group	4,757.8	3	1,585.6	**1.42** $\left(\frac{1585.6}{1114.8}\right)$	**0.261** $(P(F_{3,24} > 1.42))$
Gender × group interaction[a]	2,262.1	3	754.0	**0.68** $\left(\frac{754.0}{1114.8}\right)$	**0.575** $(P(F_{1,24} > 0.68))$
Error	26,755.0	24	1,114.8		
Corrected total	43,644.0	31			

[a]It is conventional notation to use the multiplication sign for interaction, but the multiplication should not be taken literally.

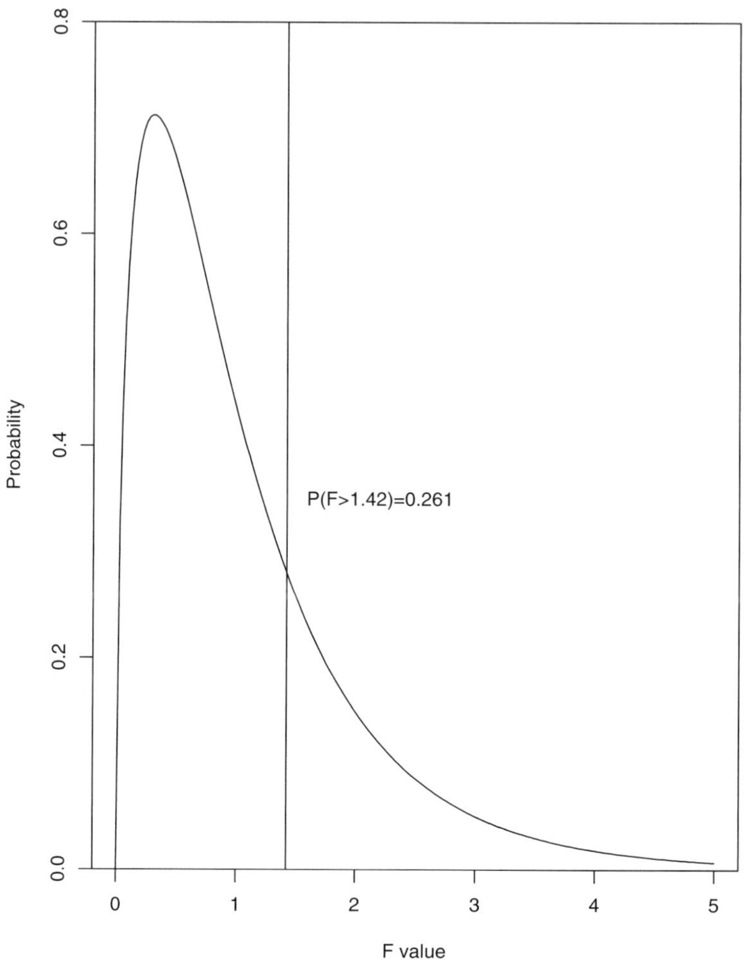

Fig. 2. The F distribution with degrees of freedom (3, 24). The probability $P(F > 1.42)$ is indicated by the area underneath the curve on the right-hand side of the vertical line $F = 1.42$.

Table 17
Interpreting *P* Value from an *F* Test at $\alpha = 0.05$

Magnitude of P value	$P > 0.05$	$P \leq 0.05$
Conclusion	The effect is not significant	The effect is significant
Meaning in terms of hypothesis	$\beta_1 = \beta_2 = \beta_3$	Some of the β_j's are not equal

4. Testing of Nested Hypotheses

Researchers often find themselves in a situation in which they have collected data on a large number of predictor variables, all of which seem to contribute in some way to the outcome, but they want to identify a parsimonious set of factors that best explains the outcome. The full model, which uses all the relevant predictor variables, always achieves a higher percentage R^2 (coefficient of multiple determination; see **Chapter 9**). However, the full model is more complex in that it uses more predictors, so a trade-off exists between how much variance is explained and model complexity. An ANOVA table is a powerful tool to help resolve the trade-off issue.

A very general ANOVA approach in **Table 18** shows how individual effects of factors can be summarized by presenting only the contribution of the entire set of factors (i.e., the model) relative to the error in terms of the SS. The notation here is such that y_i denotes the data value of the ith observation, and $\hat{y}_i$ denotes the estimated value under the model, which contains an intercept and $p - 1$ independent predictors. For two-way ANOVA with interaction, SSM includes SS due to **factor A**, **factor B**, and interaction.

To compare the reduced and full models, the following statistic, which can be obtained from ANOVA tables for the full and reduced models, is called the general linear F test:

$$F = \frac{(\text{SSE}(reduced) - \text{SSE}(full))[\text{d.f.}_{full} - \text{d.f.}_{reduced}]}{\text{MSE}(full)}. \tag{8}$$

The Venn diagram in **Figure 3** illustrates the various pieces of SS when a reduced model is compared with a full model. It can be seen that the expression SSE(*reduced*) – SSE(*full*) is the additional contribution in SS of the full model,

Table 18
ANOVA Table (Mean Corrected)

Source of variation	Sum of squares (SS)	Degrees of freedom (d.f.)	Mean square (MS)
Model	$\text{SSM} = \sum_{\text{observations}} (\hat{y}_i - \bar{y})^2$	$p - 1$	$\frac{\text{SSM}}{p-1}$
Residual	$\text{SSE} = \sum_{\text{observations}} (y_i - \hat{y}_i)^2$	$n - p$	$\frac{\text{SSE}}{n-p}$
Corrected total	$\text{SST} = \sum_{\text{observations}} (y_i - \bar{y}..)^2$	$n - 1$	

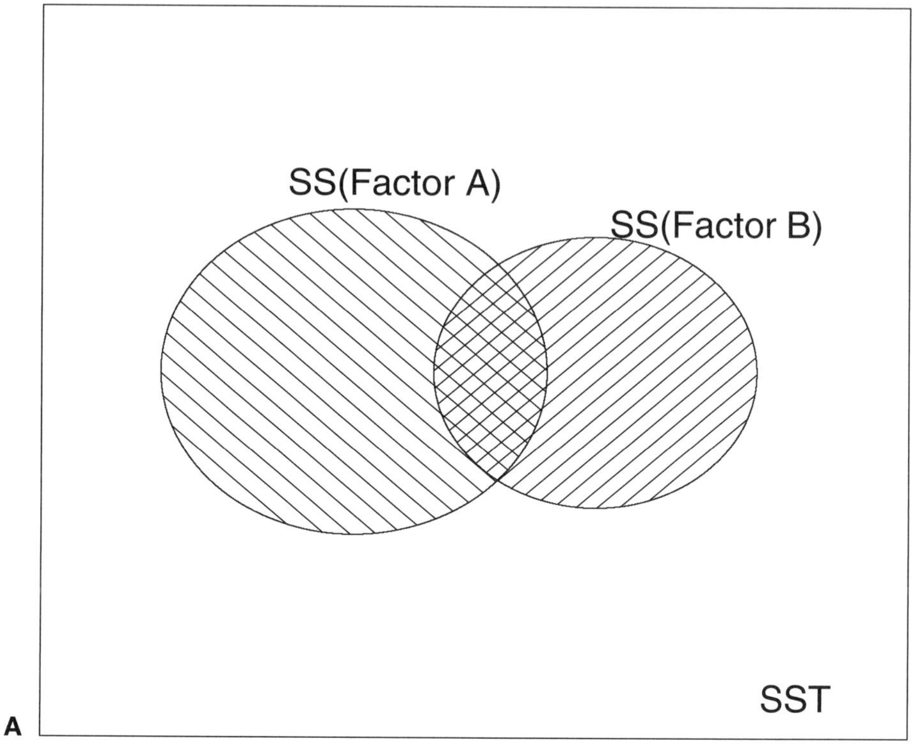

Fig. 3. **(a)** Venn diagram showing SSM and SST of a reduced model of 2 factors, **A** and **B**. The rectangular area represents SST, the shaded area SSM, and the unshaded area SSE. **(b)** Venn diagram showing SSM and SST for the full model for 3 factors, **A, B,** and **C**. The area shaded with only horizontal lines represents the additional SS contributed by **factor C**. The unshaded area, now diminished, represents SSE.

and its magnitude is represented by the darkened area in **Figure 3b**. The symbols $d.f._{full}$ and $d.f._{reduced}$ denote, respectively, the degrees of freedom of the full and reduced models. The F statistic follows an F distribution with the respective degree of freedom $d.f._{full} - d.f._{reduced}$ and d.f. in MSE(*full*).

In the phosphate example, suppose the full model contains gender, group, and **gender** × **group**, and the reduced model contains only **gender** and **group**. The ANOVA table for the full model is presented in **Table 19** and the reduced model is presented in **Table 20**. We can use these ANOVA tables to test the null hypothesis: there is no interaction effect versus the alternative hypothesis: there is interaction effect.

Applying **Equation 8**, SSE(*reduced*) − SSE(*full*) = 29017.1− 26755.0 = 2261.1, $d.f._{full} - d.f._{reduced} = 7 - 4 = 3$, we have

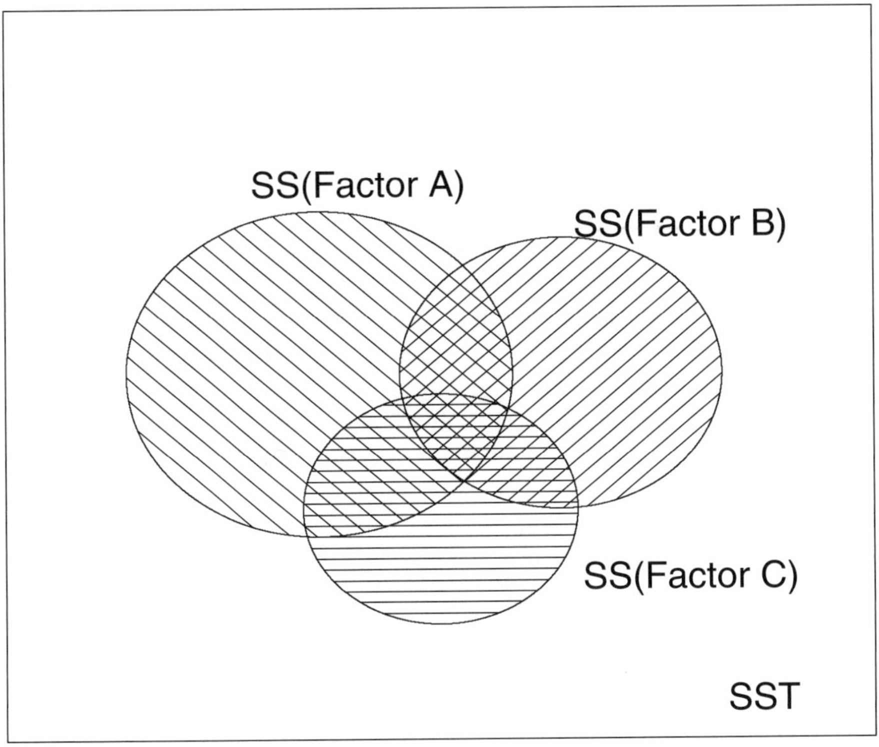

B

Fig. 3. (*continued*)

Table 19
ANOVA Table for the Full Model: $y_{ijk} = \mu + \alpha_i + \beta_j + \alpha\beta_{ij} + \varepsilon_{ijk}$

Source of variation	Sum of squares (SS)	Degrees of freedom (d.f.)	Mean square (MS)
Full model	16,889	7	2412.7
Residual	26,755.0	24	1114.8
Corrected total	43,644.0	31	

Table 20
ANOVA Table for the Reduced Model: $y_{ijk} = \mu + \alpha_i + \beta_j + \varepsilon_{ijk}$

Source of variation	Sum of squares (SS)	Degrees of freedom (d.f.)	Mean square (MS)
Reduced model	14,627	4	3656.8
Residual	29,017.1	27	1074.7
Corrected total	43,644.0	31	

$$F = \frac{2261.1/3}{1114.8} = 0.68,$$

which is distributed as an F distribution with d.f. = (3, 24) under the null hypothesis. The P value is 0.575, suggesting that the reduced model may be sufficient.

The test to determine which factor to include in the final model may seem straightforward for balanced designs in two-way ANOVA—we only need to look at which factor has significant P values. This is only possible in a balanced design because all factors are orthogonal—the magnitude of the SS of a factor does not depend on when it enters the model. Therefore, one can say such things as "The drug effect explains 30% of the variance, and the gender effect explains 10% of the variance." Unfortunately, for unbalanced designs and many other models in the general linear model class, such as the multiple regression model, the orthogonality property does not exist. In other words, the SS associated with a factor depends upon the sequence in which the factor is added or removed from the model. This is due to the "overlap" between the SS explained by various factors. For a 3-factor model, the SS for **factor C** appears in several forms: SS(C), SS(C|A), SS(C|B), SS(C|A,B), where SS(C|A) denotes the SS for **C** after **A** has entered the model, and so on. The fact that all these SSs are the same in a balanced design ANOVA is more of an exception than a rule. A way to visualize the overlapping explanatory power can be found in Ip *(8)*. A discussion of this is also provided in **Chapter 9**.

In summary, the F test can be used to scientifically determine if a reduced model suffices when evaluated against a more complex model that contains a larger number of relevant predictors. Although the method for testing nested hypotheses is rather general in nature, except in special cases, such as balanced design in ANOVA, the SS of a predictor depends upon the sequence in which it enters the model. Caution should be exercised in interpreting results when multiple steps are taken to sequentially remove or add predictor variables to a model.

5. Conclusion

The F test is useful for testing a broad range of hypotheses involving linear models. The methods described in this chapter are generalizations of the t-test, ANOVA, linear regression, and multiple linear regression discussed in previous chapters. The general linear model can be used to compare nested models that allow questions such as "does the effect of **A** remain significant after adjusting for **B**?"

References

1. Scheffe, H. (1959) *The Analysis of Variance.* New York: John Wiley & Sons.
2. Graybill, F. A. (1961) *An Introduction to Linear Statistical Models*, Vol. 1. New York, McGraw-Hill.
3. Searle, S. R. (1971) *Linear Models.* New York, John Wiley & Sons.
4. McCullagh, P., and Nelder, J. (1989) *Generalized Linear Models.* London, Chapman and Hall.
5. Schmid, C. H. (1991) Value splitting: taking the data apart. In: Hoaglin, D. C., Mosteller, F., and Tukey, J. W., eds. *Fundamentals of Exploratory Analysis of Variance.* New York, John Wiley & Sons.
6. Bradstreet, T. E. (1991) Some favorite data sets from early phases of drug research. *Proceedings of the Section on Statistical Education of the American Statistical Association* Alexandria, American Statistical Association.
7. Northway, W. H., Jr., Petrickeks, R., and Shahinian, L. (1972) Quantitative aspects of oxygen toxicity in the newborn: inhibition of lung DNA synthesis in the mouse. *Pediatrics* **50**, 67–72.
8. Ip, E. H. (2001) Visualizing multiple regression. *Journal of Statistics Education* **9**(1). Available at http://www.amstat.org/publications/jse/v9n1/ip.html.

11

Linear Mixed Effects Models

Ann L. Oberg and Douglas W. Mahoney

Summary

Statistical models provide a framework in which to describe the biological process giving rise to the data of interest. The construction of this model requires balancing adequate representation of the process with simplicity. Experiments involving multiple (correlated) observations per subject do not satisfy the assumption of independence required for most methods described in previous chapters. In some experiments, the amount of random variation differs between experimental groups. In other experiments, there are multiple sources of variability, such as both between-subject variation and technical variation. As demonstrated in this chapter, linear mixed effects models provide a versatile and powerful framework in which to address research objectives efficiently and appropriately.

Key Words: Fixed effects; mixed models; random coefficient models; random effects; two-stage analysis.

1. Introduction

A statistical model provides a mathematical description of how data are produced. An underlying goal of statistical analysis is to describe the process generating the data at hand while accounting for all sources of variation. George E. P. Box, a well-known statistician, once said "All models are wrong, but some are useful." Albert Einstein said, "Make your theory as simple as possible, but no simpler." As these quotes indicate, this process requires a balance of simplicity and proper representation of the biological system at hand. On the surface, linear mixed effects models may seem like a complex analytical approach with difficult concepts to grasp. In actuality, it is a simple extension of linear models with the added benefit of accounting for features in the data that aid in the interpretation and conclusions of the research study.

From: *Methods in Molecular Biology, vol. 404: Topics in Biostatistics*
Edited by: W. T. Ambrosius © Humana Press Inc., Totowa, NJ

Regression (**Chapter 8** and **Chapter 9**) and general linear models (**Chapter 10**) are extremely useful and versatile statistical modeling tools. They can be used to examine hypotheses on correlation, treatment effects and interactions, and for estimation of means. These basic models assume the residuals are independently and identically distributed as $N\ (0,\sigma^2)$. This assumption implies that the residuals must be independent of each other with mean 0 and constant variance σ^2 over the entire range of the response variable. In addition, it is assumed that residual or unexplainable error is the only source of random variability. However, in practice, not all experiments or studies may satisfy these assumptions. The data collection process or research design may induce correlation between observations or introduce multiple sources of random variation beyond that of the residual error. Each of these characteristics (correlation and multiple sources of error) would lead to inefficient and potentially misleading conclusions to a research study if the standard or classical methods of analysis were brought to bear on the problem. Hence, some additional tools are needed in order to make inferences.

Linear mixed effects models are powerful and useful approaches to many applications and may be used to address several study objectives. For example, they may be used in split plot experiments to account for varying sizes of experimental units (e.g., if tilling method is applied to the entire field while a variety is planted on only half of the field). They are useful in repeated measures studies to account for correlations between multiple observations per experimental unit and to specify or investigate different structures of correlation. In multilocation clinical trials, the results of fitting a mixed effects model enable researchers to broaden the inference space to the entire population of clinic locations, rather than just those participating in the trial. They can be used to model spatial variability on microchips or in geological studies. In genetics, they allow researchers to estimate and test for the heritability of a given trait.

Throughout this chapter, we discuss fixed and random effects, and it is useful at this point to discuss these terms. A key component to developing a statistical model is identifying factors that contribute to the total variation observed in a data set. This total variation is then partitioned among these explainable or identified sources of variation, and any unexplained or unidentified sources of variation are attributed to factors representing residual error. The importance of this exercise is that if an explainable source of variation is not accounted for in the statistical model, the residual error component will be inflated, resulting in inefficient analyses and inferences. The drawback of this exercise is that if too many factors are identified for a particular data set, the generalization of the research project comes into question because the researcher may be identifying factors that are uniquely associated with his or her study. The challenge then becomes a balance between maintaining generalization of the data while accounting for sources of variation.

This leads to what is referred to as the inference space of a study. The inference space, or population of interest, refers to the population about which the researchers will make conclusions based on the study at hand. For example, in a multilocation clinical trial for colon cancer treatment, one may wish to make inferences about treatment differences at one of the participating locations, overall participating locations in the study, or beyond the participating locations to all similar locations that would treat people with the disease of interest. In this context, location is an important source of variation to the data because the geographical makeup of the populations can vary greatly, but within a location a homogeneous group of subjects can be identified. These three inference spaces are very different and are called narrow, intermediate, and broad inference spaces, respectively, in the literature ***(1–3)***.

For narrow and intermediate inference spaces, a researcher would consider center location as a fixed effect. A factor (e.g., center location) is considered as a fixed effect if the levels studied represent all possible levels (or cover the expected range of levels) about which inference is to be made. For research projects conducted using similar procedures and selection criteria for the same centers, the contribution of a fixed effect factor to the variation in data can be anticipated from study to study. Within the statistical model, fixed effects (center location) represent a mean shift in the data. In this example, treating location as a fixed effect will result in a narrower confidence interval because its contribution to variation would not be added to the variability of the estimated treatment effects.

For a broad inference space, a researcher would consider center location as a random effect. A factor (e.g., center location) is considered as a random effect if the levels used in the study represent a random sample of a larger set of potential levels. In this case, if different levels of the factor (or different centers) were selected for a similarly conducted study, it would not be possible to anticipate *a priori* its contribution to the variation in the data. For this example, it may not be reasonable to assume that similar mean shifts in the data would occur if different centers take part in the study. Random effects allow the researcher to account for an important source of variation by adding an additional source of random variation to the statistical model beyond that of residual error. In this example, treating location as a random effect widens the confidence intervals of the estimated treatment effects across locations by adding its contribution of variation to the residual error.

2. Random Block Design

To begin to understand the impact of declaring an experimental factor as a random effect in a mixed model, consider the following example of a simulated multilocation clinical trial with balanced data. It is of interest to test a new chemotherapy regimen (**A**) against the standard of care (**B**) for reduction of tumor size prior to surgery for breast cancer in a multilocation clinical trial. The research plan is to conduct this trial at three different centers, and it is assumed that the randomization process within each center controls for extraneous factors that would confound the results of the study. The data are given in **Table 1**. The statistical model to describe the change in tumor size is given by

$$y_{ijk} = \mu + \tau_i + c_j + \varepsilon_{ijk}$$
$$i = \text{A,B}$$
$$j = 1,2,3$$
$$k = 1,2,3,4,5 \tag{1}$$

where, μ is the overall mean change in tumor size, τ_i is deviation from the overall change in tumor size due to treatment, c_j is the effect due to center, and ε_{ijk} is residual error, which is assumed to be $N\,(0,\sigma^2)$. In this model, c_j is a blocking factor. Blocking is used in statistical models to remove a known source of variation in the data from the residual error of the model, thus allowing for a more efficient comparison of treatment differences. In this example, varying influences such as referral patterns or geographic location of the centers may influence the outcome variable and should be accounted for in the model.

For this model, τ_i is treated as a fixed effect because the research aim is to directly compare two particular treatment regimens. However, c_j can be treated as either a fixed effect or a random effect depending on the inferences that the researcher wishes to make. In a balanced data design such as **Table 1**, the best estimate of the mean tumor size reduction due to the ith treatment regimen is

Table 1
Data from a Simulated Example of a Multilocation Clinical Trial for Change in Tumor Size

	Center			
	1 (cm)	2 (cm)	3 (cm)	Treatment mean (cm)
Treatment regimen				
A	2.0, 0.0, 0.0, 1.0, –1.0	–4.5, –1.0, –3.0, 0.0, –2.0	0.0, –3.5, –4.5, –2.0, 1.0	–1.17
B	0.0, 3.0, 1.0, 2.5, 0.5	–1.5, 1.5, 2.5, 3.5, 0.5	0.5, 0.5, 1.0, –2.5, –3.0	0.67
Center mean	0.9	–0.4	–1.25	–0.25

simply the average of all observations that received the *i*th treatment regimen. In terms of the statistical **Model 1**, this is expressed as

$$\bar{y}_{i\cdot\cdot} = \mu + \tau_i + \bar{c}_{\cdot} + \bar{\varepsilon}_{i\cdot\cdot} \tag{2}$$

where the dot denotes a sum over the respective subscript. If c_j is treated as a fixed effect, then inference made on the mean tumor size reduction due to the *i*th treatment is limited to only those centers that participated in the study. To see this, the statistical expectation and variance of the *i*th treatment mean in **Model 2** is given by

$$E(\bar{y}_{i\cdot\cdot}) = \mu + \tau_i + \bar{c}_{\cdot}$$

with

$$\mathrm{var}(\bar{y}_{i}\cdot\cdot) = \frac{\sigma_\varepsilon^2}{3\times 5}.$$

Here the average center effect is added to the estimate of the *i*th treatment mean $\mu + \tau_i$. Also note that the variance and corresponding confidence interval of this estimate would be based solely on the residual error. In order to make inferences of treatment-specific mean tumor size reduction beyond that of the participating centers under a fixed effects model, an assumption must be made that the average center effect would be similar if different participating centers took part in the study. This may be a difficult assumption to justify depending on the nature of the study, and caution is warranted when such an inference is made.

However, if c_j is treated as a random effect with a distribution of $N(0,\sigma_c^2)$, the corresponding statistical expectation and variance of the mean change given in **Model 2** is given by

$$E(\bar{y}_{i\cdot\cdot}) = \mu + \tau_i$$

with

$$\mathrm{var}(\bar{y}_{i}\cdot\cdot) = \frac{\sigma_c^2}{3} + \frac{\sigma_\varepsilon^2}{3\times 5}.$$

Here, the estimate for the mean tumor size reduction due to the *i*th treatment is free of the center effect. Under the random effects assumption, the model accounts for center effects by treating them as an additional source of variation that is added to the variance of the estimated treatment effect and not the mean. It is clear to see that the variance and corresponding confidence intervals will be larger than that of the fixed effect counterpart if the variance between centers, σ_c^2, is non-zero. In essence, the uncertainty in reproducibility of the mean center effect in a fixed effect model is attributed to the variance of the estimate.

Returning to **Table 1**, two features stand out regarding the tumor size reduction. First, treatment **regimen A** appears to have a larger mean reduction in tumor size relative to **regimen B**. Also, the mean change in tumor size is quite varied when considering treatment center. Treating c_j as a fixed or random effect in this example results in the same estimated treatment specific means of − 1.17 (**regimen A**) and 0.67 (**regimen B**). This is a special feature of balanced designs and does not hold true for designs with varying numbers of subjects per treatment combination. Under the fixed effect assumption, the variance of these estimated means is 0.2147 (a standard error of 0.4633). Under the random effect assumption, however, the variance estimate is 0.4982 (a standard error of 0.7058), which is approximately twice that of the fixed effect model.

Figure 1 gives a visual depiction of the distribution of treatment-specific mean change in tumor size with corresponding 95% confidence intervals under the two assumptions for the center effect c_j. Notice that each curve is centered about the respective estimated means and that the distribution under the random effects model is wider than that of the fixed effects model. Also, the 95% confidence intervals are wider for the random effects model. For these data, the overall F-test for significant treatment differences is the same for either assumption ($F = 7.83$, $P = 0.0096$), which is again a feature unique to balanced data.

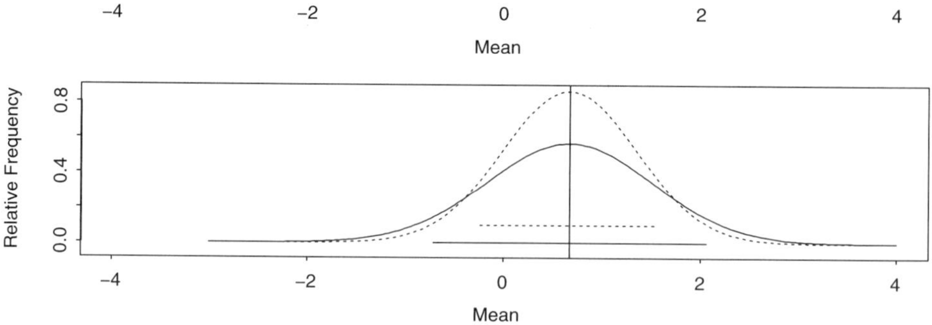

Fig. 1. The curves show the distribution of mean change due to **regimen A** (top panel) and **regimen B** (bottom panel) under the assumption of fixed (dashed lines) or random (solid lines) center effect c_j. The horizontal lines indicate the width of the respective 95% confidence intervals.

Although the effect of treatment **regimen A** on reducing tumor size is greater than that of **regimen B**, the fixed effect model and random effect model result in different interpretations on whether or not the change in tumor size for **regimen A** is significantly different from zero. When data are balanced, as they are here, the blocking factor cancels out when testing and estimating the difference between two treatments. However, when estimating or testing hypotheses regarding one treatment mean, the blocking factor does not cancel out and impacts the calculation. Under the fixed effects model, the test would indicate that the change in tumor size due to **regimen A** is significantly different from zero ($P = 0.0183$) whereas the random effects model would indicate that this is not the case ($P = 0.1104$). This can be seen in **Figure 1** by inspection of the 95% confidence intervals under each assumption and noting whether or not they contain the value zero. The conclusion from this data under the random effects model is that tumor growth would be suppressed for **regimen A** but that there is not necessarily a statistically significant overall reduction in tumor size.

The fixed effects hypothesis is considered to be a "narrow" hypothesis because extrapolation to other studies would only be valid if the same experimental factors were evaluated. That is, if another study was conducted using the *same* set of clinical centers, the expected results of the new study would be the same as the current study. The random effects hypothesis is a "broad" hypothesis in that it makes inferences to the entire population of experimental factors, not just the ones evaluated in the study. That is, if another study was conducted using a *different* set of clinical centers, the expected results would be similar to the current study.

3. Multiple Sources of Variation

In designs with multiple sources of variation (e.g., repeated or subsampling of experimental units or nested experimental factors), specifying an experimental factor as a fixed or random effect has a dramatic impact on the test of significance. Consider the following example. Redfield and others *(4)* conducted a study to estimate the prevalence of left ventricular dysfunction in the general population. A random sample of 2042 men and women age 45 and older was obtained from Olmsted County, Minnesota. Subjects were given a detailed physical exam, and echocardiographic imaging of the heart was performed to estimate ejection fraction (EF), which is the fraction of blood pumped out of the heart. Echocardiographic assessment of EF is not considered the gold standard method because of its technical variation, but it is less invasive and more cost effective than MUGA (multigated acquisition) scan, which requires the use of radioactive isotopes. For this reason, three assessments of EF were taken at different points of the cardiac cycle. Note that there are two sources of

Table 2
Ejection Fraction Data

	Male			Female			
Age (years)	Individual data	Mean	Strata mean	Individual data	Mean	Strata mean	Age strata mean
45–54	67.0, 65.8, 64.9	65.9	58.3	54.6, 61.9, 55.6	57.4	63.4	60.8
	66.4, 66.4, 64.0	65.6		65.1, 63.4, 58.7	62.4		
	42.4, 37.7, 37.1	39.1		61.3, 62.9, 62.9	62.4		
	64.0, 64.0, 64.0	64.0		55.6, 55.6, 55.6	55.6		
	64.0, 58.5, 59.9	60.8		75.0, 75.0, 75.0	75.0		
55–64	46.6, 44.5, 46.6	45.9	60.6	69.9, 70.1, 75.0	71.7	64.6	62.3
	60.0, 58.3, 60.0	59.4		57.5, 60.9, 68.6	62.3		
	64.6, 64.6, 64.6	64.6		50.7, 53.6, 50.7	51.7		
	69.1, 68.0, 62.1	66.4		65.1, 66.6, 71.6	67.8		
	65.0, 58.3, 66.0	63.1		73.0, 71.8, 71.8	72.0		
65–74	62.1, 60.4, 62.1	61.5	60.7	66.9, 66.5, 70.8	68.1	70.4	66.1
	68.0, 68.0, 69.1	68.4		75.0, 75.0, 75.0	75.0		
	58.5, 48.0, 55.6	54.0		72.0, 72.0, 72.0	72.0		
	60.9, 67.3, 64.0	64.1		63.1, 63.1, 63.1	63.1		
	55.6, 53.4, 53.4	54.1		73.9, 72.8, 75.0	73.9		
75+	60.9, 60.9, 65.0	62.3	61.8	63.4, 61.7, 63.4	62.8	55.4	58.8
	57.5, 62.3, 63.5	61.1		70.8, 75.0, 70.8	72.2		
	65.3, 58.0, 60.4	61.2		70.2, 70.2, 68.8	69.7		
	61.2, 63.5, 61.2	62.0		32.0, 23.1, 29.8	28.3		
	64.0, 63.1, 58.1	61.7		60.6, 59.0, 64.0	61.2		
Sex-specific mean	60.3			64.2			Grand mean 62.2

Ejection fraction was measured at three points of the cardiac cycle in a random sample of the Olmsted County, Minnesota population. A balanced subset of the data is displayed here stratified by age and sex where each line of data pertains to a subject.

random variation in the process giving rise to the data: that due to variation between subjects and that due to the technical variation of echocardiography itself. The goals of this study were to determine whether or not the distribution of EF depended on age and gender and to determine how much of the variability in EF was due to variation between subjects and how much was due to the technical variation of echocardiography itself. For illustration, **Table 2** displays the data on a balanced subset of the enrolled subjects stratified by age and sex.

The following linear mixed effects model can be used to address each of the researcher's objectives:

$$EF_{ijkl} = \mu + gender_i + age_j + gender_i * age_j + subject_{ijk} + \varepsilon_{ijkl}$$
$$i = 1,2$$
$$j = 1, \ldots, 4$$
$$k = 1, \ldots, 5$$
$$l = 1,2,3 \quad (3)$$

where $subject_{ijk} \sim N(0,\sigma_p^2)$ and $\varepsilon_{ijkl} \sim N(0,\sigma_\varepsilon^2)$.

In this model, EF_{ijkl} is the lth EF measurement on the kth subject from the ith gender and the jth age group, and μ is the overall grand mean. $Gender_i$ represents the effect of the ith gender, age_j represents the effect of the jth age group, and $gender_i * age_j$ allows for a potential interaction between age and gender. Both gender and age are considered fixed effects for this study because they represent the complete range of values of gender and age. $Subject_{ijk}$ is a measure of the random variation due to the kth subject from the ith gender in the jth age group, and ε_{ijkl} is the random variation or technical variation between the technical replicates on the same subject. It is assumed that $subject_{ijk}$ and ε_{ijkl} are independent of one another. $Subject_{ijk}$ and ε_{ijkl} are considered to be random effects because if the study were to be repeated again, different subjects would likely be selected in the random sample.

Table 3 summarizes the ANOVA table from fitting **Model 3** to the EF data. *P* values are presented assuming subject as a fixed and random effect for comparison. Notice the drastic differences in the *P* values! The fixed effect column tests whether or not the variation due to age and sex are significantly greater than what would be expected based on *technical* variation of the measuring device alone. Conversely, the random effect column tests whether or not the variation due to age and sex are significantly greater than what would be expected based on variation *between subjects*.The between-subject variability is estimated to be 85.67, and the technical replicate variability is estimated to

Table 3
Ejection Fraction Results

Source	SS	DF	MS	*P* value (R)	*P* value (F)
Gender	473.62	1	473.62	0.1895	0.0001
Age	484.85	3	161.62	0.6115	0.0001
Gender*age	629.87	3	209.96	0.5049	0.0001
Subject*gender*age	8,434.56	32	263.58	0.0001	0.0001
Error	526.39	80	6.58		
Total	10,549.29				

The *P* value (R) column contains results from a model considering subject as a random effect, and the *P* value (F) column contains results from a model considering subject a fixed effect.

be 6.58. Hence, the variation between subjects accounts for 85.67/(85.67 + 6.58) × 100% = 93% of the variability observed in the EF data, and the technical variability is only 7% of the total variability, which is why the two models give such different results. From these results, these data do not provide evidence that the systematic variation due to age and gender are significantly greater than the random biological variability between subjects. However, when incorrectly specifying a subject as a fixed effect, it appears that age and gender have large effects.

Combining the three EF observations per person by averaging or some other means is a common way for data such as these to be analyzed. By averaging the replicates, the amount of variability due to technical error is absorbed into the overall error term of the model and cannot be recovered directly from the ANOVA table. By averaging the replicates, the technical and between-subject sources of variation are combined into one error term, resulting in imprecise tests and confidence intervals.

4. Correlated Data and Random Effects Regression

A special case of repeated measurements on a subject or experimental unit occurs in longitudinal studies and dose-response studies. For longitudinal studies, the same response variable for the same experimental unit is repeatedly measured over some time domain (e.g., hours, years). For dose-response studies, a prespecified gradient of doses of the experimental factor under investigation is applied to the same experimental unit, and the same response variable is measured at each level of the gradient (e.g., pressure, temperature). The main objective for either study is to estimate the rate of change that occurs within an experimental unit in the response variable over the time or other gradient sampled. In addition, the association of this rate of change with other experimental factors may also be of interest to the researcher (e.g., age and gender of the subject, active drug versus placebo).

Classic repeated measures analysis techniques can be viewed as either univariate or multivariate methods. For either method, it is assumed that the sampling domain is the same across all experimental units. For example, on the time domain scale, each experimental unit is measured at exactly the same time points. However, the multivariate methods are generally restrictive in the presence of missing data or when sampling times differ between subjects, so our focus here will be on univariate techniques. Because the same response is measured repeatedly over some sampling domain on the same experimental unit, one cannot assume that these responses are independent of one another. This induces a correlation between responses on the same experimental unit, and this needs to be adequately addressed within the statistical model.

Consider an experiment in which the researcher is interested in determining if a new buffer significantly increases the growth rate of bacteria in a culture. The researcher applies the new buffer to 10 wells and leaves 10 additional wells untreated, incubates the bacteria, and measures the population size of the bacteria at hours 1, 2, 4, 8, 16, and 24. A model to describe the data for the *k*th well treated with the *i*th buffer at the *j*th time point is given by

$$y_{ijk} = \mu + buffer_i + time_j + buffer_i * time_j + \varepsilon_{ijk} \tag{4}$$

where μ is the overall grand mean, $buffer_i$ is the effect due to the *i*th buffer, $time_j$ is the effect due to the *j*th time, $buffer_i * time_j$ is the interaction term between buffer and time, and ε_{ijk} is the residual error, which we can no longer assume to be independently distributed as $N(0,\sigma^2)$ because of the six repeated measurements on each well.

For this type of situation, there are two approaches that can be used to account for the correlation. The first and most flexible approach is to specify the structure or pattern of correlation in the data when fitting the model. This can be done by calculating the correlation between time points on the residuals from **Model 4**, which assumes that the errors are independent of one another. After determining a structure that is reasonable for the data, the model is refit using a specific correlation structure. Most software packages that handle repeated measures data are suited for such a purpose. Alternatively, a random effect term can be added to the model that captures to some degree the fact that observations are correlated with one another. To achieve this, the above model would be

$$y_{ijk} = \mu + buffer_i + time_j + buffer_i * time_j + well_{ik} + \varepsilon_{ijk},$$

where everything is the same as before, but now the term $well_{ik}$ is added to the model and is assumed to be independently distributed $N(0,\sigma^2_{well})$, and the residual term is assumed to be independently distributed $N(0,\sigma^2_\varepsilon)$. The result of these assumptions is that the correlation between any two observations on the *k*th well treated with the *i*th buffer is equal to

$$corr\left(y_{ijk}, y_{ilk}\right)\frac{\sigma^2_{well}}{\sigma^2_{well} + \sigma^2_\varepsilon}.$$

This is simply the intraclass correlation coefficient. Regardless of the spacing in the sampling domain, this structure implies that all observations are equally correlated with one another. That is, responses measured 1 hour apart would have the same correlation as those measured 1 apart.

However, these approaches have limitations and cannot accommodate the following. First, the sampling times per subject may vary across the subjects in a longitudinal study. For example, a study where it was intended to measure subjects every 2 months will have some subjects with interval sampling times of a few weeks to several months. Also, the number of times the response was measured may vary from subject to subject. That is, a protocol that calls for five sampling time points may have subjects with anywhere between 1 and 5 measured time points. Lastly, in the case of the intraclass correlation model, the implied assumption that all responses are equally correlated with one another regardless of the spacing in sampling of the response is generally not reasonable. Responses measured farther apart from one another tend to be less correlated than those measured closer in time. A random effects regression model (a special case of a mixed effects model) can be used to efficiently account for the correlation between repeated measurements while allowing unequal spacing between visits and for the spacing pattern to vary across individuals. In addition, subjects with missing data are included in the analysis and are weighted according to the amount of information they provide.

Melton and others *(5)* report bone mineral density (BMD) data in a random sample of 270 (107 premenopausal, 163 postmenopausal) Olmsted County, Minnesota, women with mean age 57 ranging from 21 to 93 years. BMD was measured on each woman approximately annually, with 1 to 4 visits per woman, with 75% of them having 4 visits. The wrist BMD measurements for pre- and postmenopausal women are shown in **Figure 2** for a random subset of the study

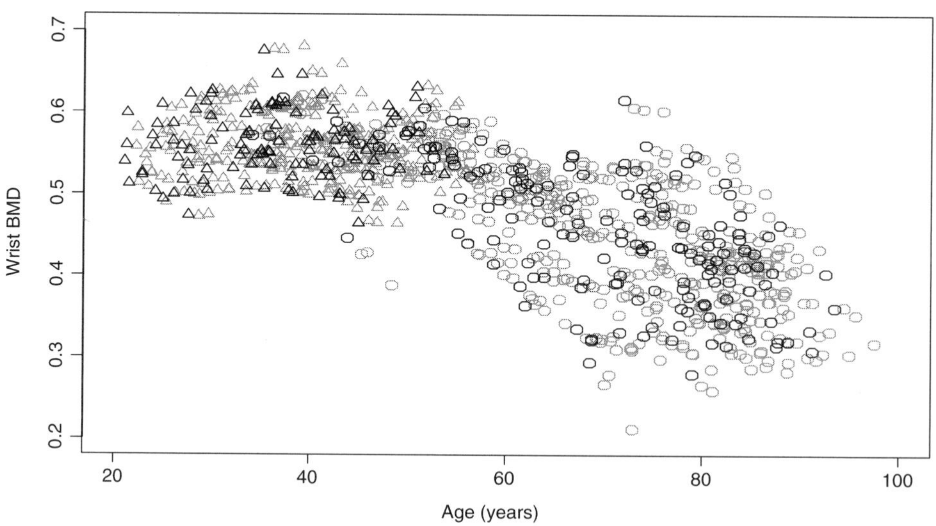

Fig. 2. Bone mineral density at the wrist versus baseline age for a random sample of women from Olmsted County, Minnesota. Triangles denote premenopausal women; circles denote postmenopausal women. First observations are highlighted in black; follow-up observations are in gray.

Table 4
Estimates Corresponding with Cross-sectional Parameters in Model 5 Fit to BMD Data Presented in Figure 2

Parameter	Estimate	Standard error	*P* value
β_0	0.5668	0.01306	
β_1	0.1723	0.01806	<0.0001
β_2	–0.0002	0.00033	<0.5927
β_3	–0.0039	0.000373	<0.0001
ε_i	~N(0, 0.0031)		

participants. In studies of osteoporosis, BMD is a measure of bone frailty with lower values of BMD indicating a higher susceptibility to bone fracture. Low bone mass in postmenopausal women is defined as a value 1 SD below the young normal mean (or 0.51 g/cm^2), whereas osteoporosis by World Health Organization criteria is defined as a value 2.5 SD or more below the mean (or 0.44 g/cm^2). **Figure 2** clearly shows that as the population ages, the mean BMD lowers indicating greater risk of bone fractures for elderly women. Our objectives here are to determine what age-related changes occur in BMD at the wrist and whether these changes are significantly higher in postmenopausal women than in premenopausal women.

An estimate of the age-related change allowing for different rates of change in the two menopausal groups can be obtained from this sample of women spanning the spectrum of age by regressing BMD on baseline age and menopausal status with an interaction term between age and menopausal status to test for differential slopes:

$$\text{BMD}_i = \beta_0 + \beta_1(meno_i) + \beta_2(age_i) + \beta_3(age_i \times meno_i) + \varepsilon_i. \quad (5)$$

Here, β_0 is the overall population average of BMD, $meno_i$ is an indicator variable with value 1 for menopausal women and value 0 for premenopausal women, β_1 is the mean shift in BMD due to menopause, β_2 is the cross-sectional or population rate of change in BMD per year of age, β_3 is the change in slope from β_2 for menopausal women, and the residual error ε_i is assumed to be independently distributed as $N(0,\sigma^2)$. This is a standard linear model, and the results of fitting **Model 5** to this data are presented in **Table 4**. The estimation of the model parameters for this model was covered in **Chapter 9**. Thus, the regression model for premenopausal women is estimated to be

$$\text{BMD}_i = 0.5668 - 0.0002(age_i)$$

and the regression model for the postmenopausal women is estimated to be

$$\mathrm{BMD}_i = 0.7391 - 0.0041(age_i).$$

These results indicate that the cross-sectional rate of change in BMD with age in menopausal women is –0.0041 g/cm^2 per year of age, significantly greater than the rate observed in premenopausal women of –0.0002 g/cm^2 per year of age.

Figure 3 presents the full longitudinal data as in **Figure 2**, but here the points for an individual subject are connected to indicate that the same subject was measured across time and to highlight the subject specific trajectories over time. **Figure 3** does little to change our initial claim about the cross-sectional decline in BMD across age groups. In addition, it appears that the variability in the rate of change may be larger in the postmenopausal group than in the premenopausal group.

Various strategies for modeling the correlation present in longitudinal data such as these have been proposed in the literature. One popular strategy is a two-stage analysis. The first stage consists of fitting the linear regression model

$$\mathrm{BMD}_{ij} = \beta_{0i} + \beta_{1i}(age_{ij}) + \varepsilon_{ij} \tag{6}$$

separately for each individual in the study. The subscript i indicates that this is the regression model associated with the ith subject. After fitting **Model 6** to

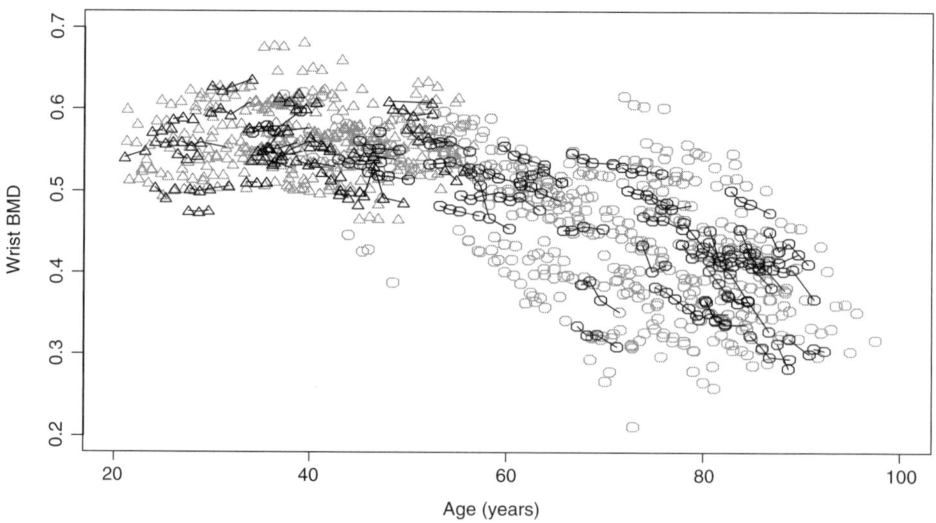

Fig. 3. Bone mineral density data as shown in **Figure 2**. Here, lines connect the observations for given subject over time. Trajectories are shown in black for a random subset of the sample in order to ease viewing.

each subject, the individual estimates of β_{1i} (the estimates of the subject specific slopes) are saved and carried forward to the second stage of analysis.

Recall that the goals of the study are to estimate the rate of change for both the premenopausal and postmenopausal women, to test whether these rates of change are significantly different from zero, and to test whether the rate of change is different for the two groups. Hence, the second stage of analysis consists of constructing confidence intervals and conducting multiple *t*-tests. The first hypotheses to test are whether the slopes for each menopausal group are significantly different from zero:

$$H_0: \beta_1^{pre} = 0 \text{ versus } H_1: \beta_1^{pre} \neq 0$$

and

$$H_0: \beta_1^{post} = 0 \text{ versus } H_1: \beta_1^{post} \neq 0$$

where β_1^{pre} and β_1^{post} are the population mean subject-specific slopes for premenopausal and postmenopausal women, respectively. The second hypothesis to test is whether or not the slopes differ between premenopausal and postmenopausal women:

$$H_0: \beta_1^{pre} - \beta_1^{post} = 0 \text{ versus } H_1: \beta_1^{pre} - \beta_1^{post} \neq 0.$$

Results for these hypothesis tests are shown in **Table 5**. This analysis indicates that there is no significant within-subject change in BMD over time for the premenopausal women, but that the postmenopausal women do have a significant decline. Confidence intervals for the quantities of interest can be constructed as well.

This is a reasonable first approach to this analysis. However, some women have more observations than others causing the estimates $\hat{\beta}_{1i}$ to have varying levels of precision. In addition, the variation in the subject-specific slopes is of interest. Knowing the variability in the women's trajectories is as important as

Table 5
Results of a Two-Stage and Mixed Effects Model Analysis of the BMD Data

Null hypothesis	Analysis	Estimate	Standard error	*P* value
$H_0: \beta^{pre} = 0$	Two-stage	–0.00001	0.0003	0.7321
	Mixed model	0.0001	0.0002	0.7897
$H_0: \beta^{post} = 0$	Two-stage	–0.0050	0.0005	<0.0001
	Mixed model	–0.0045	0.0004	<0.0001
$H_0: \beta^{pre} - \beta^{post} = 0$	Two-stage	–0.0050	0.0006	<0.0001
	Mixed model	–0.0046	0.0003	<0.0001

knowing the average trajectory. Random effect models provide a general framework, integrating these issues (the population average slope, the variability between subjects, the variability within a subject about her own slope, and the variation in precision for each subject) into one analyses method.

Linear mixed effects models are an extremely versatile tool for addressing these hypotheses efficiently. They can model the within-subject correlation, and the estimation process is weighted so that subjects providing more information and less variable information give more to the analysis than subjects with less information or more variable information.

A model containing two random effects, as follows, allows the correlation between observations to decrease as the time between them increases.

$$\mathrm{BMD}_{ij} = \beta_0 + \beta_1(age_{ij}) + b_{0i} + b_{1i}(age_{ij}) + \varepsilon_{ij}$$

where

$$b_{0i} \sim N\,(0,\sigma^2_{b_0})$$

$$b_{1i} \sim N\,(0,\sigma^2_{b_1})$$

$$cov(b_{0i},b_{1i}) = \sigma_{b_0,b_1}$$

$$\varepsilon_{ij} \sim N\,(0,\sigma^2_i). \tag{7}$$

The fixed portion of **Model 8** expresses the mean BMD across all subjects as

$$E(\mathrm{BMD}_{ij}) = \beta_0 + \beta_1(age_{ij}),$$

and the random effects portion

$$b_{0i} + b_{1i}(age_{ij}) + \varepsilon_{ij}$$

expresses each individual subject's deviation from the overall mean. That is, each subject is allowed to have their own unique regression line with intercept $(\beta_0 + b_{0i})$ and slope $(\beta_1 + b_{1i})$ but a common error distribution, ε_{ij}.

The assumptions on the random effects for **Model 8** can be evaluated and different correlation structures can be explored within the random effects regression model. The correlation and variance structure that is specified for **Model 8** was investigated using the methods outlined by Diggle and others ***(6)***. In short, the residuals from fitting the fixed effects portion of **Model 8** to the entire data set were obtained. Then, for each individual subject, the half-squared differences between residuals for age_{ij} and age_{ik} (for all $age_{ij} \leq age_{ik}$) were created and plotted against the difference in sampling times (in this case $age_{ik} - age_{ij}$). **Figure 4** displays the results of this procedure for the BMD data. An interesting aspect of this plot is that when averages of the half-squared difference in residuals are taken over small windows of time, one obtains an estimate of the variance incorporating the correlation between observations on the same subject. Also, the average across all half-squared residuals provides

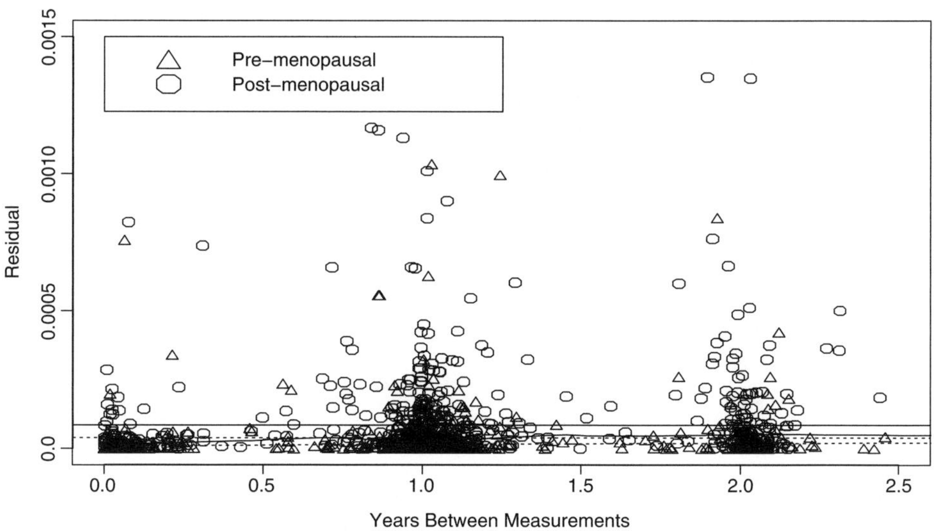

Fig. 4. Evaluation of correlation structure in residuals. The y-axis is the half-squared difference between a subject's residuals at time j and k [i.e., $\frac{1}{2}$ $(residual_{ij} - residual_{ik})^2$, for $age_{ij} \leq age_{ik}$], and the x-axis is the difference between the subject's ages at time j and time k (i.e., $age_{ij} - age_{ik}$, the amount of time between a subject's jth and kth visit). The horizontal lines are the estimated variances for the pre- (dashed) and postmenopausal (solid) women without accounting for the correlation present in the data. The smoothed curves are the estimated variances for the pre- (dashed) and postmenopausal (solid) women incorporating the correlation present in the data. A blown up portion of this plot is presented in **Figure 5**.

an estimate of the overall variance. **Figure 5** is a blow-up of **Figure 4** in order to highlight some features in the data. The horizontal lines for the pre- and postmenopausal women are highlighted to represent the overall variance in the data for these respective groups when the correlation between multiple observations on a subject is not incorporated into the model. The curves represent the estimated variances at each sampling time interval when incorporating the correlation. Notice that as the time between samples increases, the local variance approaches the overall variance in the data, indicating that the correlation is decreasing with time. Note for **Model 8** that (1) the variance increases as the spacing between sampling time increases and (2) the correlation between BMD measurements depends on the observed spacing in sampling points (and the actual sampling times across individuals may vary). All three observations made here are reasonable assumptions when dealing with longitudinal data.

Recall that our initial objective in this study was to estimate and compare the rates of change between pre- and postmenopausal women. Also, **Figure 4** and **Figure 5** indicate differences between pre- and postmenopausal women in amount of variation associated with the longitudinal rates of change.

The following model accommodates menopausal status and the different variances easily:

$$\begin{aligned}\text{BMD}_{ij} &= \beta_0 + \beta_1(meno_i) + \beta_2(age_{ij}) + \beta_3(meno_i \times age_{ij}) \\ &\quad + b_{0i} + b_{2i}(age_{ij}) + \varepsilon_{ij} \\ &\text{where } b_{0i} \sim N(0,\sigma^2_{b_0}) \\ &b_{2i} \sim N(0,\sigma^2_{b_2}) \\ &cov(b_{0i},b_{2i}) = \sigma_{b_0,b_2} \\ &\text{and } \varepsilon_{ij} \sim N(0,\sigma^2_{\varepsilon pre}), \text{ premenopausal} \\ &\varepsilon_{ij} \sim N(0,\sigma^2_{\varepsilon post}), \text{ postmenopausal.}\end{aligned} \tag{8}$$

Here, the residual error variances of $\sigma^2_{\varepsilon pre}$ and $\sigma^2_{\varepsilon post}$ are allowed to differ between premenopausal and postmenopausal women. For **Model 9**, β_1 represents the

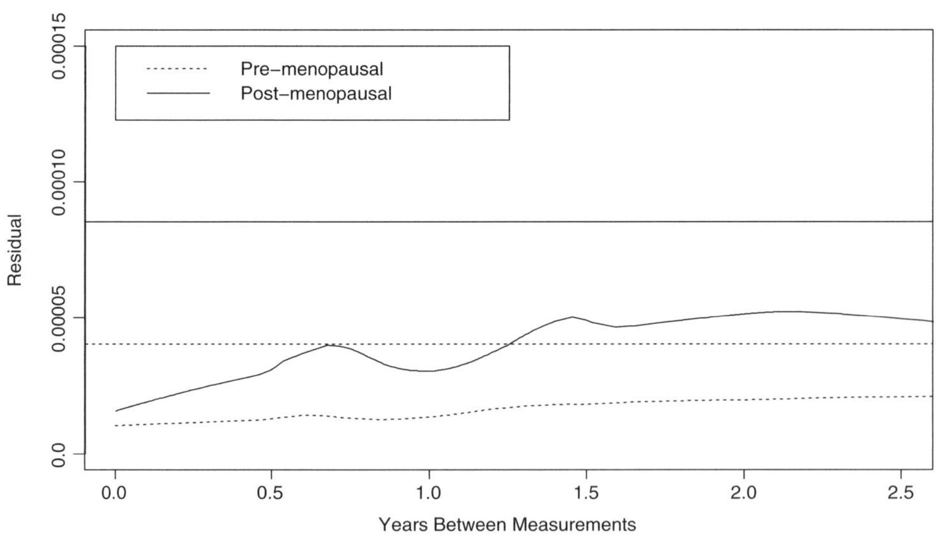

Fig. 5. Here the y-axis of **Figure 4** is expanded in order to more easily see the smoothed correlation structure for the pre- and postmenopausal groups. The horizontal lines are the estimated variances for the pre- (dashed) and postmenopausal (solid) women without accounting for the correlation present in the data. The smoothed curves are the estimated variances for the pre- (dashed) and postmenopausal (solid) women incorporating the correlation present in the data.

Table 6
Results of Fitting Model 9 to the BMD Data in Figure 3 Allowing for Different Amounts of Variation in the Subject Specific Slopes Between the Two Groups of Women

Parameter	Estimate	Standard error	*P* value
β_0	0.5611	0.0095	<0.0001
β_1	0.2075	0.0169	<0.0001
β_2	0.0001	0.0002	<0.7897
β_3	–0.0046	0.0003	<0.0001
ε_i	~N(0, 0.0003)	Premenopausal	
	~N(0, 0.0007)	Postmenopausal	
b_{0i}	~N(0, 0.0046)		
b_{2i}	~N(0, 0.0000002)		
$Cov\ (b_{0i}, b_{2i})$	–0.0001		

overall mean shift in BMD measurements for postmenopausal women, and β_3 represents the deviation from the cross-sectional association of age with BMD measurements for postmenopausal women. **Table 6** summarizes the results of fitting **Model 9** to the data presented in **Figure 3**.

The estimated models for pre- and postmenopausal women are

$$\text{BMD}_{ij} = 0.5611 + 0.00006(age_{i0}) \quad \text{and}$$

$$\text{BMD}_{ij} = 0.7686 - 0.0045(age_{i0}) \tag{10}$$

respectively. For premenopausal women, the cross-sectional rate of change in BMD due to aging is negligible ($P = 0.7897$). However, for postmenopausal women, there is clearly a negative cross-sectional rate of change in BMD due to aging. As indicated by the tests of significance, pre- and postmenopausal women have significantly different rates of change in BMD measurement. Note that the estimate of variability is twice as high in the postmenopausal group than the premenopausal group. **Table 5** shows the results of the two-stage approach and the random effects regression. Although the estimates are very similar using either approach, the random effects regression approach provides for smaller standard errors of the estimates and thus more precise confidence interval estimates.

5. Model Fitting

There are several ways to approach the estimation of random effects in a linear mixed effects model. All give equivalent results if the data are balanced with no missing data points. However, when data are unbalanced, the results differ. There are few experiments finished where a subject has not dropped out, failed to show up for a scheduled visit, or some unforeseen event causes some of the data to be missing.

Model fitting in mixed effects models has been the subject of a large amount of research. In standard linear regression and ANOVA as discussed in **Chapters 7** to **10**, ordinary least squares (OLS) methods are used to estimate parameters and construct test statistics, and these are easily calculated by hand. Though the theory for more complex situations was developed in the mid-1900s *(7)*, these models did not see significant use until the 1990s with the development of reliable and flexible software such as the MIXED procedure in SAS *(3)*.

For all but the simplest problems, the best solution to mixed models involves the maximization of a nonlinear function by an iterative algorithm. Currently, maximum likelihood (ML) and restricted maximum likelihood (REML) are considered to be the best methods available, with REML the default in most modern software packages. The numerical difficulty of the problem has led to a large number of simple approximate methods [such as method of moments (MOM), minimum norm quadratic unbiased estimation (MINQUE), and minimum variance quadratic unbiased estimation (MIVQUE)], which are still available in some systems. More powerful computers have rendered them unnecessary for almost all problems encountered by the average user. It is very important to understand what methods are being used by your software. [For further reading, see ***(3,6,8,9)***]. These references also provide discussions of the impact of missing data on mixed effects model fitting in addition to the discussion of missing data in **Chapter 17.**

6. Power and Sample Size

As discussed in **Chapter 12** and **Chapter 19**, carefully planning and designing the optimal study to address the research objective maximizes the amount of information gained and the efficiency with which the objectives are addressed. Designing the optimal study involves power and sample size calculations in order to assess feasibility of various designs. It is possible to perform power calculations for linear mixed effects models, and methods for doing so with the aid of software are demonstrated in Stroup ***(10)***. However, this requires the researcher to provide estimates for any variances and covariances that will be in the model, as well as an educated guess as to the mean responses in each treatment group. Obtaining the needed information may be daunting and require more assumptions than a researcher is willing to make.

Care should be taken to think about aspects such as dropout in a longitudinal or repeated measures study and the possible effects on the sample size and modeling process. (See **Chapter 17** for a discussion.) In addition, Verbeke and Molenberghs give a nice discussion of this and the impact on linear mixed effects models in **Chapter 23** of their book *(8)*. For more on power and sample size, see **Chapter 19**.

7. Extensions

In some cases, the data of interest cannot be assumed to be normally distributed. For example, outcomes such as presence or absence of disease that follow a binomial distribution, or count data such as number of infections for a subject that follow a Poisson distribution. In other cases, the outcome may not be linearly related to the predictors. For example, growth data frequently follow a nonlinear logistic growth curve. However, these types of data may require some of the tools for properly modeling correlation structures, or allowing for random effects as well. Mixed effects models have been extended to these situations. Generalized linear mixed effects models *(11)* have been developed for the situation of nonnormally distributed data, and nonlinear mixed effects models have been developed for the situation where a response is nonlinearly associated with the predictors *(9)*.

With these tools, as for linear mixed effects models, it is extremely important for the researcher to have an in-depth understanding of the software and what it is doing in order to perform an analysis correctly. It is also very important to understand the differences between population level and subject level inference. In the generalized linear models framework, these two levels of inference result in different models. The same model cannot be used for both as is possible in linear mixed effects models. Further reading on generalized linear mixed models can be found in McCulloch *(11)* and for nonlinear mixed effects models in Davidian and Giltinan *(9)*.

Acknowledgments

The authors wish to thank Dr. L. Joseph Melton III for his mentoring; Dr. Terry M. Therneau for his insightful suggestions regarding this chapter; Dr. Sundeep Khosla for his permission to use the bone mineral density data; Dr. Richard Rodeheffer for his permission to use the ejection fraction data set; Ms. Sara Achenbach for compiling the BMD data set and plot construction; and Mrs. Rhonda Larsen for typing the chapter.

References

1. Zeger, S. L., Liang, K. Y., and Albert, P. S. (1988) Models for longitudinal data: a generalized estimating equation approach. *Biometrics* **44**, 1049–1060.
2. McLean, R. A., Sanders, W. L., and Stroup, W. W. (1991) A unified approach to mixed linear models. *Am. Stat.* **45**, 54–64.

3. Littell, R. C., Milliken, G. A., Stroup, W. W., and Wolfinger, R. D. (2002) *SAS ® System for Mixed Models.* Cary, SAS Institute Inc.
4. Redfield, M. M., Jacobsen, S. J., Burnett, J. C. Jr., Mahoney, D. W., Bailey, K. R. and Rodeheffer, R. J. (2003) Burden of systolic and diastolic ventricular dysfunction in the community: appreciating the scope of the heart failure epidemic. *JAMA* **289**(2), 194–202.
5. Melton, L. J. III, Atkinson, E. J., O'Connor, M. K., O'Fallon, W. M., and Riggs, B. L. (1998) Bone density and fracture risk in men. *J. Bone Miner. Res.* **13**, 1915–1923.
6. Diggle, P. J., Liang, K., and Zeger, S. L. (1995) *Analysis of Longitudinal Data.* New York, Oxford University Press.
7. Rao, C. R. (1965) The theory of least squares when the parameters are stochastic and its application to the analysis of growth curves. *Biometrika* **52**, 447–468.
8. Verbeke, G., and Molenberghs, G. (2000) *Linear Mixed Models for Longitudinal Data.* New York, Springer-Verlag.
9. Davidian, M., and Giltinan, D. M. (1995) *Nonlinear Models for Repeated Measures Data.* New York, Chapman & Hall.
10. Stroup, W. W. (1999) Mixed model procedures to assess power, precision, and sample size in the design of experiments. *ASA Proceedings of the Biopharmaceutical Section.* Alexandria, American Statistical Association, pp. 15–24.
11. McCulloch, C. E. (2003) Generalized linear mixed models. *NSF-CBMS Regional Conference Series in Probability and Statistics* 7. San Francisco, Institute of Mathematical Statistics.

12

Design and Analysis of Experiments

Jonathan J. Shuster

Summary

This chapter is primarily devoted to experiments that compare 2 treatments with respect to an outcome measure. Six design scenarios are discussed: (a) completely randomized designs (treatments are assigned completely at random); (b) randomized block designs (experimental units are subdivided into blocks of like subjects, with one subject in each block randomly assigned to each treatment); (c) stratified designs (subjects are categorized into subpopulations called strata, and within each stratum, a completely randomized design is conducted); (d) crossover designs (each subject gets both treatments, but order is completely at random); (e) 2 × 2 factorial designs [design can be in any of the formats (a)–(d) but there are 4 not 2 treatments representing 2 types of treatment interventions, each with 2 levels]; and (f) randomized designs with "random" effects. This is much like the stratified design, except there is only 1 sample, at least conceptually, from the *strata*. Examples might be litters of laboratory animals, surgical practices, or batches of a therapeutic agent. The desire is to make inferences about treatments in the population as a whole, not just in the strata that were actually sampled.

Key Words: Completely randomized design; crossover design; optimal allocation; randomized block design; stratified design; 2 × 2 factorial design.

1. Introduction

In order to answer important biomedical questions, we conduct investigations called experiments. Unfortunately, when we repeat the same process, we may not get the same exact answer. For example, if one does 2 replicates of a quantitative assay of the concentration of calcium in bone samples in mice, the 2 results will generally differ. Part of the difference may be due to a true difference in the concentrations from different bone samples and part of the difference due to measurement error. If the bone samples come from different

From: *Methods in Molecular Biology, vol. 404: Topics in Biostatistics*
Edited by: W. T. Ambrosius © Humana Press Inc., Totowa, NJ

mice, we would expect a larger variation than we would if the bone samples came from different bones from the same mouse, and this in turn would be expected to have a larger variation than if the 2 samples came from the same bone of the same mouse. In experimental design, it is always a good idea to identify the *sources of variation.* In this example, we have variation between mice, variation between bones of the same mouse, variation within different parts of the same bone of the mouse, and measurement error. These sources of variation are an obstacle to our learning the truth about the study questions we pose. Accurate answers to study questions can be obtained by using designs that control extraneous variations and/or increase the sample size of the study.

The theme of this chapter is to look at the comparison of "treatments" under various randomized design scenarios. Please refer to **Chapter 2** for a discussion of observational study designs. In addition, readers are referred to **Chapter 17** and **Chapter 19** for more details on issues relevant to both randomized and observational studies. The major emphasis is on comparing 2 treatments, but multiple treatment studies are considered. The major topics are (a) completely randomized designs, (b) randomized block designs, (c) stratified designs, (d) crossover designs, (e) 2×2 factorial designs, and (f) randomized designs with random effects.

For additional reading on study design, see Shuster *(1)*, Hinkelmann and Kempthorne *(2)*, Cochran and Cox *(3)*, Frigon and Mathews *(4)*, Heath *(5)*, and Weber and Skillings *(6)*.

The *completely randomized design* assigns experimental units (research subjects or specimens) at random to 1 of 2 treatments, without regard to any other factors about the subject. For 2 treatment studies, the probability of assignment is nearly always intended to be 50% assigned to each treatment.

The *randomized block design* randomizes a *block of subjects* to treatments, typically 1 member of the block per treatment. For example, suppose you wish to test for antitumor activity of 2 agents plus a placebo control in mice. Human frozen tumor tissues from 12 patients are each split into 3 subsamples, thawed, and injected into 36 genetically equivalent laboratory mice, as blocks of 3. Within each block of 3 mice, the tumor tissue comes from the same person, and 1 is treated with each of the treatments. This design eliminates potential random variation caused by differences in individual tissue sources when it comes to comparing the treatments. A completely randomized design would use tissue from 36 individuals and randomly assign the tissues to treatments. Hence, a randomized block design can save precious tissue on the one hand while eliminating considerable variation between subjects on the other.

The *stratified design* can give some of the benefits of randomized block designs but can be almost as simple to conduct as completely randomized designs. Subpopulations (strata) are defined, and approximately equal numbers of subjects within each stratum are assigned to each treatment. Stratified designs are very effective in laboratory research where observations are done in batches (strata) over a relatively long period of time. If the treatment assignments are balanced within each batch, good control over the batch-to-batch variation and good control over the impact of changes in support personnel over time are achieved.

The *crossover design* can be an efficient method to study a biomedical question. The idea is to expose research subjects to both treatments in random order, with a *washout* in between. For example, if one wishes to study the cholesterol-lowering effect of a food additive in mice, one could randomly assign mice to the additive and to no additive in random order, controlling for other aspects of the diet. A washout segment between the 2 treatment periods would be recommended. If feasible, this design will get considerably better than 2 for 1, in terms of animals required versus a completely randomized design. Not only does each animal give 2 observations, but it does a better job of controlling a major source of variation, namely that between animals, because each serves as its own control in a crossover design.

One downside of a crossover design occurs when there is treatment by period interaction (crossover effects). If the true target treatment difference is dependent upon the order of the treatments, then the crossover design will estimate a population parameter that differs from the one estimated by the completely randomized design. In addition, there may be practical reasons not to conduct a crossover design, including end points that take a long time to collect or higher rates of dropout expected in a crossover design compared with a completely randomized design. Although one may legitimately think of the design as a special case of a randomized block design, to do so would discard vital information about the order of treatment assignment.

The *2 × 2 factorial design* can basically be viewed as conducting 2 studies at the same time. They can be grafted onto completely randomized designs, randomized block designs, or even a mixture of a completely randomized design with a crossover design (some would call this design a "split plot," as borrowed from agriculture, not borrowed from a great novelist). For example, in the Physician's Heart Study *(7)*, subjects were assigned to either aspirin or placebo as well as carotene or placebo to study protective effects against cardiovascular disease and cancer, respectively. Twenty-five percent were assigned to each of the 4 treatment scenarios. These designs are attractive to many scientists because with the expense of running clinical trials and experimental studies with animals, they afford the opportunity to answer 2 questions for the price of 1. However, the results can be confusing if the effect size of one factor is highly dependent upon which other factor the individual subject was assigned to (interaction between the treatments). The physician study seems ideal for this, because the interventions targeted different medical problems. When the 2 interventions target the same medical problem, the best experimental question may be more complicated than simply asking if the 2 treatments within each factor differ. It may also be about whether 2 active treatments are better than 1. The ISIS II Study *(8)* studied 2 clot-busting techniques (aspirin and streptokinase) in patients hospitalized for myocardial infarction. One might expect an interaction, as the 2 treatments target similar medical mechanisms. A key question, in addition to the efficacy of each therapy, might be, "Is the combination therapy better than either monotherapy?"

The *randomized design with random effects* can be used in multicenter surgical trials or in the laboratory where a design issue might be how many clusters (e.g., physician practices, litters, or divisible tissue specimens) do we utilize and how many sampling units within the cluster do we employ? The study is planned for each *tissue sample* cluster putting an equal number of *aliquots* (sampling units extracted from the tissue sample) on each of the 2 treatments, with the tissue samples forming, at least in concept, a random sample from a target population of hypothetical tissue donors that could participate.

Limitations of the Chapter

This chapter will be confined to *single-stage designs*; that is, the methods have no allowance for interim decision-making. However, most of these designs can be incorporated into multistage designs called *group sequential designs* by using the methods of this chapter in conjunction with software such as EaST *(9)*. In addition, the outcome measures considered herein will be restricted to quantitative or binary outcomes. Censored survival outcomes are not part of this chapter. For the most part, we shall concentrate on a single outcome variable. In our design considerations, we shall handle multivariate outcome data via a Bonferroni bound, rather than by a formal multivariate analysis.

2. The Completely Randomized Design

Generally, the design calls for a sample size of N subjects, half randomly assigned to each treatment without regard to any other factors. The value of N is obtained from a power analysis or a sample size calculation (as described in **Chapter 19**) to obtain a desired sensitivity to a given difference. If the 2 treatments are designated as **A** and **B**, the actual treatment assignment represents an N letter "word" of **A**s and **B**s with each word having the same probability of occurring. For example, if $N = 4$, one of the following six assignments is drawn: **AABB**, **ABAB**, **ABBA**, **BAAB**, **BABA**, **BBAA**. Say the actual draw was **ABBA**. The first subject receives **A**, the second **B**, the third **B**, and the fourth **A**.

2.1. Permutational Basis of Inference

In order to calculate a *P* value for a completely randomized design, one can consider the null hypothesis that there is no true target population treatment effect, and that the subject outcomes are predestined. One must select a statistic before collecting the data, such as the absolute difference between the sample means, and calculate the *P* value as the answer to the following question: If we went through all possible rerandomizations of the subjects, half assigned to each treatment, what fraction of these rerandomizations will have the absolute difference in means at least as large as that observed? This makes sense, because the larger the actual observed difference between the sample means, the more confidence one might have that indeed the population means are different. To actually do this in practice is feasible only for relatively small studies. For example, if *N* is as small as 20 (10 assigned to each treatment), there are 184,756 different assignments (20 letter words with 10 **A**s and 10 **B**s). Randomization tests can legitimately be done by sampling the reruns, rather than doing all. For example, we can take a sample of say 9999 rerandomizations (taken *with* replacement as tracking those already done is more trouble than it is worth), and see how the actual result stacks up among the 10,000 randomizations including the actual one. The *P* value is defined as the fraction among this 10,000 (including the actual one) that is at least as large as that observed. Not only is this a good approximation to the test that looks at this exhaustively, but also it is a legitimate permutation test in its own right. If indeed the null hypothesis is true, then, for example, there is less than a 5% chance that the *P* value is below 0.05. (The degree less depends on the probability of tying the observed value. In a continuous data situation, this chance is virtually nil.) This method is *nonparametric*, valid whatever the underlying true null distribution is. Some practitioners convert the data to ranks. The test that does this is called the Wilcoxon–Mann–Whitney test ***(10)***, which can be analyzed with exact methods via StatXact ***(11)***.

2.2. Parametric Basis of Inference: Student's t*-Test*

Strictly speaking, Student's *t*-test applies to the completely randomized design where the target population has a normal distribution under each treatment, and the target population standard deviations are the same for both treatments. However, it is also a large sample approximation to the permutational test on the means cited above. In this author's experience, the *t*-test, with modest sample sizes as low as 15 per group, gives results very close to the permutational tests except in the presence of *outliers* (observations "far" from the center of the data). See **Chapter 4** and **Chapter 7** for alternate developments of the *t*-test.

To perform this test, one computes the following statistics:

$$\bar{x}_j = \frac{\sum_{i=1}^{n_j} x_{ij}}{n_j},$$

where i represents the ith subject for treatment j, $j = 1$ or 2 for treatment 1 or 2, and n_j is the sample size for treatment j. This formula also applies when there are more than 2 treatments. The variance is estimated as

$$s^2 = \frac{\sum_{j=1}^{K}\sum_{i=1}^{n_j}(x_{ij} - \bar{x}_j)^2}{N - K} \quad (1)$$

where N is the total sample size and K is the number of treatments ($K = 2$ in the classical t-test, but **Equation 1** can be used in multitreatment settings with $K = 3$, 4, or 5).

The t-statistic is defined as:

$$t = \frac{\bar{x}_2 - \bar{x}_1}{\text{SE}} \quad (2)$$

where the *standard error for the difference* (SE) satisfies:

$$\text{SE} = s\sqrt{\frac{1}{n_1} + \frac{1}{n_2}}. \quad (3)$$

The SAS procedure PROC TTEST performs the calculation, provided one uses the "Equal Variance" output. Alternately, one can use the central t-distribution with $N - 2$ degrees of freedom to obtain the P value. For the 2-sided test, the P value is given by $2 \times P\ (t_{n-2} \leq -|t|)$ for 2 samples and $2 \times P\ (t_{N-K} \leq -|t|)$ for an unadjusted t-test to compare 2 treatments in a K-sample experiment. Here, t_{df} represents the random variable with a central t-distribution with degrees of freedom d.f. (see **Chapter 7** for a discussion).

2.3. Large Sample Inference

When the sample sizes are large (usually $n > 30$), it is recommended that the unpooled version of the standard error be employed and that the P value be defined by the normal distribution, rather than the t-distribution. Define

$$\text{SE}_u = \sqrt{\frac{s_1^2}{n_1} + \frac{s_2^2}{n_2}} \quad (4)$$

where

$$s_j^2 = \frac{\sum_{i=1}^{n_j} (x_{ij} - \overline{x}_j)^2}{n_j - 1} \tag{5}$$

and let

$$z = \frac{\overline{x}_2 - \overline{x}_1}{\mathrm{SE_u}}. \tag{6}$$

Note that if the 2 sample sizes are equal, $\mathrm{SE_u}$ and SE as defined in **Equation 3** are identical and hence $t = z$ when the sample sizes are equal. The 2-sided *P* value is defined as $2 \times P\ (Z \leq -|z|)$, where *Z* is a standard normal random variable.

For large samples, the advantage of the use of *z* over *t* is that it does not require the assumption of equal population variances to be a valid test for the difference in means. For large samples, neither the *t*-test nor the *z*-test require that the populations be normally distributed to be a valid test. The *t*-test is a valid large-sample test for equal populations, as under equal populations, the variances are indeed equal. Furthermore, it represents a large-sample approximation to the permutation test described above.

2.4. Allocation of the Sample to the Two Treatments: Is 50-50 Best?

Applied researchers often ask if the samples for the completely randomized design should be allocated 50-50 to the 2 treatments. The evasive answer is, "That depends." Here are the major considerations:

1. Is the cost of sampling approximately equal for the two "treatments"? If no, oversample from the less expensive treatment.
2. Is there substantial anticipated difference in the 2 population variances? If yes, oversample from the more variable population.
3. Are there ancillary questions within 1 of the treatment groups? For example, if a new drug is being tested against a standard, they may wish to allocate more to the new drug to get more safety data. If yes, oversample from that group.
4. Is one treatment expected to be more efficacious than the other? This one is sticky, because if you are really convinced one is better, you need to question the ethics of the study in the first place. But there are situations where the pharmaceutical industry is required to study a drug against a placebo, even when there is a known effective control. If yes, oversample from the active treatment.

Consider the first 2 questions above. We can optimize the sample size under the following assumptions:

Table 1
Values of $z_{\alpha/2}$ for Use in Equation 7

Dimensions	1	2	3	4
α	0.050	0.05/2 = 0.025	0.05/3 = 0.0167	0.05/4 = 0.0125
$z_{\alpha/2}$	1.96	2.24	2.39	2.50

1. The population variances are σ_1^2 and σ_2^2 for treatments 1 and 2.
2. The costs for each sampling unit are C_1 and C_2 for treatments 1 and 2.
3. The planning difference for the two treatment means is $\Delta = \mu_2 - \mu_1$.
4. The study is planned for a type I error of α (2-sided) and power of $1 - \beta$, where β is the type II error.

The variance of $\bar{x}_2 - \bar{x}_1$ is $\sigma_1^2/n_1 + \sigma_2^2/n_2$, where n_1 and n_2 are the sample sizes. If we denote the total sample size by $n = n_1 + n_2$ and set $\theta = n_1/n$ and $1 - \theta = n_2/n$, it can be shown [see Section A130 of Shuster ***(12)***], that the required total sample size satisfies:

$$n = \frac{(\sigma_1^2/\theta + \sigma_2^2/(1-\theta))(z_{\alpha/2} + z_\beta)^2}{\Delta^2} \tag{7}$$

where z_γ is the $100(1 - \gamma)$ percentile of the standard normal distribution. Please see **Chapter 19** for an in-depth discussion of power. Values of $z_{\alpha/2}$ to utilize for univariate and up to 4-dimensional multivariate analysis using Bonferroni bound appear in **Table 1**. Use $z_\beta = 0.842$ for 80% power.

To minimize the total cost $C = n[\theta C_1 + (1 - \theta)C_2]$, we set

$$\theta = \theta_{opt} = \frac{\sigma_1\sqrt{C_2}}{\sigma_1\sqrt{C_2} + \sigma_2\sqrt{C_1}}. \tag{8}$$

If the costs are equal and the anticipated standard deviations are equal, then the optimal allocation is 50-50.

Two designs can be compared by comparing the *relative efficiency* of the two. The relative efficiency of 2 designs is the ratio of sample sizes required to achieve a given precision. It can also be considered as the inverse of the ratios of the variances for a given cost.

Example 1

Equal sampling costs but unequal variance. The costs are proportional to the total sample size, and hence the relative efficiency (defined as the ratio of the sample size required under optimal allocation to the sample size required under inverse of the ratio of variances for a fixed cost) is simply the ratio of the n's from **Equation 7**. From **Equation 8**, for equal costs, $\theta_{opt} = \frac{\sigma_1}{\sigma_1 + \sigma_2}$, and the relative efficiency is $RE = \frac{n_{opt}}{n_{50\text{-}50}} = \frac{1}{2\theta_{opt}^2 + 2(1-\theta_{opt})^2} \leq 1$. The fact that the relative efficiency can be at most 1 means that the optimal sample size can be at most the sample size for the 50-50 split. The relative efficiency for a range of variances is given in **Table 2**. In terms of ratios of variances, for equal costs, it takes a very large disparity to make a major difference.

Example 2

Binomial sampling for a rare trait under equal costs. In a recent grant application, this author reviewed a plan to allocate observations in a 1:2 ratio, because the anticipated failure rates were anticipated to be 2% versus 5% so the allocation might favor the group with the higher anticipated rate, which will have the higher variance. If one plans for this rate to compare the groups at $\alpha = 0.05$ (2-sided) and power of $1 - \beta = 0.8$ (β is the type II error), then the following ingredients for **Equation 7** need to be used. The detectable difference is $\Delta = 0.05 - 0.02 = 0.03$ (5% minus 2%). Because the variance of a binomial random variable is $p(1 - p)$, the anticipated variances are $\sigma_1^2 = 0.02\ (1 - 0.02) = 0.0196$ and $\sigma_2^2 = 0.05\ (1 - 0.05) = 0.0475$. Using **Equation 8** (equal costs), $\theta_{opt} = \frac{\sigma_1}{\sigma_1 + \sigma_2} = \frac{\sqrt{0.0196}}{\sqrt{0.0196} + \sqrt{0.0475}} = 0.39$. The required sample sizes are 1133 (actual with a 1:2 allocation to the group anticipated to have a 2% or 5% failure rate respectively); 1169 for an equal allocation; or 1116 for the optimal allocation presuming equal costs. In short, for binomial experiments, equal allocation is close to optimal for equal costs, even in low-probability experiments where the anticipated standard deviations may be markedly different. In

Table 2
Relative Efficiencies (RE) of Equal to Optimal Allocation for Equal Costs per Sampling Unit

σ_1/σ_2	σ_1^2/σ_2^2	θ_{opt}	RE
1	1	0.5	1.00
1.5	2.25	0.6	0.96
2	4	0.667	0.90
3	9	0.75	0.80
4	16	0.80	0.73

For $\sigma_1/\sigma_2 < 1$, results can be obtained by reversing subscripts.

the middle of the probability range, the standard deviations are nearly equal. For example, $\sqrt{0.7(1-0.7)} = 0.46$ while $\sqrt{0.5(1-0.5)} = 0.50$. The optimal allocation would be 0.48 : 0.52 in such an experiment, virtually 50-50.

Example 3

A device company wishes to test a new device to control intraocular pressure in patients with ocular hypertension. They randomize patients to 2 treatments: surgery or medication (a beta-blocker). The standard deviation for change in pressure (6 months minus baseline) is anticipated to be $\sigma = 3$ millimeters mercury (mm Hg). The planning sensitivity is to a difference in means of $\Delta = 1.0$ mm Hg at $\alpha = 0.05$ (2-sided type I error, $z_{\alpha/2} = 1.96$) and power of $1 - \beta = 0.8(z_\beta = 0.842)$. The device company's research costs (payment to its subcontracting ophthalmology service) will be \$6200 for each surgical patient and \$2400 for each beta-blocker patient. From **Equation 8**, the optimal allocation is $\theta_{opt} = 0.3835$. From **Equation 7**, $n = \dfrac{(3^2/0.3835 + 3^2/(1-0.3835))(1.96+0.842)^2}{1^2} = 299$, allocated 115 (38.35%) to surgery and 184 (61.65% to medication). The total cost for the research would be C = 115(\$6200) + 184(\$2400) = \$1,154,600. Had the investigators used an equal allocation, they would have used $n = \dfrac{(3^2/0.5 + 3^2/(1-0.5))(1.96+0.842)^2}{1^2} = 283$ subjects (say 141 to surgery and 142 to medication). As expected, the sample size is smaller, but the cost would be C = 141(\$6200) + 142(\$2400) = \$1,215,000 (\$60,400 higher) for the same power.

3. Randomized Block Designs

These designs are excellent tools to get good mileage from limited research resources. The basic idea is to match the subjects assigned to the treatments as well as possible so that any difference can be attributed to the treatments, and not to the luck of the draw in assigning a disproportionate number of good actors to one of the treatments. Where feasible, randomized block designs get multiple observations from the same subject and the subject serves as its own control, thereby reducing extraneous variability. There are also potential drawbacks. For example, there may be "sympathetic effects" in ocular studies where one eye is treated and the other serves as a control. If the mechanism of action (not always well understood) is at least in part through systemic effects rather than local effects, both eyes may be affected causing regression to the mean. A loss of a real treatment effect, that could have been detected by a completely randomized design, might therefore occur. When using a randomized block design, carefully consider whether this design will answer the same question as a completely randomized design.

The simplest of all randomized block designs is the *matched pair design.* (See **Chapter 7** for the paired t-test.) One example might be to utilize 2 laboratory mice of the same gender from each of n litters, treating one with an experimental treatment and the other with a standard treatment. This will control for genetic variation in the animals.

Let us assume we have n blocks, with each block having 1 subject randomly assigned to each treatment. Let y_{ij} denote the outcome of the subject in block i ($i = 1,2 \ldots, n$) treatment j ($j = 1,2$). If we denote $x_i = y_{i2} - y_{i1}$ as the paired difference, we can treat this as a single-sample problem, for example testing for the mean against a null value of zero using a 2-tailed t-test. Specifically, define the following:

$$\bar{x} = \frac{\sum_{i=1}^{n} x_i}{n} \tag{9}$$

and

$$s^2 = \frac{\sum_{i=1}^{n} (x_i - \bar{x})^2}{n-1}. \tag{10}$$

The t-statistic (with $n - 1$ degrees of freedom) is defined as $t = \bar{x}/(s/\sqrt{n}) = \sqrt{n}\bar{x}/s$, which can be compared against the t-distribution with $n - 1$ degrees of freedom.

We shall present the "large sample" size requirements (validated by the central limit theorem) for a 2-sided type I error α and power $1 - \beta$ (type II error β) and compare this to that required by a completely randomized approach. The required number of pairs is

$$n = \frac{\sigma_x^2 (z_{\alpha/2} + z_\beta)^2}{\Delta^2} \tag{11}$$

where z_γ is the upper 100γ percentile of the standard normal distribution, σ_x^2 is the target population variance of the paired differences x_i, and Δ is the planning difference from zero of the population mean x_i under the alternative hypothesis. In contrast, the sample size required for a completely randomized design with equal sample sizes (**Equation 7**), assuming the effect size would be the same in either design, would be

$$n_{cr} = \frac{2(\sigma_1^2 + \sigma_2^2)(z_{\alpha/2} + z_\beta)^2}{\Delta^2} \tag{12}$$

subjects.

Example 4

An investigator has the choice of conducting a study on 2 femurs of the same rat or as a completely randomized design. She is willing to assume that the anticipated effect size (Δ) would be the same in either design, and $\sigma_{y1} = \sigma_{y2} = \sigma_y$. Based on elementary analysis of linear combinations of random variables, it can be shown that

$$\sigma_x = \sigma_Y \sqrt{2(1-\rho)} \tag{13}$$

where ρ is the correlation between y_{i1} and y_{i2}. A positive correlation would be expected, because an animal whose measure is higher than expected on one treatment tends to be higher than expected on the other. For example, the animal may tend to have a high mineral content in its bones, which would affect both femurs. A conservative planning value for ρ, absent pilot data, might be $\rho = 0.5$. (If in reality it is higher, we will have overestimated our sample-size needs.) This means that knowledge of the outcome of treatment 1 explains ρ^2, or just 25% of the variance of treatment 2. Under this planning scenario ($\rho = 0.5$), we use **Equation 13** to obtain $\sigma_{y1} = \sigma_{y2} = \sigma_x$. If a completely randomized study required 400 animals (400 assays), the randomized block study needs only 100 animals (200 assays) to get the same power. Under the worst-case scenario, $\rho = 0$ (there is no matching effect), the randomized block study would require 200 animals (400 assays). Also note that in the equal standard deviation case, 100ρ represents the percentage reduction in sample size of a randomized block design (matched pairs) over the corresponding completely randomized design (which can be thought of as having $\rho = 0$).

4. Stratified Designs

Operationally, we think of the target population as subdivided into subpopulations (strata), and independent, completely randomized substudies are conducted within each stratum, with the combined results of these substudies answering the primary study questions. For more on estimates from stratified studies, see Chapter 5 of Cochran ***(13)***.

As an example, you want to test the impact of an experimental antibiotic against a standard antibiotic in mice challenged by a strain of bacterium. You worry that there may be a different response according to gender, but that the overall mean effect is of interest. A completely randomized design might result in unequal proportions of females (and males) in the 2 treatment groups. By stratifying the study by gender, the randomization will ensure that the proportion of females (and males) will be the same within each treatment group. We shall demonstrate that this will lead to a more efficient study design than a completely randomized design where gender is ignored.

The following notation will be utilized. There will be 2 treatments labeled $j = 1,2$ and K strata labeled $k = 1,2, \ldots, K$. The probability that a random member of the population falls in stratum k will be denoted as W_k.

For treatment j and stratum k, the mean and variance will be denoted by μ_{jk} and σ^2_{jk}. For the total population, the mean and variance are

$$\mu_j = \sum_{k=1}^{K} W_k \mu_{jk}$$

and

$$\sigma_j^2 = \sum_{k=1}^{K} W_k [\sigma^2_{jk} + (\mu_{jk} - \mu_j)^2]. \tag{14}$$

Within stratum k, denote the sample mean and variance for each treatment by

$$\bar{x}_{jk} = \frac{\sum_{i=1}^{n_{jk}} x_{ijk}}{n_{jk}} \quad \text{and} \quad s^2_{jk} = \frac{\sum_{i=1}^{n_{jk}} (x_{ijk} - \bar{x}_{jk})^2}{n_{jk} - 1}. \tag{15}$$

The stratified estimates of treatment means are

$$\mu_j^* = \sum_{k=1}^{K} W_k \bar{x}_{jk} \quad \text{for} \quad j = 1, 2. \tag{16}$$

We now consider two competing designs.

(A) Assign a total of n patients to each treatment, proportionately allocated to each stratum, $n_{jk} = nW_k$ (for both treatments). By elementary properties of linear combinations of random variables, μ_j^* as defined in **Equation 16** is unbiased for μ_j and has variance

$$\mathrm{Var}(\mu_j^*) = \sum_{k=1}^{K} W_k^2 \frac{\sigma^2_{jk}}{n_{jk}} = \frac{1}{n} \sum_{k=1}^{K} W_k \sigma^2_{jk}. \tag{17}$$

To compare the 2 treatments, we would utilize the approximate standard normality of

$$z = \frac{(\mu_2^* - \mu_1^*) - (\mu_2 - \mu_1)}{\sqrt{\text{SE}_1^2 + \text{SE}_2^2}} \tag{18}$$

where SE_j^2 is obtained by substituting s_{jk}^2 from **Equation 15** for σ_{jk}^2 in **Equation 17** to yield $\text{SE}_j^2 = \sum_{k=1}^{K} W_k s_{jk}^2$.

This can be used to produce a confidence interval for $\mu_2 - \mu_1$ or for testing the null hypothesis that $\mu_2 - \mu_1 = 0$.

(B) The alternate design is the completely randomized design comparing the 2 treatments, based on n subjects per treatment and allowing them to fall into the strata at random. From the section on completely randomized designs and **Equation 14**,

$$\text{Var}(\mu_j^*) = \frac{\sigma_j^2}{n} = \frac{1}{n}\sum_{k=1}^{K} W_k[\sigma_{jk}^2 + (\mu_{jk} - \mu_j)^2]. \tag{19}$$

Note: The stratified study has a smaller standard deviation for each treatment unless, for both treatments, there is no variation in mean response between the strata.

4.1. How to Plan the Sample Size of a Stratified Study

Method 1: This is the most common approach. Because we rarely know enough planning detail within strata, plan the sample size as if it was an unstratified study, knowing the stratified study will have greater power than expected from the unstratified study. It is simply a matter of being unable to quantify how much greater.

Method 2: Presumes planning values for all stratum means and standard deviations are known. First calculate the sample size as if it was an unstratified study. Say this requires n_u randomized to each treatment. Let us calculate the "fudge factor" as the ratio of the variances in **Equation 17** and **Equation 19**:

$$\text{FF} = \frac{\sum_{j=1}^{2}\sum_{k=1}^{K} W_k \sigma_{jk}^2}{\sum_{j=1}^{2}\sum_{k=1}^{K} W_k[\sigma_{jk}^2 + (\mu_{jk} - \mu_j)^2]}. \tag{20}$$

Allocate $n_s = n_u \times FF$ subjects to each treatment. (For example, if the unstratified study required 100 subjects per treatment, and FF = 0.8, allocate 80 per treatment.) The individual stratum sample sizes would be $W_j n_s$ per treatment in stratum j and the W_j are the natural stratum weights (proportion of subjects in the target population falling in stratum j).

It is the experience of this author that stratification will help power, but reasonable choices as to the large number of extra planning ingredients are almost never available at the onset of the study to use **method 2**. If one uses **method 1**, one recognizes that stratification is not saving in terms of sample size over a completely randomized design, but one is comforted by knowing that the power is higher (by an unknown amount) than expected from a non-stratified study.

4.2. Poststratification in a Completely Randomized Design

As noted by Peto and others ***(14)***, nearly all of the power benefits of stratified studies can be obtained by analyzing a nonstratified study as if it were a stratified study. The key is that the role of W_k is played by $W_k^* = (n_{1k} + n_{2k})/n$, where n_{jk} is the sample size in group j and stratum k, and n is the total sample size allocated to each treatment. Like a truly prospectively assigned stratified study, this forces the contribution from each stratum to be the same for both treatments and removes the nonstratified study's tendency to assign differing effective weights in a given stratum to the 2 treatments.

Caution on Poststratification: Some investigators will conduct the study as a nonstratified study but then subject the study to a large number of exploratory analyses to detect imbalance between the arms in various substrata. A corrected analysis may or may not be done according to the results of these analyses. If you ignore how the decision to do a poststratified analysis was made, you may alter the operating characteristics. Poststratification is valid if it is prospectively planned as the primary analysis. However, the major conclusions should be based on the single primary analysis. Poststratified analyses are welcome supplements that may help to explain any difference or lack thereof.

Comment: We have treated the stratum weights as natural (i.e., in the proportion that exists in the target population). However, there is no reason that this is a requirement. For example, one might wish to oversample females in a predominately male disease. Alternately, you may wish to oversample in a stratum that has an excess of anticipated variability than expected in others. If you wish to make inference about the natural population, then indeed, stratified studies can accomplish this, but the relationship between the W_k and n_{jk} have changed over what is presented here.

Example 5

This somewhat simplified example compares sample-size requirements for a completely randomized design, a stratified design, and randomized block design. Let us consider a population with 2 strata: males and females (50% falling into each). Subjects will be randomized to 2 antibiotic treatments: **A** versus **B**. The planning means and standard deviations are presented in **Table 3**. We will assume 80% power and a 2-sided test at the 5% level.

For a completely randomized study using **Equation 7** we have

$$n_{cr} = \frac{4 \times 25^2 (1.96 + 0.842)^2}{15^2} = 88$$

with 44 allocated to each treatment. For the stratified study using **Equation 20** we have

$$n_{\text{strat}} = n_{cr} \left(\frac{20}{25} \right)^2 = 56$$

with 14 allocated to each of the 4 treatment by stratum combinations. For the randomized block study, matching only on gender, the correlation is calculated as follows. Let y_{iA} and y_{iB} denote the responses for members of matched-pair i on treatments **A** and **B** and let $x_i = y_{iB} - y_{iA}$ denote the difference. The variance of the difference is

$$\sigma_x^2 = \sigma_{yA}^2 + \sigma_{yB}^2 - 2\rho\sigma_{yA}\sigma_{yB} = 25^2 + 25^2 - 2 \times 0.36 \times 25 \times 25 = 800$$

and the expected difference is $\Delta = 30 - 45 = -15$. From **Equation 11** for a 2-sided P value of 0.05 and 80% power,

$$n_{\text{randomized block}} = \frac{800\,(1.96 + 0.842)^2}{15^2} = 28 \text{ pairs}$$

or 56 total subjects.

Both matching and stratification give you the same advantage (36% reduction) over a completely randomized design, when only matching on gender. If further important matching variables can be introduced, the benefit of matching will be greater. However, the more matching criteria one has, the

Table 3
Planning Means and Standard Deviations for Example 5

Stratum\treatment	A	B
Female (50%)	Mean = 30 (SD = 20)	Mean = 15 (SD = 20)
Male (50%)	Mean = 60 (SD = 20)	Mean = 45 (SD = 20)
Total	Mean = 45 (SD = 25)	Mean = 30 (SD = 25)

We also assume that the within-subject correlation is 0.36 between measurements on treatments **A** and **B**.

greater the chance that a subject will not be able to be conveniently matched to another.

5. Crossover Designs

Consider a 2-treatment design (**A** vs. **B**) where both treatments are given to each subject, but the order is determined by a completely randomized design (50% allocated to each ordering). Generally, it is advisable to have a washout period between the treatment periods to minimize any *carryover effects* of the first period upon the second period. Although there are multitreatment analogs, we shall restrict the discussion to a randomized study of 2 treatments. This design can be legitimately considered as a randomized block design yielding 1 sample (matched pair) analysis, as given in **Section 3** or **Chapter 7**. This approach ignores potentially useful data on the treatment ordering. We shall advocate a 2-sample approach where the dependent variable will be the period 2 value less the period 1 value irrespective of the treatment ordering, but the orderings will be compared via a completely randomized design approach as given in **Section 2**. The estimated treatment effect from this analysis will represent the average of the treatment difference (**B** minus **A**) when **A** is given first and when **B** is given first.

To illustrate the advantage of overcoming carryover effects, we can model the outcome as follows. Let Y_{ij} represent the period 2 minus the period 1 difference for patient i and order j (j = 1 for **A** then **B** and j = 2 for **B** then **A**). The model is parameterized as $Y_{i1} = -\mu + \tau + \varepsilon_{i1}$ for $j = 1$ and $Y_{i2} = \mu + \tau + \varepsilon_{i2}$ for $j = 2$. We assume that the ε_{ij} are independent, have mean zero, have a common standard deviation σ (a reasonable assumption for a crossover study), and are normally distributed.

Two-sample approach: The 2-sample t-test comparing the Y_{i1} to the Y_{i2} is the least squares solution for testing the null hypothesis that $\mu = 0$. The 2-sample t-test is also an approximation to the permutational t-test (nonparametric). See Shuster *(1)* and **Section 2** for further details. In summary, μ is the main effect of treatment, and τ is the carryover effect. These may be estimated using simple linear regression with Y_{ij} as the response and with an independent predictor variable X that is -1 when $j = 1$ and 1 when $j = 2$. The intercept estimates the carryover effect τ and the slope term estimates the treatment effect μ.

One-sample approach: The most common method of analysis is the 1-sample approach, which bases the inference on $\frac{\sum_i Y_{i2} - \sum_i Y_{i1}}{n}$, where n is the combined sample size. This can be obtained from linear model considerations for the same model cited above. Let $X_i = -Y_{i1}$ (**A** then **B**) and $X_i = Y_{i2}$ (**B** then **A**). Simply perform a 1-sample t-test and compare the mean of the X_i to zero.

Substitution in the linear model for the X_i above yields $X_i = \mu + Z_i + \varepsilon_i$, where Z_i are independent binary random variables taking on values $+\tau$ if the order is **BA** and $-\tau$ if the order is **AB**. The +/– each occur with probability 0.5 (assuming 50-50 unconditional randomization to each order **AB** or **BA**) and are independent of the ε_i. The ε_i are $\pm\varepsilon_{ij}$ from the 2-sample model and have mean 0 and the same common standard deviation σ as the ε_{ij}. Unconditionally, $E(Z_i) = 0$, so the sample mean of the X_i is unconditionally unbiased for μ. However, conditional on the actual sample sizes, the sample mean is biased unless the sample sizes happen to be equal or if $\tau = 0$. In reality, a different number may be randomized to each ordering, and when that happens, there will be a different number of $+\tau$'s and $-\tau$'s in the calculation of the sample mean of the X_i's.

Comparison of the 1- and 2-sample approaches: (1) When there is no carryover effect, the 2 methods are essentially the same. You lose 1 degree of freedom for error in the 2-sample analysis. (2) If τ is not zero and the sample sizes are equal, both methods still provide unbiased estimates of effect size, but the 1-sample method overestimates the variance on average. The 1-sample estimate of variance has expected value of $\sigma^2 + \tau^2$, whereas the 2-sample estimate of variance has expected value σ^2. The point estimates match when the sample sizes are equal. (3) Perhaps the worst scenario occurs when the sample sizes are unequal. Conditional on the actual sample sizes to each ordering, the 1-sample estimate is biased, unless we are lucky enough to have $\tau = 0$.

The 2-sample method prevails. Item (2) above demonstrates superior power of the 2-sample method under a non-zero carryover, for equal sample sizes. The relative efficiency of the 2-sample version under equal sample sizes is approximately $(\sigma^2 + \tau^2)/\sigma^2$. Item (3) tells us that whenever the sample sizes assigned to the 2 orderings differ, the point estimate is biased for the 1-sample approach, but not for the 2-sample approach. This further will erode the power over and above the impact on the variance. Finally, the 1-sample dependent variable has a discrete component ($\pm\tau$) and this would violate normality. For large samples, that is not a major factor. But the 1-sample method discards valuable information and as a consequence suffers with inferior performance. Sample size derivations follow the randomized block designs (1-sample approach) or completely randomized design (2-sample approach). For more on crossover designs, see Jones and Kenward ***(15)*** and Senn ***(16)***.

6. Two-by-Two Factorial Designs

A 2×2 factorial design investigates 2 questions simultaneously in the same study. One classical example is the U.S. Physician Study ***(7)***. Once the effort to study 1 intervention in healthy adults was undertaken, it seemed reasonable to introduce a second intervention and get answers to 2 research questions for

Table 4
Treatment Assignments (and Target Population Means) for a 2 × 2 Factorial Design

Treatments	B1	B2
A1	**A1B1** (μ_{11})	**A1B2** (μ_{12})
A2	**A2B1** (μ_{21})	**A2B2** (μ_{22})

little more than the cost of each one individually. Physicians were equally randomized to 4 groups (double placebo, aspirin plus placebo carotene, placebo aspirin plus carotene, or aspirin plus carotene). The aspirin question dealt with preventing heart disease and improving total mortality, while the carotene question dealt with reducing cancer incidence. For the purpose of analysis, each treatment was compared as if the study was stratified for the companion question. The parameter to be estimated for the aspirin effect size was the average of (a) the aspirin effect size within placebo carotene and (b) the aspirin effect size within carotene. Two-by-two factorial studies allow for the examination of an interaction between treatments. This is not possible with separate randomized studies.

There are four treatments groups that are shown in **Table 4**. This type of study can be performed using several types of designs including completely randomized designs, stratified designs, or randomized block designs. At the simplest level, when comparing **A1** versus **A2** (the main effect of **A**), one can consider **B** a stratification variable. If 25% of subjects are randomly assigned to each cell in **Table 4**, the comparison asks the question if the mean outcome of treatment **A1** (averaged over the levels of **B**) differs from the corresponding mean outcome of treatment **A2**. This is a valid question, whether or not there is an interaction between **A** and **B**. That is, the mean difference between treatments **A1** and **A2** differ according to whether **B1** or **B2** was assigned.

The major experimental questions, expressed as null hypotheses, that can be asked in a 2 × 2 factorial study are

1. Is there a difference between treatments **A1** and **A2**, stratified for **B** (main effect of **A**)? The null hypothesis states $\frac{1}{2}(\mu_{11}+\mu_{12})-\frac{1}{2}(\mu_{21}+\mu_{22})=0$.
2. Is there a difference between treatments **B1** and **B2**, stratified for **A** (main effect of **B**)? The null hypothesis states $\frac{1}{2}(\mu_{11}+\mu_{21})-\frac{1}{2}(\mu_{12}+\mu_{22})=0$.
3. Is there an interaction between **A** and **B**? The null hypothesis states $(\mu_{11} - \mu_{12}) - (\mu_{21} - \mu_{22}) = 0$.

4. Does **A2B2** differ from **A1B1**? This may be a safety concern where **A1B1** are both placebo and **A2B2** are both active treatments. The null hypothesis states $\mu_{11} - \mu_{22} = 0$.
5. Does **A2B2** differ from **A1B2** and **A2B1**? This is a multiple comparison of 2 treatments against a control, per Dunnett *(17)*. This may be an additivity question where **A1** and **B1** are placebos and **A2** and **B2** are active treatments. Is the combination of both "active" treatments different from only one active treatment? The null hypothesis states $\mu_{12} = \mu_{21} = \mu_{22}$.
6. Does **A1B1** differ from **A1B2** and **A2B1**? This is also a multiple comparison of the Dunnett type *(17)*. This may be an individual efficacy question where **A1** and **B1** are placebos and **A2** and **B2** are active treatments. Is either single agent different from the double placebo? The null hypothesis states $\mu_{11} = \mu_{12} = \mu_{21}$.

As an example, the ISIS II trial *(8)* randomized patients with a myocardial infarction to placebo aspirin versus aspirin (**A1** vs. **A2**) and placebo streptokinase versus streptokinase (**B1** vs. **B2**). Of course, questions 1 and 2 are of primary interest. While the study is ongoing, concern over the double placebo group is high, and an effective way to monitor this is to compare this group to the double drug group (question 4). Further, if efficacy is established in questions 1 and 2, then question 5 becomes very important; namely, is there a difference between the group with both active drugs than each monotherapy (1 active drug and 1 placebo)? The quantitative interaction question (question 3) asks whether the effect of aspirin is different depending upon whether streptokinase or placebo streptokinase was used.

In general, the methods for obtaining sample sizes cited above can be utilized for 2 × 2 factorial studies for all of the hypotheses except question 3, the interaction hypothesis. Although this chapter takes a frequentist approach, an interesting Bayesian approach can be found in Simon and Freedman *(18)*.

7. Randomized Designs with Random Effects

The design situation covered in this section is a 2-treatment study where high-cost treatment units can each produce unlimited experimental subunits at a relatively low cost. For a fixed total cost, there is a trade-off between obtaining smaller numbers of the larger units and more subunits per large unit compared with getting more of the larger units, each with fewer of the smaller units. Consider the following example. Suppose you wish to compare 2 surgical treatments with respect to a quantitative outcome (e.g., 6-minute walking distance). You will recruit surgeons from all over the country to participate, pay for their travel to come to a central location for training, and pay them a capitation fee for providing data on each patient they treat. It is recognized that the true effect size will vary from physician to physician. Each has a target population (or long-term) difference between treatments. The overall goal of the study is to assess whether the average effect size in the population of surgeons (of which the participants are, at least in concept, a random sample), is greater or less than zero. The analysis will be very similar to a *meta-analysis* [see Hedges and Olkin ***(19)***, Lipsey and Wilson ***(20)***, and Arthur and others ***(21)***], but the design has control over how many physicians to recruit and how many patients to randomize for each physician. This is an application of a mixed model as described in **Chapter 11** of this book. See Khuri and others ***(22)*** or Chapter 9 of Cochran ***(13)*** for a more detailed look at this topic.

Consider the surgery example described above. The total cost of the study is

$$C = C_1N + 2C_2NK \tag{21}$$

where N is the number of clusters (surgeons), $2K$ is the number of patients per cluster (K planned to be assigned to each treatment), C_1 is the cost for recruiting and training 1 surgeon, and C_2 is the marginal cost of recruiting and treating a patient. The response model will be defined as

$$X_{ijk} = \tau_{ij} + \varepsilon_{ijk} \tag{22}$$

where τ_{ij} is the target population mean outcome for surgeon i and treatment j ($j = 1,2$), ε_{ijk} is the random error (assumed independent of the τ_{ij}) with common variance σ_ε^2, and $k = 1,2, \ldots, K$ represents the patient number within the physician and treatment. The term σ_ε^2 is known as the *within-surgeon variance*. The target population effect size for physician i is represented by

$$\gamma_i = \tau_{i2} - \tau_{i1} \tag{23}$$

and is estimated by the difference in treatment means for the individual physician as

$$\gamma_i^* = \frac{1}{K}\sum_{k=1}^{K}(X_{i2k} - X_{i1k}). \tag{24}$$

The overall effect size $E(\gamma_i)$ is estimated by the sample mean of the γ_i^* as

$$\gamma^* = \frac{1}{NK}\sum_{i=1}^{N}\sum_{k=1}^{K}(X_{i2k} - X_{i1k}). \tag{25}$$

We will let σ_γ^2 denote the variance of the γ_i, which is known as the *between-surgeon variance in effect size*. The variance of the effect size γ^* is

$$Var(\gamma^*) = \frac{\sigma_\gamma^2}{N} + \frac{2\sigma_\varepsilon^2}{NK}. \tag{26}$$

Under the cost function in **Equation 21**, the values that minimize the variance of the effect size for a given total cost are

$$K = \frac{\sigma_\varepsilon}{\sigma_\gamma}\sqrt{\frac{C_1}{C_2}} \tag{27}$$

and

$$N = \frac{C}{C_1 + 2C_2K}. \tag{28}$$

Note that all other things being equal, if within-surgeon to between-surgeon variability is high, K, the number of patients per treatment per surgeon, tends

Table 5
Designs Keeping the Total Cost Under $1,000,000

N	K	Var (γ*)	Cost (thousands of dollars)
20	30	29.28	1000
21	27	27.94	987
22	25	26.71	990
23	23	25.59	989
24	21	24.57	984
25	20	23.62	1000
26	18	22.77	988
27	17	21.96	999
28	15	21.26	980
29	14	20.57	986
30	13	19.94	990
31	12	19.35	992
32	11	18.82	992
33	10	18.33	990
34	9	17.88	986
35	8	17.49	980
37	7	16.68	999
38	6	16.42	988
40	5	15.84	1000
41	4	15.80	984
43	**3**	**15.63**	**989**
45	2	16.00	990
47	1	18.38	987

N is the number of surgeons, $2K$ is the number treated per surgeon (K on each regimen), and the cost is $20,000 per surgeon plus $500 per patient. Var ($\gamma$*) is the variance of estimate of treatment effect. The optimal design is highlighted in **boldface**.

to be larger than it would be under a lower ratio. If the relative fixed cost to train a surgeon (C_1) is low relative to the fixed cost per patient (C_2), then K tends to be low. Of course, sample-size numbers have to be rounded to whole numbers.

Example 6

A device maker wants to compare its experimental device against a standard device. Based on a pilot study, the following planning parameters were used: C = \$1,000,000 (the total cost allocated), C_1 = \$20,000 (the cost to provide travel and hands-on training to a surgeon and data manager), C_2 = \$500 (capitation cost for data management per patient), $\sigma_\varepsilon = 12$ (within surgeon standard deviation), and $\sigma_\gamma = 24$ (between-surgeon standard deviation of effect size). Using **Equation 27**, $K = 3.16$, and using **Equation 21**, $N = 43.2$. With $K = 3$ and $N = 43$, $\text{Var}(\gamma^*) = 15.63$. Other design choices that keep the costs just below or at C = \$1,000,000 are given in **Table 5**, along with variance and actual costs. The ratio of variances represent the relative efficiency for a fixed cost. The curve is rather flat, in that it takes a substantial variation from the optimal choice to make a major impact on the variance. For example, if one failed to do these calculations, either analytically or by trial and error as in **Table 5**, and one arbitrarily decided to enroll 20 patients per surgeon, $K = 10$ allocated to each treatment, one would need 40 surgeons (not 33) to match the precision of the optimal design of 43 surgeons with 6 patients per surgeon ($K = 3$ allocated to each treatment). The cost would escalate by 20% to \$1.2 million.

8. Conclusion

In this chapter, we examined several design strategies that can be deployed in randomized experiments. Randomization is the gold standard to obtain fair comparisons of competing treatments, because any other form of treatment assignment may be associated with selection bias. As noted in this chapter, there are many options as to how to design randomized experiments, from the simple, completely randomized design to the fairly complex stratified or block designs. Simpler designs have the benefit of being easier to explain to the public and are generally easier to conduct than more complex designs. On the other hand, when feasible, more complex designs often give more efficient estimates of treatment effects than the simpler ones, either in costs for a given precision or in precision for a given cost.

Acknowledgments

This work was partially supported by General Clinical Research Center grant M01 RR00082 from the National Center for Research Resources.

References

1. Shuster, J. J. (2005) Diagnostics for assumptions in moderate to large simple clinical trials: Do they really help? *Stat. Med.* **24**, 2431–2438.
2. Hinkelmann, K., and Kempthorne, O. (1994) *Design and Analysis of Experiments, Volume I. Introduction to Experimental Design.* Somerset, Wiley-Interscience.
3. Cochran, W. G., and Cox, G. M. (1992) *Experimental Designs*, 2nd ed. New York, Chichester, John Wiley & Sons.
4. Frigon, N. L., and Mathews, D. (1997) *Practical Guide to Experimental Design.* New York, Chichester, John Wiley & Sons.
5. Heath, D. (1998) *An Introduction to Experimental Design and Statistics for Biology.* London, University College London Press.
6. Weber, D., and Skillings, J. H. (2000) *A First Course in the Design of Experiments: A Linear Models Approach.* Boca Raton, CRC Press.
7. Stampfer, M. J., Buring, J. E., Willett, W., Rosner, B., Eberlein, K., and Hennekens, C. H. (1985) The 2×2 factorial design: Its application to a randomized trial of aspirin and carotene in U.S. physicians. *Stat. Med.* **4**, 111–116.
8. ISIS #2 Collaborative Group. (1988) Randomised trial of intravenous streptokinase, oral aspirin, both, or neither among 17,187 cases of acute myocardial infarction: ISIS 2. *Lancet* **2**, 349–360.
9. Mehta, C., Pampallona, S., Kale, M., Ghanekar, A., Kulthe, S., and Sathe, A. (2004) EaST 3 (Early Stopping in Clinical Trials). Cambridge, Cytel Software.
10. Lehmann, E. L. (1975) *Nonparametrics: Statistical Methods Based on Ranks.* San Francisco, Holden-Day, Ch. 1–2.
11. Mehta, C., and Patel, N. (2003) StatXact 6. Cambridge, Mass.: Cytel Software.
12. Shuster, J. J. (1992) *Practical Handbook of Sample Size Guidelines for Clinical Trials.* Boca Raton, CRC Press, pp. 197–198.
13. Cochran, W. G. (1977) *Sampling Techniques.* New York, Chichester, John Wiley & Sons.
14. Peto, R., Pike, M. C., Armitage, P., Breslow, N. E., Cox, D. R., Howard, S. V., Mantel, N., McPherson, K., Peto, J., and Smith, P. G. (1979) Design and analysis of randomized clinical trials requiring prolonged observation of each patient. II. analysis and examples. *Br. J. Cancer* **35**, 1–39.
15. Jones, B., and Kenward, M. G. (1989) *Design and Analysis of Cross-Over Trials.* London, New York, Chapman & Hall.
16. Senn, S. (1993) *Cross-Over Trials in Clinical Research.* New York, Chichester, John Wiley & Sons.
17. Dunnett, C. W. (1955) A multiple comparison procedure for comparing several treatments with a control, *J. Am. Stat. Assoc.* **50**, 1096–1121.
18. Simon, R., and Freedman, L. S. (1997) Bayesian design and analysis of 2×2 factorial clinical trials, *Biometrics* **53**, 456–464.
19. Hedges, L. V., and Olkin, I. (1985) *Statistical Methods for Meta-analysis.* New York, London, Academic Press.
20. Lipsey, M. W., and Wilson, D. B. (2000) *Practical Meta-analysis.* Newbury Park, Calif., London, Sage Publications.

21. Arthur, W., Bennett, W., and Huffcutt, A. I. (2001) *Conducting Meta-analysis Using SAS*. Hillsdale, Lawrence Erlbaum Associates.
22. Khuri, A. I., Mathew, T., and Sinha, B. K. (1998) *Statistical Tests in Mixed Linear Models*. New York, Chichester, John Wiley & Sons.

13

Analysis of Change

James J. Grady

Summary

When the same subjects or laboratory animals are observed across a set of different conditions or over time, we are usually interested in studying change. In these study designs, each subject serves as its own control. In this chapter, we consider different ways to assess change over time, for example, analyses for evaluating changes from a baseline condition. Study designs and analyses for single group studies and studies with two groups are discussed in detail. Examples come from published data. Statistical methods used in the examples include paired *t*-tests and analysis of covariance. The use of difference scores is discussed relative to analysis of covariance.

Key Words: ANCOVA; baseline values; change scores; difference scores; paired t-test; pre- and post scores; rank sum test.

1. Introduction

When the same group of experimental units (i.e., human subjects, laboratory animals, etc.) is observed across a set of different conditions or over time, we are usually interested in studying change. In these study designs, each subject serves as its own control. The different measures can be before and after an imposed experimental intervention, or they can be purely observational, such as measures made during a baseline period and at later time points. The simplest experimental study design of this type is the pretest versus posttest situation, where measurements are made before and after a single intervention. A common characteristic of this type of design is that every subject experiences the same condition or intervention: the study has one group and one intervention. These designs can get more complicated by adding more conditions (e.g., several doses of a drug) or by adding different groups of subjects (e.g., different age

From: *Methods in Molecular Biology, vol. 404: Topics in Biostatistics*
Edited by: W. T. Ambrosius © Humana Press Inc., Totowa, NJ

Table 1
Average Urinary 2-Hydroxyestrone Excretion Levels for a One-Group Study, with Two Diets

Subject	With (+) isoflavones*	Without (−) isoflavones*
1	34.2	18.3
2	24.0	18.1
3	20.0	19.2
4	13.1	2.5
5	13.0	12.8
6	12.9	9.0
7	12.8	8.3
8	9.4	8.1

*Expressed as amount excreted (nmol) in 12 hours.

groups). Extensions of these designs lead to repeated measures designs, which are described in **Chapter 11**.

2. The One-Group Study, Pretest and Posttest Design

The data in **Table 1** are from Lu and others *(1)*, who studied the effects of 2 different soy diets (approximately 1 month of soy ingestion with isoflavones, followed by 1 month of soy diet without isoflavones) on female hormone levels among 8 premenopausal women. The statistical approach to analyze this data is the same as for a pre- and posttest design, or a study with baseline measure and a single follow-up. The data in **Table 1** are average urinary excretion levels for 2-hydroxyestrone. The data collected from an experiment involving change is typically entered in a spreadsheet in this fashion. Each subject has a row of data with a subject identifier followed by 2 or more columns for data collected across time or under different conditions. Notice there is no variable for group membership, because all subjects receive both diets and undergo the same experimental conditions.

2.1. Graphical Displays and Other Data Summaries

As in most data analyses, a graphical display of change data can relate a compelling story about the data. **Figure 1** shows perhaps the best way to display change data across 2 conditions, or time points. These simple line plots are easy to make in Excel and other software. One can quickly see that 6 out of 8 subjects had lower values during the diet without isoflavones, and 2 subjects had little or no change. These types of line plots, however, can get visually unattractive for large data sets.

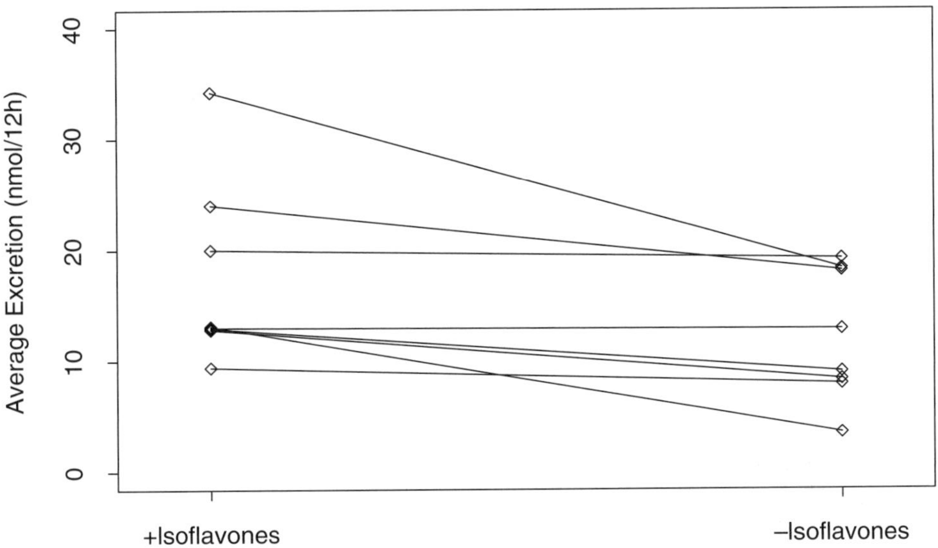

Fig. 1. Excretion levels for each subject.

Another way to summarize the data would be to report the average change. For **Table 1**, the average change in excretion for the 8 women was –5.3 with a standard deviation of 5.29. The interpretation of this biological effect is that excretion levels of these women were 5.3 units lower on average on the diet without isoflavones. The null value is zero, which represents the expected mean for "no change."

2.2. Assessing Statistical Significance

For the urinary excretion data in **Table 1**, there are two basic approaches for assessing statistical significance. The first would be to use a parametric approach, which would involve the paired *t*-test to test the null hypothesis that the mean change was equal to zero. The second would be to use nonparametric tests, the Wilcoxon signed rank test (also called the Mann-Whitney *U* test), or the sign test. These tests are discussed in **Chapter 7**. The results are given in **Table 2**.

Table 2
Results of Data from Table 1

Test	*P* value	One might report as
Paired *t*-test	0.0261	0.026 or 0.03
Wilcoxon signed rank test	0.0078	0.008 or $P < 0.01$
Sign test	0.0078	0.008 or $P < 0.01$

Notes

1. For a small data set, it is sometimes difficult to defend the assumptions needed for applying a parametric test. For a paired t-test, the main assumption is that the difference scores are normally distributed. This makes some analysts favor non-parametric tests for small data sets, although this generally results in a loss of power. On the other hand, the t-test has been shown to be robust, in that it still gives valid inference even when the assumptions are not met. It is worth noting that some journal reviewers will not accept parametric tests for small data sets.

3. A More Complicated Design: One-Group Study with Baseline and Two Follow-Up Times

The next type of design involves a study with a baseline measure and 1 or more subsequent measures. The data in **Table 3** are measures of mean arterial pressure from a study of fluid therapy in sheep before the study *(2)*, immediately after (0 min) and 20 min after a bolus of dopamine. Assume that the main comparisons of interest are the extent to which measures at 0 and 20 min vary from the prestudy period.

Analysis of this small data set might proceed as follows.

1. Using a nonparametric approach, test whether there is a time effect across the 3 time points. This can be accomplished with Friedman's test, which is a nonparametric test that compares a set of related measures. For this data set, each subject has a set of 3 correlated measures. The hypothesis is that the 3 measures at prestudy, 0 and 20 min have identical effects. The results of Friedman's test gives $P = 0.003$, indicating that there is a difference among the three measures across the 3 time periods.
2. Having established that there are differences in the measures across time, use a basic parametric or nonparametric test for paired data to compare 0 and 20 min to the prestudy measure. Parametric *t*-test: Compute 2 paired *t*-tests comparing each time point with prestudy. **Table 4** indicates that statistical significance might depend on whether one makes an alpha correction to account for multiple tests.

Table 3
Mean Arterial Pressure (mm/Hg) Prestudy, 0 and 20 Minutes

Sheep	Prestudy	0 min	20 min
1	97	152	148
2	86	96	119
3	86	98	94
4	79	97	124
5	94	106	123
6	91	99	95

Table 4
Results from Paired *t*-Tests and Wilcoxon Signed Rank Tests

	P value	Adjusted *P* value[a]
Paired *t*-tests		
(0 min–prestudy)	0.047	$0.047 > 0.025$, NS
(20 min–prestudy)	0.015	$0.015 < 0.025$
Wilcoxon signed rank tests		
(0 min–prestudy)	0.03	$0.03 > 0.025$, NS
(20 min–prestudy)	0.03	$0.03 > 0.025$, NS

[a]If an alpha correction of $\alpha/2$ (i.e., $0.05/2 = 0.025$) was applied to account for the 2 pairwise tests being conducted, the only significant result to report would be at 20 min via the paired *t*-test. Discrepancies between statistical significance tests happen more often with small data sets and require thoughtful reporting.

Nonparametric Wilcoxon signed rank test: Using a nonparametric approach and calculating the Wilcoxon signed rank tests for paired data, comparing each time to prestudy. The results are also shown in **Table 4**.

A more sophisticated approach to these data: It should be noted that if the data set was larger, an alternative parametric approach would be to conduct a repeated measures analysis of variance (ANOVA) to test for an overall effect over time, followed by pairwise comparisons of each time point to the prestudy period.

4. Repeated Measures Designs

The data in **Table 5** are from the same animal laboratory experiment described above in **Section 1.3** and represent an extension of the data matrix described in **Table 3**. Analysis of this data is more complicated and falls under the topic of repeated measures analysis of variance. Methods of analyses for these data are described in **Chapter 11**.

Table 5
Mean Arterial Pressure (mm/Hg)

Sheep	Baseline	0 min	20 min	60 min	120 min	180 min
1	97	152	148	119	118	98
2	86	96	119	111	132	177
3	86	98	94	84	118	125
4	79	97	124	99	130	129
5	94	106	123	125	129	132
6	91	99	95	101	105	108

Table 6
Average Excretion Levels, Two Age Groups, Two Diets

Subject	Age group (years)	With (+) isoflavones*	Without (–) isoflavones*	Difference
1	19–25	34.2	18.3	15.9
2	19–25	24.0	18.1	5.9
3	19–25	20.0	19.2	0.8
4	19–25	13.1	10.5	2.6
5	19–25	13.0	12.8	0.2
6	19–25	12.9	9.0	3.9
7	19–25	12.8	8.3	4.5
8	26+	33.5	32.1	1.4
9	26+	29.0	29.0	0
10	26+	27.0	25.4	1.6
11	26+	16.5	17.5	−1.0
12	26+	13.2	12.2	1.0
13	26+	11.7	12.1	–0.4
14	26+	12.0	13.5	−1.5

*Expressed as amount excreted (nmol) in 12 hours.

5. Comparison of Change Among Subgroups in a One-Group Study

Most studies, even with only one overall group, will proceed to compare change among subgroups. We will examine approaches to analyzing data among subgroups from a 1-group study with a pretest versus posttest design. Consider the data set in **Table 6**. The study design is the same as in **Section 2** with average excretion measurements on 2 isoflavone diets: a standard diet with isoflavones, followed by a diet without isoflavones. The grouping or stratification variable of interest is age, which is categorized as 19 to 25 versus 26+ years. We have added a column to display the difference scores of the 2 diets.

The data are displayed in line graphs in **Figures 2 and 3**. By comparing the slopes of the lines between the 2 groups, we notice that the slopes in the 19- to 25-year-old group are generally steeper than the slopes in the 26 and older group, some of which are flat. This suggests a greater effect in the 19- to 25-year-old group.

A study with this design often asks three questions:

1. Did the 19- to 25-year-old group have a significant change between diets?
2. Did the 26+ year-old group have a significant change between diets?
3. Were the changes in the 2 groups the same or different?

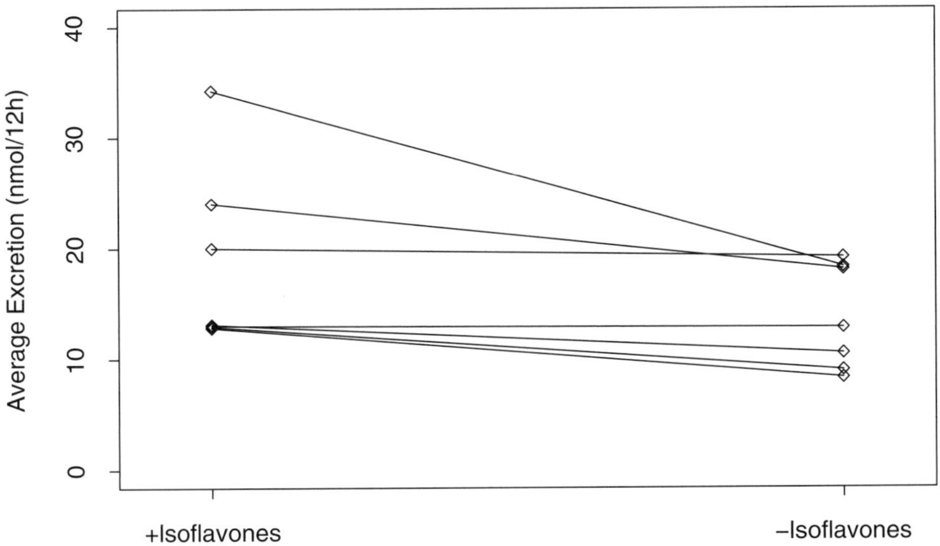

Fig. 2. Excretion levels for each subject, age 19 to 25 years.

We will describe 2 general statistical approaches for analyzing this data. The first approach involves creating difference scores and using t-tests (or nonparametric alternatives) to compare group mean changes. The second utilizes a statistical model called analysis of covariance to evaluate group differences. Each method will be demonstrated.

5.1. Analysis Using Tests for Paired Data

Investigators will often test for a diet effect in the 2 groups separately. The difference scores in each group can be analyzed with tests for paired data as shown in **Table 7**. Although the excretion levels across the diets are statistically significant for the 19- to 25-year-old group and not for the 26+ year-old group, this does not formally test whether the average differences in excretion are significantly different between the 2 groups. To accomplish that comparison,

Table 7
Within Group Tests

Age (years)	Mean difference	Paired *t*-test *P* value	Wilcoxon signed rank test *P* value
19 to 25	4.8	0.05	0.02
26+	0.2	0.74	0.75

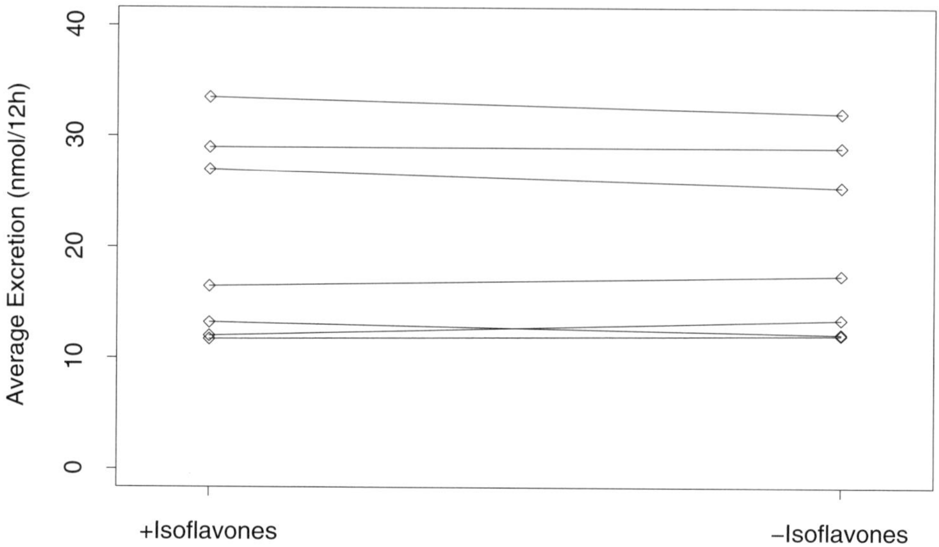

Fig. 3. Excretion levels for each subject, age 26+ years.

we apply a 2-sample *t*-test to the difference scores. This gives the results shown in **Table 8**. This demonstrates in a formal test that the average change in excretion level was larger in the younger age group compared with the older age group ($P = 0.04$ or $P = 0.02$).

Note of caution for one group studies: One of the assumptions for the analysis of change scores is that the 2 groups being compared need to have similar distributions for pretest values, here the excretion values in the isoflavone-free diet. In clinical trials where subjects are randomized to treatment groups, pretest values are usually evenly balanced across groups through the randomization process. In studies such as this one, however, where grouping factors are created after the study design and sample are established, one must be concerned with the balance of pretest values across any created groups. In this example, if age was related to baseline excretion levels, it could invalidate the results. For example, if women aged 19 to 25 years had lower excretion rates

Table 8
Between-Group Tests

	Mean Age 19 to 25 years	Age 26+ years	2-sample *t*-test *P* value	Wilcoxon rank sum test *P* value
Mean difference	4.8	0.2	0.04	0.02

Table 9
Excretion (nmol/12 h)During the Diet with Isoflavones

	Mean	SD
Age 19 to 25 years	18.6	8.19
Age 26+ years	20.4	9.15

compared with women aged 26+ at baseline, comparison of change scores could lead to spurious results. This dilemma has its basis in the phenomenon known as regression to the mean and has been discussed in length by many authors *(3,4)*. Subjects with pretest scores greater than the mean will tend to have smaller difference scores, and subjects with pretest scores lower than the mean will tend to have larger differences scores. As a result, the assessment of change in a study with 1 group, by analyzing change in the whole group or in subgroups, must proceed with caution.

One quick way to check this assumption is to compare the means and standard deviations of the pretest, or baseline values for the 2 subgroups. Assessment of this assumption can be difficult for small data sets. If the distributions appear similar, the analysis of change scores using a 2-sample *t*-test will most likely give valid results. The mean and standard deviations for the 2 age groups are given in **Table 9** for excretion during the diet with isoflavones. It appears that age group is not related to excretion during the diet with isoflavones.

5.2. Analysis of Change Using Analysis of Covariance

Another way to test for differences between the groups for urinary excretion after the groups are switched over to the isoflavone-free diet is to use a statistical modeling approach called analysis of covariance (ANCOVA). Let's assume that diet with flavones (with isoflavone) is the control or standard diet, and the goal is to test if there are differences between the age groups after the subjects are switched over to the isoflavone-free diet (without isoflavone). To accomplish this, we would fit a regression model in which the response variable is the excretion measure during the isoflavone-free period (i.e., the "second" diet). The explanatory variables in the model would be (1) an indicator variable for age group (e.g., 1 = 19 to 25, 0 = 26 or older) and (2) a covariate measure of excretion measure during the diet with isoflavones (i.e., the "first" diet). The model equation for the ANCOVA model to test for an age effect in the isoflavone free diet is

$$Y_{-ISO} = \beta_0 + \beta_1 A + \beta_2 X,$$

where A is an indicator variable for "age group" and takes on the values of 1 or 0; and X is the excretion level during the diet with isoflavones. The results

Table 10
Analysis of Covariance Model

Parameter	Estimate	SE	*t* value	*P* value
β_0 Intercept	5.71	2.35	2.42	0.034
β_1 Age 19–25 years	–5.20	1.63	–3.18	0.009
Age 26+ years	Reference			
β_2 Excretion + isoflavones	0.71	0.10	7.06	<0.001

of the analysis are given in **Table 10**. The age effect is significant indicating that the 2 age groups had different excretion levels once switched over to the isoflavone-free diet. The covariate, excretion during the diet with isoflavones, is also a significant predictor.

Notes

1. For these data, both the grouping variable age and the covariate excretion during the diet with isoflavone were significant ($P < 0.01$), indicating that the 26+ age group had higher levels of urinary excretion compared with the 19- to 25-year-olds, adjusting for excretion values collected during the baseline diet with isoflavone.
2. The analysis of covariance estimate has the advantage of having smaller variance than that for change scores, and for this reason it is favored by many statisticians. It can be shown that the relative efficiency in terms of a change score analysis versus analysis of covariance, as measured by the ratio of variances is $2(1 + \rho)$ in favor of ANCOVA *(5)*. Despite this, analysis of change scores to assess change remains popular, perhaps because it does not require using regression techniques and is easier to interpret.
3. There is an underlying assumption that any chosen covariate, whether it is a premeasure for the outcome or some other prognostic variable, is at least moderately correlated with the outcome variable. The Pearson correlation between excretion level on the 2 diets was 0.85, indicating that analysis of covariance is well suited for this data set.
4. The analysis of covariance can be inappropriate and lead to spurious results in nonrandomized designs in which the grouping variables are related to the covariate. This was summarized by Miller and Chapman *(6)* recently and by others in the past *(7,8)*. For this study, a problem would have arisen if age was correlated with excretion during the diet with isoflavones. This problem arises in studies with nonrandom group assignment, such as cohort studies, and can lead to invalid interpretations.
5. Analysis of covariance assumes that the relationship between the response and the covariate be the same for the 2 groups being compared. This is sometimes called the assumption of parallelism. This gets its name from imagining 2 separate but parallel regression lines, one above the other, for the 2 groups being compared. The

distance between the lines is what is being tested against the null hypothesis of no difference. For more on this, see **Chapter 8** and **Chapter 9**.

6. The covariates in ANCOVA may or may not be statistically significant in these models. In many ANCOVA models, a baseline measurement serves as the covariate.

References

1. Lu, L-J. W., Cree, M., Shylaja, J., Nagamani, M., Grady, J. J., and Anderson, K. E. (2000) Increased urinary excretion of 2-hydroxyestrone but 16α-hydroxyestrone in premenopausal women during a soy diet containing isoflavones. *Cancer Res.* **60**, 1299–1305.
2. Vane, L. A., Prough, D., Kinsky, M. A., Williams, C. A., Grady, J. J., and Kramer, G. C. (2004) Effects of different catecholamine on dynamics and volume kinetics of crystalloid infusion. *Anesthesiology* **101**, 1136–1144.
3. Fleiss, J. L. (1986) *The Design and Analysis of Clinical Experiments*. New York, John Wiley & Sons.
4. Bonate, P. L. (2000) *Analysis of Pretest-Posttest Designs*. New York, Chapman &Hall/CRC Press.
5. Bock, R. D. (1975) *Multivariate Statistical Methods in Behavioral Research*. New York, McGraw-Hill.
6. Miller, G. A., and Chapman, J. P. (2001) Misunderstanding analysis of covariance. *J. Abnormal Psychol.* **110**(1), 40–48.
7. Fleiss, J. L., and Tanur, J. M. (1972) The analysis of covariance in psychology. In: Hammer, M., Salzinger, K., and Sutton, S., eds. Psychopathology: Contributions from the Social, Behavioral, and Biological Sciences. New York, John Wiley & Sons, pp. 509–527.
8. Huitema, B. (1980) *Analysis of Covariance and Alternatives*. New York, John Wiley & Sons.

14

Logistic Regression

Todd G. Nick and Kathleen M. Campbell

Summary

The Medical Subject Headings (MeSH) thesaurus used by the National Library of Medicine defines logistic regression models as "statistical models which describe the relationship between a qualitative dependent variable (that is, one which can take only certain discrete values, such as the presence or absence of a disease) and an independent variable." Logistic regression models are used to study effects of predictor variables on categorical outcomes and normally the outcome is binary, such as presence or absence of disease (e.g., non-Hodgkin's lymphoma), in which case the model is called a binary logistic model. When there are multiple predictors (e.g., risk factors and treatments) the model is referred to as a multiple or multivariable logistic regression model and is one of the most frequently used statistical model in medical journals. In this chapter, we examine both simple and multiple binary logistic regression models and present related issues, including interaction, categorical predictor variables, continuous predictor variables, and goodness of fit.

Key Words: Interaction; logit; odds ratio; predictive accuracy; sample size.

1. Introduction

Logistic regression models, which will be explained in this chapter, were developed from other seminal works on the analysis of binary data ***(1–3)***. The Medical Subject Headings (MeSH) thesaurus used by the National Library of Medicine for indexing articles for the Medline/PubMED database introduced logistic models as a term in 1990. Logistic regression models are defined as "statistical models which describe the relationship between a qualitative dependent variable (that is, one which can take only certain discrete values, such as the presence or absence of a disease) and an independent variable" ***(4)***. Synonymous terms are logistic regression, logistic models, and logit models.

From: *Methods in Molecular Biology, vol. 404: Topics in Biostatistics*
Edited by: W. T. Ambrosius © Humana Press Inc., Totowa, NJ

Logistic regression models are used to study effects of predictor variables on categorical outcomes. Normally, the outcome is binary, such as presence or absence of disease (e.g., non–Hodgkin lymphoma), in which case the model is called a binary logistic model. When there is only one predictor variable in a logistic regression model, the model is referred to as a simple logistic regression. When there are multiple predictors (e.g., risk factors and treatments), including categorical and continuous variables as predictors, the model is referred to as a multiple or multivariable logistic regression.

Logistic models are regularly applied when studying the relationships between risk factors and the occurrence of disease in epidemiologic studies. These models are frequently used in medical journals not specializing in epidemiology and public health. Of medical journals with high impact in their medical specialty, the most frequently used complex statistical model (models that adjust for confounding) is the logistic model *(5)*. The next most frequently used models in these high-impact journals are the Cox proportional hazards model *(6)* followed by multiple linear regression. This result is somewhat surprising, because before 1985, no more than 400 papers appeared in the bibliographic database Medline (Medical Literature Analysis and Retrieval System Online) using the search term "logistic regression." However, coinciding with the introduction of the procedure LOGIST in the SAS supplemental library (later replaced by Proc Logistic in SAS) *(7)* and other user-friendly statistical software that is available today, the frequency of use of logistic regression in scientific papers has exponentially increased (**Fig. 1**), thus improving the analysis of data on binary variables.

In this chapter, we examine the most common logistic regression model, the binary logistic model. We then describe both the simple and multiple binary logistic regression models and present related issues including interaction, categorical predictor variables, continuous predictor variables, and goodness of fit.

2. Example: Effect of TGF-β1 Gene Polymorphism on Renal Dysfunction After Liver Transplantation in Children

To illustrate various aspects of modeling a binary variable with a logistic regression model, consider the following example involving the effects of a particular gene polymorphism on renal dysfunction in children after liver transplantation.

Ojo and others *(8)* demonstrate that by 3 years after transplant, renal failure develops in 16.5% of all nonrenal solid organ recipients. Among liver transplant recipients, renal failure develops in 13%. The incidence of renal failure increases with time since transplant and is also associated with increased age and a number of other clinical characteristics.

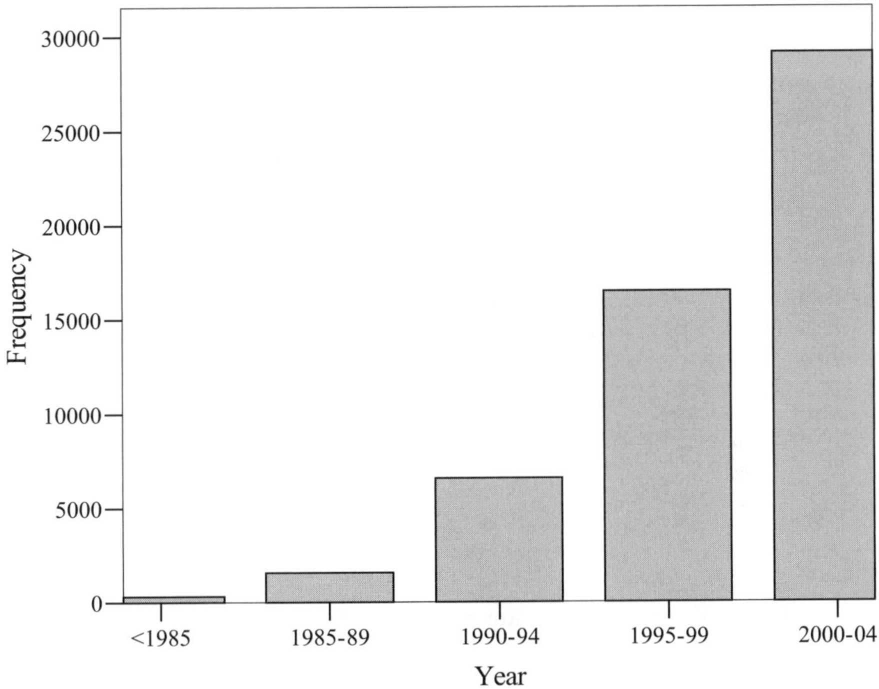

Fig. 1. The frequency of papers with logistic models or logistic regression as search terms appearing in Medline from 1965 to 2004.

To address the possible role of genetic factors on posttransplant renal dysfunction, Baan and others *(9)* examine the association between renal dysfunction after heart transplantation and the cytokine transforming growth factor (TGF)-$\beta 1$ codon 10 polymorphism in a predominately male population with a mean age of 45 years. An association is found between the presence of the **C** allele, in which case the amino acid coded for is proline (Pro), and renal dysfunction at 7 years after transplant. At this locus the major allele is **T**, and leucine (Leu) is the amino acid coded. The patients with the Pro/Pro (**CC**) or Pro/Leu (**CT**) genotype have more than a fourfold increase in the odds of renal dysfunction versus the Leu/Leu (TT) genotype.

To assess the association of the TGF-$\beta 1$ codon 10 polymorphism with posttransplant renal dysfunction among children, after controlling for sex and time since transplantation, a study involving pediatric liver transplant recipients is ongoing. Hypothetical data from this study is presented in **Table 1** and is sorted by sex, genotype, age, and disease. The data includes 60 children, 50% female, ages 5 to 20 years, who received a liver transplant. Renal dysfunction is defined as a serum creatinine $\geq$250 µmol/L *(9)* and is presented as a binary outcome.

Time since liver transplantation is measured in years and, in the study sample, ranges from 4 to 17 years, with a median of 10 years and lower and upper quartiles of 8 and 12 years. The allele frequencies of **T** and **C** are 55% and 45%, respectively. The frequencies of the genotypes are 35% homozygous for Leu, 40% heterozygous, and 25% homozygous for Pro. The proportion of subjects with renal dysfunction is given in **Table 2**. The traits female sex, shorter

Table 1
Hypothetical Data from 60 Subjects on Sex, Distribution of TGF-β1 Codon 10 Genotype (Genotype), Time Since Transplant (TST), and Renal Dysfunction (Disease)

ID	Sex	Type	TST	Disease	ID	Sex	Type	TST	Disease
1	Female	Leu/Leu	4	Absent	31	Male	Leu/Leu	5	Absent
2	Female	Leu/Leu	6	Absent	32	Male	Leu/Leu	5	Absent
3	Female	Leu/Leu	7	Absent	33	Male	Leu/Leu	7	Absent
4	Female	Leu/Leu	8	Absent	34	Male	Leu/Leu	7	Absent
5	Female	Leu/Leu	9	Absent	35	Male	Leu/Leu	11	Absent
6	Female	Leu/Leu	9	Absent	36	Male	Leu/Leu	11	Present
7	Female	Leu/Leu	10	Present	37	Male	Leu/Leu	15	Present
8	Female	Leu/Leu	11	Absent	38	Male	Leu/Leu	17	Present
9	Female	Leu/Leu	12	Absent	39	Male	Leu/Leu	17	Present
10	Female	Leu/Leu	13	Absent	40	Male	Leu/Pro	5	Absent
11	Female	Leu/Leu	13	Absent	41	Male	Leu/Pro	6	Absent
12	Female	Leu/Leu	15	Present	42	Male	Leu/Pro	7	Absent
13	Female	Leu/Pro	4	Absent	43	Male	Leu/Pro	8	Present
14	Female	Leu/Pro	5	Absent	44	Male	Leu/Pro	8	Present
15	Female	Leu/Pro	6	Absent	45	Male	Leu/Pro	9	Present
16	Female	Leu/Pro	8	Absent	46	Male	Leu/Pro	10	Present
17	Female	Leu/Pro	8	Absent	47	Male	Leu/Pro	11	Present
18	Female	Leu/Pro	8	Present	48	Male	Leu/Pro	12	Absent
19	Female	Leu/Pro	10	Absent	49	Male	Leu/Pro	12	Present
20	Female	Leu/Pro	10	Present	50	Male	Leu/Pro	12	Present
21	Female	Leu/Pro	10	Present	51	Male	Leu/Pro	15	Present
22	Female	Leu/Pro	14	Present	52	Male	Pro/Pro	4	Absent
23	Female	Leu/Pro	16	Present	53	Male	Pro/Pro	8	Absent
24	Female	Leu/Pro	17	Present	54	Male	Pro/Pro	8	Absent
25	Female	Pro/Pro	8	Absent	55	Male	Pro/Pro	9	Present
26	Female	Pro/Pro	8	Present	56	Male	Pro/Pro	10	Present
27	Female	Pro/Pro	9	Absent	57	Male	Pro/Pro	12	Absent
28	Female	Pro/Pro	10	Present	58	Male	Pro/Pro	13	Absent
29	Female	Pro/Pro	11	Present	59	Male	Pro/Pro	14	Present
30	Female	Pro/Pro	17	Present	60	Male	Pro/Pro	17	Present

Table 2
Proportion of Subjects with Renal Dysfunction

Predictor	*n*	*N*	% Renal dysfunction
Overall	**28**	**60**	**47**
Sex			
Female	12	30	40
Male	16	30	53
Time since transplant			
4–8 years	4	24	17
9–11 years	11	17	65
12–18 years	13	19	68
Genotype distribution			
Leu/Leu	6	21	29
Leu/Pro	14	24	58
Pro/Pro	8	15	53

time since transplantation, and homozygosity for Leu are associated with less renal dysfunction.

3. Measures of Effect for Categorical Outcomes

3.1. Odds and Odds Ratio

The logistic model uses the odds ratio to determine the effect a predictor variable has on the outcome variable ***(10)***. When there is only 1 predictor variable, the odds ratio is called a crude, or unadjusted, odds ratio. When there are at least 2 predictor variables, an adjusted odds ratio quantifies the effect a predictor has on an outcome while holding the other predictors constant ***(11)***.

An odds ratio (OR) ***(12)*** is simply the ratio of 2 odds and is used extensively in medical studies as a measure of effect for categorical data. Odds are usually expressed in terms of probability of an event. If the probability of an event is p, then an odds can be defined as $p/(1 - p)$. For example, if the probability of an event is 1/3, then the odds ratio of that event is 1 to 2, or 1/2. Odds can be converted to probabilities by taking $p = \text{odds}/(1 + \text{odds})$. For example, an odds of 2 would have $2/(1 + 2) = 0.67$ probability of occurring. As probability goes from 0 to 1, odds vary from 0 to ∞. An odds less than 1 would have probability less than 0.50, and an odds equal to 1 has a probability of 0.50, or 50/50 chance. If the odds ratio is 2.0, there is a twofold increase in the odds of an event occurring, or double the odds. To illustrate the calculation of odds and odds ratios, examples from the data in **Table 1** are provided below. Further discussion of odds is given in **Chapter 2**.

To determine the odds and odds ratio for a binary predictor and outcome variable, a 2×2 contingency table can be constructed as in **Table 3**. To illustrate the

Table 3
2 × 2 Table Showing Hypothetical Association Between Sex and Renal Dysfunction (RD)

Sex	Renal dysfunction Yes	No	Total
Male	16 (a)	14 (b)	30 ($a + b$)
Female	12 (c)	18 (d)	30 ($c + d$)
Total	28 ($a + c$)	32 ($b + d$)	60 (n)

calculation, the risk and odds of renal dysfunction overall and separately for females and males are computed based on the frequencies in **Table 3**.

The overall probability, or risk, that a child will have renal dysfunction (RD), P(RD), is $[(a + c)/n] = 28/60 = 0.467$, or 0.47 when rounding to 2 decimal places. The odds that a child will have renal dysfunction, odds(RD), is estimated by $P(\text{RD})/[1 - P(\text{RD})] = 0.875$. Note that rounding in intermediate steps of the calculation may lead to slight differences.

To compute an odds ratio, the probability and odds are calculated separately for males and females. Females have a risk of having renal disease of 12/30 = 0.40 and males 16/30 = 0.53. These conditional probabilities are typically denoted P(RD|Female) and P(RD|Male), respectively. That is, the probabilities are computed conditioned on each sex.

The odds of renal dysfunction (RD) given a female child, denoted Odds(RD|Female), is expressed as

$$\text{Odds(RD|Female)} = \frac{P(\text{RD|Female})}{1 - P(\text{RD|Female})} = \frac{c/(c+d)}{d/(c+d)} = c/d = 12/18 = 0.667.$$

Similarly, the odds of RD given a male is expressed as

$$\text{Odds(RD|Male)} = \frac{P(\text{RD|Male})}{1 - P(\text{RD|Male})} = \frac{a/(a+b)}{b/(a+b)} = a/b = 16/14 = 1.143.$$

The odds of RD given a female is 0.67 (0.67 to 1), or 2 to 3. The odds of having RD for males is 1.14, or 1.14 to 1. This is equivalent to saying the odds of RD for a male is 8 to 7, which is slightly higher than 1 to 1, or even, odds.

To compare the 2 odds for males and females, an odds ratio is used. Odds ratios, denoted OR, are one of many ways to compare 2 groups with a binary outcome. The ratio of the 2 odds are calculated, or the formula OR = ad/bc can be used. The odds ratio is given by

$$\text{OR} = \frac{\text{Odds(RD | Male)}}{\text{Odds(RD | Female)}} = \frac{1.143}{0.667} = 1.71.$$

Males are 33% more likely to have RD and, in absolute terms, 13% more males have RD. On the other hand, the odds of RD occurring is 71% more for males than females.

For large sample sizes, the natural log of the odds ratio is approximately normally distributed, and a 95% confidence interval (CI) can be calculated. The natural logarithm of the odds is denoted ln(OR) with standard error, SE[ln(OR)], equal to

$$\text{SE}[\ln(\text{OR})] = \sqrt{\frac{1}{a} + \frac{1}{b} + \frac{1}{c} + \frac{1}{d}}. \quad (1)$$

For the OR above, the SE[ln(OR)] = 0.522. The antilog of a number is denoted exp and the 100 × (1 – α)% CI of the OR is then calculated as

$$\exp[\ln(\text{OR}) \pm z_{1-\alpha/2} \times \text{SE}[\ln(\text{OR})]]$$

where $z_{1-\alpha/2}$ is the 100 × (1 – α/2) percentile of the normal distribution. Common values for $z_{1-\alpha/2}$ are 1.645, 1.96, and 2.33 for 90%, 95%, and 99% CIs. For example, the 95% CI for the odds of RD for a male compared with a female is

$$\begin{aligned} 95\%\text{CI(OR)} &= \exp[\ln(1.71) \pm 1.96 \times 0.522] = \exp[0.536 \pm 1.023] \\ &= \exp(-0.487, 1.559) = 0.61, 4.75. \end{aligned}$$

The 95% CI of the OR is 0.61 to 4.75. Because the 95% CI includes 1, statistical significance using the chi-square test would not be achieved. That is, there is not sufficient evidence of an association between sex and RD. An odds ratio of 1, the null hypothesis value, would occur if the odds of RD in males and females were exactly the same.

Odds ratios are used extensively in medical studies, and their use in molecular biology is increasing. For example, Slattery and others ***(13)*** uses odds ratios to describe the associations among a cytochrome P-450 gene (CYP1A1), cigarette smoking, and cancer. The greatest colon cancer risk among men is observed for current smokers who have any CYP1A1 variant allele [odds ratio = 2.4; 95% confidence interval (CI) = 1.3 to 4.8]. Hishida and others ***(14)*** recently used logistic regression to determine relationships between polymorphisms and non–Hodgkin lymphoma. A possible association between the p53 Pro72 allele and non–Hodgkin lymphoma in a Japanese population is demonstrated (OR = 1.59; 95% CI = 0.99 to 2.57).

3.2. Relative Risk and Absolute Risk Measures

Although logistic regression uses odds ratios, two other useful statistics that measure effects in a 2 × 2 table are the absolute risk reduction (ARR) and relative risk (RR) statistics. The frequencies in **Table 3** are used to illustrate the calculations.

The simplest measure is the ARR, which is the difference between the 2 absolute risks. The ARR for the data in **Table 3** is given by

$$\text{ARR} = \frac{a}{a+b} - \frac{c}{c+d} = 0.53 - 0.40 = 0.13.$$

The ARR is 0.13, or 13%. That is, females have a 13% increase in absolute risk of having renal disease compared with males. The RR is a ratio similar to the odds ratio and is sometimes referred to as the risk ratio. The RR is given by

$$\text{RR} = \frac{a/(a+b)}{c/(c+d)} = \frac{0.53}{0.40} = 1.33.$$

Based on the RR, females are 33% more likely to have renal disease than males.

4. Logistic Regression

4.1. Formulating a Model

The OR and the 95% CI may be calculated using a simple logistic regression model for a predictor of any type, dichotomous or continuous. Define the outcome variable to be Y and let $Y = 1$ denote the occurrence of an event such as renal dysfunction, and let $Y = 0$ denote no occurrence. Define the predictor variable to be X_1. The subscript 1 is used to generalize when multiple predictors are present. Then the logistic model can be stated in terms of the probability that the event occurs given the value of the predictor which is denoted $P(Y = 1 | X_1)$.

The fundamental assumption is that the log of the odds that $Y = 1$ occurs is linearly related to the predictor variable(s) ***(15)***. The odds below is defined as the odds of the event or disease occurring ($Y = 1$) given the predictor variable, X_1. This can be written as

$$\log \text{odds}[Y = 1 | X_1] = \log\left[\frac{P(Y = 1 | X_1)}{1 - P(Y = 1 | X_1)}\right] = \beta_0 + \beta_1 X_1, \tag{2}$$

where β_0 is the intercept and β_1 is the regression coefficient of X_1. The coefficients are on a logarithmic scale, and the log of the odds is known as the logit transformation. From **Equation 2**, the model is a linear regression model in the log odds that $Y = 1$.

The logistic probability function can then be expressed as

$$P(Y=1 \mid X_1) = \frac{1}{1+\exp[-(\beta_0+\beta_1 X_1)]} = \frac{\exp(\beta_0+\beta_1 X_1)}{1+\exp(\beta_0+\beta_1 X_1)}. \quad (3)$$

Equation 3 is useful when determining the predicted probability of an occurrence of an event given the value of the predictor(s).

4.2. Relationship Between Logit and Probability Scale

To illustrate the relationship between the logit and probability scale, consider the relationship between TST and occurrence of renal dysfunction (**Fig. 2**). The relationship is linear on the log odds (logit) scale and S-shaped on the probability scale *(16)*. On the probability scale, the logistic function is constrained between 0 and 1 and is one of the reasons it is so popular today. As shown in the figure, in its simplest form the logistic model assumes that TST is linearly related to the log odds of RD, the outcome. As time since transplant increases,

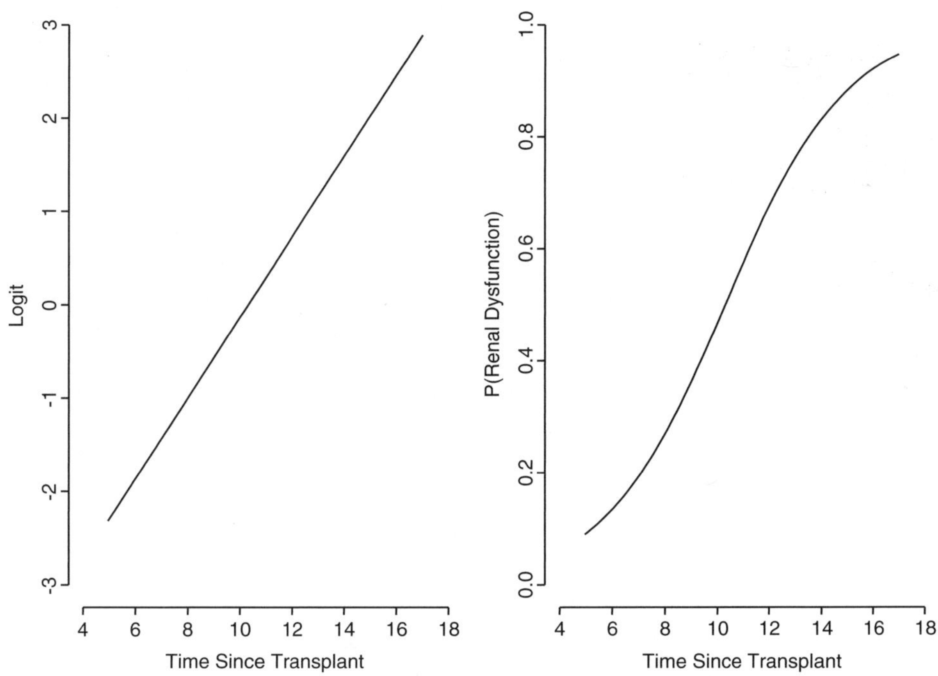

Fig. 2. Fitted log odds (left plot) and logistic curve for time since transplant on renal dysfunction (right plot).

the probability of renal dysfunction increases as well. When TST is relatively short, TST < 8, chance of RD is not likely. As TST increases, chance of RD increases and levels off around TST = 16.

4.3. Interpretation of Coefficients

The size of the coefficient β_1 controls the rate of change in the probability and can be referred to as the slope. The slope gives the change in the log odds for an increase of 1 unit in X_1 (e.g., $X_1 = 1$ vs. $X_1 = 0$ or $X_1 = 21$ vs. $X_1 = 20$). When the slope is positive, the curve increases from probability 0 to 1, and the odds ratio is greater than 1. When the slope is negative, the curve decreases from probability 1 to 0 and the odds ratio is less than 1.

The intercept β_0 is the log odds when $X_1 = 0$. When there are 1 or more predictors, β_0 is the log odds when all the predictors are at 0 and is often not meaningful. For a given value of the slope, β_0 controls the location of the curve. For a detailed discussion of coefficients in a logistic regression model, see Dupont ***(17)***.

4.4. Odds Ratio

To compare the odds of RD for 2 values of a predictor variable, **Equation 3** can be used to construct an odds ratio. For example, to compare the odds of RD when $X_1 = 1$ to the odds when $X_1 = 0$, the odds ratio becomes

$$\mathrm{OR} = \frac{\dfrac{P(Y=1 \mid X_1=1)}{1-P(Y=1 \mid X_1=1)}}{\dfrac{P(Y=1 \mid X_1=0)}{1-P(Y=1 \mid X_1=0)}} = \frac{\exp(\beta_0+\beta_1)}{\exp(\beta_0)} = \exp[(\beta_0+\beta_1)-\beta_0] = \exp(\beta_1). \quad (4)$$

The log of the odds ratio is given by $\ln[\exp(\beta_1)] = \beta_1$. The coefficient, β_1, or the odds ratio, $\exp(\beta_1)$, gives the change in the log odds or ratio of odds for an increase of 1 unit in X_1.

For the OR above, the $100 \times (1 - \alpha)\%$ CI of OR in a logistic regression model is written similar to the $100 \times (1 - \alpha)\%$ CI for the basic OR above. The $100 \times (1 - \alpha)\%$ CI is given as

$$100 \times (1-\alpha)\%\mathrm{CI} = \exp[\beta_1 \pm z_{1-\alpha/2} \times \mathrm{SE}[\beta_1]],$$

where $z_{1-\alpha/2}$ is the $100 \times (1 - \alpha/2)$ percentile of the normal distribution. For the 95% CI, $z_{1-\alpha/2} = 1.96$.

Although the formulas above are complex, the simple relationship between β_1 and the odds ratio is the main reason why logistic models are a proven tool to model relationships between predictors and a categorical outcome ***(18)***.

Table 4
Logistic Model Output with TST as a Predictor

					Odds ratio		
Predictor	Coef.	SE	Wald	*P* value	Estimate	95% CI	
Intercept	$\beta_0 = -4.455$	1.205	13.5	<0.001	—	—	—
TST	$\beta_1 = 0.432$	0.118	13.7	<0.001	1.54	1.22	1.94

5. Simple Logistic Regression Model

5.1. Results of Fitting a Simple Logistic Regression Model

Output from statistical software packages are consistent in that they give coefficients (Coef.), standard error of coefficients (SE), Wald statistics and corresponding *P* values, and estimated odds ratio with a 95% CI. The Wald chi-square statistic is presented below. Some packages give the square root of the Wald chi-square statistic, which is referred to as the Wald *z*-statistic ***(10)***. The relevant null hypothesis is $H_0 : \beta_1 = 0$, which is equivalent to $\mathrm{H_0 : OR = 1.0}$, where OR is the odds ratio in the population. This can be stated as there is no association or relationship between the predictor and the occurrence of the outcome.

Using the data in **Table 1**, the output for a simple logistic regression model predicting the occurrence of RD with the continuous predictor TST is given in **Table 4**. The output for a model predicting RD with the dichotomous predictor Sex is given in **Table 5**. From **Table 4**, TST was significantly associated with occurrence of RD (OR 1.54, 95% CI = 1.22 to 1.94; $P < 0.001$). The odds increases 54% for an increase of 1 year in TST. From **Table 5**, we infer that sex was not significantly associated with occurrence of RD (OR 1.71, 95% CI = 0.62 to 4.77).

Table 5
Logistic Model Output with Sex as a Predictor

					Odds ratio		
Predictor	Coef.	SE	Wald	*P* value	Estimate	95% CI	
Intercept	$\beta_0 = -0.405$	0.373	1.184	0.28	—	—	—
Male Sex[a]	$\beta_1 = 0.539$	0.522	1.065	0.30	1.71	0.62	4.77

[a]Coded 1 for male and 0 for female.

5.2. Coding Categorical Predictors

5.2.1. Binary Predictor Variable

Males and females are included in the model presented in **Table 5** by constructing a design (indicator or dummy) variable (X_1 = 1 if male; X_1 = 0 if female). Design variables are used because regression models typically cannot handle character strings such as "Male" or "Female." From **Table 5**, the odds of RD for males is 71% greater than the odds for females. Because unity is included in the 95% CI, there is no evidence for an association. Estimates of the intercept and slope are $\beta_0 = -0.405$ and $\beta_1 = 0.539$. Plugging those values into **Equation 3** gives

$$P(Y = 1 | X_1 = 1) = 0.53.$$

That is, a male has a 53% chance of having RD, and this is simply the proportion of males that have RD as shown above. The odds ratio, although given in **Table 5**, can be derived using **Equation 4** or from $\exp(\beta_1) = \exp(0.539) = 1.71$. The 95% CI of the estimated OR is

$$95\%\text{CI} = \exp[0.539 \pm 1.96 \times 0.522] = \exp(1.562, -0.484) = (0.62, 4.77).$$

It is important to note that the coefficients given in **Table 5** for predicting RD from sex would change if values other than 0 and 1 were used to encode sex. For example, if we switched the values of X_1 for males and females the coefficients would be $\beta_0 = 0.134$ and $\beta_1 = -0.539$. Further details are given in Hosmer and Lemeshow ***(18)***. Regardless of the values used, the estimated probability of renal disease occurring will remain the same for each sex.

5.2.2. Nominal and Ordinal Predictor Variables

The number of design variables in a model required to represent a nominal predictor is 1 less than the number of categories of the predictor. For binary predictors, only 1 design variable is needed, and a column of 0s and 1s are created. For predictors with more than 2 categories, more design variables are required. For example, the TGF-β1 codon 10 polymorphism listed in **Table 1** has 3 genotypes. A contingency table is given in **Table 6** showing the association between genotype and renal dysfunction.

Because there are 3 genotypes, 2 design variables with the values 0 and 1 are created, and a reference category needs to be selected for comparison purposes to calculate regression coefficients and odds ratios. Here, Leu/Leu will be the reference. There are several ways to code categorical predictors. The most common is the nominal codings for either nominal or ordinal predictors. The nominal codings are usually sufficient for ordinal variables with up to 5 categories. These codings compare the odds of the reference category to the other categories. An alternative for ordinal predictors is to use ordinal codings

Table 6
3 × 2 Table Showing Hypothetical Relationship Between Genotype and Renal Dysfunction

	Renal dysfunction		
Genotype	Yes	No	Total (%)
Leu/Leu	6	15	21 (35%)
Leu/Pro	14	10	24 (40%)
Pro/Pro	8	7	15 (25%)
Total	28	32	60 (100%)

(19,20). Using ordinal codings allows determination of the amount of change occurring from one category to the next. **Table 7** applies both coding schemes to the TGF-β1 codon 10 polymorphism (Genotype) variable.

To demonstrate the interpretation of regression coefficients and odds ratios and to compare the coding schemes in **Table 7**, the genotype variable is used alone in a logistic model. **Table 8** presents the output using the 2 coding schemes. To interpret **Table 8** for the nominal codings, we see the estimated OR and 95% CI for the first design variable, GT_1, is 3.50 (1.01 to 12.18) and is statistically significant (*P* value = 0.05). The odds of RD occurring in a patient with genotype Leu/Pro is estimated to be 3.5 times that of a patient with genotype Leu/Leu, the reference. Recall that an OR of 1 would imply no association between genotype and RD, or equivalent odds for the 2 genotypes being compared. The odds of a patient with genotype Pro/Pro is estimated to be 2.86 times that of Leu/Leu (95% CI = 0.71 to 11.44; *P* value = 0.14). Based on the *P* value = 0.14, there is insufficient evidence to conclude that there is a difference in the odds for the 2 genotypes. For the ordinal coding scheme, the first design variable (GT_1) has the same interpretation as that of the nominal coding (OR = 3.50, 95% CI = 1.01 to 12.18; *P* value = 0.05). The interpretation of GT_2 is not the same. The coefficient and odds ratio compare the second and third

Table 7
Nominal and Ordinal Coding of Design Variables for Genotype

	Nominal		Ordinal	
Genotype	GT_1	GT_2	GT_1	GT_2
Leu/Leu	0	0	0	0
Leu/Pro	1	0	1	0
Pro/Pro	0	1	1	1

Table 8
Coefficients and Odds Ratio of Nominal and Ordinal Coding of Design Variables

	Nominal			Ordinal		
	Wald	*P* value	OR (95% CI)	Wald	*P* value	OR (95% CI)
Constant	3.60	0.06	—	3.60	0.06	—
GT_1	3.88	0.05	3.50 (1.01, 12.18)	3.88	0.05	3.50 (1.01, 12.18)
GT_2	2.20	0.14	2.86 (0.71, 11.44)	0.09	0.76	0.82 (0.22, 2.99)

genotypes, Leu/Pro and Pro/Pro. The odds are 18% less for homozygous Pro compared with heterozygous individuals (OR = 0.82, 95% CI = 0.22 to 2.99) and is not significant (*P* value = 0.76). Stated another way, the odds significantly increases when Leu/Pro is compared with homozygous Leu but does not significantly change when Pro/Pro is compared with Leu/Pro.

Note that before coding the categorical predictors, it is important to inspect the frequencies of the categories to determine whether a reduction of the categories is needed ***(19)***. For example, there are only 15 patients with genotype Pro/Pro. Merging this genotype with Leu/Pro would reduce the number of design variables to 1 and lead to more precise estimation. In general, it is better to merge categories into "similar" groups instead of keeping them separate before the analysis phase of the data. Basing the reduction of the genotypes on an inspection of the regression modeling is not appropriate and can lead to biased models. For categorical predictors, the categories are compared with each other depending on the coding scheme applied. For continuous predictors, the odds ratio has a different interpretation.

5.3. Continuous Predictor Variables

5.3.1. Interpretation of Odds Ratios

In computer output and subsequently in medical papers, the coefficient and the odds ratio are presented as the increase (or decrease) in the log odds or odds of disease (RD) occurring for a 1-unit increase in the continuous predictor. For example, for a 1-unit increase in TST, the log odds of RD increases 0.432 and the odds increases 1.54-fold (see **Table 4**). Although coefficients and odds ratios are often reported in terms of a 1-unit increase in the predictor variable, as in **Equation 4**, a useful description would be to determine a meaningful change in a predictor, such as a 5- or 10-unit change. That is, changing the scale of the predictor may aid in interpreting meaningful clinical effects. For example, using logistic regression, Ford and others ***(21)*** found that after adjustment for covariates, a 10-unit change in the Healthy Eating Index (HEI) reduced the odds of an elevated C-reactive protein concentration by 8% (OR = 0.92, 95% CI = 0.86 to 0.99). If the standard 1-unit change was used for the calculation, a 1% reduction would have been reported (OR = 0.99, 95% CI = 0.985 to 0.995). Regardless of the unit-change used in the calculation, statistical significance (or insignificance) will remain the same. However, by using a clinically useful change in the predictor, a more meaningful interpretation of the change in odds can be made.

For any c-unit change, the log odds ratio in a predictor X_1 is simply $c\beta_1$ and the odds ratio is $\exp(c\beta_1)$ (18). A 95% CI for a c-unit change in a predictor X_1 is written as

$$\begin{aligned} 95\%\text{CI(OR)} &= \exp[c\beta_1 \pm z_{1-\alpha/2} \times c\text{SE}[\beta_1]] \\ &= \exp[c]\exp[\beta_1 \pm z_{1-\alpha/2} \times \text{SE}[\beta_1]]. \end{aligned} \quad (5)$$

For example, there is a 1.54-fold (95% CI = 1.22 to 1.94) increase in odds for a 1-year increase in TST (see **Table 4**). However, it may be more meaningful to compute the increase in the odds for any 2-year increase in TST. The odds ratio is calculated as $\exp(c\beta_1) = \exp[2(0.432)] = 2.37$. Using **Equation 5**, the 95% CI for a 2-year increase in TST is

$$\begin{aligned} 95\%\text{CI(OR)} &= \exp[2(0.432) \pm 1.96(2)(0.118)]] \\ &= \exp[0.158, 1.570) = (1.49, 3.77). \end{aligned}$$

Other useful strategies are to use some standardized measure for c, such as 1 standard deviation or the difference between the outer quartiles (0.25 and 0.75 quantiles). If the predictor is normally distributed, then standardized logistic regression coefficients could be used to indicate the effect of a 1-unit standard deviation difference in the predictor on the outcome ***(22)***. However, we recommend using the change from the outer quartiles as a measure for c, because predictors are not always normally distributed. Using the difference in the outer quartiles as a measure for c, the odds ratio is called the interquartile range (IQR) or half-sample odds ratio ***(23)***. For example, the 25th and 75th percentiles of TST are 8 and 12. The difference between the quartiles is 4. Let $c = 4$ and the OR is $\exp[4(0.432)] = 5.6$. The 95% CI is 2.2 to 14.2. That is, patients at the 75th percentile of TST (TST = 12) had a 5.6-fold (95% CI = 2.2 to 14.2) higher odds of renal dysfunction compared to those at the 25th percentile (TST = 8). Examples on the reporting and interpreting of logistic models and specifically the half-sample OR can be found in Refs. ***24–26***.

An alternative is to categorize a continuous predictor and use the nominal coding scheme. In fact, of the articles reviewed in epidemiological journals by Ottenbacher and others ***(27)***, 65% of the 99 articles reviewed that used logistic regression only utilized binary predictor variables. Although practiced in the medical literature, we do not recommend categorizing a continuous predictor in a logistic regression model unless the number of subjects is large. Categorizing continuous variables may result in a significant loss of information. In addition to Ottenbacher, the guidelines presented by Lang and Secic ***(28)*** and Bagley and others ***(29)*** are helpful.

5.3.2. Relaxing the Linearity Assumption

In logistic regression models, in their simplest form, it is assumed that continuous or ordinal predictor variables are linearly related to the log odds. If the linearity assumption is not relaxed, exploration should be done to ensure conformity with the linear gradient ***(29,30)***. However, information, or indication of, verifying this assumption is usually not provided ***(27,29)***. For example, in Figure 2, TSA is constrained to be linearly related to the log odds of renal dysfunction. To fit more flexible curves, nonlinear effects should be allowed by using either polynomials or restricted cubic splines ***(15,31)***. Polynomials are also used in multiple linear regression models as discussed in **Chapter 9**. Alternatively, transformations may be made on the predictor (e.g., log) because log of the predictor may be linearly related to the log odds. **Figure 3** displays

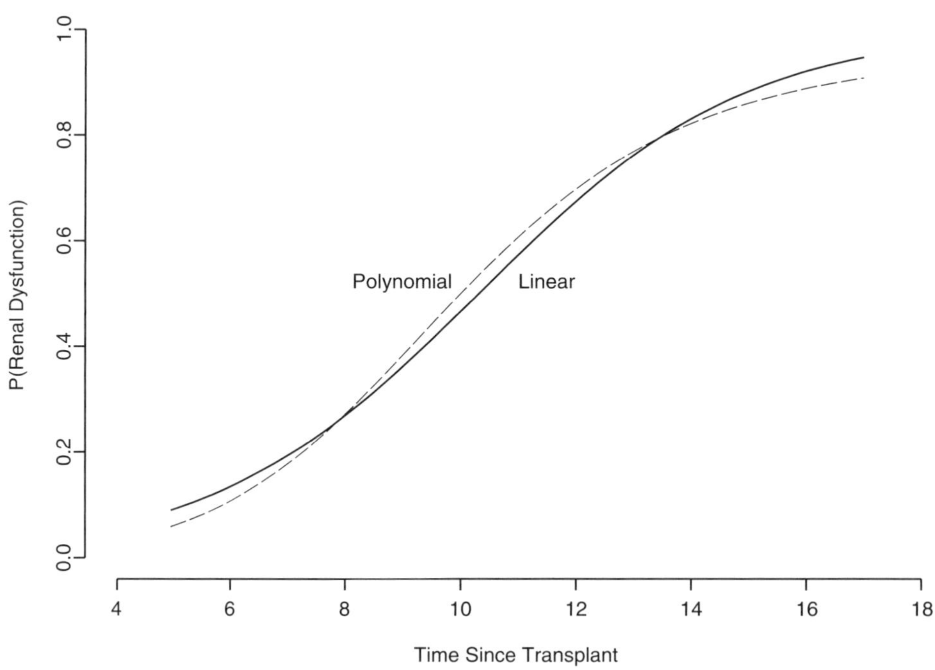

Fig. 3. Relationship between time since transplant and probability of renal dysfunction assuming linear (solid line) and relaxing linearity using polynomial (dashed line).

the relationship TST has on probability of renal disease. Only small differences are seen between the 2 lines. The relationship appears to conform to the linear gradient. Nick and Hardin ***(19)*** give an example when the relationship does not conform to the linear gradient.

Typically, a log transformation may satisfy the linearity assumption. However, the transformation that satisfies the linearity assumption is often not known. Polynomials are incorporated in a regression model by including terms that are powers of the predictor variable. For example, the polynomial in **Figure 3** was created by including TST and TST^2 as predictor variables in the model. This allowed the linearity assumption to be relaxed.

Splines fit a wider variety of functions and can allow for threshold effects. The number of knots, or inflection points in the curve, needs to be specified. A greater number of knots will result in a regression model that more closely fits the data. Three knots should be used with samples with less than 30 subjects and 4 knots with samples up to 100. Usually 5 knots are sufficient for large samples ***(10)***. If splines or polynomials are used, linearity is not assumed on the log odds so the odds ratio will depend on the value of the predictor. It is then very important to use the half-sample odds ratio to describe the effects the predictor has on outcome.

6. Logistic Regression with Multiple Predictors

6.1. Introduction

In **Section 5**, predictors were related to the occurrence of an event with respect to an outcome, but only 1 predictor was handled at a time. It is important to demonstrate the handling of many predictors and various different types of predictors using multivariable models. When there are multiple predictors and a categorical outcome, it is important to examine the predictors simultaneously using a multiple logistic regression model. The multiple logistic regression coefficients and odds ratio are adjusted for the other predictors in the model. For example, Hishida and others ***(14)*** reported the association of polymorphisms and risk of non–Hodgkin lymphoma, adjusted for age and sex. Age and sex were included as predictors in their multiple logistic regression models and held constant when studying the association of polymorphisms with occurrence of lymphoma.

6.2. Model Assuming Additivity

The multivariable model involves a linear combination of the predictors. Consider the case of 3 predictor variables, X_1, X_2, and X_3. The log of the odds can be represented by

$$\log \text{odds}(Y = 1 \mid X_1, X_2, X_3) = \beta_0 + \beta_1 X_1 + \beta_2 X_2 + \beta_3 X_3. \quad (6)$$

The above model could be used to study the effects that TST and Genotype have on occurrence of renal dysfunction, simultaneously. Recall TST is in years and there are 3 genotypes (Leu/Leu, Leu/Pro, Pro/Pro). The model is written as

$$\log \text{odds}(Y = \text{RD}|\text{TST}, \text{GT}_1, \text{GT}_2) = \beta_0 + \beta_1\text{TST} + \beta_2\text{GT1} + \beta_3\text{GT2}, \qquad (7)$$

where GT1 and GT2 represent the design variables using the nominal coding scheme given in **Table 7**. **Table 9** is output from a standard statistical package using TST and Genotype. The standard output determines the coefficient for a 1-unit change as stated above. Therefore, as TST increases 1-year the odds increase 1.74-fold (95% CI = 1.30 to 2.33). The design variable GT1 compares Leu/Pro with Leu/Leu. Based on the statistics in the row labeled GT1, subjects with genotype Leu/Pro have an increase in the odds of renal dysfunction occurrence compared with genotype Leu/Pro (OR = 1.54, 95% CI = 1.84 to 83.2). Although the confidence intervals for comparing genotypes are wide and imprecise, significant differences are observed for one of the genotype comparisons. See **Section 6.4.5** for a discussion of sample sizes for logistic regression models.

Standard computer output is usually not presented in medical research papers. We suggest restructuring **Table 9** so a more meaningful change in TST is used and the genotype comparisons are clear. For example, a half-sample odds ratio is presented for TST. Additionally, we perform a statistical test to assess the overall contribution of genotype (which is explained in **Section 6.4.1**). Table 10 is a suggested presentation of a multiple logistic regression model. Typically, only the odds ratio and 95% CI are reported. A prediction formula with coefficients may be given in text or in a footnote. *P* values are reported below but are expendable if 95% CIs are given. *P* values are useful for testing the contribution of a set of variables as below. Graphical presentations, especially on the probability scale, are useful to display important relationships. **Figure 4** shows the predicted log odds and probability of renal dysfunction. The basic

Table 9
Logistic Model Output with Predictors TST and Genotype

					Odds ratio		
Predictor	Coef.	SE	Wald	*P* value	Estimate	95% CI	
Intercept	$\beta_0 = -7.2$	1.92	14.0	<0.001	—	—	—
TST (per 1-year change)	$\beta_1 = 0.55$	0.15	13.6	<0.001	1.74	1.30	2.33
GT1	$\beta_2 = 2.52$	0.97	6.7	0.01	12.37	1.84	83.2
GT2	$\beta_3 = 1.72$	0.98	3.1	0.08	5.56	0.82	37.5

Table 10
Presenting the Logistic Model with Predictors TST and Genotype

		Odds ratio		
Variable	*P* value	Estimate	95% CI	
TST (per 4-year change)[a]	<0.001	9.1	2.8	29.4
Genotype (reference, Leu/Leu)	0.01[b]			
Leu/Pro		12.4	1.84	83.2
Pro/Pro		5.6	0.82	37.5

[a]Comparison for TST is half-sample odds ratio; compares the 0.75 to 0.25 quantile.
[b]$P = 0.01$ represents a 2 degree of freedom likelihood ratio (LR) test on genotype.

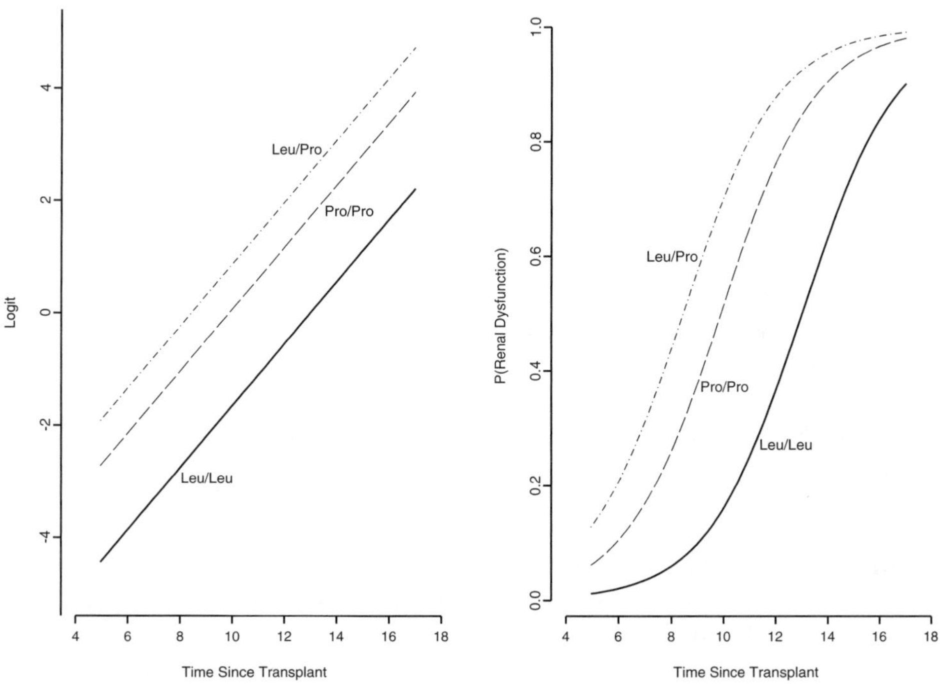

Fig. 4. Predicted log odds (left) and probability (right) of renal dysfunction for the 3 genotypes (Leu/Leu, Leu/Pro, Pro/Pro). No interaction is assumed between genotype and TST.

assumption in the model is that there is no interaction between the predictors and therefore the lines are forced to be parallel on the log odds scale. We suggest reporting the probability scale for important relationships.

6.3. Model with Interaction Among Predictors

In its simplest form, an assumption of a multiple logistic regression model is that there is no interaction between the predictor variables. Statistical interaction is present when there is nonindependence of the effect of 2 predictors on the outcome ***(32,33)***. This is referred to as a nonadditive model. That is, the effect of a predictor on an outcome does not depend on another predictor. Statistical interaction is analogous to drug interaction, where the effect of one drug is altered by another drug. In epidemiological journals, Ottenbacher and others ***(27)*** found 61% of the articles they reviewed that used logistic models did not report or discuss testing for interactions.

In statistical packages, interaction is included by taking the product of 2 or more predictor variables ***(34)***. Consider the model in **Equation 6** for the predictors X_1 and X_2, but now allow for interaction between the 2 predictors. The log odds can be represented by

$$\log \text{odds}(Y = 1 | X_1, X_2) = \beta_0 + \beta_1 X_1 + \beta_2 X_2 + \beta_{12} X_1 X_2,$$

and holds when the predictors are either continuous or binary. The coefficient β_{12} represents the interaction term. A hypothesis test is used to test if the interaction coefficient is 0 and is illustrated below.

We relax the no interaction assumption given in **Equation 7** and allow an interaction between genotype and TST. That is, we want to allow for differences in occurrence of renal dysfunction among the 3 genotypes at any year since transplant. We can determine if there is evidence of complexity due to interaction between genotype and TST. The model is written as

$$\begin{aligned}\log \text{odds}(\text{RD}|\text{TST}, \text{GT1}, \text{GT2}) = \beta_0 &+ \beta_1\text{TST} + \beta_2\text{GT1} + \beta_3\text{GT2} \\ &+ \beta_{12}\text{TST} \times \text{GT1} + \beta_{13}\text{TST} \times \text{GT2}.\end{aligned}$$

Figure 5 below relaxes the parallelism assumption based on the interaction model above and can be compared with the no-interaction model in **Equation 7** and plots in **Figure 4**. That is, we allow for interaction between TST and genotype in **Figure 5**. Comparisons can be made to determine if the effect of TST differs across the 3 genotypes. The genotype Pro/Pro appears to be similar to Leu/Pro when TST is short, less than 10 years. But for high values of TST, greater than 12 years, Pro/Pro has similar occurrence of renal dysfunction as Leu/Leu. If lines are more or less parallel, the interaction effect is not likely to be significant. See **Chapter 10** for a discussion of tests of parallelism in ANCOVA.

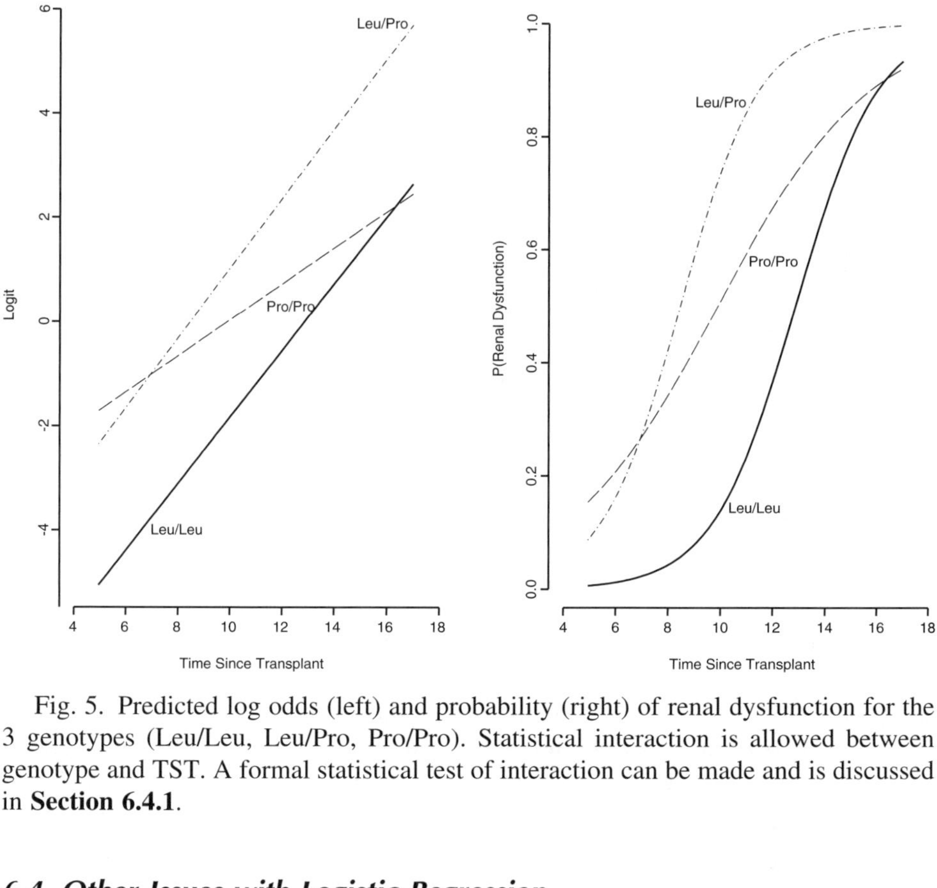

Fig. 5. Predicted log odds (left) and probability (right) of renal dysfunction for the 3 genotypes (Leu/Leu, Leu/Pro, Pro/Pro). Statistical interaction is allowed between genotype and TST. A formal statistical test of interaction can be made and is discussed in **Section 6.4.1**.

6.4. Other Issues with Logistic Regression

6.4.1. Global Test of Model and Testing a Group of Predictors

A simultaneous test of all predictors in a logistic model should be performed similar to multiple linear regression as described previously in **Chapter 9**. Before proceeding to an individual hypothesis test, a global test of all the predictors is computed using a likelihood ratio (LR) test. The LR statistic has approximately a chi-square distribution with degrees of freedom equal to the number of predictors in the model, including all design variables, interaction terms, and terms to allow for curvature. For example, a model with TST and Genotype that allows for interaction would include 5 degrees of freedom: 1 for TST, 2 for design variables for genotype, and 2 for interaction between TST and Genotype. Statistical packages call the global test the LR chi-square test or Model chi-square test. If the Model chi-square test is significant at the 0.05 level, then one proceeds to test predictors individually or subsets of predictors using Wald or LR tests. For example, the model in **Section 6.2** with 3 degrees of freedom (TST, GT1, and GT2) has an LR test chi-square = 30.7 (P value < 0.001; 3 d.f.).

LR and Wald tests can also be performed on subsets of variables or pooled tests ***(10)***. For example, we can perform a pooled test on all terms that involve genotype (e.g., GT1 and GT2 as in **Table 10**). To test the overall genotype effect, we can perform a multiple degree of freedom test by removing genotype and obtaining the LR test chi-square value. For example, the model with TST alone (1 d.f.) has an LR test chi-square = 21.9. The difference between the LR chi-square values of the 2 models, 30.7 − 21.9 = 8.8, represents the pooled effect of genotype with 2 degrees of freedom. A chi-square = 8.8 with 2 d.f. has a P value = 0.012 and this is the P value presented in **Table 10** for the overall genotype effect. Similarly, if genotype has 4 terms, such as a model with interaction (GT1, GT2, GT1 × TST, GT2 × TST), then the difference between the LR chi-square values for the overall model and a model with only TST would give the overall contribution of genotype with one pooled test.

6.4.2. Assessing Lack of Fit and Influential Data

Once allowance is made for suspected nonlinear trends in the continuous predictors and interaction effects, influential observations should be checked and goodness-of-fit tests performed using standard statistical software. One common measure to check for influential points is to use the leverage statistic ***(35)***. Leverage is a measure of the overall influence of an observation on the coefficients. The influential observations are determined by plotting the probability of the event against leverage. Other typical regression diagnostics similar to linear regression can be performed as well ***(18)***.

Measures of global goodness-of-fit tests are numerous, but the most common are the Pearson chi-square and Hosmer-Lemeshow tests ***(18)***. By the definition of *good fit*, the logistic function should be the correct function and no more additional terms are needed for nonlinear or interaction effects. However, the P value is not appropriate for the Pearson chi-square test because the expected frequencies are not large enough. The Hosmer-Lemeshow test is sensitive to how its fixed groups of estimating probabilities are formed. These tests will likely not detect interaction effects if present unless the sample size is at least 500 ***(36)***.

Using the no interaction model in **Section 6.2** and SPSS statistical software, the Hosmer and Lemeshow test has a chi-square = 8.8 (P value = 0.27, 7 degrees of freedom). For the interaction model, the test has a chi-square = 4.7 (P value = 0.69, 7 degrees of freedom). However, because the sample size is less than 100 in this hypothetical data set, the goodness-of-fit test is not very sensitive in detecting lack of fit. Harrell ***(10)*** recommends to specify alternative hypotheses, such as nonlinear trends and interaction terms, because they are more powerful than goodness-of-fit tests.

6.4.3. Assessing Predictive Accuracy

The predictive ability of linear regression models is determined by R^2, the coefficient of multiple determination. R^2 is a measure of how well an individual subject will be predicted from a model. For logistic regression models, many *discrimination* measures are used including R^2-type measures, overall rate of classification, and the *c*-statistic. Additionally, the receiver operating characteristic (ROC) curve may be used and is discussed in **Chapter 6**.

The *c*-statistic is related to Somers' *D* rank correlation [$D = (c - 0.5)/0.5$] and identical to the area under the ROC curve ***(23,37)***. These statistics measure the extent to which a model's predicted probability agrees with the observed outcome (e.g., presence or absence of an event). A *c*-statistic of 0.50 suggests no discrimination, <0.70 poor, 0.70 to 0.79 acceptable, 0.80 to 0.89 excellent, and at least 0.90 outstanding ***(18)***.

R^2-type measures include R^2 (used for continuous outcomes), R_N^2 ***(38)***, pseudo R^2, and others ***(39)***. For binary outcomes, R^2 has a useful interpretation but tends to yield low values. Pseudo R^2 is common but the maximum value will not approach 1.0 in some cases. R_N^2 can vary from 0 to 1 and is often reported in software packages.

Classification tables cross-classify the number of correct outcomes with observed outcomes into a 2×2 table. The overall probability of correct classification or misclassification is given and is commonly reported in medical papers. However, classification tables are not recommended because logistic regression is a probabilistic model and is not to be used for estimating the occurrence of an event ***(10,18)***.

Although Homser and Lemeshow ***(18)*** do not recommend reporting R^2, we agree with Ash and Shwartz ***(37)*** and encourage the reporting of both *c* and R^2 statistics for logistic models. See Riester and others ***(31)*** for an illustration of *c*-statistics and ROC curves to compare models.

To illustrate the use of discrimination statistics, we use the models without and with interaction term in **Sections 6.2** and **6.3**. The model without interaction has a *c*-statistic = 0.88, Somer's $D = 0.75$, and $R_N^2 = 0.54$. The model with interaction has a *c*-statistic = 0.88, Somer's $D = 0.76$, and $R_N^2 = 0.55$. That is, very little predictive power is added when the interaction term is included.

In addition to reporting discrimination statistics such as the *c*-statistic, a model should ideally be validated as well. Validation refers to the accuracy of a model when applied in new patient samples, but it is usually not reported ***(27,29)***. There are internal and external procedures for evaluating validity

(40,41). Internal procedures refer to the use of sophisticated statistical techniques, such as the bootstrap or other resampling method. External procedures use a subset of the original sample that is left out before modeling takes place or use an entirely new sample. Whichever procedure is used, the statistics upon validation are commonly lower than estimated from the development sample. This may be due to differences between the actual samples used or because the developed model was overfitted and thus too optimistic. The risk of overfitting after extensive modeling using many predictors is high, especially in small data sets. An optimism-adjusted statistic can then be used instead of the original discrimination statistic. To validate a regression model, the Design Library of Harrell can be used ***(42)***.

With the data in **Table 1** and using **Equation 7** with no interaction, the Somer's D = 0.75 and R_N^2 = 0.54. After internally validating the model, the Somer's D = 0.69 and R_N^2 = 0.45. With 60 subjects and only 3 degrees of freedom, we have some concern for overfitting. With the interaction model in **Section 6.3**, the Somer's D = 0.76 and R_N^2 = 0.55. Upon validation, the optimism-adjusted discrimination statistics reduce to Somer's D = 0.68 and R_N^2 = 0.38. There is more than a 10% optimism in the Somer's D statistic and more than 30% optimism in R_N^2. The original model is overly optimistic when studying an interaction effect using only 60 subjects.

6.4.4. Sample Size/Power and Automated Selection Routines

Sample size should be sufficient to produce reliable and valid models. Power determinations can be made with software such as nQuery Advisor ***(43)*** or PASS ***(44)***. (See **Chapter 19** for further discussion on power.) For example, by using nQuery Advisor, sample size can be determined to have sufficient power to detect an increase in the odds ratio for a 1 standard deviation increase in a normally distributed predictor. This assumes the predictor has a prespecified relationship with the other predictors, measured by R^2. Using an $R^2 = 0.25$, a 2-tail significance level = 0.05, and assuming the probability of the outcome event is 50% for the average value of the predictor, a sample size of at least 125 subjects is required to achieve 80% power to detect a twofold increase in the odds when the predictor increases 1 standard deviation. If the probability of the outcome event is 25% instead of 50%, the sample size needed is 165 subjects. Oftentimes, power determinations are made based on simpler models, such as the chi-square test.

Although we encourage power determinations when prior information is available, investigators should also be concerned when the number of events per variable (EPV) in a logistic model is low ***(45)***. For logistic regression models to have accurate regression coefficients, a general rule of thumb is to have at least 10 events per variable. The number of events is defined to be the number in the less frequent category. In the hypothetical data from 60 subjects in **Table 1**, 28 subjects have renal disease and 32 do not. Based on the rule of thumb, we could only examine about 2 to 3 predictors because there are 28 subjects in the less frequent category.

It is important to note that the number of variables in the model includes all the terms excluding the intercept, for example, multiple terms for nonlinear trends, interaction, and design variables. For example, in the model above that allows for genotype by TST interaction, we have 1 TST term, 2 genotype terms, and 2 interaction terms for a total of 5 variables or degrees of freedom. Based on the rule of thumb, we need 50 subjects with renal dysfunction. Because RD occurs in about 50% of cases in our sample, we would need approximately 100 subjects total. Because there are only 60 subjects in the hypothetical study sample above, regression coefficients are imprecise as indicated in the confidence intervals and the validation section. Note that if RD was expected to occur 25% of the time, then we would still need 50 subjects with RD but 150 without, for a total of 200 subjects.

If there are too many predictors or terms, data reduction should take place. One common method uses bivariable selection to reduce the number of predictors, such as the *t*-test or chi-square test. However, this method is not recommended because it does not properly control for confounding variables. Another method of data reduction is to use automated selection routines, such as stepwise regression for logistic models. Stepwise selection can result in substantial bias of regression coefficients and odds ratios and, therefore, standard errors lose their interpretation ***(46)***. Automated methods also produce models that are unstable and not reproducible ***(47)***. However, automated procedures have been shown to predict accurately when EPV was greater than 20 ***(48)***.

Another method is to use variable clustering or some similar method to help to reduce the number of predictors ***(31,49)***. Alternatively, with a model that is overfitted, adjustments can be made to the model using a shrinkage factor or penalized maximum likelihood estimation (PMLE) ***(50,51)***. Unlike automated procedures, these methods have more potential when the EPV is low.

7. Conclusion

Logistic regression is used frequently in medical research, especially because these procedures are readily available in statistical software packages. Additionally, logistic models are easy to interpret because they are probability models and their coefficients can be expressed as odds ratios. Although they have been little used in applications involving molecular biology to date, logistic regression models have enormous potential for developing models using various forms of data, both genomic and clinical, to predict class membership, or the phenotype, of individual patients. However, the EPV is typically low for applications involving molecular biology. The methods described above to adjust for overoptimism using penalized maximum likelihood estimation have tremendous potential. For example, Antoniadis *(52)* has successfully applied penalized logistic regression to microarray data to classify acute leukemia patients where there are hundreds of predictors on less than 40 subjects (EPV < 1).

Campbell discusses the statistical issues involved in genetic and genomic tests *(53)*. The challenge is to evaluate efficiently the different predictive claims that may be associated with a gene chip when predicting class membership. For example, can a microarray gene expression pattern predict which patients have cancer and which do not *(53)*? Because there are thousands of potential predictors, the EPV is very low. Campbell suggests using logistic regression but to use resampling techniques such as the bootstrap to confirm analyses or to use training sets.

Logistic regression should become a valuable tool to determine the predictive role microarrays have in predicting class membership. However, care should be taken that models not be overfitted. Validation will become paramount as well as the PMLE and shrinkage methods ***(50,51)***. We strongly encourage good reporting of models and following the guidelines suggested previously ***(27–29)***.

References

1. Cox, D. R. (1958) The regression analysis of binary sequences. *J. R. Stat. Soc. Ser. B.* **20**, 215–42.
2. Berkson, J. (1955) Maximum likelihood and minimum chi-square estimates of the logistic function. *J. Am. Stat. Assoc.* **50**, 130–62.
3. Cornfield, J., Gordon, T., and Smith W. W. (1961) Quantal response curves for experimentally uncontrolled variables. *Bull. Int. Stat. Inst.* **38**, 91–115.
4. MeSH B. (2005) Bethesda: National Library of Medicine. Available at http://www.nlm.nih.gov/mesh/MBrowser.html. Retrieved April 2, 2007.
5. Mullner, M., Matthews, H., and Altman D. G. (2002) Reporting on statistical methods to adjust for confounding: a cross-sectional survey. *Ann. Intern. Med.* **136**, 122–6.
6. Tibshirani, R. (1982) A plain man's guide to the proportional hazards model. *Clin. Invest. Med.* **5**, 63–8.
7. Harrell, F. E. (1986) *SUGI Supplemental Library User's Guide, Version 5 Edition*. Cary, SAS Institute Inc., pp. 269–93.
8. Ojo, A. O., Held, P. J., Port, F. K., Wolfe, R. A., Leichtman, A. B., Young, E. W., Arndorfer, J., Christensen, L., and Merion, R. M. (2003) Chronic renal failure after transplantation of a nonrenal organ. *N. Engl. J. Med.* **349**, 931–40.
9. Baan, C. C., Balk, A. H., Holweg, C. T., van Riemsdijk, I. C., Matt, L. P., Vantrimpont, P. J., Niesters, H. G., and Weimar, W. (2000) Renal failure after

clinical heart transplantation is associated with the TGF-beta 1 codon 10 gene polymorphism. *J. Heart Lung Transplant.* **19**, 866–72.

10. Harrell, F. E., Jr. (2001) *Regression Modeling Strategies: With Applications to Linear Models, Logistic Regression, and Survival Analysis.* New York, Springer-Verlag.
11. Katz, M. H. (1999) *Multivariable Analysis: A Practical Guide for Clinicians.* New York, Cambridge University Press.
12. Bland, J. M., and Altman, D. G. (2000) Statistics notes. The odds ratio. *BMJ* **320**, 1468.
13. Slattery, M. L., Samowtiz, W., Ma, K., Murtaugh, M., Sweeney, C., Levin, T. R., and Neuhausen, S. (2004) CYP1A1, cigarette smoking, and colon and rectal cancer. *Am. J. Epidemiol.* **160**, 842–52.
14. Hishida, A., Matsuo, K., Tajima, K. Ogura, M., Kagami, Y., Taji, H., Morishima, Y., Emi, N., Naoe, T., and Hamajima, N. (2004) Polymorphisms of p53 Arg72Pro, p73 G4C14-to-A4T14 at exon 2 and p21 Ser31Arg and the risk of non-Hodgkin's lymphoma in Japanese. *Leuk. Lymphoma* **45**, 957–64.
15. Harrell, F. E., Jr., Lee, K. L., and Pollock, B. G. (1988) Regression models in clinical studies: determining relationships between predictors and response. *J. Natl. Cancer Inst.* **80**, 1198–202.
16. Kleinbaum, D. G. (1994) *Logistic Regression: A Self-learning Text.* New York, Springer-Verlag.
17. Dupont, W. D. (2002) *Statistical Modeling for Biomedical Researchers.* Cambridge, Cambridge University Press.
18. Hosmer, D. W., and Lemeshow, S. (2000) *Applied Logistic Regression.* New York, John Wiley & Sons.
19. Nick, T. G., and Hardin, J. M. (1999) Regression modeling strategies: an illustrative case study from medical rehabilitation outcomes research. *Am. J. Occup. Ther.* **53**, 459–70.
20. Walter, S. D., Feinstein, A. R., and Wells, C. K. (1987) Coding ordinal independent variables in multiple regression analyses. *Am. J. Epidemiol.* **125**, 319–23.
21. Ford, E. S., Mokdad, A. H., and Liu, S. (2005) Healthy Eating Index and C-reactive protein concentration: findings from the National Health and Nutrition Examination Survey III, 1988–1994. *Eur. J. Clin. Nutr.* **59**, 278–83.
22. Menard, S. (2004) Six approaches to calculating standardized logistic regression coefficients. *Am. Stat.* **58**, 218–23.
23. Harrell, F. E., Jr., Lee, K. L., and Mark, D. B. (1996) Multivariable prognostic models: issues in developing models, evaluating assumptions and adequacy, and measuring and reducing errors. *Stat. Med.* **15**, 361–87.
24. Adams, R. A., Sherer, M., Struchen, M. A., and Nick, T. G. (2004) Post-acute brain injury rehabilitation for patients with stroke. *Brain Inj.* **18**, 811–23.
25. Sherer, M., Hart, T., and Nick, T. G. (2003) Measurement of impaired self-awareness after traumatic brain injury: a comparison of the patient competency rating scale and the awareness questionnaire. *Brain Inj.* **17**, 25–37.
26. Sherer, M., Hart, T., Nick, T. G., et al. (2003) Early impaired self-awareness after traumatic brain injury. *Arch. Phys. Med. Rehabil.* **84**, 168–76.

27. Ottenbacher, K. J., Ottenbacher, H. R., Tooth, L., and Ostir, G. V. (2004) A review of two journals found that articles using multivariable logistic regression frequently did not report commonly recommended assumptions. *J. Clin. Epidemiol.* **57**, 1147–52.
28. Lang, T. A., and Secic, M. (1997) *How to Report Statistics in Medicine: Annotated Guidelines for Authors, Editors, and Reviewers*. Philadelphia, American College of Physicians.
29. Bagley, S. C., White, H., and Golomb, B. A. (2001) Logistic regression in the medical literature: standards for use and reporting, with particular attention to one medical domain. *J. Clin. Epidemiol.* **54**, 979–85.
30. Concato, J., Feinstein, A. R., and Holford, T. R. (1993) The risk of determining risk with multivariable models. *Ann. Intern. Med.* **118**, 201–10.
31. Riester, K. L. A., Peduzzi, P., Holford, T. R., Ellison, R. T., 3rd, and Donta, S. T. (1997) Statistical evaluation of the role of Helicobacter pylori in stress gastritis: applications of splines and bootstrapping to the logistic model. *J. Clin. Epidemiol.* **50**, 1273–9.
32. Matthews, J. N., and Altman, D. G. (1996) Statistics notes. Interaction 2: compare effect sizes not P values. *BMJ* **313**, 808.
33. Altman, D. G., and Bland, J. M. (2003) Interaction revisited: the difference between two estimates. *BMJ* **326**, 219.
34. Farewll, V. T. (1998) Interaction, In: Armitage, P., and Colton, T., eds. *Encyclopedia of Biostatistics.* New York, John Wiley & Sons, pp. 2060–2061.
35. Pregibon, D. (1981) Logistic regression diagnositcs. *Ann. Stat.* **9**, 705–24.
36. Hosmer, D. W., Hosmer, T., Le Cessie, S., and Lemeshow, S. (1997) A comparison of goodness-of-fit tests for the logistic regression model. *Stat. Med.* **16**, 965–80.
37. Ash, A., and Shwartz, M. (1999) R2: a useful measure of model performance when predicting a dichotomous outcome. *Stat. Med.* **18**, 375–84.
38. Nagelkerke, N. J. D. (1991) A note on a general definition of the coefficient of determination. *Biometrika* **78**, 691–2.
39. Mittlbock, M., and Schemper, M. (1996) Explained variation for logistic regression. *Stat. Med.* **15**, 1987–97.
40. Steyerberg, E. W., Harrell, F. E., Jr., Borsboom, G. J., Eijkemans, M. J., Vergouwe, Y., and Habbema, J. D. (2001) Internal validation of predictive models: efficiency of some procedures for logistic regression analysis. *J. Clin. Epidemiol.* **54**, 774–81.
41. Steyerberg, E. W., Bleeker, S. E., Moll, H. A., Grobbee, D. E., and Moons, K. G. (2003) Internal and external validation of predictive models: a simulation study of bias and precision in small samples. *J. Clin. Epidemiol.* **56**, 441–7.
42. Harrell, F. E. (2005) Design Library. Available at http://biostat.mc.vanderbilt.edu/twiki/bin/view/Main/RS.
43. Elashoff, J. (2005) *nQuery Advisor Version 6.0 User's Guide*. Los Angeles, Statistical Solutions.
44. Hintze, J. (2002) *PASS*. Kaysville, NCSS Statistical Software.

45. Peduzzi, P., Concato, J., Kemper, E., Holford, T. R., and Feinstein, A. R. (1996) A simulation study of the number of events per variable in logistic regression analysis. *J. Clin. Epidemiol.* **49**, 1373–9.
46. Steyerberg, E. W., Eijkemans, M. J., and Habbema, J. D. (1999) Stepwise selection in small data sets: a simulation study of bias in logistic regression analysis. *J. Clin. Epidemiol.* **52**, 35–42.
47. Austin, P. C., and Tu, J. V. (2004) Automated variable selection methods for logistic regression produced unstable models for predicting acute myocardial infarction mortality. *J. Clin. Epidemiol.* **57**, 1138–46.
48. Ambler, G., Brady, A. R., and Royston, P. (2002) Simplifying a prognostic model: a simulation study based on clinical data. *Stat. Med.* **21**, 3803–22.
49. Harrell, F. E., Jr., Margolis, P. A., Gove, S., Mason, K. E., Mulholland, E. K., Lehmann, D., Muhe, L., Catchalian, S., and Eichenwald, H. F. (1998) Development of a clinical prediction model for an ordinal outcome: the World Health Organization Multicentre Study of Clinical Signs and Etiological Agents of Pneumonia, Sepsis and Meningitis in Young Infants. WHO/ARI Young Infant Multicentre Study Group. *Stat. Med.* **17**, 909–44.
50. Moons, K. G., Donders, A. R., Steyerberg, E. W., and Harrell, F. E. (2004) Penalized maximum likelihood estimation to directly adjust diagnostic and prognostic prediction models for overoptimism: a clinical example. *J. Clin. Epidemiol.* **57**, 1262–70.
51. Steyerberg, E. W., Borsboom, G. J., van Houwelingen, H. C., Eijkemans, M. J., and Habbema, J. D. (2004) Validation and updating of predictive logistic regression models: a study on sample size and shrinkage. *Stat. Med.* **23**, 2567–86.
52. Antoniadis, A. (2003) Penalized logistic regression and classification of microarray data. Available at http://www.bioconductor.org/workshops/2003/Milan/Lectures/anestisMilan3.pdf. Assessed April 2, 2007.
53. Campbell, G. (2004) Some statistical and regulatory issues in the evaluation of genetic and genomic tests. *J. Biopharm. Stat.* **14**, 539–52.

15

Survival Analysis

Hongyu Jiang and Jason P. Fine

Summary

This chapter introduces some fundamental results in survival analysis. We first describe what is censored failure time data and how to interpret the failure time distribution. Two nonparametric methods for estimating the survival curve, the life table estimator and the Kaplan-Meier estimator, are demonstrated. We then discuss the two-sample problem and the usage of the log-rank test for comparing survival distributions between groups. Lastly, we discuss in some detail the proportional hazards model, which is a semiparametric regression model specifically developed for censored data. All methods are illustrated with artificial or real data sets.

Key Words: Actuarial estimator; Cox model; nonparametric methods; product-limit estimator; rank testing; right censoring; semiparametric regression.

1. Introduction

Survival analysis is a branch of biostatistical methods for analyzing data representing times to the occurrence of some specific event, which are often referred to as *failure time*, *survival time*, or *lifetime*. Time to a particular event is often of interest in medical or biological studies. As prolonging survival is the ultimate goal of medicine, time from an appropriately defined origin to death is the most frequently used marker for intervention effect. For example, in a lung cancer clinical trial, time from the start of treatment to death is often used to evaluate whether new therapy is superior to the standard treatment in prolonging survival. In some other situations, a marker event of disease progression might be considered. For example, in HIV clinical trials, time from the start of treatment to viral rebound is the most frequently used end point for comparing effects of antiretroviral treatments. However, not all the events considered are negative. In this chapter, we follow the convention in survival

From: *Methods in Molecular Biology, vol. 404: Topics in Biostatistics*
Edited by: W. T. Ambrosius © Humana Press Inc., Totowa, NJ

analysis and refer to the event times as *failure times*. Statistically, we denote this time T by a nonnegative valued random variable.

Appropriately defining a time-to-event variable is sometimes not a trivial task, especially for a retrospective observational study. According to Cox and Oakes *(1)*, a well-defined failure time needs to satisfy three requirements:

1. A time origin must be unambiguously defined.
2. A scale of measuring the passage of time must be agreed.
3. The meaning of *failure* must be entirely clear.

Note that a commonly defined time origin does not mean that all subjects have to be followed from the same calendar date. Most studies allow subjects to have individual entry dates, in other words, staggered entry into the study.

In most situations, failure times observed from different individuals can be assumed to be independently distributed, while distributions of failure times under distinct treatments may differ from each other. To describe the random behavior of failure times under different treatment, we use the so-called survival function, which is defined as: $S(t) \equiv Pr(T > t)$, where t is any positive time point. $S(t)$ can be interpreted as the proportion of subjects not experiencing the event of interest at time t. Some examples of survival functions are displayed in the left panel of **Figure 1**. For example, at time 20 (say the unit is in months), the

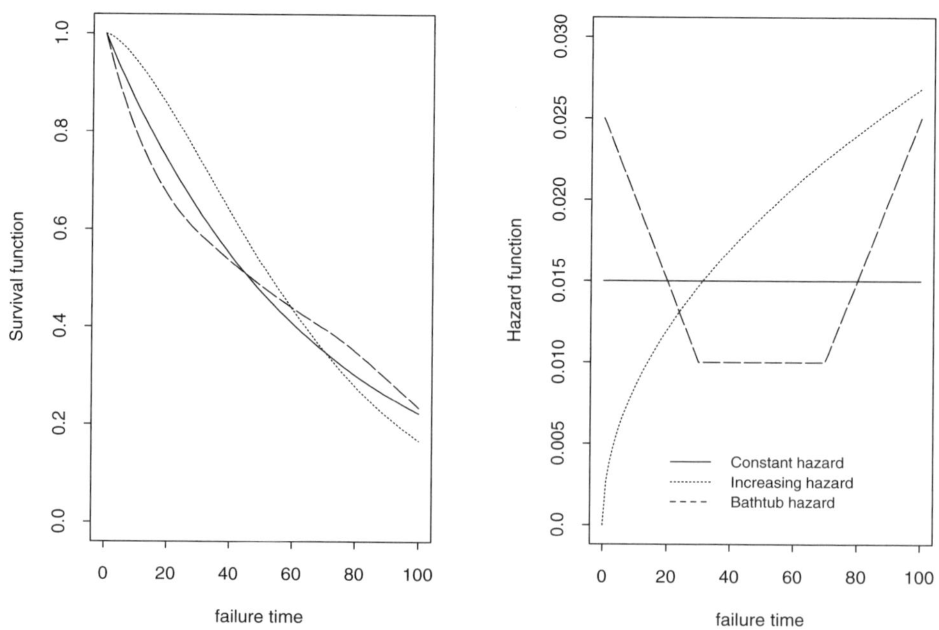

Fig. 1. Examples of survival function and hazard function.

survival probability is 85% for the dotted curve, 74% for the solid curve, and 67% for the dashed curve. However, at 80 months, the dashed curve has the largest survival probability, 35%, while the solid and dotted curves have survival probabilities of 30% and 28%, respectively. The function $S(t)$ is also referred to as survival curve or event-free probability curve. Survival functions share some common properties:

1. At time $t = 0$, $S(t) = 1$, which means no subject fails at the time origin.
2. Similar to proportions, $S(t)$ takes values between 0 and 1. Depending on the failure mechanism, $S(t)$ may reach 0 at infinity or some positive time point.
3. $S(t)$ is a nonincreasing function over time.
4. $S(t) = 1 - F(t)$, where $F(t)$ is the cumulative distribution function, which is often used in describing distribution for nonfailure time type random variables.

Although the survival curve may provide the event-free probability for a specific study population, the probabilities may not be useful in determining an individual's risk of failure at specific time points. For example, a 30% 3-year survival rate under a particular treatment may not be relevant to a subject who is still alive at 3 years after initiating the treatment. For describing this individual's risk of failure at 3 years after treatment, one may use the *hazard function*, which is usually denoted as $\lambda(t)$. The hazard function is interpreted as the instantaneous failure rate at time t given that a subject has already survived to t. In statistical notation,

$$\lambda(t) = \lim_{\Delta t \to 0} \frac{\Pr(t \le T < t + \Delta t | T \ge t)}{\Delta t},$$

where Δt denotes a small interval of time.

From the definition of the hazard function, it is obvious that the function can only take nonnegative values because probabilities must be nonnegative. But, unlike the survival curve, which is monotonically decreasing, hazard functions may take a variety of shapes, including special cases like constant over time, monotonically increasing or decreasing, and a "bathtub" shape (decreasing in the beginning, plateau in the middle, and increasing later on). The right panel of **Figure 1** presents the hazard functions corresponding with the survival curves in the left panel of **Figure 1**. The solid and dotted survival curves correspond with the constant and monotonically increasing hazard functions over time, respectively, whereas the dashed survival curve is driven by a bathtub-shaped hazard function.

Survival functions and hazard functions are related through the following identity:

$$S(t) = \exp\left(-\int_0^t \lambda(u)\,du\right).$$

Hence, knowing one of these quantities is sufficient to determine the other. For example, the exponentially distributed failure time has constant hazard function λ_0, and the survival function is $S(t) = \exp(-\lambda_0 t)$.

2. Censoring versus Failure

Fortunately in most medical or biological studies, not all subjects would fail during the limited follow-up time. When a failure time is not observed before the end of follow-up, we say the failure time is *right censored* because it is only known to lie on the right side of the end of follow-up. This phenomenon is referred to as *right censoring*, where the follow-up period is the so-called censoring time, which is often denoted by the random variable C. There are many ways of representing right-censored failure time data. One way is as in the following example, where a "+" is attached to observation times where the failure time is censored.

Example 1

In a lung cancer clinical trial with study duration of 8 months and staggered patient entry, the observed survival times in months under a particular treatment are as follows:

$$5,\ 4+,\ 7,\ 1,\ 7+,\ 5,\ 3.$$

Notice that in this example, there is an observed failure time tied with a censored failure time at 7 months.

The above method is clear and simple for displaying survival data but may not be convenient if we want to construct estimators from the data representation. A more convenient way of representing the observed survival data makes use of two additionally defined random variables. The first is the total length of observation time Y, which could represent a true failure time, or the total follow-up period without observing a failure. The second is a binary indicator δ, which takes value 1 if the observed time is the true failure time and value 0 if the observed time is the censoring time. In other words, $\delta = I\{T \leq C\}$, where C denotes the censoring time, and $I\{.\}$ is the indicator function. The observed survival data from a random sample of size n can be expressed as $\{(Y_i,\delta_i), i = 1, \ldots, n\}$. For example, the above data can be represented as in the following way:

Y	5	4	7	1	7	5	3
δ	1	0	1	1	0	1	1

Besides right censoring, there exist other types of censoring, although less frequently observed. One is left censoring, which means for some subjects, the failure time is only known to be shorter than a certain duration. Left-censored data can be easily converted to artificially right-censored data ***(2)*** and may be analyzed using the methods that will be presented in the later sections. Another type is interval censoring, which means the failure time is only known to lie inside an interval $[C_l, C_u]$ (i.e., $C_l < T < C_u$). A special case of interval-censored data is the so-called current status data, in which each subject's status is only evaluated at a single time point C_t. If the event has already occured prior to C_t, then the current status is coded 1; if the event has not yet happened, then the status is coded 0. Methods for right-censored data are not applicable to interval-censored data. Other special methods are needed but are beyond the scope of this chapter. For interested readers, methods for interval-censored data can be found in advanced textbooks ***(3,4)***.

In this chapter, we focus on inferential methods for right-censored data assuming that T and C are independent. That is, the time at which a failure time is right-censored is assumed to carry no information about when the failure event will happen. With administrative loss to follow-up occurring at the end of a study, censoring is clearly independent. However, when a patient drops off a study prior to this time, censoring may be related to disease processes and potentially informative.

In the presence of right censoring, the estimation of survival function becomes quite complicated. If the data were completely observed, we may simply use the sample proportion of subjects surviving beyond time t to estimate the survival probability $S(t)$. However, with censored data, the sample proportion is no longer computable when there are subjects whose failure times were censored prior to t. In the next two sections, two methods for estimating survival function with right-censored data will be introduced. Both methods are nonparametric, meaning that neither makes any assumption on the underlying true survival distribution. They are the standard analyses in biomedical research for censored data.

3. Life Table Methods

The *life table* method is probably the oldest method in survival analysis. It was developed to track mortality behavior in large populations using annual census data. Because the method groups survival data into time intervals, the method can be computed fairly easily by hand, which is why it was very useful in the precomputer age. The life table estimate of the survival function is also called the actuarial estimate because the method is used routinely in actuarial science.

With a given survival data set, the steps for calculating the life table estimator of survival function is as follows:

1. First, divide the time scale into m intervals, which may not be of same length, such that all the observed failure times fall into one of the intervals.
2. For the ith interval, count the number of failure events, D_i, and the number of censored observations, C_i, in that interval.
3. Determine N_i, the average number of subjects at risk (e.g., not experiencing the failure event) during the ith interval. Usually, the formula, $N_i = N_i^0 - C_i/2$ is used, where N_i^0 denotes the number of subjects at risk in the beginning of the ith interval. Note that $N_{i+1}^0 = N_i^0 - D_i - C_i$.
4. Compute the probability of surviving to the end of the ith interval given survival to the beginning of the ith interval as $(N_i - D_i)/N_i$.
5. Compute the life table estimate of survival probability at the end of ith interval:

$$\hat{S}_i^{LT} = \prod_{j=1}^{i} \frac{N_i - D_i}{N_i}.$$

The estimated survival curve can be obtained by interpolating the above estimates at the end of each interval.

The life table estimator makes implicit assumptions that death and censoring occur at uniform rates during any particular interval.

For the above **Example 1**, we consider the following intervals: [0,2), [2,4), [4,6), and [6,8), which are numbered from 1 to 4. (Note that the interval $[a,b)$ includes $a \le t < b$.) Hence, D_i, C_i, N_i^0, and N_i can be determined as follows:

	Intervals			
	1	2	3	4
D_i	1	1	2	1
C_i	0	0	1	1
N_i^0	7	6	5	2
N_i	7	6	4.5	1.5

The life table estimates of survival probabilities at the end of each interval are computed as:

$$\hat{S}_1^{LT} = \left(\frac{7-1}{7}\right) = \frac{6}{7} = 0.86$$

$$\hat{S}_2^{LT} = \hat{S}_1^{LT}\left(\frac{6-1}{6}\right) = \frac{6}{7} \times \frac{5}{6} = 0.71$$

$$\hat{S}_3^{LT} = \hat{S}_2^{LT}\left(\frac{4.5-2}{4.5}\right) = \frac{6}{7} \times \frac{5}{6} \times \frac{2.5}{4.5} = 0.40$$

$$\hat{S}_4^{LT} = \hat{S}_3^{LT}\left(\frac{1.5-1}{1.5}\right) = \frac{6}{7} \times \frac{5}{6} \times \frac{2.5}{4.5} \times \frac{0.5}{1.5} = 0.13.$$

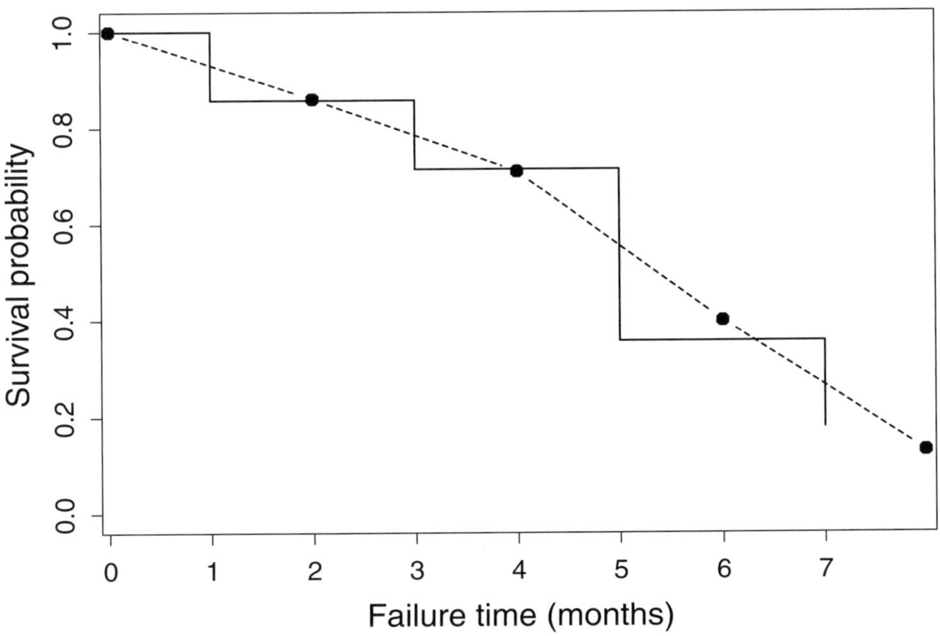

Fig. 2. Comparison of survival curves of time to relapse. The dashed line is the life table estimate, and the solid line is the Kaplan-Meier curve.

The life table estimate of $S(t)$ is as shown in the dashed line in **Figure 2**, where the solid dots indicate the point estimates at the end of each time interval.

4. Kaplan-Meier Curves

For census data that are naturally grouped, the life table estimator is a good choice for estimating the survival curve. However, for usual right-censored survival data that are not grouped, artificially grouping the data may not be efficient. A more efficient nonparametric method is the so-called Kaplan-Meier estimator, which is sometimes referred to as the product-limit estimator ***(5)***.

To discuss the Kaplan-Meier estimator, the following notation is needed. Let $t_1 < t_2 < \ldots < t_m$ be the m ordered, unique, uncensored event times, d_j be the number of subjects who have failure events at time t_j, and r_j be the number of subjects who have not failed before time t_j, that is, at risk at t_j, where $j = 1, \ldots, m$. One can easily verify that

$$d_j = \sum_{i=1}^{n} I\{Y_i = t_j \text{ and } \delta_i = 1\} \quad \text{and} \quad r_j = \sum_{i=1}^{n} I\{Y_i \geq t_j\}.$$

The Kaplan-Meier estimator of $S(t)$ is defined on the range of the observed data, $[0,\tau_n]$, where $\tau_n = \max(Y_i, i = 1, \ldots, n)$:

$$\hat{S}(t) = \begin{cases} 1 & \text{if} \quad t < t_1 \\ \prod_{j:t_j \le t} \left(1 - \frac{d_j}{r_j}\right) & \text{if} \quad t_1 \le t \le \tau_n. \end{cases} \tag{1}$$

Unlike the life table estimator, the Kaplan-Meier curve is a right-continuous step function that only jumps at the uncensored failure times. For $t > \tau_n$, if there is no censoring at the largest observed time τ_n, then $\hat{S}(t) = 0$, otherwise, $\hat{S}(t)$ is undefined. Efron ***(6)*** proposed estimating $S(t)$ beyond τ_n by 0, and Gill ***(7)*** suggested $\hat{S}(\tau_n)$. The approaches are equivalent in large samples. Klein ***(8)*** showed that in small samples, Gill's estimator has smaller bias in the tail of the distribution.

The variance of the Kaplan-Meier estimator at time t, $\sigma^2(t)$, can be estimated by *Greenwood's formula* ***(9)***:

$$\hat{\sigma}^2(t) = \hat{S}(t)^2 \sum_{j:t_j \le t} \frac{d_j}{r_j(r_j - d_j)}. \tag{2}$$

With no censoring, **Equation 2** reduces to the standard binomial variance estimator, $n^{-1}\hat{S}(t)[1 - \hat{S}(t)]$. In large samples, $\hat{S}(t)$ is approximately normally distributed with variance $\sigma^2(t)$. Using Greenwood's formula, a $100(1 - \alpha)\%$ pointwise confidence interval for $S(t)$ can be constructed using a normal approximation: $\hat{S}(t) \pm Z_{\alpha/2}\hat{\sigma}(t)$, where $Z_{\alpha/2}$ is the $(1 - \alpha/2)$th quantile of the standard normal distribution.

We can demonstrate the computation of the Kaplan-Meier estimator using **Example 1**. First we order the distinct uncensored failures times, which are 1, 3, 5, 7. We then determine how many subjects failed and are at risk at each of those time points. These are d_i and r_i, respectively. The details are provided in the following table:

	Uncensored times t_i			
	1	3	5	7
d_i	1	1	2	1
r_i	7	6	4	2

Next, we plug in the above numbers into **Equation 1** to obtain the Kaplan-Meier estimates at t_i's:

$$\hat{S}(1) = \left(1 - \frac{1}{7}\right) = \frac{6}{7} = 0.86$$

$$\hat{S}(3) = \hat{S}(1)\left(1 - \frac{1}{6}\right) = \frac{6}{7} \times \frac{5}{6} = 0.71$$

$$\hat{S}(5) = \hat{S}(3)\left(1 - \frac{2}{4}\right) = \frac{6}{7} \times \frac{5}{6} \times \frac{2}{4} = 0.36$$

$$\hat{S}(7) = \hat{S}(5)\left(1 - \frac{1}{2}\right) = \frac{6}{7} \times \frac{5}{6} \times \frac{2}{4} \times \frac{1}{2} = 0.18.$$

The Kaplan-Meier curve is illustrated in the solid line in **Figure 2**.

5. Log-Rank Test

In practice, we are frequently interested in formally comparing two survival functions rather than just estimating the curves. In a randomized clinical trial with two treatment arms, experimental versus control, the goal is to evaluate if the experimental treatment improves the survival experience compared with the control arm. To answer this question, we need to conduct hypothesis testing. (See **Chapter 4** for a complete discussion of hypothesis testing.)

Because we are comparing the entire survival experience over the observable range, the null hypothesis has to be formulated with the distribution function, not just the probability at a particular time point. One way to express the null hypothesis is in terms of the survival functions. That is, $H_0: S_1(t) = S_2(t)$ for $t \in [0, \tau_n]$. Equivalently, we may also assume that under the null hypothesis, the hazard functions in the two treatment groups are the same, that is, $H_0: \lambda_1(t) = \lambda_2(t)$ for $t \in [0, \tau_n]$.

With two-sample right-censored data, the above hypothesis can be tested using a nonparametric test called the *log-rank test* ***(10,11)***. The advantage of the log-rank test is that it does not make any specific assumption about the distribution of failure time. The idea of the log-rank test is to create a two-by-two table comparing the number of failure events relative to the numbers at risk between groups at each distinct uncensored failure time and then combine information from tables at all uncensored time points into a single chi-square distributed test statistic.

Specifically, assume there are k uncensored failure times. At the ith uncensored failure time point t_i, let r_i and r_{1i} be the number at risk in the pooled sample and in group 1, respectively; let d_i and d_{1i} be the number of failure events in the pooled sample and in group 1, respectively. Then at t_i, the observed number of failures in group 1, denoted as O_i, equals d_{1i}. Under the null hypothesis of no difference in survival experience between the two groups, we would expect the number of failure events occuring at time t_i in group 1, denoted as E_i, to be proportional to the size at risk in group 1 relative to the total size at risk. That is, conditional on (d_i, r_{1i}, r_i), we expect $E_i = d_i r_{1i} / r_i$.

The log-rank statistics is formulated as

$$\text{LR} = \frac{\left[\sum_{i=1}^{k} (O_i - E_i)\right]^2}{\sum_{i=1}^{k} \text{Var}(O_i - E_i)}, \tag{3}$$

which has chi-square distribution with 1 degree of freedom, where

$$\mathrm{Var}(O_i - E_i) = \frac{d_i(r_i - d_i)r_{1i}(r_i - r_{1i})}{r_i^2(r_i - 1)}.$$

Note that it does not matter which group we choose to compute the log-rank statistic—the results should be identical. Log-rank test can also be used to compare the failure time distributions for more than two groups. If the total number of groups is g, we simply need to select any $g - 1$ groups and sum up the log-rank statistic as in **Equation 3** for each group. The resulting statistic is

$$\mathrm{LR} = \frac{\left[\sum_{i=1}^{k}(O_{i1} - E_{i1})\right]^2}{\sum_{i=1}^{k}\mathrm{Var}(O_{i1} - E_{i1})} + \ldots + \frac{\left[\sum_{i=1}^{k}(O_{i,g-1} - E_{i,g-1})\right]^2}{\sum_{i=1}^{k}\mathrm{Var}(O_{i,g-1} - E_{i,g-1})},$$

which follows a chi-square distribution with $g - 1$ degrees of freedom.

To illustrate the log-rank test, we assume that the observations in **Example 1** are from subjects randomized to the control treatment. There are another 7 subjects who were randomized to the experimental treatment. The pooled data set is analyzed in the following example.

Example 2

In a randomized lung cancer clinical trial, the observed survival times (in months) under two study treatments are as follows:

- Experimental arm: 2+, 5, 6+, 7, 7+, 8+, 8+
- Control arm: 5, 4+, 7, 1, 7+, 5, 3

In the two arms, there are altogether 4 distinct uncensored failure times. Letting the experimental arm be group 1, the computation of the observed and expected number of failures in group 1 at each time point is illustrated below:

t_i	O_i	d_i	r_i	r_{1i}	E_i	$O_i - E_i$	$\mathrm{Var}(O_i - E_i)$
1	0	1	14	7	$\frac{1}{2}$	$-\frac{1}{2}$	0.25
3	0	1	12	6	$\frac{1}{2}$	$-\frac{1}{2}$	0.25
5	1	3	10	6	$\frac{9}{5}$	$-\frac{9}{5}$	0.56
7	1	2	6	4	$\frac{4}{3}$	$-\frac{1}{3}$	0.36
Total						−1.63	1.42

Hence, the log-rank statistic is $\frac{(-1.63)^2}{1.42} = 3.22$ with P value = 0.07.

6. Proportional Hazards Model

We need to fit a regression model on survival data for several possible reasons. First, the log-rank test can only help us test the treatment effect and cannot quantify the difference in efficacy between the groups. A regression model can help us assess the effect size. Second, many studies do not have a randomized design, which means the intervention effect could be confounded by other factors that may also affect the failure time distribution. Thus, it may be helpful to adjust the intervention effect for other possible confounding factors through a multiple regression model. Third, for some studies, the main purpose may not be to test treatment effects but to investigate the relationship among the survival outcome and some risk factors. It is of more interest to create a predictive model for survival with the prognostic factors.

With right-censored failure time data, the usual linear regression model cannot be used because the least squares method is no longer feasible. Cox developed a special semiparametric regression model for censored failure time data, which is the famous *proportional hazards (PH) model* or the *Cox model* ***(12)***. This model assumes that covariates have multiplicative effects on the baseline hazard function. Let $\mathbf{x_i}$ denote the $p \times 1$ vector of covariates for ith subject, $\mathbf{x_i} = (x_{i1}, x_{i2}, \ldots, x_{ip})^t$, and β the $1 \times p$ vector of coefficients. The PH model assumes that

$$\lambda(t;\mathbf{x_i}) = \lambda_0(t)e^{\beta\mathbf{x_i}} = \lambda_0(t)e^{\beta_1 x_{i1}+\beta_2 x_{i2}+\ldots+\beta_p x_{ip}}, \tag{4}$$

where $\lambda_0(t)$ corresponds with the hazard function at t for subject whose covariate values all equal to 0. $\lambda_0(.)$ is often referred to as the baseline hazard function.

The PH model has several important properties:

- The PH model is semiparametric because $\lambda_0(t)$ is left unspecified.
- Under a PH model, the hazard ratio of two subjects with fixed covariates is a constant over time. That is, $\dfrac{\lambda(t;\,\mathrm{x_i})}{\lambda(t;\,\mathrm{x_j})} = \dfrac{\lambda_0(t)e^{\beta\mathrm{x_i}}}{\lambda_0(t)e^{\beta\mathrm{x_j}}} = e^{\beta(\mathrm{x_i}-\mathrm{x_j})}$.
- The coefficients can be interpreted as the logarithm of hazard ratio. A positive coefficient means the hazard of failing increases with each unit increase in the covariates. A negative coefficient means the risk decreases with increasing covariate values.

Coefficients in a PH model can be estimated by maximizing the so-called partial likelihood ***(1,13)***. At each uncensored failure time t_i, let R_i be the risk set at t_i, which includes all the indexes of subjects who have not failed prior to t_i. The partial likelihood can be expressed as

$$\mathrm{PL}(\beta) = \prod_{i=1}^{n}\left[\frac{e^{\beta\mathrm{x_i}}}{\sum_{j\in R_i} e^{\beta\mathrm{x_j}}}\right]^{\delta_i}. \tag{5}$$

Each term in the partial likelihood can be interpreted as

Pr(individual with $\mathbf{x_i}$ fails at t_i|one individual from risk set R_i fails).

The maximum partial likelihood estimate of β, $\hat{\beta}$, has properties similar to those of a regular maximum likelihood estimator for a parametric model. The properties of the estimator are insensitive to the underlying baseline failure time distribution ***(1,4)***.

The original PH model does not allow ties in the uncensored failure times or covariates with changing values over time (time-dependent covariates). However, the PH model has since been modified to permit ties and time-dependent covariates, and these features have been incorporated into many commercial software packages ***(3,4)***.

With $\hat{\beta}$, we may obtain the nonparametric estimate of the baseline survival function as $\hat{S}_0(t) = \exp[-\hat{H}_0(t)]$, where $\hat{H}_0(t)$ is the Breslow estimator of the baseline cumulative hazard function ***(14)***:

$$\hat{H}_0(t) = \sum_{t_i \le t} \frac{d_i}{\sum_{j \in R_i} e^{\hat{\beta}\mathbf{x_i}}}.$$

Thus, for a subject with covariate values $\mathbf{x_i}$, the subject-specific survival curve can be estimated as $[\hat{S}_0(t)]^{\exp[\hat{\beta}\mathbf{x_i}]}$. This is very useful in graphically comparing survival curves for subjects with different covariate values.

The validity of the proportional hazards assumption should be checked for each covariate included in the model. There are in general two ways of checking the assumption. One is using diagnostic plots. For example, we may plot the log[–log] transformed Kaplan-Meier curves for different values of a covariate to see if the curves roughly have constant distances. We may also use diagnostic plots of the so-called martingale residuals (a pseudoresidual from the PH model), which allows checking the proportionality assumption adjusting for other covariates ***(15–17)***. The residual plot may be used to check the proportional hazards assumption for a particular covariate, as well as to determine whether a covariate should be included in the model and what functional form the covariate should take. A detailed review can be found in Therneau and Grambsch ***(18)***. Another way of checking the PH assumption is testing whether there is significant interaction between covariate effect and time. This statistical testing approach is less subjective but may be sensitive to small departures from proportionality when the sample size is large.

7. Application

We will finish this chapter with application of the above methods to a real clinical trial example ***(19)***.

Example 3

In a clinical trial, 21 pairs of children with acute leukemia matched by remission status were recruited. In each pair, one child was randomly assigned to the drug 6-mercaptopurine (6-MP) and the other to the control treatment. Patients were followed from randomization until their leukemia returned (relapse) or until the end of the study. Following are the observed data:

6-MP arm: 6,6,6+,6+,7,9+,10,10+,11+,13,16,17+,19+,20+,22,23, 25+,32+,32+,34+,35+.

Control arm: 1,1,2,2,3,4,4,5,5,8,8,8,8,11,11,12,12,15,17,22,23.

The Kaplan-Meier estimates of the survival curves of time to relapse for the two arms are plotted in **Figure 3**. The thick solid line is the Kaplan-Meier estimator of the survival function for the 6-MP arm, and the thin solid lines are the pointwise upper and lower 95% confidence intervals. Similarly, the thick dotted line and thin dotted lines are the Kaplan-Meier estimate and its pointwise confidence limits, respectively, for the control arm. It is easy to see that patients receiving 6-MP treatment have much higher survival probabilities than patients receiving the control treatment.

We define a treatment indicator, *X*, taking value 1 for the 6-MP arm and 0 for the control arm. The two-sample log-rank test indicates that the two

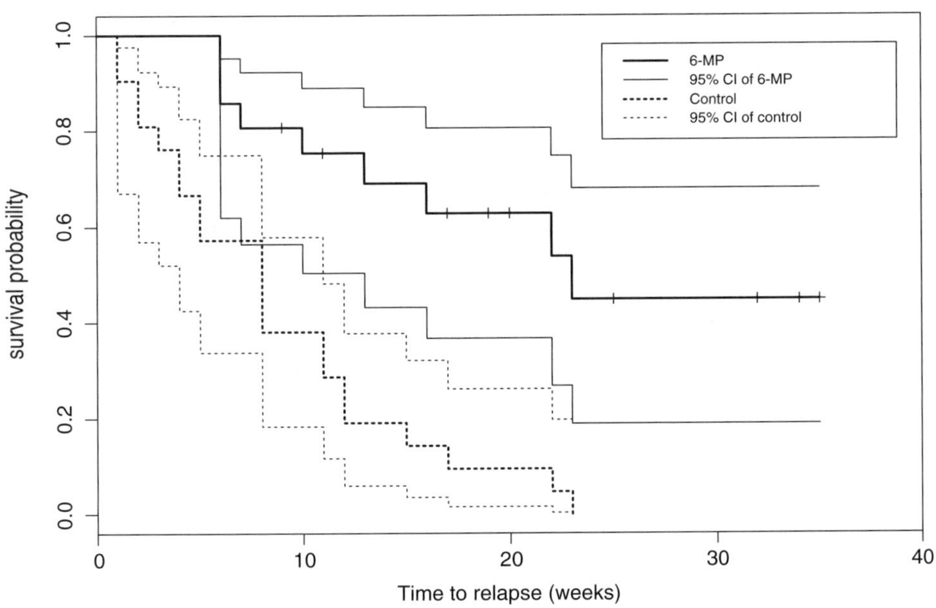

Fig. 3. Comparison of survival curves of time to relapse.

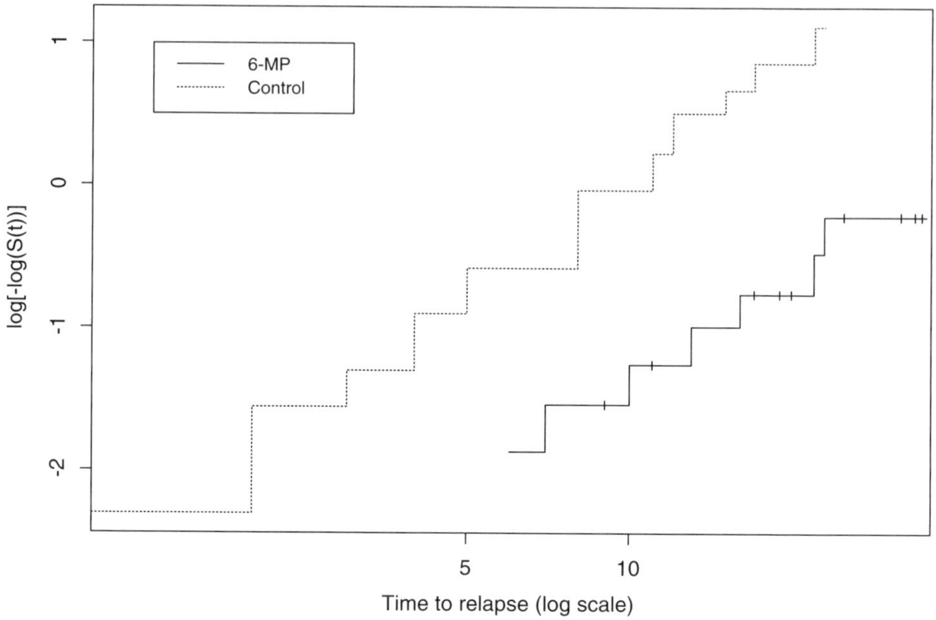

Fig. 4. Checking the proportional hazards assumptions.

treatments differ significantly in terms of survival (test statistic = 16.8, *P* value <0.0001). To quantify the difference in treatment effect, we fit a univariate proportional hazards model with *X* being the only predictor. The coefficient for *X* is estimated to be –0.786. The hazard ratio of the 6-MP arm to the control arm is exp(–0.786) = 0.46, which means 6-MP treatment reduces the hazard of death by about 54% compared with the control treatment.

As shown in **Figure 4**, we plotted the log[–log] transformed Kaplan-Meier curves for the two treatment groups versus time in log scale to check the proportional hazards assumption between the groups. The distance between the two curves is roughly constant over time. Given the sample size, this plot does not provide strong evidence for lack-of-fit of the Cox model.

8. Conclusion

In this chapter, we first introduced some fundamental concepts in survival analysis, which included the definition of failure time variable, as well as ways of describing the failure time distribution, such as the survival function and the hazard function. As failure time data are typically subject to independent right censoring, special statistical methods are needed for making inference. In the presence of censoring, nonparametric or semiparametric inferential methods that do not make any distributional assumptions are preferred in biomedical applications. Two nonparametric methods for estimating the survival curve, the life table estimator and the Kaplan-Meier estimator, were presented. For group comparisons, we discussed the log-rank test, a nonparametric test for testing the equality of survival distributions among groups. The proportional hazards model was demonstrated for evaluating intervention effect adjusting for confounding factors. The application of survival analysis methods is obviously not restricted to the medical area. The same methods can be applied in many other fields, including engineering, econometrics, sociology, and wherever censored time-to-event data are collected.

References

1. Cox, D. R., and Oakes, D. (1984) *Analysis of Survival Data*. London, Chapman & Hall, pp. 52–53.
2. Ware, J. H., and DeMets, D. L. (1976) Reanalysis of some baboon descent data. *Biometrics*. **32**, 459–463.
3. Kalbfleisch, J. D., and Prentice, R. L. (2002) *The Statistical Analysis of Failure Time Data*. New York, Chichester, John Wiley & Sons.
4. Klein, J. P., and Moeschberger, M. L. (1997) *Survival Analysis Techniques for Censored and Truncated Data*. New York, Springer-Verlag, pp. 100–103.
5. Kaplan, E. L., and Meier, P. (1958) Nonparametric estimation from incomplete observations. *J. Am. Stat. Assoc*. **53**, 457–481.
6. Efron, B. (1967) The two sample problem with censored data. In: *Proceedings of the Fifth Berkeley Symposium on Mathematical Statistics and Probability*. New York, Prentice-Hall, 4, pp. 831–853.
7. Gill, R. D. (1980) Censoring and Stochastic Integrals, *Mathematical Center Tracts* 124. Amsterdam, Mathematisch Centrum.
8. Klein, J. P. (1991) Small-sample moments of some estimators of the variance of the Kaplan-Meier and Nelson-Aalen estimators, *Scand. J. Stat.* **18**, 333–340.
9. Greenwood, M. (1926) The errors of sampling of the survivorship tables. In: *Reports on Public Health and Statistical Subjects*, no. 33. London, HMSO, Appendix 1.
10. Gehan, E. A. (1965) A generalized Wilcoxon test for comparing arbitrarily singly-censored samples. *Biometrika* **52**, 203–223.
11. Harrington, D. P., and Fleming, T. R. (1982) A class of rank test procedures for censored survival data. *Biometrika* **69**, 553–566.
12. Cox, D. R. (1972) Regression models and life-tables (with discussion). *J. R. Stat. Soc. Ser. B Methodol.* **34**, 187–220.
13. Cox, D. R. (1975) Partial likelihood. *Biometrika* **62**, 269–276.
14. Breslow, N. E. (1975) Analysis of survival data under the proportional hazards model. *Int. Stat. Rev.* **43**, 45–58.
15. Cox, D. R., and Snell, E. J. (1968) A general definition of residuals (with discussion). *J. R. Stat. Soc. Ser. B Methodol.* **30**, 248–275.

16. Schoenfeld, D. (1982) Partial residuals for the proportional hazards regression model. *Biometrika* **69**, 239–241.
17. Lin, D. Y. Wei, L. J., and Ying, Z. (1993) Checking the Cox model with cumulative sums of martingale-based residuals. *Biometrika*. **80**, 557–572.
18. Therneau, T. M., and Grambsch, P. M. (2000) Modeling survival data: extending the Cox model. Berlin, New York, Springer-Verlag.
19. Freireich, E. J., Gehan, E., Frei, E., Schroeder, L. R., Wolman, I. J., Anbari, R., Burgert, E. O., Mills, S. D., Pinkel, D., Selawry, O. S., and others. (1963) The effect of 6-mercaptopurine on the duration of steroid-induced remissions in acute leukemia: a model for evaluation of other potentially useful therapy. *Blood* **21**, 699–716.

16

Basic Bayesian Methods

Mark E. Glickman and David A. van Dyk

Summary

In this chapter, we introduce the basics of Bayesian data analysis. The key ingredients to a Bayesian analysis are the likelihood function, which reflects information about the parameters contained in the data, and the prior distribution, which quantifies what is known about the parameters before observing data. The prior distribution and likelihood can be easily combined to from the posterior distribution, which represents total knowledge about the parameters after the data have been observed. Simple summaries of this distribution can be used to isolate quantities of interest and ultimately to draw substantive conclusions. We illustrate each of these steps of a typical Bayesian analysis using three biomedical examples and briefly discuss more advanced topics, including prediction, Monte Carlo computational methods, and multilevel models.

Key Words: Monte Carlo simulation; posterior distribution; prior distribution; subjective probability.

1. Introduction

As with most academic disciplines, researchers and practitioners often choose from among several competing schools of thought. In music, for example, some composers have been guided by the rules of Romanticism, Impressionism, or Atonality in developing their work; in art, painters have at various periods followed the rules of Cubism, Expressionism, or Dadaism with widely differing results. One might assume that a scientific discipline such as statistics is immune to such philosophical divides. Interestingly, this is not the case. Statistics, as a discipline, consists of two main competing schools of thought: The *frequentist* or classical approach to statistical inference, and the *Bayesian* approach. The frequentist approach, which includes hypothesis testing and confidence intervals as two of the main modes of inference, has been the main framework for

From: *Methods in Molecular Biology, vol. 404: Topics in Biostatistics*
Edited by: W. T. Ambrosius © Humana Press Inc., Totowa, NJ

most of the techniques discussed thus far in this book. We discuss the basics of the Bayesian approach in this chapter.

The underlying difference between the Bayesian and frequentist approaches to statistical inference is in the definition of probability. A frequentist views probability as a long-run frequency. When a frequentist asserts that the probability of a fair coin tossed landing heads is $\frac{1}{2}$, he means that in the long run, over repeated tosses, the coin will land heads half the time. In contrast, a Bayesian, who will also surely say that the probability a coin lands heads is $\frac{1}{2}$, is expressing a *degree of belief* that the coin lands heads, perhaps arguing that based on the symmetry of the coin there is no reason to think that one side is more likely to come up than the other side. This definition of probability is usually termed *subjective* probability. Whereas, in practice, a frequentist uses probability to express the frequency of certain types of data to occur over repeated trials, a Bayesian uses probability to express belief in a statement about unknown quantities.

These definitions have profound impact on a framework for statistical inference. Because a Bayesian uses subjective probability, he can describe uncertainty of a statement about an unknown parameter in terms of probability. A frequentist cannot. So, for example, it is legitimate for a Bayesian to conclude as a result of a data analysis that an interval contains a parameter of interest with 95% probability. A frequentist, in contrast, will use probability to describe how often the calculations that produce an interval will cover the parameter of interest in repeated samples. For instance, frequentist 95% confidence intervals have the property that, in the long run, 95% of such intervals will cover the parameters being estimated. But, unfortunately for the frequentist, once a set of data is observed and an interval is computed, the frequentist concept of probability is no longer relevant. Further, when a Bayesian is evaluating two competing hypotheses about an unknown parameter, he can calculate the probability of each hypothesis given observed data and then choose the hypothesis with the greater probability. A frequentist, on the other hand, cannot use probability in such a direct way, and instead will approach the problem asymmetrically and ponder the long-run frequency under one of the hypotheses of sampling data as extreme or more extreme than what was observed.

This chapter describes the basics of Bayesian statistics. We begin by describing the main ingredients of a Bayesian analysis. In this discussion, we explain how to obtain the *posterior distribution* of model parameters and how to obtain useful model summaries and predictions for future data. We then demonstrate an application of the Bayesian approach to multilevel models, using *Monte Carlo simulation* as a computational tool to obtain model summaries.

2. Fundamentals of a Bayesian Analysis

A typical Bayesian analysis can be outlined in the following steps.

1. Formulate a probability model for the data.
2. Decide on a *prior distribution*, which quantifies the uncertainty in the values of the unknown model parameters *before* the data are observed.
3. Observe the data, and construct the *likelihood* function (see **Section 2.3**) based on the data and the probability model formulated in **step 1**. The likelihood is then combined with the prior distribution from **step 2** to determine the posterior distribution, which quantifies the uncertainty in the values of the unknown model parameters *after* the data are observed.
4. Summarize important features of the posterior distribution, or calculate quantities of interest based on the posterior distribution. These quantities constitute statistical outputs, such as point estimates and intervals.

We discuss each of these steps in turn in **Sections 2.1–2.4**.

The main goal of a typical Bayesian statistical analysis is to obtain the posterior distribution of model parameters. The posterior distribution can best be understood as a weighted average between knowledge about the parameters before data is observed (which is represented by the prior distribution) and the information about the parameters contained in the observed data (which is represented by the likelihood function). From a Bayesian perspective, just about any inferential question can be answered through an appropriate analysis of the posterior distribution. Once the posterior distribution has been obtained, one can compute point and interval estimates of parameters, prediction inference for future data, and probabilistic evaluation of hypotheses. Predictive inference is the topic of **Section 2.5**.

2.1. Data Models

The first step in a Bayesian analysis is to choose a probability model for the data. This process, which is analogous to the classic approach of choosing a data model, involves deciding on a probability distribution for the data if the parameters were known. If the n data values to be observed are $y_1, \ldots, y_n$, and the vector of unknown parameters is denoted $\boldsymbol{\theta}$, then, assuming that the observations are made independently, we are interested in choosing a probability function $p(y_i \mid \boldsymbol{\theta})$ for the data (the vertical bar means "conditional on" the quantities to the right). In situations where we have extra covariate information, x_i, for the ith case, as in regression models, we would choose a probability function of the form $p(y_i \mid x_i, \boldsymbol{\theta})$. When the data are not conditionally independent given the parameters and covariates, we must specify the joint probability function, $p(y_1, \ldots, y_n \mid x_1, \ldots, x_n, \boldsymbol{\theta})$.

Example 1

A random sample of 300 women aged 60–69 years whose immediate families have had histories of cancer are to be screened for breast cancer. Let y_i be 1 if woman i has a positive test, and 0 if not, for $i = 1, \ldots, 300$. Let θ be the probability that a randomly selected woman aged 60–69 years with a family history of cancer has a positive breast cancer screening. Then an appropriate model for the data is to assume that the y_i independently follow a Bernoulli distribution with probability θ, that is,

$$p(y_i \mid \theta) = \theta^{y_i}(1 - \theta)^{1-y_i}$$

for $i = 1, \ldots, 300$.

Example 2

A random sample of 50 men with a history of cardiovascular disease enters a study on LDL (low-density lipoprotein) cholesterol. Let y_i be the LDL cholesterol level (in mg/dL) for man i, $i = 1, \ldots, 50$. A reasonable probability model for LDL cholesterol levels is a normal distribution. We can assume that the y_i are independently normal with unknown common mean μ and variance σ^2. The probability function for y_i is given by

$$p\left(y_i \mid \mu, \sigma^2\right) = \frac{1}{\sqrt{2\pi\sigma^2}} \exp\left(-(y_i - \mu)^2/2\sigma^2\right)$$

for $i = 1, \ldots, 50$.

2.2. Prior Distribution

Once the data model (probability model) is chosen, a Bayesian analysis requires the assertion of a prior distribution for the unknown model parameters. The prior distribution can be viewed as representing the current state of knowledge, or current description of uncertainty, about the model parameters prior to data being observed.

Approaches to choosing a prior distribution divide into two main categories. The first approach involves choosing an *informative* prior distribution. With this strategy, the statistician uses his knowledge about the substantive problem perhaps based on other data, along with elicited expert opinion if possible, to construct a prior distribution that properly reflects his (and experts') beliefs about the unknown parameters. The notion of an informative prior distribution may seem at first to be overly subjective and unscientific. In response to this concern, it should be pointed out that the selection of a data model, which a frequentist needs to make, is also a subjective choice, so that frequentist analyses are not devoid from subjectivity either. Furthermore, it can be argued that if extra information or knowledge about the model parameters exists prior to observing data, it would be unscientific *not* to incorporate such information into a data analysis. For example, in a study measuring the weight of preterm births, it would be sensible to incorporate into the prior distribution that the "prior probability" of a mean birth weight above 15 lb is negligible. Another criticism by frequentists of using informative prior distributions is that two Bayesian statisticians are likely to use two different prior distributions, which leads to two different sets of inferences for the same scientific problem. Again, it is reasonable to respond to this criticism by pointing out that when frequentists use different data models on the same data, conclusions will be different as well. From a Bayesian point of view, a prior distribution is part of the overall statistical model, so that two Bayesian statisticians selecting different prior distributions is analogous to two frequentist statisticians choosing two different data models.

The second main approach to choosing a prior distribution is to construct a *noninformative* prior distribution that represents ignorance about the model parameters. Besides *noninformative*, this type of distribution is also called objective, vague and diffuse, and sometimes a reference prior distribution. Choosing a noninformative prior distribution is an attempt at objectivity by acting as though no prior knowledge about the parameters exists before observing the data. This is implemented by assigning equal probability to all values of the parameter (or at least approximately equal probability over localized ranges of the parameter). The appeal of this approach is that it directly addresses the criticisms of informative prior distributions as being subjectively chosen. In some cases, there is arguably a single best noninformative prior distribution for a given data model, so that this prior distribution can be used as a default option, much like one might have default arguments in computer programs. Unfortunately, noninformative prior distributions are not without their problems either. First, because there are various commonly accepted criteria for constructing noninformative prior distributions, it is rare that, for a given data model, all these criteria produce the same unique noninformative prior distribution. Second, some common methods for constructing noninformative prior distributions, such as always assuming a uniform distribution for a parameter, result in an interesting inconsistency. Any method for constructing a noninformative prior distribution ought to be invariant to the measurement scale of the parameter; if, for example, the method of constructing a noninformative prior distribution is applied to a data model with parameter θ, and then applied to the same model reparameterized with parameter $\eta = \log(\theta)$, it would be desirable that the distributions on θ and η were representing equivalent probabilistic information. It turns out that this is a difficult criterion to satisfy (one approach constructed to satisfy this invariance criterion is *Jeffrey's rule*, which works well with one-parameter data models but with mixed results for multiparameter models). Finally, many commonly used methods for constructing a noninformative prior distribution result in probability functions that integrate to infinity, usually called *improper* distributions, and are not formally probability distributions. Luckily, for many problems, having an improper prior distribution still allows for a coherent Bayesian analysis.

In general, if an objective prior distribution is desired, one defensible strategy is to construct a relatively uniform proper (i.e., integrates to 1) prior distribution. If the information contained in the data is supposed to be the main determining factor in producing statistical inferences (as it should be), then we should expect that the choice among a range of relatively flat prior distributions will not make much of a difference. On the other hand, if the choice of a relatively flat prior distribution does matter, this may be an indication that the data conveys little information about the parameter of interest, and it may be appropriate to rethink the form of the data model, or to collect additional data.

Example 1 (Continued)

Recall that θ is the probability a randomly selected woman, aged 60–69 years with a family history of cancer, has a positive breast cancer screening. According to the American Cancer Society, roughly 3.6% of women aged 60–69 years develop invasive breast cancer, so that we may form an informative prior distribution for θ that reflects this information. A flexible choice of a prior distribution for a Bernoulli probability is $\theta \sim \text{Beta}(\alpha, \beta)$, that is, θ has a Beta distribution with specified parameters α and β. The probability function is given by

$$p(\theta|\alpha,\beta) = \frac{\Gamma(\alpha+\beta)}{\Gamma(\alpha)\Gamma(\beta)}\theta^{\alpha-1}(1-\theta)^{\beta-1}$$

where $\Gamma()$ represents the Gamma function.[1] The mean of a Beta distribution is $\alpha/(\alpha + \beta)$. The value $\alpha + \beta$ has an interpretation as the amount of information about θ viewed as a sample size. For the cancer screening problem, the choice $\theta \sim \text{Beta}(0.36, 9.64)$ is sensible, as this distribution has a mean of $0.36/(0.36 + 9.64) = 0.036$, the estimate given by the American Cancer Society, and the information represented by this distribution is equivalent to that in $0.36 + 9.64 = 10$ data values. A plot of the probability function is given in **Figure 1**. Note that the greatest probability under this distribution of θ is concentrated around very low values, which is meant to reflect our initial belief that a value of θ much larger than 0.1 or 0.15 is not very plausible. With an eventual sample of 500 observations, the data is about 50 times more informative than the prior distribution.

Example 2 (Continued)

For studying LDL cholesterol levels, we assume a noninformative prior distribution for the mean μ and variance σ^2 of the normal data model. A strategy

[1] The Gamma function is closely related to the factorial function: For a positive integer n, $\Gamma(n) = (n - 1)!$. For more details about the Gamma function, see *(1)*.

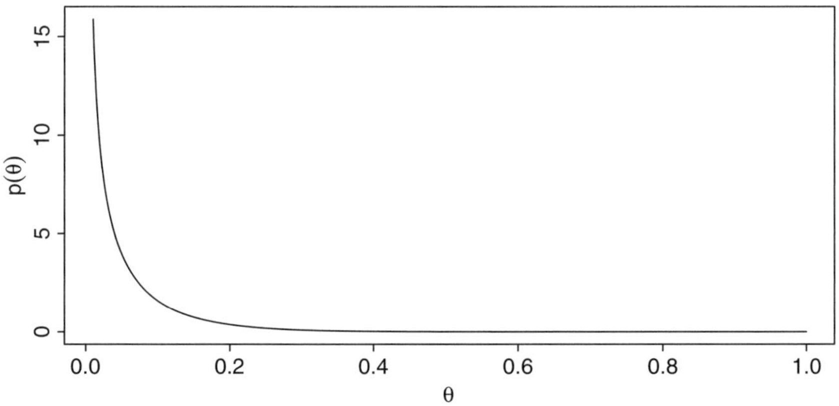

Fig. 1. Probability function for the Beta(0.36, 9.64) distribution.

that can often be employed for models with multiple parameters is to consider each parameter separately and form the joint prior distribution as a product of the several independent distributions.

The most common noninformative choice for a location parameter, such as a mean (or a regression coefficient), is to assume an *improper* uniform distribution over the entire real line. Thus we assume

$$p(\mu) = 1$$

for $-\infty < \mu < \infty$ even though this function does not integrate over the range. We further assume, independently, that the prior distribution for σ^2 is the improper probability function

$$p(\sigma^2) = 1/\sigma^2.$$

By a change-of-variables argument from elementary calculus, this distribution on σ^2 corresponds with a uniform distribution on $\log(\sigma^2)$ over the entire real line. Besides having the appeal of placing a uniform distribution over a parameter that has been transformed to take values over the entire real line, as with μ, this prior distribution also recognizes that extremely large values of σ^2 are less believable *a priori* than are small values. A uniform distribution on the untransformed variance, σ^2, in contrast, asserts that a variance between 1,000,000 and 1,000,001 is as likely *a priori* as a variance between zero and one, which is not particularly believable. We therefore assume an improper joint prior distribution for (μ, σ^2) equal to

$$p(\mu, \sigma^2) = p(\mu)p(\sigma^2) = 1 \cdot (1/\sigma^2) = 1/\sigma^2.$$

2.3. From the Likelihood to the Posterior Distribution

Once the data has been observed, the likelihood function, or simply the likelihood, is constructed. The likelihood is the joint probability function of the data, but viewed as a function of the parameters, treating the observed data as fixed quantities. Assuming that the data values, $\boldsymbol{y} = (y_1, \ldots, y_n)$ are obtained independently, the likelihood function is given by

$$L(\theta|y) = p(y_1, \ldots, y_n|\theta) = \prod_{i=1}^{n} p(y_i|\theta).$$

In the Bayesian framework, all of the information about $\boldsymbol{\theta}$ coming directly from the data is contained in the likelihood. Values of the parameters that correspond with the largest values of the likelihood are the parameters that are most supported by the data.

To obtain the posterior distribution, $p(\boldsymbol{\theta}|\boldsymbol{y})$, the probability distribution of the parameters once the data have been observed, we apply Bayes' theorem:

$$p(\boldsymbol{\theta}|y) = \frac{p(\boldsymbol{\theta})p(\boldsymbol{\theta}|y)}{\int p(\boldsymbol{\theta})p(y|\boldsymbol{\theta})d\boldsymbol{\theta}} = \frac{p(\boldsymbol{\theta})L(\boldsymbol{\theta}|y)}{p(y)} \propto p(\boldsymbol{\theta})L(\boldsymbol{\theta}|y)$$

where "$\propto$" means "is proportional to" (i.e., that the expressions are equal when the right-most term is multiplied by a normalizing constant that doesn't depend on $\boldsymbol{\theta}$). Operationally, therefore, it is straightforward in principle to obtain the posterior distribution: Simply multiply the prior distribution by the likelihood, and then determine the constant (not depending on $\boldsymbol{\theta}$) that forces the expression to integrate to 1. An effective strategy for computing the posterior distribution is to drop multiplicative constants from the prior distribution and likelihood that do not depend on $\boldsymbol{\theta}$, and then in the final step determine the normalizing constant.

Example 1 (Continued)

Suppose, for the breast cancer screening study, 14 of the 300 women had positive tests. Thus 14 women have $y_i = 1$, and the remaining 286 have $y_i = 0$. The likelihood is therefore given by

$$L(\theta|y) = \prod_{i=1}^{300} \theta^{y_i}(1-\theta)^{1-y_i} = \theta^{14}(1-\theta)^{286}.$$

The posterior distribution is proportional to the product of the Beta prior distribution (with parameters $\alpha = 0.36$ and $\beta = 9.64$) and the likelihood,

$$L(\theta\,|\,y) \propto p(\theta)L(\theta\,|\,y) \propto \left(\frac{\Gamma(10)}{\Gamma(0.36)\Gamma(9.64)}\theta^{-0.64}(1-\theta)^{8.64}\right)\cdot\theta^{14}(1-\theta)^{286}$$
$$\propto \theta^{-0.64}(1-\theta)^{8.64}\cdot\theta^{14}(1-\theta)^{286} \propto \theta^{13.36}(1-\theta)^{294.64}.$$

Note that the normalizing constant in the prior distribution was dropped as it does not depend on θ. Rather than determine the normalizing constant analytically, we notice that the final expression is proportional to a Beta distribution with parameters $\alpha = 14.36$ and $\beta = 295.64$, so that the posterior distribution must be

$$p(\theta\,|\,y) = \frac{\Gamma(330)}{\Gamma(14.36)\Gamma(295.64)}\theta^{13.36}(1-\theta)^{294.64}.$$

Thus, the posterior distribution is $\theta|\mathbf{y} \sim \text{Beta}(14.36, 295.64)$.

Example 2 (Continued)

In the LDL cholesterol study, suppose the 50 LDL cholesterol measurements are taken. The likelihood is the product of 50 normal probability functions:

$$L(\mu,\sigma^2\,|\,y) = \prod_{i=1}^{50} p(y_i\,|\,\mu,\sigma^2) = \prod_{i=1}^{50}\frac{1}{\sqrt{2\pi\sigma^2}}\exp(-(y_i-\mu)^2/2\sigma^2)$$
$$= \frac{1}{(2\pi\sigma^2)^{25}}\exp\left(-\sum_{i=1}^{50}(y_i-\mu)^2/2\sigma^2\right).$$

Letting $\bar{y} = \frac{1}{n}\sum_{i=1}^{50} y_i$ and $s^2 = \frac{1}{49}\sum_{i=1}^{50}(y_i-\bar{y})^2$ be the sample mean and variance, respectively, the likelihood can be rewritten in a more useful form as

$$L(\mu,\sigma^2\,|\,y) = \frac{1}{(2\pi\sigma^2)^{25}}\exp\left(-\sum_{i=1}^{50}(y_i-\mu)^2/2\sigma^2\right)$$
$$= \frac{1}{(2\pi\sigma^2)^{25}}\exp\left(49s^2 + 50(\mu-\bar{y})^2/2\sigma^2\right).$$

We again use the standard choice of noninformative prior distribution on the parameters of a normal model, $p(\mu, \sigma^2) = 1/\sigma^2$. With this choice of prior distribution, the posterior distribution can be computed as follows:

$$p(\mu,\sigma^2\,|\,y) \propto p(\mu,\sigma^2)L(\mu,\sigma^2\,|\,y) \propto \frac{1}{\sigma^2}\cdot\frac{1}{(2\pi\sigma^2)^{25}}\exp(-(49s^2+50(\mu-\bar{y})^2)/2\sigma^2)$$
$$\propto (\sigma^2)^{-25.5}\exp(-49s^2/2\sigma^2)\cdot\frac{1}{\sigma}\exp\left(-(\mu-\bar{y})^2/2(\sigma/\sqrt{50})^2\right).$$

The second term in the above expression, as a function of μ with the appropriate constant, is a normal distribution with mean $\bar{y}$ and variance $\sigma^2/50$. The first term, with the appropriate constant, is an *inverse-*χ^2 distribution; this means that $1/\sigma^2$ has the more familiar chi-square distribution. The posterior distribution $p(\mu, \sigma^2|\mathbf{y})$ therefore factors into a *marginal posterior* distribution of σ^2, $p(\sigma^2|\mathbf{y})$, which is inverse-χ^2, and a *conditional posterior* distribution of μ given σ^2, $p(\mu|\sigma^2,\mathbf{y})$, which is normal. A marginal posterior distribution specifies the posterior distribution for a subset of the model parameters without regard to the other parameters. A conditional posterior distribution, on the other hand, is the posterior distribution of a subset of the parameters subject to the other parameters having specified values.

In this example, the joint posterior distribution can be written

$$p(\mu, \sigma^2|\mathbf{y}) = p(\sigma^2|\mathbf{y})p(\mu|\sigma^2, \mathbf{y})$$

where $\sigma^2|\mathbf{y} \sim \text{Inv-}\chi^2(49, s^2)$ (i.e., $49s^2/\sigma^2$ has a chi-square distribution on 49 degrees of freedom), and $\mu|\sigma^2, \mathbf{y} \sim \text{N}(\bar{y}, \sigma^2/50)$. Once the sample mean and sample variance have been computed from the data, these values can be substituted in to obtain the actual distributions. It is also worth noting that σ^2 can be integrated out of the joint posterior density to obtain the *marginal* posterior density of μ, which is

$$\mu|\mathbf{y} \sim t_{49}(\bar{y}, s^2/50),$$

that is, a *t*-distribution with 49 degrees of freedom that is centered at $\bar{y}$ and rescaled by $s/\sqrt{50}$.

2.4. Posterior Summaries

Once the posterior distribution has been determined, inferential conclusions can be summarized with an appropriate analysis. Point estimates of parameters are commonly computed as the mean or the *mode* (i.e., highest point) of the posterior distribution. Interval estimates can be calculated by producing the end points of an interval that correspond with specified percentiles of the posterior distribution. For example, a 95% *central posterior interval* involves computing the 2.5%-ile and 97.5%-ile of the posterior distribution. Probabilities of competing composite hypotheses can be evaluated by calculating their posterior probability, that is, the probability of the hypotheses based on the posterior distribution.

Example 1 (Continued)

With a posterior distribution for the probability of a positive breast cancer screening of Beta(14.36, 295.64), we can compute informative inferential summaries about θ. The posterior mean and posterior mode are the two most common point estimates for a parameter. For a Beta distribution with parameters α and β, the mean is $\alpha/(\alpha + \beta)$, and the mode is $(\alpha - 1)/(\alpha + \beta - 2)$. The posterior mean estimate of θ is therefore

$$\mathrm{E}(\theta|\mathbf{y}) = 14.36/(14.36 + 295.64) = 0.0463.$$

The posterior mode estimate of θ, the most "believable" value of θ, is

$$\mathrm{Mode}(\theta|\mathbf{y}) = (14.36 - 1)/(14.36 + 295.64 - 2) = 0.0434.$$

To construct a 95% central posterior interval for θ, we need to find the appropriate percentiles of the Beta(14.36, 295.64) distribution. Analytically, this involves evaluating the integral $\int_0^c p(\theta|\mathbf{y})d\theta = 0.025$ and solving for c to obtain the lower end point of the interval, and similarly for the upper end point. Using statistical software (like R or S-Plus, SAS, Stata, SPSS, etc.), the percentiles can easily be evaluated numerically. The 2.5%-ile and the 97.5%-ile of the posterior distribution are computed to be 0.0259 and 0.0723, respectively, so that the 95% central posterior interval for θ is (0.0259, 0.0723). There is a 0.95 posterior probability that θ lies in this interval.

Suppose for health policy reasons that it is important to know whether $\theta > 0.05$. We can translate the question into a posterior probability computation of

$$\mathrm{P}(\theta > 0.05|y) = \int_{0.05}^{1} p(\theta|y)\, d\theta.$$

Rather than attempting to evaluate this Beta integral analytically, we can evaluate it numerically using statistical software. The probability from the Beta posterior distribution is computed to be 0.351, which implies that the probability $\theta < 0.05$ is 0.649. Thus we may conclude that it is more likely than not that $\theta < 0.05$.

Example 2 (Continued)

We computed the joint posterior distribution of μ and σ^2, the mean and variance of the normal model, in the LDL cholesterol study. This posterior distribution depends on the data through the sample mean and sample variance of the 50 measurements, $\bar{y}$ and s^2, respectively. Now suppose that upon observing the measurements, we compute $\bar{y} = 110$ and $s^2 = 100$. From a Bayesian perspective, the posterior distribution is a complete summary of what we know about the parameters, both from the data and—as quantified via the prior distribution—from other sources of information. In this case, we can plot the posterior distribution and use the plots to quantify what we understand about the unknown

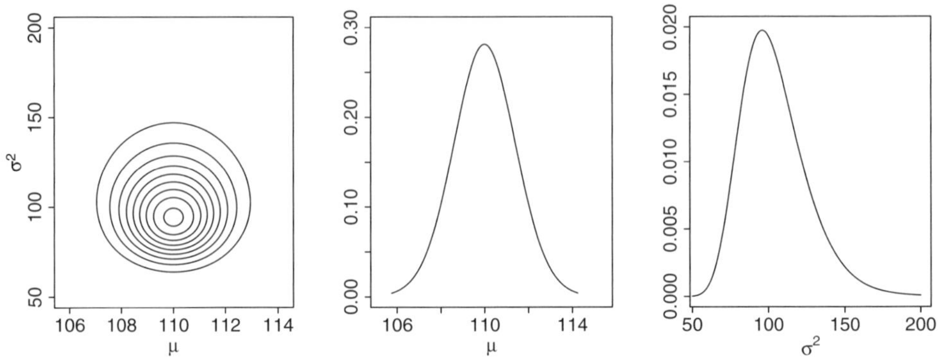

Fig. 2. The posterior distribution of parameters of LDL cholesterol levels. The three figures depict the 2-dimensional joint posterior distribution of the mean and variance of LDL cholesterol in the population of men. A contour plot of the joint distribution and plots of both of the marginal distributions are given.

parameters. A contour plot of the joint posterior distribution appears in the first panel of **Figure 2**. The next two panels represent the marginal posterior distributions of μ and σ^2, respectively. These distributions represent our knowledge about likely values of the mean and variance of LDL cholesterol levels in this particular population of men. Judging from the posterior distribution of μ, the mean LDL cholesterol level is about 110 plus or minus about four. The posterior distribution of σ^2 tells us how much the level varies among men: The variance appears to be about 100 but could be as low as 60 or as high as 175. Notice that the posterior distribution of σ^2 is slightly skewed toward the right. Looking at the joint distribution, the mean and variance appear to be uncorrelated. This means that inference about particular values of μ does not have a relationship to our inference about values of σ^2.

2.5. Predictive Distributions

One of the benefits of the Bayesian approach is that predictive inference is a straightforward computation once the posterior distribution has been obtained. Suppose we have observed data $\mathbf{y} = (y_1, \ldots, y_n)$, and we would like to make a prediction about a future observation y. From an analysis of the data, we have obtained $p(\boldsymbol{\theta}|\mathbf{y})$, the posterior distribution. We are interested in making probabilistic statements about an unobserved y, so that we want to compute the *posterior predictive distribution* of y. The posterior predictive distribution is written as $p(y \mid \mathbf{y})$. Note that we are not interested in conditioning on parameter values, but that we only want to condition on what we have observed: the previous data.

The posterior predictive distribution can be computed using the equation

$$p(y|\mathbf{y}) = \int p(y|\boldsymbol{\theta})p(\boldsymbol{\theta}|\mathbf{y})d\boldsymbol{\theta}$$

which makes the often appropriate assumption that future data is independent of past data conditional on the parameters. Thus, integrating the product of the data model distribution with the posterior distribution with respect to the model parameters produces the posterior predictive distribution, which can then be summarized for predictive inferences.

Example 2 (Continued)

Let y be an LDL cholesterol measurement taken of a man with a history of cardiovascular disease not yet sampled. We are interested in deriving the posterior predictive distribution of y, that is, $p(y|\mathbf{y})$. We must therefore evaluate

$$\begin{aligned} p(y|\mathbf{y}) &= \iint p(y|\mu,\sigma^2)p(\mu,\sigma^2|\mathbf{y})\,d\mu d\sigma^2 \\ &= \iint p(y|\mu,\sigma^2)p(\sigma^2|\mathbf{y})p(\mu|\sigma^2,\mathbf{y})\,d\mu d\sigma^2. \end{aligned}$$

It can be shown that with the normal distribution for y, the normal conditional posterior distribution for μ given σ^2, and the inverse-χ^2 marginal posterior distribution for σ^2, the integral is evaluated to

$$p(y|\mathbf{y}) \propto \left(1+\frac{50(y-\bar{y})^2}{49s^2(1+1/50)}\right)^{-25}$$

which is a t-distribution on 49 degrees of freedom centered at $\bar{y}$ and with a scale parameter of $\sqrt{s^2(1+1/n)}$. (In our example, $\bar{y} = 110$, $s^2 = 100$, and $n = 50$.)

3. Application to Multilevel Models

3.1. Monte Carlo Methods

The examples above illustrate how statistical summaries of scientific interest can be expressed as integrals of the posterior distribution. Although in simple cases these integrals can sometimes be computed analytically, in more complex realistic examples, numerical methods are required. Even computing a 95% central posterior interval for the probability of breast cancer, θ, in **Example 1** required numerical methods. In this section, we describe Monte Carlo methods, which have revolutionized applied Bayesian data analysis over the past 20 years. Monte Carlo methods are so important because they are often relatively easy to understand and implement, yet are powerful enough to enable us to compute relevant statistical summaries even when fitting highly structured models.

As an introduction to Monte Carlo methods, we return to the LDL cholesterol study.

Example 2 (Continued)

Monte Carlo methods are simulation-based methods. With a specified probability distribution, a typical Monte Carlo simulation involves a computer program generating multiple plausible values from the distribution. In Bayesian data analysis, this generally involves acquiring a sample from the posterior (or posterior predictive) distribution. In **Figure 3**, we compare a Monte Carlo sample from the posterior distribution with the three plots of the posterior distribution given in **Figure 2**. The key here is that we can draw the same inferences regarding μ and σ^2 from either the plots of the Monte Carlo sample or from the plots of the posterior distribution itself. In addition to the qualitative descriptions discussed in **Section 2.3**, we can compute posterior means by averaging over the Monte Carlo sample or compute a 95% central interval, by computing the 2.5%-ile and 97.5%-ile of the Monte Carlo sample.

Example 2 is a simple illustration with only two parameters. This makes it easy to visually examine the joint posterior distribution and to compute the marginal posterior distributions of the parameters of interest. In more complex settings, however, the dimension of the unknown parameter may be much larger. In image analysis (e.g., functional magnetic resonance imaging), for example, there may be an unknown image intensity in each of a large number of pixels or voxels. In such settings, there may be hundreds or thousands of unknown parameters. It is in such settings that Monte Carlo methods are so

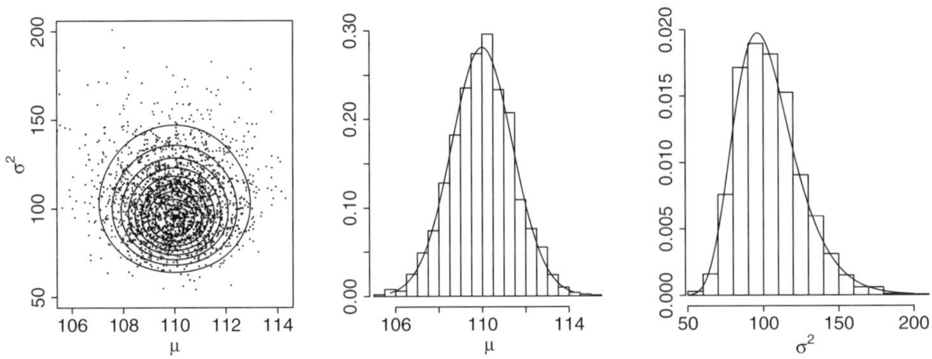

Fig. 3. A Monte Carlo sample from the posterior distribution of parameters of LDL cholesterol levels. A Monte Carlo sample is compared with each of the 3 plots given in Figure 2. The Monte Carlo sample carries the same information about the posterior distribution as the analytically computed plots.

useful. Although we cannot plot the joint posterior distribution or even compute the high-dimensional integrations that are required to evaluate the marginal posterior distributions of low-dimensional quantities of scientific interest, we may be able to acquire a Monte Carlo sample from the posterior distribution. That is, although we cannot produce plots analogous to those in **Figure 2**, we can produce scatter plots and histograms analogous to those in **Figure 3**. From these representations of the Monte Carlo sample, we can construct statistical inferences for unknown quantities of scientific interest, even in highly complex models. This strategy is illustrated in a more complex setting in **Section 3.2**.

There are a variety of techniques available for acquiring a Monte Carlo sample from a given posterior distribution. Perhaps the most important class of such techniques is known as *Markov chain Monte Carlo* (MCMC). It was the development of MCMC in the statistical literature, starting in the late 1980s, that greatly expanded the class of models that can be fit using Monte Carlo techniques. An important example of MCMC is the *Gibbs sampler*. Rather than directly acquiring a Monte Carlo sample from the posterior distribution, the Gibbs sampler cycles through a set of conditional posterior distributions, sampling from each distribution conditional on the most recent draw of the remaining parameters. Because the conditional distributions involve a smaller number of unknown parameters, they tend to be simpler to simulate. Carefully designed Gibbs samplers allow highly complex models to be divided into a sequence of simpler more standard models, all of which can be fit using standard Bayesian statistical techniques. The iterative nature of the Gibbs sampler (and other MCMC techniques) means that it can be sensitive to starting values, and its Monte Carlo nature means that convergence diagnostics can be subtle. Here, we have only scratched the surface of the numerous technical issues involved in designing, implementing, and detecting convergence of MCMC samplers. Nonetheless, interpreting the scientific results is done in much the same way as with the Monte Carlo methods described here. Readers interested in learning more about this important class of Bayesian computational methods are directed to the references in **Section 4** and the citations therein.

3.2. Multilevel Models

The power of Monte Carlo sampling in conjunction with Bayesian methodology is that it allows us to fit models that are explicitly designed to capture the complexity of any given data generation mechanism. We often accomplish this by hierarchically combining a series of simple models into a single more appropriate model. In this section, we illustrate this strategy in an extended example. Although this example is relatively simple by current standards, we hope that it will give the reader a flavor for how multilevel models are constructed and for the power of combining Monte Carlo sampling with Bayesian methods.

Table 1
Data for the 16 Litters of Rats in the Treatment Group

	Litter															
	1	2	3	4	5	6	7	8	9	10	11	12	13	14	15	16
Size	12	11	10	9	11	10	10	9	9	5	9	7	10	6	10	7
Surviving	12	11	10	9	10	9	9	8	8	4	7	4	5	3	3	0

The litter sizes and the number of pups surviving the 21-day lactation period are recorded.

Example 3

In an experiment described by Weil *(2)*, 32 pregnant female rats were divided into 2 groups. In the control group, the mothers were fed a control diet during pregnancy and lactation. In the second group, the mothers' diets were treated with a chemical. The number of pups in each litter that survived 4 days was recorded as the litter size. Of these, the number that survived the 21-day lactation period were also recorded. For our purposes, we consider only the treatment group and investigate how the probability of 21-day survival varies among the litters in this population and fit the probability of survival for each of the 16 observed treatment litters. The data for the treatment litters appear in **Table 1**, which records the size of each litter (number of pups that survive for 4 days) and number of these that survive for 21 days.

We begin by formulating a probability model for the data. For each litter, let n_i be the size of the litter and y_i be the number of pups that survive the 21-day lactation period. We assume the pups within each litter have equal probability of survival and use a binomial distribution to model the number that survive. In particular, we assume $y_i | \theta_i \sim \text{Binomial}(n_i, \theta_i)$, that is,

$$p(y_i|\theta_i) = \binom{n_i}{y_i} \theta_i^{y_i} (1-\theta_i)^{(n_i - y_i)}.$$

Because we believe the survival rates vary among the litters, we allow θ_i to depend on i. The distribution of the θ_i is of primary interest in this study (in particular, we may be interested in how the distribution is affected by the treatment). Therefore we introduce a probability model for the θ_i. As discussed in **Example 1**, the Beta distribution is particularly well suited for modeling probabilities. Thus, we assume $\theta_i \sim \text{Beta}(\alpha, \beta)$. The parameters α and β determine the shape, mean, and variability of the Beta distribution and thus of the survival probabilities among litters in the treatment group.

In **Example 1**, we used prior information as to the probability of breast cancer to set the values of α and β. In this case, however, α and β are fit to the data to describe the distribution of the survival probabilities. Because α and β, both restricted to be positive, are treated as model parameters, we must decide on prior distributions for these 2 parameters. Here we choose independent noninformative prior distributions that are uniform on $\log(\alpha)$ and $\log(\beta)$. As in **Example 2**, this corresponds with prior distributions that are proportional to the reciprocal, that is, $p(\alpha, \beta) \propto 1/\alpha\beta$.

Combining the two parts of the specification of the data model with the prior distribution leads to a 3-level model. In particular, the statistical model can be formulated as a Beta-binomial model *(3)* with noninformative prior distribution as follows:

Level 1: $y_i \mid \theta_i \sim \text{Binomial}(n_i, \theta_i)$ for $i = 1, \ldots, 16$.
Level 2: $\theta_i \mid \alpha, \beta \sim \text{Beta}(\alpha, \beta)$ for $i = 1, \ldots, 16$.
Level 3: $p(\alpha, \beta) \propto 1/\alpha\beta$.

Level 1 specifies the 16 within-litter distributions, **level 2** describes the variability among the litters in the treatment population, and **level 3** is the (noninformative and improper) prior distribution. This is a simple illustration of how standard probability distributions can be combined hierarchically to form more complex and more appropriate models—models that can more fully describe the richness of the data generation mechanism.

With the data model, prior distribution, and observed data in hand, we construct and compute the posterior distribution as described earlier. We acquire a Monte Carlo sample from the joint posterior distribution of $(\theta_1, \ldots, \theta_{16}, \alpha, \beta)$. **Figure 4** represents the Monte Carlo sample from the marginal posterior

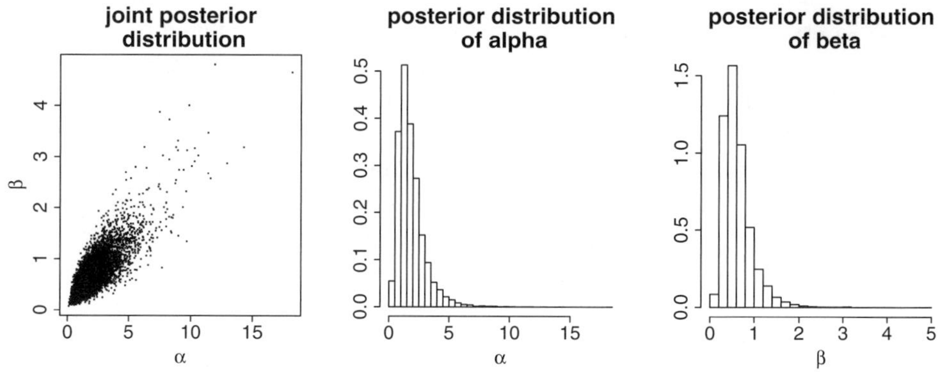

Fig. 4. A Monte Carlo sample from the joint posterior distribution of α and β.

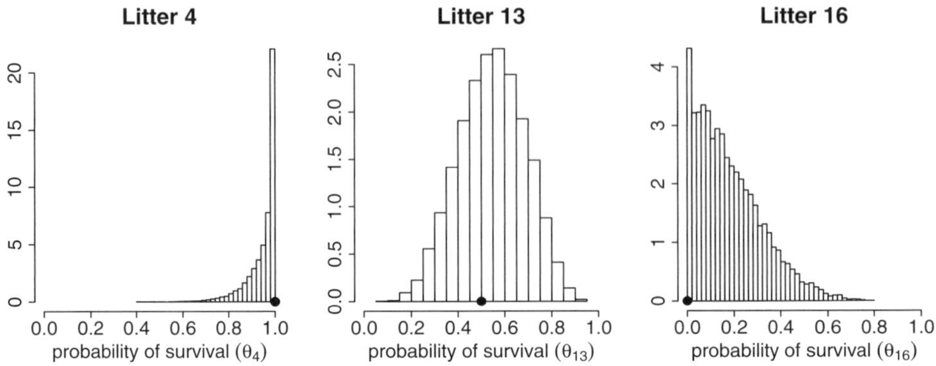

Fig. 5. Histograms of the Monte Carlo sample of the survival probabilities of 3 of the litters. The solid circles on the horizontal axis of each of the histograms represent the sample proportion of the pups that survived in that litter. Notice that in all 3 cases the histograms have their centers of mass a bit off of the sample proportion, in the direction of the fitted population mean of 0.74. This is known as *shrinkage*: the posterior mean "shrinks" from the sample proportion toward the fitted population mean.

distribution of α and β, and **Figure 5** represents a sample from the marginal posterior distributions of θ_4, θ_{13}, and θ_{16}. In this case, the plots in **Figure 5** are more relevant because the parameters are more easily interpreted: they are the marginal posterior distributions of the survival probabilities for 3 of the litters.

Comparing the three plots in **Figure 5**, it is clear that the survival probabilities vary among the litters. To explore this further, we can acquire a Monte Carlo sample from the predictive distribution of the survival probability of another litter. A histogram of this Monte Carlo sample appears in the first panel of **Figure 6**. This distribution accounts for both the variability among the litters and the uncertainty in the distribution of the survival probabilities. These two variance components correspond with the variability among the histograms in **Figure 5** and the uncertainty in α and β illustrated in **Figure 4**, respectively. The final histogram in **Figure 6** is a Monte Carlo sample from the posterior predictive distribution of the number of surviving pups for an additional litter of size 10. This distribution accounts for both the variability in θ as represented by the first histogram in **Figure 6** and for the binomial variation of pup survival.

We can fit the survival probabilities of each of the 16 litters by averaging over the Monte Carlo sample of each of these 16 parameters. The results, along

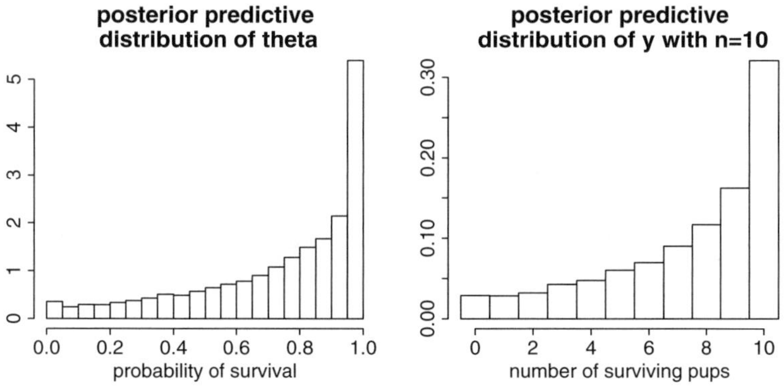

Fig. 6. Monte Carlo samples from the posterior predictive distribution. The first histogram represents a sample from the predictive distribution of the survival probability of another litter from this population. The second histogram corresponds with a sample from the predictive distribution of the number of surviving pups from this additional litter, given that the litter is of size 10.

with the sample proportion of surviving pups in each litter, appear in **Table 2**. Notice that in each case, the fitted probability is between the sample proportion and the expected survival probability of a new litter, 0.74. Although the sample proportion is the standard estimate of the survival probability for a single litter, like all statistical estimates, these have error because of the variable nature of binomial data. Because we are simultaneously fitting the population distribution of survival probabilities, we have some information as to the direction of the estimates' error. The Bayesian estimate is an average of the population mean and the sample proportion. As the size of the litter increases, this average is weighted more heavily toward the sample proportion. These fitted values are often called *shrinkage estimates* because they "shrink" the fitted probability from the sample proportion toward the population mean. Shrinkage is automatic

Table 2
Shrinkage

	Litter															
	1	2	3	4	5	6	7	8	9	10	11	12	13	14	15	16
Sample	1.00	1.00	1.00	1.00	0.91	0.90	0.90	0.89	0.89	0.80	0.78	0.57	0.50	0.50	0.30	0.00
Fitted	0.96	0.96	0.95	0.95	0.88	0.87	0.87	0.86	0.86	0.78	0.77	0.61	0.55	0.57	0.38	0.18

The sample proportion of surviving pups and the fitted probability of survival are recored for each of the 16 litters. Each of the fitted values is between the population mean (0.74) and the sample proportions for the particular litter.

when the Bayesian posterior distribution is used to generate statistical estimates.

4. Other Resources

In this chapter, we have introduced only the most basic aspects of Bayesian modeling, methods, and computation. There are a number of accessible treatises on Bayesian methods that interested readers might refer to, including Gelman and others ***(4)*** and Carlin and Louis ***(5)***, both of whom offer excellent introductions.

References

1. Abramowitz, M., and Stegun, I. A. (1972) *Handbook of Mathematical Functions with Formulas, Graphs, and Mathematical Table*, 9th ed. New York, Dover Publications.
2. Weil, C. S. (1970) Selection of the valid number of sampling units and a consdieration of their combination in toxicological studies involving reproduction, teratongenesis or carcinogensis. *Food and Cosmetics Toxicology* **8**, 177–182.
3. Williams, D. A. (1975) 394: The analysis of binary responses from toxicological experiments involving reproduction and teratogenicity. *Biometrics* **31**, 949–952.
4. Gelman, A., Carlin, J. B., Stern, H. S., and Rubin, D. B. (2003) *Bayesian Data Analysis*, 2nd ed. London, Chapman & Hall.
5. Carlin, B. P., and Louis, T. A. (2000) *Bayes and Empirical Bayes Methods for Data Analysis*, 2nd ed. Boca Raton, Chapman & Hall.

17

Overview of Missing Data Techniques

Ralph B. D'Agostino, Jr.

Summary

Missing data frequently arise in the course of research studies. Understanding the mechanism that led to the missing data is important in order for investigators to be able to perform analyses that will lead to proper inference. This chapter will review different missing data mechanisms, including random and non-random mechanisms. Basic methods will be presented using examples to illustrate approaches to analyzing data in the presence of missing data.

Key Words: Imputation, missing data mechanism, MAR, MCAR, nonignorable missing data

1. Introduction

In the previous 16 chapters, you have been presented with a variety of methods and techniques for analyzing data in order to make valid inference. In these previous chapters, a common assumption concerning the validity of the techniques has been that there is complete data available on all units measured in the experiment or study. The goal of this chapter is to present an overview of what can be done when this assumption is violated and missing data occurs on observations in an experiment or study.

Missing data frequently arise in the course of research studies. This phenomenon, though rarely intended, can have varying impact on the ability of investigators to draw proper conclusions concerning the relevance of their data. Often, the existence of missing data itself is not the issue that is of most importance, but rather understanding the mechanism that led to data being missing is most relevant. If one can understand the mechanisms that led to data being missing, then often appropriate analytical strategies can be used to handle its occurrence. This chapter will introduce basic concepts concerning approaches

From: *Methods in Molecular Biology, vol. 404: Topics in Biostatistics*
Edited by: W. T. Ambrosius © Humana Press Inc., Totowa, NJ

to analyzing data in the presence of missing data. General concepts and terminology will be presented as well as descriptions of analytical tools that can be used to address the missing data.

The focus of this chapter, as well as this book, is on analyses that researchers face often in laboratory settings. One may wonder if missing data is really an issue in the laboratory setting when most if not all aspects of an experiment are under the control of the investigator. Although this may appear to be the case, the reality is that missing data may often exist yet the investigator may not recognize this fact. Consider the following examples.

Example 1

An investigator wishes to test the impact of a new therapy on an outcome of cell proliferation measured in mice that are to be euthanized after 15 weeks of exposure. The investigator breeds 30 mice (that are genetically predisposed to develop prostate cancer) for the experiment and assigns 15 to receive the new treatment and 15 to receive a placebo treatment. Fourteen weeks into the experiment, 4 mice in the placebo group have died, whereas 1 mouse in the treatment group has died. In the next week (week 15), when the primary outcome assessment is to be made on the mice, there are now 11 mice on placebo and 14 mice on treatment. What data should be used for the 5 mice that died prior to week 15 and therefore do not have a week 15 assessment?

Example 2

Consider a laboratory experiment where one wishes to study the cleavage of certain proteins to different enzymes. The whole experiment is conducted using test tubes, and the goal of the experiment is to expose one set of proteins to a new compound and to expose a different set to an existing compound. At 6 fixed time points (0, 15, 30, 45, 60, and 75 min), a fixed amount of protein is to be extracted from the test tube and measured for its cleavage to a different enzyme. What happens if a test tube is dropped after the first 3 measurements are made but prior to the last 3 assessments? What happens if the investigator cannot be present for the 60-min assessment for some of the tubes and it does not take place until 73 min? How should these data be handled?

The remainder of this chapter will present methods that may be useful in confronting situations similar to these as well as other ones that may occur in laboratory research settings. **Section 2** will introduce notation and terminology useful for describing the missing data mechanism. **Section 3** discusses missing data mechanisms. **Section 4** will present an overview of *ad hoc* procedures that have been suggested for handling missing data and, where appropriate, describe their shortcomings. **Section 5** will present 2 model-based approaches to missing

Table 1
Illustration of Notation and Missing Data

Unit	X	Y
1	Observed	Observed
2	Observed	Observed
k	Observed	Observed
$k + 1$	Observed	Missing
$k + m = n$	Observed	Missing

data including a brief overview of the concept of multiple imputation. Finally, **Section 6** will present a summary and list appropriate references to examine for more thorough handling of missing data.

2. Notation

To illustrate the concepts of missing data, consider a simple example where there are 2 variables, one outcome, referred to as Y, and one predictor variable, referred to as X. In **Table 1**, there are n total observations, k with complete data and m with the outcome Y missing. Both X and Y can be partitioned into their observed and missing parts as follows: $Y = (Y_{obs}, Y_{mis})$ and $X = (X_{obs}, X_{mis})$, where Y_{obs} and X_{obs} represent the observed parts of Y and X, respectively, and Y_{mis} and X_{mis} represent the missing parts of Y and X. In the above example, $X = X_{obs}$, because X has no missing data, whereas Y_{obs} is represented by the first k observations of Y and Y_{mis} is represented by the m observations where Y is missing. In addition, we define the response indicator, R_i, as an indicator variable that equals 1 if the ith observation ($i = 1, \ldots, n$) is observed and 0 if the ith observation is missing. For each variable in a data set (i.e., in the above X and Y), there is a corresponding response indicator. Thus, in our example, the response indicator corresponding with X is equal to 1 for all observations, whereas the response indicator for Y is equal to 1 for observations 1 through k and 0 for observations $(k + 1)$ to $(k + m) = n$.

3. Missing Data Mechanisms

In most discussions concerning how to handle missing data, it is generally agreed that the most important step to decide how to handle missing data is to determine what mechanism led to the data being missing in the first place. Intuitively, one can imagine that one would analyze the same set of data differently if the mechanism that led to the data being missing were different. For instance, for the example introduced at the beginning of the chapter where levels of cell proliferation are to be compared between 2 groups after a fixed period of time, consider these 2 scenarios. **Scenario 1**: From the original 30 animals (15 per group), you are given data on 25 (14 from **group A** and 11 from **group B**) and told that the missing data on the 5 animals occurred because when performing the measurements, there was a power outage and the data on those 5 animals was lost. **Scenario 2**: The 5 animals with missing data all died prior to the time point when cell proliferation measures could be made.

Clearly, the mechanism by which the data came to be missing in these 2 scenarios is different, and in fact knowing the mechanism in this example is likely to influence how you would analyze the observed data. Intuitively, you would imagine that the observed data from **Scenario 1** would represent a relatively accurate description of the difference between the 2 groups because the missing data seemed to have been generated by a "random" mechanism. However, you may not feel the same about the data generated in **Scenario 2**. Here you have one group with less data (**group B** has 11 observations vs. 14 in **group A**) and the fact that data is missing may in fact be linked to the outcome of interest. In this case, one might argue that the mechanism that led to missing data was not random.

In order to describe the missing data mechanism, we must examine the probability (Pr) that response indicator equals 1 for a variable, conditional on the values for that variable taken on by the observed and missing observations. We denote this by

$$\Pr(R \mid Y_{obs}, Y_{mis}).$$

3.1. Missing Completely at Random

In the first scenario just described, the probability that a particular animal is missing its measure of cell proliferation does not depend on observed or unobserved measurements of cell proliferation (if we assume that the power outage itself was not related to the experiment being performed). This can be written as

$$\Pr(R \mid Y_{obs}, Y_{mis}) = \Pr(R).$$

When this relationship exists, we refer to the missing data mechanism as *missing completely at random* (MCAR) ***(1,2)***. In other settings, this mechanism may be referred to as *uniform nonresponse* (i.e., in the sample survey literature).

When data are MCAR, we can perform analyses using complete data techniques, described in the previous 16 chapters, and still arrive at consistent results. In general, there will be a loss of power (and information) in these studies because of the missing data. This occurs because the variability of statistics used in treatment comparisons is reduced as the sample size increases, and thus because missing data reduces the sample size available, this in turn increases variability. Still, when MCAR is present, one can perform analyses using only those units with complete data (often referred to as *completers*) and still be confident that valid inferences can be made.

3.2. Missing at Random

Although MCAR may occur in many settings, one can clearly imagine there are settings where data are not missing completely at random, yet the missing data are not linked directly to the values of the unobserved data as would be the implication of **Scenario 2** above. A question that arises is whether there exist other mechanisms (other than MCAR) where one can perform valid analyses in the presence of missing data, and the answer is yes as long as the missing data mechanism itself does not depend on the unobserved data. This can be written as

$$\Pr(R \mid Y_{obs}, Y_{mis}) = \Pr(R \mid Y_{obs}),$$

and when this condition is met, we refer to the missing data as *missing at random* (MAR) *(1,2)*. In words, this means that if several characteristics are measured on units in a study, then the statistical relationship among these characteristics (variables) remains the same whether or not they are observed or missing for each unit. For example, see **Table 2**. There is a clear relationship among **units 1, 2, 3,** and **4** when all four are observed. Under MAR, **variables 3** and **4** for **unit 4** will have the same distribution as **variables 3** and **4** from **units 1, 2,** and **3**.

At times, the word *random* in MAR causes confusion because the probability of a value being missing often does depend on observed values of the data and therefore does not appear to be "random." This confusion is justified. As the nomenclature for missing data has used this term for many years, one must understand its particular meaning and remember here that the important concept is that when data are MAR, the missing value mechanism can be described solely in terms of the observed observations.

It should be noted that in general, it is difficult to prove that data from a particular study are in fact MAR, however there are many situations where

Table 2
Illustration of Data Where the Missing Data Are MAR

	Variables			
Unit	1	2	3	4
1	2	3	4	5
2	5	10	15	20
3	8	15	22	29
4	10	22	?	?

given the description of how data were collected, one can infer that the missing data mechanism is likely to be MAR. An example that may arise in many experiments or trials is that a characteristic is to be measured on a unit twice at the same visit (i.e., systolic blood pressure). Many protocols state that if the 2 measurements differ by a certain amount, a third measurement will be taken as well, however for most units this third measurement will be missing at each visit. This missing data mechanism is clearly not MCAR but it would be considered to be MAR.

One should be aware that situations may arise where data is MCAR within classes and MAR overall. For instance, consider an experiment where 2 observations are made: a mouse's overall weight and a mouse's level of a certain biomarker that is known to exist and is related to the mouse's weight. The sensitivity of the instruments used to detect the biomarker is such that for many small mice, the actual value of the biomarker cannot be recorded by the instrument and appears as a missing value. In this example, if one were to look at the data from the biomarker and take a simple average of the values from those mice with data, the mean would be upwardly biased because many of the small mice would have missing data for the biomarker.

However, if one were to group mice by weight, then conditional on a mouse's weight the missing data for the biomarker is random. Then, the biomarker data is missing at random because the mechanism that led to the missing data depends on the mouse's weight. In other words, once we know a mouse's weight, the missingness does not depend on the value of the biomarker itself.

We can then estimate the overall mean of the biomarker in the experiment by taking averages of the observed biomarker data within groups of mice with similar weights. We then combine these stratified estimates together, weighting each by the proportion of total mice in each group. For instance, consider the following illustration. In an experiment there were 20 total mice: 10 with high weight and observed biomarker data with a mean value for the biomarker data of 15, and 10 with low weight of which 6 have observed biomarker data with a mean value of the biomarker data of 25 and 4 have missing data for the biomarker data. If we had taken an average of the observed data for the biomarker, we would get 18.75 {i.e., $[(10 \times 15) + (6 \times 25)]/16 = 18.75$}. However, if we average the 2 strata, we get the mean for the biomarker data to be 20 (the average of 15 in the high-weight group with 25 in the low-weight group). In this way, the 6 low-weight mice with observed biomarker data are contributing information into the estimate as if they were 10 mice.

This example illustrates how simple summary statistics (the average of the 16 observed biomarker values) would be biased, whereas by using a model that appropriately takes into account the MAR structure of the data, by conditioning on the mouse's observed weight, an unbiased estimate is derived. In general, likelihood-based methods, which have been discussed in the previous chapters, are valid when data are MAR.

3.3. Missing Not at Random/Non-ignorable Missing Data

MAR and MCAR describe two possible ways that data can be missing; the third and final way we describe missing data is when $\Pr(R \mid Y_{obs}, Y_{mis})$ depends on both the observed and missing parts of Y and therefore cannot be explained or modeled easily. When this type of data arises we often refer to the missing data mechanism as being *non-ignorable* because the distribution of R, the response indicator, cannot be explained or modeled easily and therefore cannot be ignored when using likelihood methods for inference. When data are MAR, the missing data mechanism was considered *ignorable* because one could ignore it when making likelihood-based inference once one conditioned on the observed data.

The implication of nonignorable missing data is that the reason some observations are missing depends on the unobserved observations themselves. In the biomarker example above, had the researcher not recorded the weight of the mice in the experiment but had only recorded the values of the biomarker data, then the mechanism would have not been ignorable and the data would not have been MAR. In order for inference to be made when the missing data mechanism is non-ignorable, one has to model both Y and R (the data and response indicator) jointly to make inference about Y.

In general, there is no easy way to prove that data are MAR, MCAR or missing not at random (MNAR). Usually one can make reasonable assumptions concerning the origin of missing data and based on those assumptions determine which mechanism led to the missing data. If data are determined to be MNAR, it is often difficult to determine an appropriate model for the response indicator.

4. Ad Hoc Methods for Handling Missing Data

There are many *ad hoc* methods that investigators use to handle missing data, and in general these methods are not advisable to be used except in certain specific problems or circumstances. We now describe a few of these methods and point out their shortcomings. In particular, we describe analyses that use completers only, analyses that use a method called *last observation carried forward*, and simple mean or regression imputation.

4.1. Complete Case Analysis

Complete case analysis, as its name implies, is an analysis that uses only units that have complete data. All units that have missing data are removed from the analyses. As pointed out above, if data are MCAR, then analyses based

Table 3
Hypothesized Data from 18 Mice

Mouse	Biomarker 1	Biomarker 2	Biomarker 3	Weight	Treatment
1	1.4	4.4	9.1	101	A
2	2.9	5.4	8.4	121	B
3	1.2	4.2	7.7	118	A
4	2.6	3.4	9.4	141	B
5	2.1	4.4	8.5	131	B
6	1.2	6.2	6.7	125	A
7	1.9	7.1	9.1	108	A
8	0.8	3.1	6.1	132	B
9	2.4	8.0	9.3	128	A
10	1.1	5.4	7.2	140	B
11	1.9	5.6	8.8	102	A
12	2.0	5.3	7.5	109	B
13	3.2	?	9.9	121	A
14	?	?	?	114	B
15	2.4	6.1	?	125	A
16	1.5	?	?	110	B
17	?	4.6	8.3	131	A
18	?	?	9.1	139	B

on completers will not be biased and therefore can be used; however, even in this situation, one must proceed with caution. When a completer's analysis is performed, one has reduced the sample size from the original intent of the study. This implies that the planned study power is no longer being maintained, and thus treatment effects may not be detected with this reduced power.

In experiments that require examining the relationship among several variables in a sequential manner, one is often confronted with fitting a series of models with different units included in each model. For instance, consider the hypothetical data measured in an experiment using 18 mice and presented in **Table 3**. The question of interest is whether a particular treatment (**A** or **B**) predicts the weight of mice after controlling for the effects of three biomarkers. To look at this, using completers only, a series of models would be fit. The first could include all 18 mice to see the effect of treatment on weight. Next, a model that includes the first 13 mice and **mouse 15** and **mouse 16** could look at the effect of treatment on weight while adjusting for **biomarker 1**. A third model could look at the impact of treatment and **biomarker 2** on weight, but this would exclude **mice 13, 14, 16,** and **18**. Additional models could be fit to look at **biomarker 3** or some combination of **biomarkers 1, 2,** and **3**. Each model would include a different subset of mice, until finally the full model would only include the first 12 mice, because they are the only mice with totally complete data. In general, even if one assumes MCAR, one might be concerned that results from analyses using data such as this may be influenced as much by which subset of mice are included in each analysis as by a real scientific relationship between treatment and weight. If the data are not MCAR, then analyses based on completers only will be biased and not result in valid inferences.

4.2. Last Observation Carried Forward

Last observation carried forward (LOCF) is a method that can only be used in experiments with longitudinal follow-up. Essentially, this method takes the last observation measured in a longitudinal study and uses it to impute future missing observations. For instance, if the data in **Table 4** were observed on 4 units in an experiment, the LOCF method would consider the **visit 4** and **5** values for **unit 2** to be 7, because that was the last observed value for **unit 2**. Likewise, the **visit 4** and **5** values for **unit 3** would be considered 12, because that was the last observed value. For **unit 4**, we imputed **visits 3** and **5** with the data from **visits 2** and **4**.

Using this method, all missing data are filled in, and analyses are then performed as if there were no missing data. This method will nearly always produce incorrect treatment effect estimates and measures of variability, regardless of whether the data are MCAR, MAR, or MNAR. The severity of the potential bias introduced by using this method depends on the actual mechanism that led to the missing data and the particular treatment effect estimate over time. For instance, in the simple example above, one can see clearly that there appears to be a predictable reduction in the variable measured over time for each of the 3 units. Using the LOCF method, it appears at **time 5** that there is a large difference among the 4 units (**3, 7, 12,** and **9**).

Although it may seem clear that the LOCF method will almost always present incorrect information for inference, it is still widely used in many settings. One reason is because it is easy to implement as no models need to be fit, and it allows all units to be included in the final analysis. Often, when LOCF

Table 4
Illustration of Data for LOCF Example

	Visit				
Unit	1	2	3	4	5
1	10	8	6	5	3
2	11	9	7	? (7)	? (7)
3	20	16	12	? (12)	? (12)

is presented in analyses, it is shown as one of several approaches for handling missing data and is presented as part of a *sensitivity* analysis. Thus, investigators who use LOCF usually understand that the results based on the analysis are not correct, but they argue that the results may give some understanding concerning a range of what outcomes are possible. Much research has been done to suggest that even in this case, using LOCF is not an optimal technique *(3)*.

4.3. Mean/Regression Imputation

Additional *ad hoc* methods that are often used in practice are to perform some form of simple imputation in order to fill in the missing data that is based on the observed data. The easiest version is referred to as *mean* imputation, and this method estimates the mean based on observed data for each variable in a study. Using the data from **Table 3**, we see that the mean value for the observed data in **biomarker 1** is 1.91. This estimated mean is then imputed into all missing values for that variable (**mice 14, 17,** and **18**). This process is repeated for **biomarkers 2** and **3** as well. The intuitive appeal of this method is that the overall mean in each group does not change after the imputation is performed, and this method results in a complete data set to be analyzed using all of the original participants. The disadvantages are that this method will underestimate the variability of each of the individual variables where imputation occurs because imputation of a mean value will not increase the variance estimate for a variable by definition (as the variance estimate calculates the squared distances of each observation from the mean and averages these over the data, and therefore observed values at the mean contribute no extra variability to this estimate). A second disadvantage to this approach is that it ignores any possible relationship among variables measured in a study. Therefore, it is likely to bias any analyses that involve estimating associations or regression models.

A second simple imputation method that is used by investigators is referred to as regression mean imputation. In this approach, participants with complete data are used to estimate regression equations describing the relationship between variables with missing data to the remaining variables. For instance, consider the data from **Table 3** again. There is data missing for **biomarkers 1, 2,** and **3**. The first 12 observations have complete data, thus a regression model predicting **biomarker 1** based on **biomarkers 2** and **3**, and treatment would be fit using these 12 observations. The resulting regression equation is

$$\text{(Biomarker 1)} = -2.00 + 0.161\text{(biomarker 2)} + 0.461\text{(biomarker 3)} - 0.003\ \text{Weight} - 0.807\ \text{(treatment A)}.$$

Thus, using this equation, one would use the observed values for **mouse 17** (**biomarker 2** = 4.6, **biomarker 3** = 8.3, **weight** = 131, **treatment** = A) to predict **biomarker 1** to be 1.37. To determine the predicted value for units that have more than 1 missing data point (i.e., **mouse 16** missing **biomarker 2** and **biomarker 3**), one can proceed in 1 of 2 ways. The first would be to fit a model predicting **biomarker 2** using **biomarker 1** and **treatment** and then imputing the predicted value for this variable. Next, a model using **biomarker 1**, **biomarker 2**, and **treatment** would be fit to predict **biomarker 3** and the observed values for **biomarker 1** and **treatment** and the *imputed* value for **biomarker 2** will be used to predict **biomarker 3**. (Note: This process could have been performed in reverse order predicting **biomarker 3** first and then **biomarker 2**.) An alternative approach would be to fit 2 models both with **biomarker 1** and **treatment** as predictors to determine predicted values for **biomarker 2** and **biomarker 3**, respectively.

This imputation approach is better than a simple mean imputation approach because it can maintain unbiased estimates of statistics such as means, correlations, and regression coefficients. However, the variance associated with this statistic will be underestimated using this approach *(3)*.

5. Model-Based Approaches to Missing Data

5.1. Likelihood-Based Modeling

A common theme of the *ad hoc* methods discussed above was to develop a method to handle the missing data that created a "complete data" set on which to perform analyses (either by imputation or restricting the analysis to completers). We now describe methods that focus on estimating summary statistics to be used to make inference without actually *imputing* any values into the data set. These approaches do not propose to impute a specific value of a measurement into missing data but rather propose to estimate an appropriate summary statistic that incorporates all information from units with both observed and missing data.

To implement this method, one needs to have a statistical model specified for the complete data. Using this model, one then determines what the likelihood for the model is and then based on assumptions concerning the missing data mechanism, the likelihood is maximized.

Recall, $Y = (Y_{obs}, Y_{mis})$ where Y_{mis} denotes the missing values and Y_{obs} denotes the observed values of Y. If we denote $f(Y|\theta)$ as $f(Y_{obs}, Y_{mis}|\theta)$ as the density function of the joint distribution of Y_{obs} and Y_{mis} and θ are the unknown parameters for the distribution of Y, then the marginal density of Y_{obs} is

$$f(Y_{obs}|\theta) = \int f(Y_{obs}, Y_{mis}|\theta)dY_{mis}.$$

Using this, we can define the likelihood of θ based on Y_{obs} ignoring the missing data mechanism to be any function of θ proportional to $f(Y_{obs}|\theta)$ because

$L(\theta | Y_{obs}) \propto f(Y_{obs} | \theta)$ where L is the likelihood. When the missing data mechanism is ignorable (i.e., if the missing data is MAR or MCAR), then inference can be based on the likelihood using the observed data only $L(\theta | Y_{obs})$ *(2)*.

To show this more clearly, consider the joint distribution of Y and R (the response indicator). We can express this as the distribution of Y and the conditional distribution of R given Y:

$$f(Y, R | \theta, \psi) = f(Y | \theta) f(R | Y, \psi)$$

where ψ is the unknown parameter from the distribution of the missing data indicator.

The observed data are only (Y_{obs}, R), and thus the distribution of the observed data can be found if one integrates Y_{mis} out of the joint density of (Y, R) where $Y = (Y_{obs}, Y_{mis})$,

$$f(Y_{obs}, R | \theta, \psi) = \int f(Y_{obs}, Y_{mis} | \theta) f(R | Y_{obs}, Y_{mis}, \psi) dY_{mis}.$$

Using this joint distribution, we can see that the likelihood for the parameters θ and ψ is any function of θ and ψ proportional to $f(Y_{obs}, R | \theta, \psi)$:

$$L(\theta, \psi | Y_{obs}, R) \propto f(Y_{obs}, R | \theta, \psi).$$

If the distribution of the missing data indicator does not depend on $Y_{mis}(f(R | Y_{obs}, Y_{mis}, \psi) = f(R | Y_{obs}, \psi))$, then it can be shown that

$$\begin{aligned} f(Y_{obs}, R | \theta, \psi) &= f(R | Y_{obs}, \psi) \int f(Y_{obs}, Y_{mis}, \theta) dY_{mis} \\ &= f(R | Y_{obs}, \psi) f(Y_{obs} | \theta). \end{aligned}$$

When the above is satisfied, then the missing data mechanism is considered to be ignorable and thus inference concerning the parameter θ can be made using the likelihood based only on the observed data, Y_{obs} *(2)*.

In many real data applications, one assumes that the missing data mechanism is ignorable and such common statistical modeling techniques that use linear mixed models (models that have both fixed and random effects included, see **Chapter 11**) are able to handle missing data in their estimation approach. The mixed models are valid when the data are MAR.

5.2. Stochastic Imputation: Single and Multiple

Above, we discussed some *ad hoc* imputation strategies (mean and regression imputation), but each of these had limitations because they would underestimate the true variability that would have existed in the data had the missing data been observed. A better imputation strategy is to impute values that are chosen randomly from an appropriate distribution of potential values for the missing data *(3)*.

For instance, in the biomarker example above, the mean value for the observed data for **biomarker 1** was 1.91 with a standard deviation of 0.706. If we assumed that the distribution of values that this data comes from was a Normal distribution, then we could generate a Normal distribution with mean 1.91 and standard deviation 0.706 (using a statistical software package such as SAS) and then choose 3 random values from this distribution to impute into the missing values from the data. The expected value for any of these random draws would be 1.91, but the actual values chosen would likely vary around that value.

We can perform a similar imputation using the regression imputation strategy described earlier, where instead of imputing the predicted value from the regression equation, we would define the distribution of the predicted values (mean and standard deviation at the particular set of observed predictors) and use this distribution to choose random observations to impute into the data set in place of missing observations.

Both of these imputation methods provide better estimates of the variability of the data than the *ad hoc* methods described earlier (recall that the *ad hoc* methods did provide unbiased estimates of the mean values for the variables with missing data but had underestimated the variability). However, estimating the correct variance for a variable that contains missing data after using an imputation strategy is still somewhat complicated. Imagine if several investigators were given the same data set that contained missing values and were told to impute the values using an imputation strategy that takes random draws from an appropriate distribution. Each investigator will likely get a different set of imputed values and thus a different estimate of the variability of the variable that had contained missing data. Because each of these individual variability estimates are in fact estimates of the true variability estimates, if one were to examine the variability among each investigator's imputed data set, an estimate of the true variability could be derived.

This concept was then formalized to say rather than asking several investigators to each impute data into the same data set, each investigator should perform this "multiple" imputation on his or her own. Thus, several sets of complete data would be generated and analyzed as complete data to make inference. The variance estimate from an analysis that uses a multiple imputation approach combines the *within-imputation* and *between-imputation* components of variance to get an overall variance estimate. Details of this can be found in Rubin *(4)*.

6. Conclusion

Missing data is a reality in research. Despite the best efforts of researchers to control their experiments, unforeseen events can (and do) occur. Therefore, one must be prepared to handle the presence of missing data in research. Much of this chapter has focused on describing approaches to handling missing data if the data are MCAR or MAR. The reason for this is because under those scenarios, principled approaches can be applied to handling the missing data, and research has shown that valid inference can be made. There is a growing statistical literature for handling the more difficult situation of a nonignorable missing data mechanism. The difficulties that arise in this situation are that one must carefully specify what the nonignorable mechanism is in order to make inference, and yet often there is little opportunity to validate whether the correct specifications have been made.

This chapter has introduced the reader to some of the concepts that surround the handling of missing data in applications. An introduction of terminology and a brief overview of methods to handle missing data have been presented to meet the goal of educating the researcher about missing data problems. Ultimately, an appropriate analysis that incorporates missing data often needs to be performed in collaboration with a trained statistician, and we would encourage the applied researcher to seek such collaboration when missing data problems arise. In addition to this chapter, there are several useful references ***(1–7)*** that describe much of the methodology in this chapter, particularly a Web page maintained by James Carpenter and Mike Kenward *(3)* that provides useful descriptions of missing data techniques presented in an easily accessible format as well as an extensive bibliography of related materials.

References

1. Rubin, D. B. (1976) Inference and missing data. *Biometrika* **63**, 581–592.
2. Little, R. J. A., and Rubin, D. B. (1987) *Statistical Analysis with Missing Data*. Chichester, John Wiley & Sons.
3. Kenwood, M., and Carpenter, J. (2006) Missing data. Available at http://www.lshtm.ac.uk/msu/missingdata/index.html.
4. Rubin, D. B. (1987) *Multiple Imputation for Nonresponse in Surveys*. New York, John Wiley & Sons.
5. Allison, P. D. (2001) *Missing Data*. Thousand Oaks, Sage Publications.
6. Schafer, J. L. (1997) *Analysis of Incomplete Multivariate Data*. London, Chapman & Hall.
7. Carpenter, J., Pocock, S., and Lamm, C. J. (2002) Coping with missing data in clinical trials: a model based approach applied to asthma trials. *Stat. Med.* **21**, 1043–1066.

18

Statistical Topics in the Laboratory Sciences

Curtis A. Parvin

Summary

This chapter concerns statistical concepts and procedures that are applicable to diagnostic testing performed in the clinical laboratory. Three important laboratory issues are addressed: the estimation of analytical imprecision, the design of an effective laboratory quality control strategy, and the establishment of population reference ranges. These three topics were selected because each demonstrates a valuable statistical principle. Estimation of analytical imprecision highlights the important role of study design. Evaluating laboratory quality control strategies emphasizes the importance of choosing appropriate statistical models. The estimation of population reference ranges demonstrates that there can be many different approaches to developing good statistical estimators.

Key Words: Quality control; reference limits; variance components.

1. Introduction

This chapter concerns statistical concepts and procedures that are applicable to diagnostic testing performed in the clinical laboratory. The life cycle of a laboratory test is generally divided into preanalytical, analytical, and postanalytical phases. The preanalytical phase encompasses patient preparation, sample collection, and transport to the laboratory. The analytical phase involves the measurement of quantities in tissues and body fluids. The postanalytical phase relates to reporting and interpretation of laboratory test results. There are many interesting statistical issues that arise within each of these 3 phases; more than can be addressed in a single chapter. This chapter will address 3 common statistical problems faced by most laboratories: characterizing the analytical imprecision of an assay, determining a quality-control testing strategy to ensure that accurate results are produced, and establishing population reference ranges.

From: *Methods in Molecular Biology, vol. 404: Topics in Biostatistics*
Edited by: W. T. Ambrosius © Humana Press Inc., Totowa, NJ

These 3 topics demonstrate the role that statistical thinking can play in the laboratory with respect to issues such as study design, statistical modeling, and robust estimation.

2. Estimating Analytical Imprecision

2.1. The Precision Performance Study

One of the important characteristics of a laboratory assay that determines its medical usefulness is its inherent analytical imprecision. The purpose of a precision performance study is to estimate the total analytical variability of an assay, and additionally to estimate the relative magnitudes of the different variance components that contribute to the total. The analytical imprecision of an assay often depends on the concentration of the analyte being measured. Therefore, imprecision should generally be estimated at multiple concentration levels throughout the concentration range of an assay.

The different sources of variability constituting total analytical imprecision will depend on the particular assay and testing environment, but a common approach to variance estimation divides total imprecision into 3 components: day-to-day variability, batch-to-batch variability within a day, and within-batch variability. The statistical model representing this situation can be given as

$$Y_{ijk} = \mu + \alpha_i + \beta_{j(i)} + \varepsilon_{k(ij)}.$$

This is a nested random effects analysis of variance (ANOVA) model *(1)*. Y_{ijk} represents the kth measured value from the jth batch on the ith day, μ is the true concentration, α_i is the random error component for the ith day, $\beta_{j(i)}$ is the random error component for the jth batch within the ith day, and $\varepsilon_{k(ij)}$ is the random error associated with the kth measurement in the jth batch on the ith day. It is generally assumed that α_i, $\beta_{j(i)}$, and $\varepsilon_{k(ij)}$ are independently distributed Normal random variables with zero means and variances σ_α^2, σ_β^2, and σ_ε^2, respectively. Assuming this model, the total analytical variance associated with a single measurement, Y_{ijk}, is

$$\sigma_Y^2 = \sigma_\alpha^2 + \sigma_\beta^2 + \sigma_\varepsilon^2.$$

In order to compute estimates of the variance components, data are obtained by performing multiple assays per batch, with multiple batches per day, over multiple days. A commonly employed precision performance experiment will use stable pools of test material at a minimum of 2 concentration levels (one low and one high) and will analyze 2 samples at each concentration level in a batch, with 2 batches performed each day for at least 20 days *(2)*.

There are a number of different methods available for estimating variance component models. The more sophisticated methods require advanced statistical software such as SAS *(3)*. If the experiment is performed using a balanced design, meaning that the number of replicates within each batch and the number of batches performed each day are constant, then the classic *method of moments* estimators are relatively easy to calculate without the need for advanced software, and the equations provide some useful insights regarding variance estimation.

Assume data are collected for n_1 days with n_2 batches per day and n_3 measurements per batch at each concentration level. Four different sample variance estimates can be defined:

between day:

$$S_\alpha^2 = \frac{\sum_{i=1}^{n_1} (\bar{y}_{i}.. - \bar{y}...)^2}{n_1 - 1},$$

between batch within day:

$$S_\beta^2 = \frac{1}{n_1} \sum_{i=1}^{n_1} \left(\frac{\sum_{j=1}^{n_2} (\bar{y}_{ij}. - \bar{y}_{i}..)^2}{n_2 - 1} \right),$$

between replicates within batch:

$$s_\varepsilon^2 = \frac{1}{n_1 n_2} \sum_{i=1}^{n_1} \sum_{j=1}^{n_2} \left(\frac{\sum_{k=1}^{n_3} (y_{ijk} - \bar{y}_{ij}.)^2}{n_3 - 1} \right), \text{ and}$$

total:

$$s_Y^2 = \frac{\sum_{i=1}^{n_1} \sum_{j=1}^{n_2} \sum_{k=1}^{n_3} (y_{ijk} - \bar{y}...)^2}{n_1 n_2 n_3 - 1}.$$

In these equations, $\bar{y}_{ij}.$ denotes the average of the n_3 measurements in the jth batch performed on the ith day, $\bar{y}_{i}..$ denotes the average of all $n_2 n_3$ measurements assayed on the ith day, and $\bar{y}...$ denotes the average of all $n_1 n_2 n_3$ measurements. The between-day formula computes the sample variance of the daily averages. The between-batch formula averages the n_1 sample variances computed from the batch averages within a day. The between-replicates formula averages the $n_1 n_2$ sample variances computed from the replicate measurements within each batch. The total formula calculates the sample variance using all the individual measurements.

What do these 4 sample variances estimate? Phrased another way, what are their expected values? The expected value of the average of the $n_1 n_2$ within-batch sample variances is σ_ε^2. So s_ε^2 provides an unbiased estimate of σ_ε^2. However, the expected value of s_β^2 is not equal to the batch-to-batch variance σ_β^2 but rather it is equal to $\sigma_\beta^2 + \sigma_\varepsilon^2/n_3$. The reason for this is that the batch averages, $\bar{y}_{ij.}$, used in the computation of s_β^2, still possess a component of within-batch variability equal to σ_ε^2/n_3. Thus, the expected value of the between-batch sample variance equals the between-batch variance plus this additional component of the within-batch variance that depends on the number of replicates per batch. However, an unbiased estimate of σ_β^2 can be obtained by computing $s_\beta^2 - s_\varepsilon^2/n_3$. The same reasoning can be used to show that the expected value of the between-day sample variance s_α^2 is not the day-to-day variance component σ_α^2, but rather the between-day variance with an additional component of between-batch variance that depends on the number of batches per day and an additional component of within-batch variance that depends on the number of batches per day and the number of measurements per batch, $\sigma_\alpha^2 + \sigma_\beta^2/n_2 + \sigma_\varepsilon^2/n_2n_3$. An unbiased estimate of σ_α^2 can be obtained by computing $s_\alpha^2 - s_\beta^2/n_2$.

Thus unbiased estimates for each variance component can be obtained as

$$\hat{\sigma}_\varepsilon^2 = S_\varepsilon^2$$

$$\hat{\sigma}_\beta^2 = s_\beta^2 - s_\varepsilon^2/n_3$$

$$\hat{\sigma}_\alpha^2 = s_\alpha^2 - s_\beta^2/n_2,$$

and an unbiased estimate for total analytical imprecision is then

$$\hat{\sigma}_Y^2 = \hat{\sigma}_\alpha^2 + \hat{\sigma}_\beta^2 - \hat{\sigma}_\varepsilon^2.$$

Note that when computing $\hat{\sigma}_\alpha^2$ and $\hat{\sigma}_\beta^2$, it is possible to obtain values that are less than zero. If a value less than zero is obtained, it is standard practice to set it to zero when computing $\hat{\sigma}_Y^2$. Many advanced statistical software packages implement more sophisticated methods such as restricted maximum likelihood (REML) estimation that will always produce nonnegative variance component estimates ***(3)***. Additionally, most statistical software packages can estimate variance components when the study data are not completely balanced (n_2 and n_3 not necessarily constant).

Lastly, it is interesting to examine the expected value of the sample variance computed from all of the individual measurements, s_Y^2. A simple way to determine the expected value of s_Y^2 is to make use of the algebraic identity

$$\sum_{i=1}^{n_1}\sum_{j=1}^{n_2}\sum_{k=1}^{n_3}(y_{ijk} - \bar{y}_{...})^2 = \sum_{i=1}^{n_1}\sum_{j=1}^{n_2}\sum_{k=1}^{n_3}\left\{(y_{ijk} - \bar{y}_{ij.})^2 + (\bar{y}_{ij.} - \bar{y}_{i..})^2 + (\bar{y}_{i..} - \bar{y}_{...})^2\right\}$$

or

$$(n_1n_2n_3 - 1)s_Y^2 = n_1n_2(n_3 - 1)s_\varepsilon^2 + n_1n_3(n_2 - 1)s_\beta^2 + n_2n_3(n_1 - 1)s_\alpha^2.$$

Thus

$$(n_1n_2n_3 - 1)E(s_Y^2) = n_1n_2(n_3 - 1)E(s_\varepsilon^2) + n_1n_3(n_2 - 1)E(s_\beta^2) + n_2n_3(n_1 - 1)E(s_\alpha^2),$$

where $E()$ denotes expected value. Substituting for the expected values of s_ε^2, s_β^2, and s_α^2, dividing by $(n_1n_2n_3 - 1)$ and combining terms gives

$$E\left(s_Y^2\right) = \sigma_\varepsilon^2 + \frac{(n_1n_2n_3 - n_3)}{(n_1n_2n_3 - 1)}\sigma_\beta^2 + \frac{(n_1n_2n_3 - n_2n_3)}{(n_1n_2n_3 - 1)}\sigma_\alpha^2.$$

Note that if the number of measurements per batch, n_3, and the number of batches per day, n_2, are >1, then $E(s_Y^2) < \sigma_\alpha^2 + \sigma_\beta^2 + \sigma_\varepsilon^2$. This implies that $E(s_Y^2) < \sigma_Y^2$ unless both σ_α^2 and σ_β^2 are equal to zero. Therefore, estimating total analytical variance, σ_Y^2 using the sample variance estimate, s_Y^2, will tend to produce estimates that are, on average, too small.

2.2. Confidence Interval for Total Imprecision

Although estimates of the variance components are important for the laboratory to understand the different sources of variability that contribute to total analytical imprecision, in the end it is the estimate of total analytical imprecision that is most important. The reliability of the estimate of total variance will depend on the study design and the number of measurements. Computing a confidence interval for total imprecision is a useful way of conveying the reliability of the estimate of total analytical imprecision.

Assuming the measurement error distributions are approximately normal, Satterthwaite *(4)* describes a method for approximating the distribution of a linear combination of independent mean squares. Let $\hat{\sigma}^2 = a_1\text{MS}_1 + a_2\text{MS}_2 + \ldots$, and let the degrees of freedom associated with each mean square be denoted d.f._1, $\text{d.f.}_2, \ldots$, respectively. Compute

$$\widehat{\text{d.f.}} = \frac{(a_1\,\text{MS}_1 + a_2\,\text{MS}_2 + \ldots)^2}{\dfrac{(a_1\,\text{MS}_1)^2}{\text{d.f.}_1} + \dfrac{(a_2\,\text{MS}_2)^2}{\text{d.f.}_2} + \ldots}.$$

Then the distribution of $\hat{\sigma}^2$ is approximately related to a chi-square distribution with $\widehat{\text{d.f.}}$ degrees of freedom. For the precision study design described above, the sources of variation and associated mean squares are given in Table 1. The mean squares (MS) are defined as SS/d.f. They are related to the sample variance estimates defined earlier: $s_\alpha^2 = \text{MS}_D/(n_2n_3)$, $s_\beta^2 = \text{MS}_B/n_3$, and $s_\varepsilon^2 = \text{MS}_R$. The estimate of total analytical imprecision can be written in terms of the mean squares as: $\hat{\sigma}_Y^2 = \text{MS}_D/(n_2n_3) + (n_2 - 1)\text{MS}_B/(n_2n_3) + (n_3 - 1)\text{MS}_R/n_3$. This is a linear combination of independent mean squares with $a_1 = 1/(n_2n_3)$, $a_2 = (n_2 - 1)/(n_2n_3)$, $a_3 = (n_3 - 1)/n_3$ and therefore, by Satterthwaite's approximation

Table 1
Analysis of Variance (ANOVA) Table for a Nested Random Effects Design

Source	SS	d.f.	MS	E(MS)
Day	$n_2 n_3 \sum_{i=1}^{n_1} (\bar{y}_{i}.. - \bar{y}...)^2$	$n_1 - 1$	MS_D	$\sigma_\varepsilon^2 + n_3\sigma_\beta^2 + n_2 n_3 \sigma_\alpha^2$
Batch	$n_3 \sum_{i=1}^{n_1} \sum_{j=1}^{n_2} (\bar{y}_{ij}. - \bar{y}_{i}..)^2$	$n_1(n_2 - 1)$	MS_B	$\sigma_\varepsilon^2 + n_3\sigma_\beta^2$
Replicate	$\sum_{i=1}^{n_1} \sum_{j=1}^{n_2} \sum_{k=1}^{n_3} (y_{ijk} - \bar{y}_{ij}.)^2$	$n_1 n_2(n_3 - 1)$	MS_R	σ_ε^2

$$\widehat{\text{d.f.}} = \frac{(a_1\,MS_D + a_2\,MS_B + a_3\,MS_R)^2}{\dfrac{(a_1\,MS_D)^2}{n_1 - 1} + \dfrac{(a_2\,MS_B)^2}{n_1(n_2 - 1)} + \dfrac{(a_3\,MS_R)^2}{n_1 n_2(n_3 - 1)}}.$$

An approximate 95% confidence interval for total analytical imprecision can be computed as;

$$\left(\hat{\sigma}_Y \sqrt{\frac{\widehat{\text{d.f.}}}{\chi^2_{0.975}\left(\widehat{\text{d.f.}}\right)}},\ \hat{\sigma}_Y \sqrt{\frac{\widehat{\text{d.f.}}}{\chi^2_{0.025}\left(\widehat{\text{d.f.}}\right)}}\right)$$

where χ_p^2(d.f.) denotes the pth percentile of the chi-square distribution with d.f. degrees of freedom. Many modern software packages will provide a function that computes the percentiles of the chi-square distribution. In the Stata statistical software package (StatCorp. College Station, Tex.), the function invchi2 (d.f., p) will return the pth percentile of the chi-square distribution with d.f. degrees of freedom.

As an example of the above calculations, assume a precision study was carried out over 20 days with 2 batches run per day, 2 replicates assayed within each batch, and the estimated sample variances are $s_\alpha^2 = 2.8$, $s_\beta^2 = 3.4$, and $s_\varepsilon^2 = 3.6$. Then

$$\hat{\sigma}_\varepsilon^2 = 3.6$$
$$\hat{\sigma}_\beta^2 = 3.4 - 3.6/2 = 1.6$$
$$\hat{\sigma}_\alpha^2 = 2.8 - 3.4/2 = 1.1,$$

and an estimate of total analytical imprecision is $\hat{\sigma}_Y = \sqrt{1.1 + 1.6 + 3.6} = 2.51$. Additionally, $MS_D = 4(2.8) = 11.2$, $MS_B = 2(3.4) = 6.8$, $MS_R = 3.6$, $a_1 = 1/4$, $a_2 = 1/4$, $a_3 = 1/2$, and

$$\widehat{\text{d.f.}} = \frac{(11.2/4 + 6.8/4 + 3.6/2)^2}{\dfrac{(11.2/4)^2}{(20-1)} + \dfrac{(6.8/4)^2}{20(2-1)} + \dfrac{(3.6/2)^2}{20(2)(2-1)}} = 62.20.$$

Using Stata's invchi2 function, $\chi^2_{0.975}(62.20) = 85.89$, $\chi^2_{0.025}(62.20) = 42.29$ and a 95% confidence interval for total analytical imprecision is

$$\left(2.51\sqrt{\frac{62.20}{85.89}}, 2.51\sqrt{\frac{62.20}{42.29}}\right) = (2.14, 3.04).$$

3. Designing a Laboratory Quality Control Strategy

3.1. Defining Laboratory Quality

Before an assay is implemented in the laboratory, a thorough evaluation of its analytical performance characteristics is undertaken. This includes estimation of analytical imprecision, determination of the linear range of the assay, and identification of potential interfering substances. Once the assay is put into production, it is necessary to have a process in place to ensure that the expected performance characteristics of the assay are maintained. Particularly, any undetected "drift" in an assay will result in an additional component of measurement error in the reported results that could adversely impact the clinical management of patients. The purpose of laboratory quality control (QC) procedures is to detect clinically important changes from the stable operating characteristics of an assay in order to prevent results with medically important errors from being produced and reported. Statistical models of a laboratory's analytical measurement processes during the presence of various out-of-control error states provide a mechanism for evaluating and comparing the performance of alternative QC strategies.

Let the in-control testing process be modeled as

$$X_i = \mu_X + \varepsilon_i,$$

where X_i is the measured result for the ith patient. The parameter μ_X is the true unknown concentration of the patient's sample, and ε_i is the random measurement error due to analytical imprecision of the assay. Assume ε_i is normally distributed with mean = 0 and variance = $\sigma^2 g^2(\mu_X)$. If $g(\mu_X) = 1$, then the assay has constant variance independent of the concentration of the sample. If $g(\mu_X) = \mu_X$, then the assay has a constant coefficient of variation (CV). The coefficient of variation is defined as the standard deviation divided by the mean, so if $g(\mu_X) = \mu_X$, then CV = $\sigma\mu_X/\mu_X = \sigma$. These are the 2 most common assumptions regarding analytical imprecision.

Let the out-of-control testing process be modeled as

$$X_i = \mu_X + \delta(\mu_X) + \theta\varepsilon_i.$$

The parameters $\delta(\mu_X)$ and θ represent possible out-of-control states. When $\delta(\mu_X) = 0$ and $\theta = 1$, the assay is operating in its stable in-control state. If $\delta(\mu_X) \neq 0$, it implies that the assay has shifted and is producing biased results. If $\theta > 1$, the assay's variability has increased and is greater than expected during stable operation. Out-of-control shifts are generally modeled as $\delta(\mu_X) = \beta_0 + \beta_1\mu_X$. If $\beta_0 \neq 0$, $\beta_1 = 0$, then a constant shift has occurred across all concentration levels. If $\beta_0 = 0$, $\beta_1 \neq 0$, then the out-of-control shift is proportional to the true concentration of the sample.

The quality of a laboratory measurement is related to the magnitude of the error in the result. The larger the difference between the measured value and the true concentration of the sample, the poorer the quality of the result. There are a number of possible ways to quantify the quality of the results produced by an assay. The most common way in laboratory medicine has been based on the concept of total allowable error. Let $E_i = X_i - \mu_X$ represent the error in a measured result. Let $E_a(\mu_X)$ represent the total allowable error specification for the assay. If the error in a reported result exceeds $E_a(\mu_X)$, the result is considered to be of unacceptable quality; the error is large enough to potentially cause harm to the patient. The increase in the probability of producing unacceptable results due to an undetected out-of-control error state can be computed as $P_E = P(|E_i| > E_a \mid \delta, \theta) - P(|E_i| > E_a \mid \delta = 0, \theta = 1)$ for any given out-of-control state. **Figure 1** and **Figure 2** graphically depict the concepts. Using this model, a laboratory's quality goals and QC performance are defined in terms of the total allowable error specified for each analyte, the ability of the laboratory's QC procedures to detect out-of-control error states when they occur, and the probability of producing and reporting unacceptable patient results when the QC procedures are implemented.

3.2. Quality Control Performance Measures

3.2.1. Batch Mode Testing

The most common form of laboratory quality control is based on the testing of quality control samples. A control sample is a stable substance with a known concentration. For out-of-control states that affect the analytical testing process, it is assumed that the control samples will reflect the error state in the same way that patient samples do. Because the concentrations of the control samples are known, statistical methods can be used to assess the true state of an assay based on the measured values obtained from the control samples *(5)*.

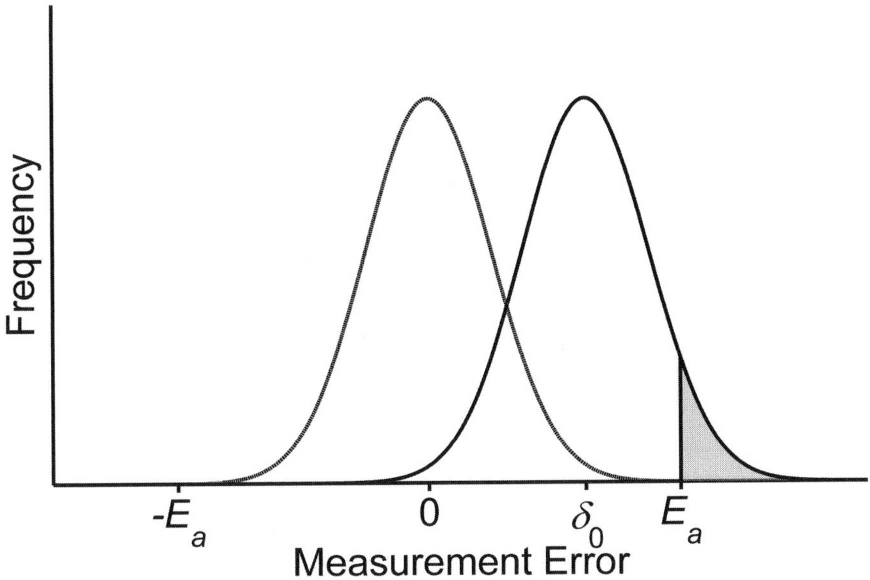

Fig. 1. The probability of producing an unacceptable test result during an out-of-control error condition. The dashed curve represents the in-control frequency distribution of analytical measurement error. The solid curve shows an out-of-control condition that has caused a shift equal to δ_0 in the frequency distribution. E_a is the allowable error specification for the assay. The shaded area represents the probability of producing an unacceptable result for the out-of-control error state.

There are two basic modes of laboratory testing: batch mode and continuous mode. In the batch testing mode, a group of patient samples, along with some number of control samples, are processed and measured as a batch. It is assumed that all samples in the batch are in the same control state. The control sample results are used to decide whether the batch is in-control (and can be accepted) or is out-of-control (and should be rejected). In the continuous testing mode, there is no physically defined batch. Rather, patient samples are tested in a continuous stream. An out-of-control error condition may occur at any point in the testing stream. Periodically, one or more control samples are inserted into the testing stream. The control sample results are used to decide whether the testing process is in-control (and can continue operating) or is out-of-control (and should be stopped).

For batch mode laboratory testing, the probability of batch rejection, P_R, is a natural outcome measure that can be used to evaluate the performance of a QC procedure. The probability of rejecting a batch that is in-control is the false-

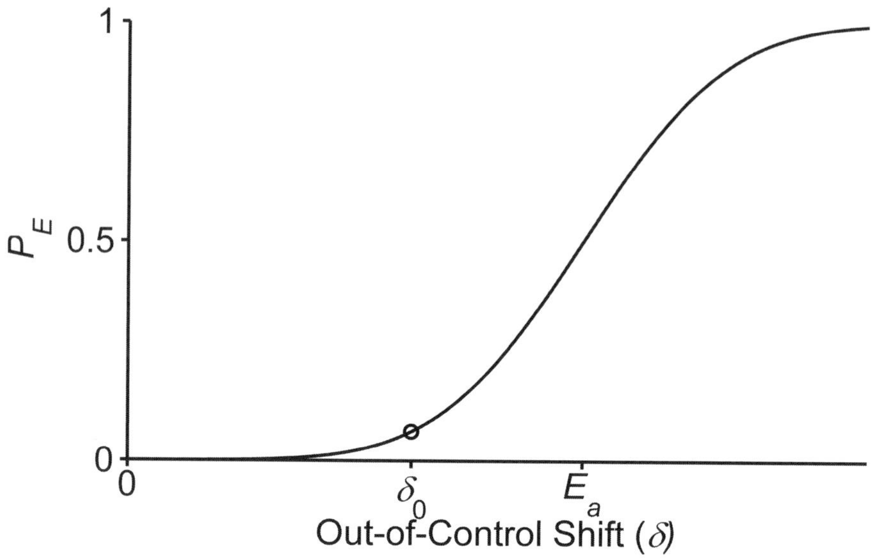

Fig. 2. The increase in probability, P_E, that the total error in a measured result will exceed the allowable error specification, E_a, as a function of the magnitude δ, of an out-of-control shift in the analytical testing process. The value δ_0 is the out-of-control shift illustrated in **Figure 1**.

rejection probability. A good QC procedure should have a low false-rejection probability. The probability of rejecting a batch that is out-of-control is the error detection probability. The probability of error detection will depend on the magnitude of the out-of-control error condition. Plotting the probability of error detection versus the magnitude of the out-of-control error condition is referred to as a power function graph. For many batch mode QC procedures, the probabilities of false rejection and error detection can be mathematically derived. In other cases, computer simulations are used *(6)*. **Figure 3** shows an example of a power function graph that was derived mathematically.

The probability of reporting unacceptable patient results, P_U, due to an out-of-control batch will depend both on the magnitude of the out-of-control error condition, the probability of generating unacceptable patient results in the presence of the error condition, and the probability that the QC procedure fails to detect the error condition *(7)*. This is illustrated in **Figure 4**. The behavior of the curve representing the increase in probability of reporting unacceptable results due to an undetected out-of-control batch can be explained as follows. For very small out-of-control error conditions, the probability of the QC procedure rejecting the batch is low, but the probability of generating unacceptable results in the presence of the error condition is also low. For very large out-of-control error conditions, the probability of generating unacceptable results is high, but the probability of rejecting the batch is also high. The worst case, in the sense of being associated with the largest increase in the probability of reporting unacceptable results due to an undetected out-of-control error condition, will be somewhere in between the 2 extremes. The worst-case out-of-control error condition and its associated increase in probability of reporting unacceptable results will depend on the total allowable error specification, the analytical imprecision of the assay, the number of control observations included in the batch, and the QC rule applied to the control observations.

Different batch-testing QC procedures can be compared by evaluating their power functions and their probabilities of reporting unacceptable results. The 2 main features of a batch QC procedure that can be modified are the number of control samples in the batch and the test statistics computed from the control sample results. The greater the number of control samples in the batch, the better the QC performance will be. In general, the most powerful test statistics for batch QC procedures are based on the sample mean and variance of the control results in the batch ***(8,9)***.

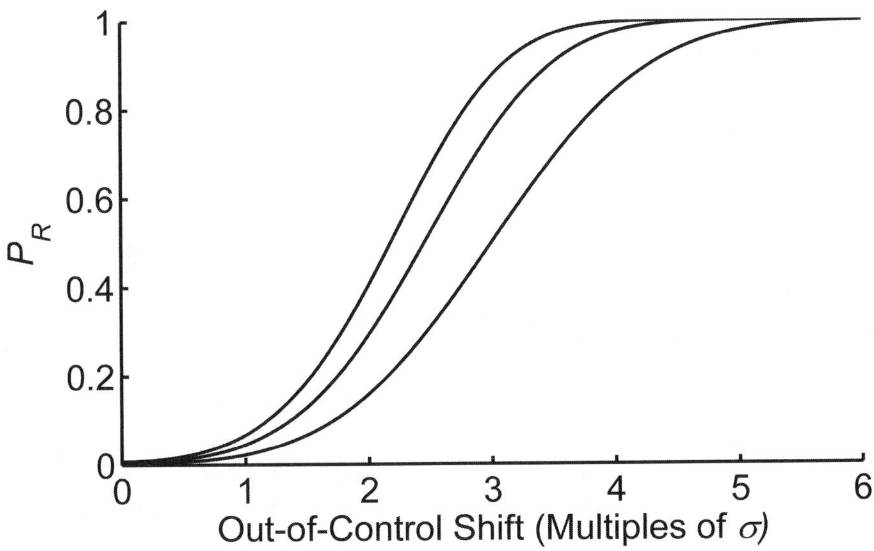

Fig. 3. The probability of a quality control rejection, P_R, based on the magnitude of the out-of-control error condition. For the QC procedure evaluated here, the QC rule rejects if any control sample's measured result is more than 3 analytical standard deviations from the control sample target value. The 3 curves, in increasing order, are the power functions when 1, 2, and 3 control samples are evaluated per batch.

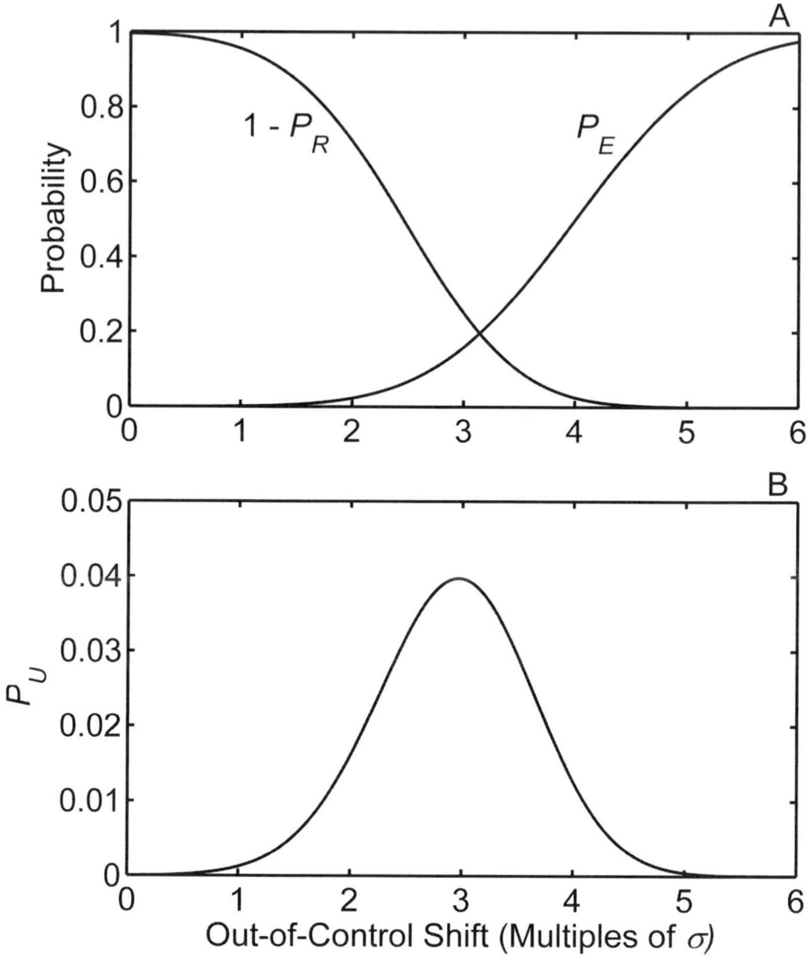

Fig. 4. (**A**) The probability of producing an unacceptable result (increasing curve) and the probability the QC procedure fails to reject the batch (decreasing curve) as a function of the magnitude of out-of-control error condition. The probability of producing an unacceptable result curve is based on a total allowable error specification equal to 4σ. The probability of accepting the batch is based on a QC procedure with 3 standard deviation limits and 2 control samples per batch (see **Fig. 3**). (**B**) The probability of reporting unacceptable patient results, P_U, due to an undetected out-of-control error condition.

The mean rule is designed to detect shifts in the testing process, and the variance rule is designed to detect increases in analytical imprecision. If there are n control samples tested in a batch, then define

$$y_i = \frac{x_i - \mu_i}{\sigma_i}, \quad \bar{y} = \frac{1}{n}\sum_{i=1}^{n} y_i, \quad s_y^2 = \sum_{i=1}^{n}(y_i - \bar{y})^2/(n-1),$$

where x_i, is the ith control sample measurement, μ_i is the true concentration level of the ith control sample, and σ_i is the stable analytical imprecision at the ith control sample concentration. A QC rule based on the sample mean rejects the batch if $\sqrt{n}\,|\bar{y}| > z_{1-\alpha_1/2}$, and a QC rule based on the sample variance rejects the batch if $(n - 1)s_Y^2 > \chi^2_{1-\alpha_2}(n - 1)$. The symbol $z_{1-\alpha_1/2}$ denotes the 100(1 – α_1/2) percentile of the standard normal distribution, and α_1 determines the false-rejection rate for the mean rule. Likewise, $\chi^2_{1-\alpha_2}(n - 1)$ denotes the 100(1 – α_2) percentile of the chi-square distribution with $(n - 1)$ degrees of freedom, and α_2 determines the false-rejection rate for the sample variance rule. These two QC tests are statistically independent of one another, so the overall false rejection probability if both rules are applied is computed as $1 - (1 - \alpha_1)(1 - \alpha_2)$.

3.2.2. Continuous Mode Testing

For laboratory testing processes that operate in continuous mode, the probability of accepting or rejecting a "batch" is a difficult performance measure to interpret because well-defined batches don't exist. In this situation, a more meaningful and interpretable outcome measure is the expected number of patient samples that are produced between the point at which an out-of-control error condition occurs and the point at which the QC procedure detects the out-of-control error condition ***(10)***. Let a QC event be defined as the point where control samples are tested and a QC acceptance/rejection decision is made. The number of patient samples that are produced during an undetected out-of-control error condition will depend on where the out-of-control condition begins relative to the next scheduled QC event, the power of the QC procedure to detect the out-of-control condition when a QC event occurs, and how frequently QC events are scheduled.

A QC strategy based on periodic QC events can only detect an out-of-control error condition at the points where QC events occur. Rather than computing the probability of rejection at a QC event, more typically the expected number of QC events until rejection is computed. This is commonly referred to as the average run length (ARL) to rejection. When the process is in-control, ARL should be large. When an out-of-control condition exists, ARL should be small. ARL will attain its minimum value of 1 when the probability of rejection at the first QC event after the error occurs is 1. There is a simple inverse relationship between ARL and the probability of rejection for QC procedures that only use control sample results from the current QC event: ARL = $1/P_R$. For QC procedures that combine control sample results from previous QC events with the control sample results from the current QC event, this simple inverse relationship no longer holds. In these cases, though ARL is still a meaningful and interpretable measure, the probability of rejecting a QC event is more difficult to interpret because it is not constant from one QC event to the next. Rather, it varies depending on how long an out-of-control error condition has existed without detection by previous QC events ***(11)***.

The expected number of patient samples produced during an out-of-control error condition can be computed as $E(N_P) = E(N_0) + E(N_Q)(\text{ARL} - 1)$, where $E(N_0)$ is the expected number of patient samples from the onset of the out-of-control error condition until the next scheduled QC event, and $E(N_Q)$ is the expected number of patient samples tested between scheduled QC events. The expected number of unacceptable patient results, $E(N_U)$, due to an out-of-control error condition will be the product of the expected number of patient samples produced during the error condition and the probability of generating unacceptable patient results given the magnitude of the out-of-control error condition. **Figure 5** demonstrates these performance measures.

The behavior of the curve representing the increase in expected number of unacceptable patient results due to an undetected out-of-control condition can be explained as follows. For very small out-of-control error conditions, the ARL for the QC procedure is large, but the probability of producing unacceptable results in the presence of the error condition is low. For very large out-of-control error conditions, all of the patient results, from the time the error condition occurs until the next scheduled QC event, are unacceptable, and the probability is near 1 that the QC event will detect the error condition. If out-of-control error conditions can occur with equal probability anywhere within the laboratory testing stream then, on average, the number of patient samples processed between the occurrence of the error condition and the next scheduled QC event will be one-half the expected number of patient samples tested between QC events: $E(N_0) = E(N_Q)/2$. Therefore, the expected number of unacceptable patient results due to an undetected out-of-control error condition will depend on the analytical imprecision of the assay, the total allowable error specification, the frequency of QC events, the number of control observations assayed during each QC event, and the QC rules that are applied. Clearly, for continuous-mode laboratory operations, in addition to the power of a QC procedure, the frequency of QC events is an important consideration when evaluating alternative QC strategies.

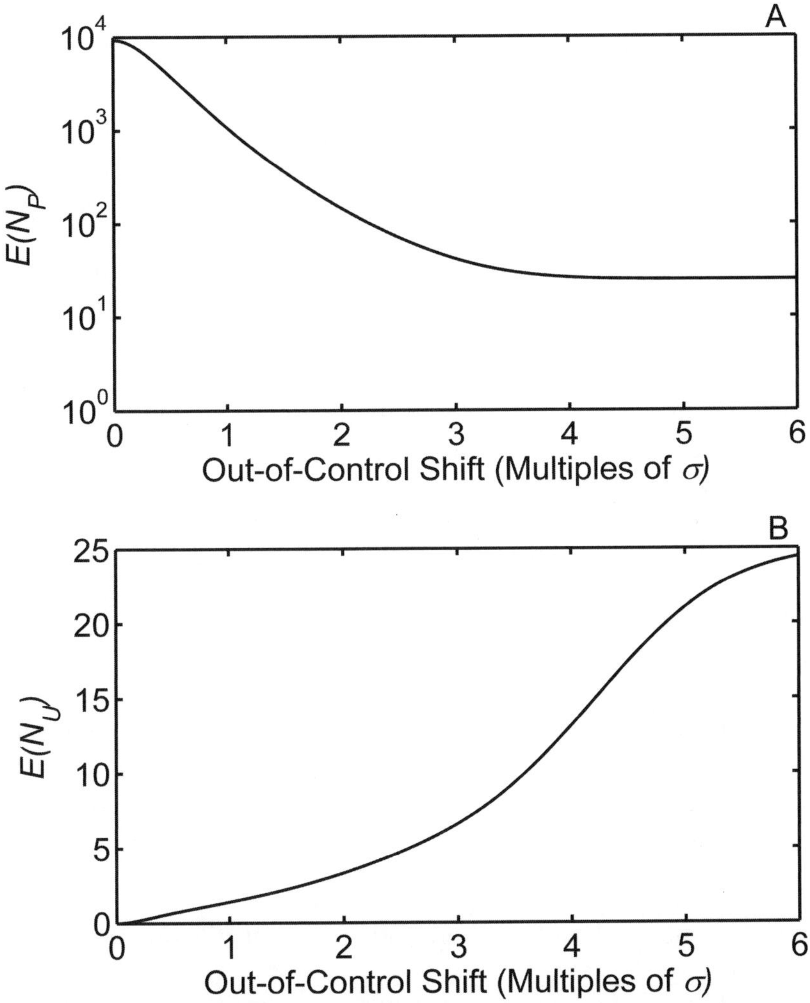

Fig. 5. (**A**) The expected number of patient results produced before detection of an out-of-control error condition, $E(N_P)$, as a function of the magnitude of the out-of-control error condition. For this case, the expected number of patient samples between scheduled QC events is $E(N_Q) = 50$. The QC rule applied at each QC event uses 3 standard deviation limits and 2 control samples (see **Fig. 3**). (**B**) The expected number of unacceptable patients results, $E(N_U)$, due to an undetected out-of-control error condition. The probability of producing an unacceptable result is based on a total allowable error specification equal to 4σ (see **Fig. 4A**).

4. Establishing Reference Ranges

4.1. Reference Limit Estimation

A reference range (or reference interval) is the interval between, and including, 2 reference limits. In most situations, reference limits are defined as the 2.5th and 97.5th percentiles of the distribution of test results from individuals representing the reference population. Reference ranges are generally derived for "healthy" populations, but may be determined for any well-defined clinical state. A laboratory test result is usually compared with a reference interval to assist in making a diagnosis or other medical management decision.

There are many different methods that have been described in the literature for estimating percentiles of a distribution. They can be divided into two basic categories: parametric and nonparametric methods. Parametric methods assume that after suitable transformation, the reference population can be represented by a normal (Gaussian) distribution. Nonparametric methods don't make any assumptions about the distributional form of the reference population. Although there are still disagreements about which approach is preferable, most often the nonparametric approach is recommended for routine use ***(12)***.

Nonparametric methods for estimating a percentile from a reference distribution are based on the theory of order statistics ***(13)***. Assume measurements from n subjects representing the reference population have been obtained. The measured values, sorted in order from smallest to largest, are called order statistics and are denoted by $x_{(1)}, x_{(2)}, \ldots, x_{(n)}$. Let ξ_p denote the pth population quantile. We are generally interested in estimating $\xi_{0.025}$ and $\xi_{0.975}$. A number of estimates of ξ_p based on different functions of the order statistics have been proposed. Parrish ***(14)*** compares 10 different nonparametric quantile estimators, all based on order statistics.

The simplest and most popular nonparametric method estimates ξ_p as the observation with rank $r = p(n + 1)$. If p and n are such that r is an integer, then the estimate is simply $\hat{\xi}_p = x_{(r)}$. If r is not an integer, then a number of different possibilities exist. The most common approach is to linearly interpolate between the 2 order statistics with integer ranks on either side of the value of $p(n + 1)$; $\hat{\xi}_p = (1 - h)x_{(k)} + hx_{(k+1)}$, where k is the largest integer value less than $p(n + 1)$ and $h = p(n + 1) - k$ is the fractional part of $p(n + 1)$. As an example, if measurements from 100 subjects have been obtained, then the estimates for the 2.5th and 97.5th percentiles would be

$$(0.025)(101) = 2.525 \Rightarrow k = 2, h = 0.525 \Rightarrow \hat{\xi}_{0.025} = 0.475x_{(2)} + 0.525x_{(3)}$$

$$(0.975)(101) = 98.475 \Rightarrow k = 98, h = 0.475 \Rightarrow \hat{\xi}_{0.975} = 0.525x_{(98)} + 0.475x_{(99)} \,.$$

The above estimator can be thought of as a weighted average of 2 order statistics. An alternative weighted average estimator that incorporates all n order statistics has been proposed by Harrell and Davis ***(15)***. It can be expressed as

$$\hat{\xi}_p = \sum_{i=1}^{n} w_i x_{(i)},$$

where the weights, w_i, are computed as the difference between two incomplete beta functions:

$$w_i = \mathrm{I}_{i/n}(p(n+1), (1-p)(n+1)) - I_{(i-1)/n}(p(n+1), (1-p)(n+1)).$$

The heaviest weights are applied to order statistics with ranks near the value *of* $p(n + 1)$, with the magnitude of the weights dropping off for order statistics with ranks more distant from $p(n + 1)$. For the above example, **Table 2** gives the weights for order statistics with weights that are at least 0.0001. Most modern statistical programming packages will provide the capability to compute the incomplete beta function. The values in **Table 2** were obtained using Stata's betainc function. Parrish found the Harrell-Davis estimator to be the most precise among the estimators he evaluated ***(14)***.

Table 2
Harrell-Davis Order Statistic Weights, *w*, to Estimate the 2.5th and 97.5th Percentiles Based on a Sample Size, *n* = 100

Quantile			
0.025		0.975	
Rank	*w*	Rank	*w*
1	0.1457	88	0.0001
2	0.2994	89	0.0002
3	0.2474	90	0.0005
4	0.1532	91	0.0014
5	0.0820	92	0.0034
6	0.0401	93	0.0081
7	0.0184	94	0.0184
8	0.0081	95	0.0401
9	0.0034	96	0.0820
10	0.0014	97	0.1532
11	0.0005	98	0.2474
12	0.0002	99	0.2994
13	0.0001	100	0.1457

4.2. Confidence Intervals for a Reference Limit

The reference limits computed from a sample of subjects randomly selected from the reference population are estimates of the true population percentiles. Another random sample selected from the reference population would produce similar, but not identical, estimates. A useful way of characterizing the variability in the estimates is to compute a confidence interval for the population percentile being estimated. Confidence intervals for reference limits can be defined in at least 3 ways. The most common approach is to estimate a lower and upper bound for the true percentile, ξ_p. Alternatively, one can estimate a lower and upper bound for the percentage, p, of the distribution excluded by the estimated percentile, $\hat{\xi}_p$. Finally, one can estimate a lower and upper bound for the true fraction of the reference distribution contained between the upper and lower reference limits. Only the first type of confidence interval will be described here.

Like $\hat{\xi}_p$, nonparametric confidence interval estimates for ξ_p can be obtained from the sample order statistics. That is, a $100(1 - \alpha)$ percent confidence interval for ξ_p can be obtained as $(x_{(r)}, x_{(s)})$, where the rank values r and s are determined such that $P(x_{(r)} \leq \xi_p \leq x_{(s)}) \geq 1 - \alpha$. It is known ***(13)*** that

$$P\left(x_{(r)} \leq \xi_p \leq x_{(s)}\right) = \sum_{i=r}^{s-1} \binom{n}{i} p^i (1-p)^{n-i} = \gamma, \quad r < s.$$

The summation in the above equation can be computed from either the binomial distribution function or the incomplete beta function. For instance, in Stata, the functions Binomial(n, k, p) and ibeta(k, $n - k + 1$, p) both compute $\sum_{i=k}^{n} \binom{n}{i} p^i (1-p)^{n-i}$. Therefore, given values of r, s, n, and p, the above summation could be computed in Stata as Binomial(n, r, p) – Binomial(n, s, p) or as ibeta(r, $n - r + 1$, p) – ibeta(s, $n - s + 1$, p).

Any choice of r and s that makes $\gamma \geq 1 - \alpha$ will produce a confidence interval with confidence coefficient $\geq 1 - \alpha$. In general, there will be multiple combinations of r and s that make $\gamma \geq 1 - \alpha$ and no choice that will make γ exactly equal to $1 - \alpha$. **Table 3** gives an example of 3 different (r, s) pairs that produce 90% confidence intervals for the 2.5th percentile when $n = 180$. The first case ($r = 1$, $s = 9$) meets the added restriction that both $P(\xi_{0.025} < x_{(r)})$ and $P(\xi_{0.025} \geq x_{(s)})$ be less than or equal to $\alpha/2 = 0.05$. The second ($r = 1$, $s = 8$) and third ($r = 2$, $s = 9$) cases provide narrower intervals that still meet the required coverage probability, but each has one-tail probability exceeding 0.05.

Tables listing the rank numbers r and s that provide a 90% confidence interval for the 2.5th percentile for different sample sizes, n, have appeared in a number of laboratory medicine publications ***(12,16)***. The tables may also be used to obtain 90% confidence intervals for the 97.5th percentile by subtracting

Table 3
Distribution-Free Confidence Intervals for the 2.5th Percentile Based on a Sample Size of 180

Condition	P	Alternative 90% confidence interval limits		
$\xi_{0.025} < x_{(1)}$	0.0105	0.0105	0.0105	0.0589
$x_{(1)} \le \xi_{0.025} < x_{(2)}$	0.0484			
$x_{(2)} \le \xi_{0.025} < x_{(3)}$	0.1111			
$x_{(3)} \le \xi_{0.025} < x_{(4)}$	0.1691			
$x_{(4)} \le \xi_{0.025} < x_{(5)}$	0.1918	0.9513	0.9055	
$x_{(5)} \le \xi_{0.025} < x_{(6)}$	0.1731			0.9029
$x_{(6)} \le \xi_{0.025} < x_{(7)}$	0.1295			
$x_{(7)} \le \xi_{0.025} < x_{(8)}$	0.0825			
$x_{(8)} \le \xi_{0.025} < x_{(9)}$	0.0458			
$x_{(9)} \le \xi_{0.025} < x_{(10)}$	0.0224	0.0382	0.0840	0.0382
$\xi_{0.025} \ge x_{(10)}$	0.0158			

the tabled values for r and s from $n + 1$. Apparently, the rank values given in these tables have been derived with the restriction that neither tail probability exceeds 5% (as demonstrated in the first case in **Table 3**). Thus, in a number of cases the actual coverage probability is considerably greater than 90%.

Beran and Hall *(17)* proposed an interpolation scheme to obtain confidence interval estimates that more closely achieve the desired $1 - \alpha$ coverage probability. To estimate an equal-tailed two-sided $1 - \alpha$ confidence interval, the lower limit is computed by determining r such that

$$\sum_{i=0}^{r-1} \binom{n}{i} p^i (1-p)^{n-i} = \gamma_{r-1} < \alpha/2$$

$$\sum_{i=0}^{r} \binom{n}{i} p^i (1-p)^{n-i} = \gamma_r \ge \alpha/2.$$

Then compute

$$\pi = \frac{\alpha/2 - \gamma_{r-1}}{\gamma_r - \gamma_{r-1}}.$$

The interpolated lower limit is computed as $(1 - \pi)x_{(r)} + \pi x_{(r+1)}$. Likewise, for the interpolated upper limit, determine s such that

$$\sum_{i=0}^{s-1} \binom{n}{i} p^i (1-p)^{n-i} = \gamma_{s-1} < 1 - \alpha/2$$

$$\sum_{i=0}^{s} \binom{n}{i} p^i (1-p)^{n-i} = \gamma_s \ge 1 - \alpha/2$$

with

$$\pi = \frac{(1-\alpha/2)-\gamma_{s-1}}{\gamma_s - \gamma_{s-1}}$$

and compute the interpolated upper limit as $(1 - \pi)x_{(s)} + \pi_{x(s+1)}$.

The computations for a 90% confidence interval for the 2.5th percentile when $n = 180$ are

$$\gamma_0 = 0.0105 < 0.05, \quad \gamma_1 = 0.0589 \geq 0.05, \quad \pi = \frac{0.05 - 0.0105}{0.0589 - 0.0105} = 0.8161$$

$$\text{lower limit} = 0.1839x_{(1)} + 0.8161x_{(2)}$$

$$\gamma_7 = 0.9160 < 0.95, \quad \gamma_8 = 0.9618 \geq 0.95, \quad \pi = \frac{0.950 - 0.9160}{0.9618 - 0.9160} = 0.7424$$

$$\text{upper limit} = 0.2576x_{(8)} + 0.7424x_{(9)}.$$

4.3. Sample Size Considerations

At least 39 samples are required in order to estimate the 2.5th and 97.5th percentiles using the $p(n + 1)$ order statistic. In this case, the reference limits are estimated by the minimum and maximum values in the sample, $(x_{(1)}, x_{(39)})$. The Harrell-Davis estimator, being a weighted average of all the order statistics, can be estimated with smaller sample sizes.

Larger sample sizes are required to obtain nonparametric confidence intervals for the 2.5th and 97.5th percentiles. For sample sizes less than 91, 90% rank-based confidence intervals aren't possible. For sample sizes less than 119, 90% confidence intervals with ≤5% in each tail cannot be obtained. **Table 4** gives minimum sample sizes required for nonparametric rank-based confidence intervals with different coverage probabilities.

Table 4
Minimum Sample Sizes, *n*, Necessary to Obtain Rank-Based Confidence Intervals for the 25 Percentile with a Specified Coverage Probability

			Probability		
Confidence interval	n	(r, s)	Below $x_{(r)}$	Within limits	Above $x_{(s)}$
90%	91	(1, 10)	0.0999	0.9000	0.0001
≤5% in each tail	119	(1, 7)	0.0492	0.9204	0.0304
95%	119	(1, 11)	0.0492	0.9506	0.0002
≤2.5% in each tail	146	(1, 9)	0.0248	0.9636	0.0116
99%	182	(1, 16)	0.0100	0.9900	0.0000
≤0.5% in each tail	210	(1, 13)	0.0049	0.9924	0.0026

The precision of a reference interval estimate will depend on the number of subjects randomly sampled from the reference population. Various recommendations have been made regarding the preferred sample size to estimate reference limits ***(18)***. One common recommendation is that at least 120 samples be used. This recommendation originated in one of the early laboratory medicine publications describing nonparametric confidence intervals for a reference limit ***(19)***. It was based on the minimum number of samples necessary to obtain 90% confidence intervals for the 2.5th and 97.5th reference limits that ensured ≤5% in each tail (row 2 in **Table 4**).

Some argue that it is unwise to have one of the confidence interval limits be represented by the minimum or maximum sample values, because if any outliers are present in the sample, then the confidence interval estimate will be adversely affected ***(20)***. A sample size of 188 is the smallest n that meets this criterion when computing 90% confidence intervals for the 2.5th and 97.5th percentiles that exclude ≤5% in each tail. With a sample size of 188, the confidence intervals for the 2.5th and 97.5th percentiles are $(x_{(2)}, x_{(9)})$ and $(x_{(180)}, x_{(187)})$, respectively. Others have recommended that sample sizes for establishing reference limits be as large as 700 or more, especially for highly skewed distributions ***(21)***.

It is difficult to predict in advance the width of a confidence interval such as $(x_{(2)}, x_{(9)})$. Its width will depend on the particular values obtained for the second and ninth smallest reference sample values, which in turn will depend on the tail characteristics of the reference population. An alternative consideration would be to take into account the precision of the population percentage excluded by a reference limit. For example, in a sample of 119 subjects the nonparametric rank-based estimate of the 2.5th percentile is $x_{(3)}$. The 90% confidence interval for the true percentage below $x_{(3)}$ when $n = 119$ is (0.7%, 5.2%). If n is increased to 199, the nonparametric estimate of the 2.5th percentile is $x_{(5)}$, and the 90% confidence interval for the true percentage below $x_{(5)}$ narrows to (1.0%, 4.5%). These confidence intervals have the advantage that they do not depend on the distribution of the reference population. Thus, with n in the neighborhood of 200 samples, the nonparametric rank-based 90% confidence intervals for the 2.5th and 97.5th reference limits do not include the extreme sample values, and the 90% confidence intervals for the true percentages excluded by the estimated reference limits will be around (1%, 4.5%) and (95.5%, 99%), respectively.

5. Conclusion

Laboratory medicine has steadily been evolving into an "information" science. As this trend continues, the role of statistics and statistical thinking becomes more significant and essential. This chapter has addressed only 3 of the many statistical issues that arise in the clinical laboratory. These topics were selected because they address important laboratory questions and because they each demonstrate a valuable statistical principle. The estimation of an assay's analytical imprecision demonstrates the impact of study design on the appropriate estimation and interpretation of an assay's total analytical imprecision. The design of an effective quality control strategy for an assay demonstrates the key role that statistical modeling plays in determining how to ask and answer the most relevant questions. The establishment of population reference ranges demonstrates that many different factors can come into play when trying to develop a sound estimator. The contributions of statistical principles such as these will continue to play an increasingly important role in the evolution of the modern laboratory.

References

1. Neter, J., Kutner, M. H., Wasserman, W., and Nachtsheim, C. J. (1996) *Applied Linear Statistical Models*, 4th ed. Homewood, Irwin.
2. Clinical and Laboratory Standards Institute. (2004) *Evaluation of Precision Performance of Quantitative Measurement Methods; Approved Guideline—Second Edition.* EP5-A2. Villanova, Clinical and Laboratory Standards Institute.
3. Littell, R. C., Milliken, G. A., Stroup, W. W., Wolfinger, R. D., and Schabenberger, O. (2006) *SAS for Mixed Models*, 2nd ed. Cary, SAS Institute.
4. Satterthwaite, F. E. (1946) An approximate distribution of estimates of variance components. *Biometrics* **2**, 110–114.
5. Clinical and Laboratory Standard Institute. (1999) *Statistical Quality Control for Quantitative Measurements: Principles and Definitions; Approved Guideline—Second Edition.* C24-A2. Villanova, Clinical and Laboratory Standards Institute.
6. Westgard, J. O., and Klee, G. G. (2006) Quality management. In: Burtis, C. A., Ashwood, E. R., and Bruns, D. E., eds. *Tietz Textbook of Clinical Chemistry and Molecular Diagnostics*, 4th ed. St. Louis, Elsevier Saunders, pp. 485–532.
7. Parvin, C. A. (1997) Quality-control (QC) performance measures and the QC planning process. *Clin. Chem.* **43**, 602–607.
8. Linnet, K. (1991) Mean and variance rules are more powerful or selective than quality control rules based on individual values. *J. Clin. Chem. Clin. Biochem.* **29**, 417–424.
9. Parvin, C. A. (1993) New insight into the comparative power of quality-control rules that use control observations within a single analytical run. *Clin. Chem.* **39**, 440–447.
10. Parvin, C. A., and Gronowski, A. M. (1997) Effect of analytical run length on quality-control (QC) performance and the QC planning process. *Clin. Chem.* **43**, 2149–2154.
11. Parvin, C. A. (1991) Estimating the performance characteristics of quality-control procedures when error persists until detection. *Clin. Chem.* **37**, 1720–1724.

12. Clinical and Laboratory Standards Institute. (2000) *How to Define and Determine Reference Intervals in the Clinical Laboratory; Approved Guideline—Second Edition.* C28-A2. Villanova, Clinical and Laboratory Standards Institute.
13. David, H. A., and Nagaraja, H. N. (2003) *Order Statistics*, 3rd ed. New York, John Wiley & Sons.
14. Parrish, R. S. (1990) Comparison of quantile estimators in normal sampling. *Biometrics* **46**, 247–257.
15. Harrell, F. E., and Davis, C. E. (1982) A new distribution-free quantile estimator. *Biometrika* **69**, 635–640.
16. Solberg, H. E. (2006) Establishment and use of reference values. In: Burtis, C. A., Ashwood, E. R., and Bruns, D. E. eds. *Tietz Textbook of Clinical Chemistry and Molecular Diagnostics*, 4th ed. St. Louis, Elsevier Saunders, pp. 425–448.
17. Beran, R., and Hall, P. (1993) Interpolated nonparametric prediction intervals and confidence intervals. *J. Roy. Statist. Soc. B* **55**, 643–652.
18. Harris, E. K., and Boyd, J. C. (1995) *Statistical Bases of Reference Values in Laboratory Medicine.* New York, Marcel Dekker.
19. Reed, A. H., Henry, R. J., and Mason, W. B. (1971) Influence of statistical method used on the resulting estimate of normal range. *Clin. Chem.* **17**, 275–284.
20. Horn, P. S., and Pesce, A. J. (2003) Reference intervals: an update. *Clin. Chim. Acta.* **334**, 5–23.
21. Linnet, K. (1987) Two-stage transformation systems for normalization of reference distributions evaluated. *Clin. Chem.* **33**, 381–386.

19

Power and Sample Size

L. Douglas Case and Walter T. Ambrosius

Summary

In this chapter, we discuss the concept of statistical power and show how the sample size can be chosen to ensure a desired power. Power is the probability of rejecting the null hypothesis when the null hypothesis is false, that is the probability of saying there is a difference when a difference actually exists. An underpowered study does not have a sufficiently large sample size to answer the research question of interest. An overpowered study has too large a sample size and wastes resources. We will show how the power and required sample size can be calculated for several common types of studies, mention software that can be used for the necessary calculations, and discuss additional considerations.

Key Words: Clinically meaningful effect; hypothesis test; power; sample size; type I error; type II error.

1. Introduction

Consideration of power and sample size is crucial in the design of most research studies and should be addressed early in the planning stage. During this stage, the investigator can estimate the number of participants needed to test a specific hypothesis, the power available to detect a specific alternative given a fixed sample size, or the difference that can be detected with a specified power with a given sample size. Power is the probability of rejecting the null hypothesis when the null hypothesis is false; that is, the probability of saying there is a difference when a difference actually exists. Having an adequately powered study is important, of course, because the goal of most investigations is to show a treatment effect. It is a waste of time, money, and participant resources for an investigator to conduct a study that is unlikely to show what he wants to show, even when he is correct in his beliefs. On the other hand, an overpowered study,

From: *Methods in Molecular Biology, vol. 404: Topics in Biostatistics*
Edited by: W. T. Ambrosius © Humana Press Inc., Totowa, NJ

one that uses many more subjects than is necessary to adequately answer the question, also wastes time, money, and participant resources. Early consideration of power and sample size can answer the question of whether the study is feasible given the resources available and the expected treatment effect. Addressing these issues early helps the investigators focus on crucial design elements, leading to a tighter and more rigorous study.

In this chapter, we consider the concept of power in some detail and show how it can be calculated in several common situations. We then show how to calculate a sample size to ensure an appropriate power to test a particular hypothesis. Research studies come in a wide assortment of designs (e.g., cross-sectional, pre-post, longitudinal, cross-over, etc.), involving one or more groups, with outcomes that may be continuous, dichotomous, censored, or correlated (or some combination of these), and sample size formulae can be derived for each of these. Indeed, entire books have been written detailing sample size calculations for a multitude of designs and outcomes [see, for example, Cohen *(1)*, Kraemer and Thiemann *(2)*, Murphy and Myors *(3)*, and Chow and others *(4)*]. However, we are most interested in the ideas involved, and we will focus mainly on single-arm and 2-arm studies involving continuous outcome measures. Programs are mentioned that will allow the reader to calculate a power or sample size for studies using other designs or testing other outcomes. Other considerations in power and sample size determination are then discussed.

2. Power of a Test

Hypothesis testing was discussed in detail in **Chapter 4**. In brief, a hypothesis is a statement made about a population (or alternatively, a statement made about the distribution of a random variable). This statement may or may not be true, and, in reality, we will never know which. However, based on sample data and some decision rule, we either reject or fail to reject the hypothesis. This process is called hypothesis testing.

Consider a simple example. Suppose a physician knows that historically weight in 15-year-old boys has been normally distributed with a mean ± SD of 120 ± 20 lb. She believes, however, that the weight of these young boys today has increased (perhaps owing to current diets and the overuse of video games), and she wants to show that the mean weight is now greater than 120 lb. Measuring weight on every 15-year-old boy is physically impossible, so the physician will not be able to determine with certainty whether the true mean weight is now less than or greater than 120. However, she wants to reach a decision based on weight measurements made on a sample of boys. In terms of hypothesis testing, she wants to decide between 2 competing hypotheses, one that the true mean is less than or equal to 120 and another that it is greater than 120. That is, based on sample data, she wants to decide between

$$H_0: \mu_{wgt} \leq 120 \text{ versus } H_1: \mu_{wgt} > 120,$$

where H_0 is called the null hypothesis and H_1 the alternative hypothesis (sometimes denoted by H_a). Typically, the current state of affairs is taken as the null hypothesis, and the statement that a researcher wants to demonstrate is called the alternative hypothesis, although this choice is sometimes a bit arbitrary. The hypotheses above are called 1-sided hypotheses because we are interested in alternatives in only 1 direction. Were the researcher interested in alternatives in either direction, she would use a 2-sided hypothesis test. For example, she may hypothesize that the mean weight is no longer 120 lb. That is

$$H_0: \mu_{wgt} = 120 \text{ versus } H_1: \mu_{wgt} \neq 120.$$

We will talk about 1- versus 2-sided hypothesis tests in more detail below.

The original hypotheses are also called composite hypotheses because the state of nature is incompletely specified. For example, considering $H_1: \mu_{wgt} > 120$, the true mean could be 121 or 125 or any other value greater than 120. For the 2-sided hypothesis test, the null hypothesis is called a simple hypothesis because it completely specifies the state of nature ($\mu_{wgt} = 120$).

The physician researcher now has to develop a decision rule to help her choose between the 2 competing hypotheses based on the data she will collect. Suppose she decides to accept H_1 if the sample mean is greater than 120, a choice that might seem natural at first because that is the value that separates the null and alternative hypotheses. Values of the test statistic (mean weight in our case) that lead to rejection of the null hypothesis (sample mean weights >120) are called the rejection region or the critical region, whereas values of the test statistic that lead to acceptance of the null hypothesis (sample mean weights ≤120) are called the acceptance region. The value of the test statistic that separates the acceptance and rejection regions is called the critical value, which we will denote by C.

In deciding between H_0 and H_1, the researcher can decide correctly or incorrectly. The 2 types of incorrect decisions would be to choose H_1 when H_0 is actually true or to choose H_0 when H_1 is actually true. The first type of incorrect decision is called a type I error, and the probability of a type I error, P(reject $H_0|H_0$ true), is denoted by α. The probability of a type I error is sometimes referred to as the level of significance, the size of the test, or the size of the critical region. The second type of error is called a type II error, and the probability of a type II error, P(accept $H_0|H_0$ false), is denoted by β. These probabilities are summarized in **Table 1** and are illustrated in **Figure 1**.

The complement of the type II error $(1 - \beta)$ is called the power of a test and is the probability of rejecting H_0 when H_0 is false; that is, P(reject $H_0|H_0$ false). It is important to note that the probability of a type I or II error depends on the true state of nature (in our example the true weight), which we never know, as well as the decision rule. That is, there is not just one α or one β but rather an α for every point in the acceptance region and a β for every point in the rejection region.

Table 1
Possible Decisions in Hypothesis Testing

		Decision	
		H_0	H_1
Truth	H_0	Correct decision	Incorrect decision (type I error)
	H_1	Incorrect decision (type II error)	Correct decision

The researcher should naturally be concerned about the probability of making a type I or II error. Because weight is normally distributed with a standard deviation of 20, we know that the mean weight is also normally distributed with a standard deviation of $20/\sqrt{n}$, where n is the number of boys sampled by the physician. Thus, we can calculate the probability of rejecting the null hypothesis for any value of the true mean weight. This probability is given by

$$P(\bar{X}_{wgt} > 120 \mid \mu_{wgt}) = 1 - \Phi\left(\frac{120 - \mu_{wgt}}{20/\sqrt{25}}\right) = 1 - \Phi\left(\frac{120 - \mu_{wgt}}{4}\right),$$

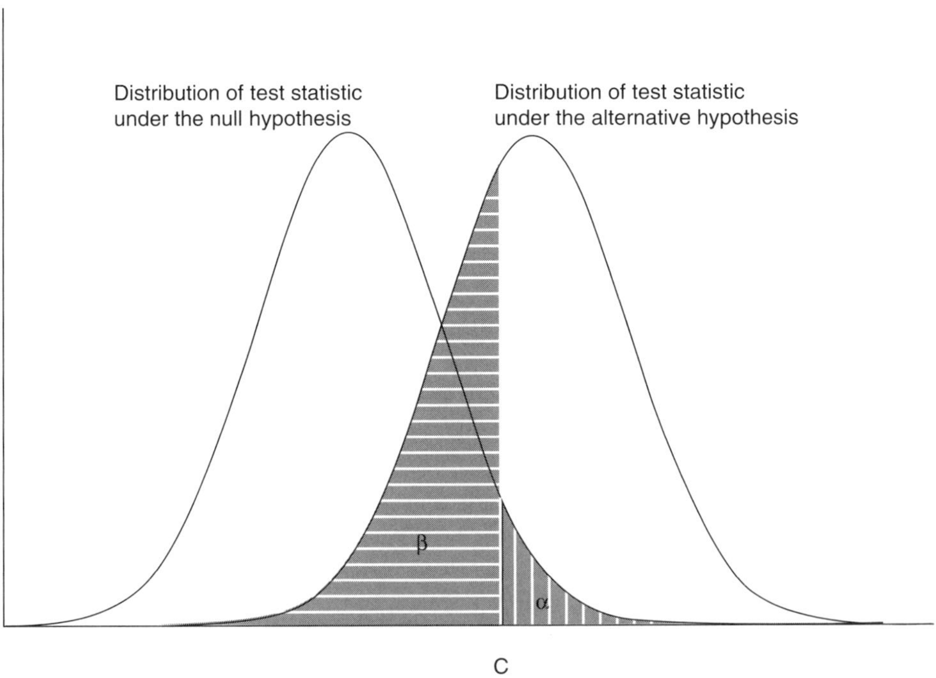

Fig. 1. Illustration of type I (α) and II (β) errors.

where Φ is the cumulative standard normal distribution function. If the true mean is 120, the probability is 50% that the null hypothesis would be rejected in favor of the alternative hypothesis. This would be a type I error. It is unlikely that a probability of 50% would be acceptable for such an error. Note that the probability of rejecting H_0 when the true mean is 115 is 0.106. This again is a type I error. As mentioned above, there are multiple probabilities of type I error (an infinite number for a composite null hypothesis), corresponding with all the points in the acceptance region. By convention, the probability of a type I error is usually reported as the maximum type I error across all those possible in the acceptance region. So another definition for the type I error is the maximum power of the test under H_0. Probabilities of rejecting H_0 for other choices of the true mean weight are shown in **Table 2**. We see that the probability of rejecting H_0 (power) is 0.894 when $\mu_{wgt} = 125$ and is 0.994 when $\mu_{wgt} = 130$.

Indeed, the researcher would not be satisfied with a 50% chance of making a type I error (or at least the scientific community would not be satisfied). Thus, she would need to define a new decision rule (i.e., a new critical region, or equivalently, a new test). Suppose she decides to reject H_0 if $\bar{X}_{wgt}$ is greater than 126. With this new decision rule (new test), one sees that the probabilities of rejecting H_0 for μ_{wgt} equal to 120, 125, and 130 are 0.067, 0.401, and 0.841. One notes that the probability of a type I error has been reduced from 0.5 to 0.067, but the power has also been reduced. For example, at μ_{wgt} equal to 125 the power is decreased from 0.894 to 0.401, and for μ_{wgt} equal to 130, the power

Table 2
Probabilities of Rejecting H_0 for Various Tests (Critical Values) and Various True Weights

Critical value	True mean weight				
	115	120	125	130	135
120	0.106	0.500	0.894	0.994	0.999
121	0.067	0.401	0.841	0.988	0.999
122	0.040	0.309	0.773	0.977	0.999
123	0.023	0.227	0.691	0.960	0.999
124	0.012	0.159	0.599	0.933	0.997
125	0.006	0.106	0.500	0.894	0.994
126	0.003	0.067	0.401	0.841	0.988
127	0.001	0.040	0.309	0.773	0.977
128	0.000	0.023	0.227	0.691	0.960
129	0.000	0.012	0.159	0.599	0.933
130	0.000	0.006	0.106	0.500	0.894

is decreased from 0.994 to 0.841. This will be true, in general. For a fixed sample size, reducing the probability of a type I error will reduce the power of the test for any specific alternative. Probabilities of rejecting H_0 for additional choices of the critical value are also shown in **Table 2**.

Suppose the investigator wants the probability of a type I error to be fixed at 5%. What should she choose as the critical value? We see from **Table 2** that a 5% type I error corresponds with a critical value between 126 and 127. Because, under H_0, $\bar{X}_{wgt}$ is normally distributed with a mean of 120 and a standard deviation of $20/\sqrt{25} = 4$, we know that the area to the right of 120 + 1.645 × 4 = 126.58 equals 0.05. Thus, our critical value for rejection would be

$$C = \mu_0 + z_{1-\alpha}\,\sigma/\sqrt{n} = 120 + 1.645 \times 20/\sqrt{25} = 126.58,$$

where $z_{1-\alpha}$ is the 100(1 − α)th percentile of the standard normal distribution function. This is illustrated in **Figure 2**. Thus the area to the right of 126.58 in

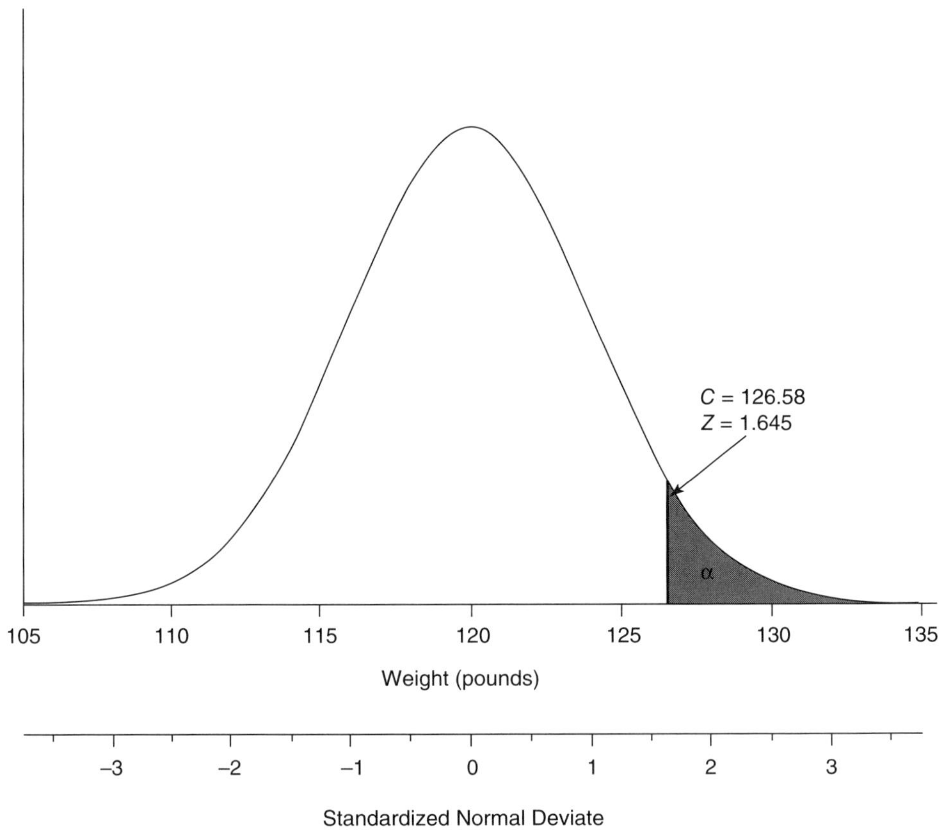

Fig. 2. Sampling distribution of mean weight with 0.05 1-sided critical region highlighted on the original scale and corresponding standardized normal deviate scale.

Figure 2 ($P(\bar{X} > C)$), which corresponds with the area to the right of 1.645 for a standardized normal deviate ($P(Z > 1.645)$), is 0.05.

So clearly, our test statistic could be, and is usually, written as the standardized normal Z statistic; that is,

$$Z = \frac{\bar{X} - \mu_0}{\sigma / \sqrt{n}}.$$

The advantage of using the Z statistic is that the percentiles of the standardized normal distribution are well tabulated. For this test statistic, the critical value is 1.645 for a 1-sided type I error of 0.05.

Now we can calculate the area to the right of the critical value (126.58 in our example) for various choices of the mean under the alternative hypothesis. This gives us the power of the test at that particular choice of the mean. That is

$$\text{Power} = P(\mu) = 1 - \Phi\left(\frac{C-\mu}{\sigma/\sqrt{n}}\right) = 1 - \Phi\left(z_{1-\alpha} - \frac{\mu - \mu_0}{\sigma/\sqrt{n}}\right),$$

where Φ represents the cumulative standard normal distribution function. This is done in **Figure 3** for several choices of μ. As the reader can easily see, the power increases as μ_{wgt} gets further and further away from μ_0. That is, the power increases as the difference in the hypothesized means increases. For example, if the true mean is 124 lb, the power is 0.26, if $\mu_{wgt} = 128$, the power is 0.64, and if $\mu_{wgt} = 132$, the power is 0.91. The probability of rejecting H_0 can be plotted as power versus the true mean in what is called a power curve. The power curve for this example is illustrated in **Figure 4**. One can then easily approximate the power for any particular value of μ_{wgt} by drawing a vertical line up from the μ_{wgt} axis until it intersects the power curve; then draw a horizontal line from the point of intersection to the power axis, where the approximate power can be read. For example, the reader can see that when the true mean is 130, the probability of rejecting the null hypothesis at the 5% 1-sided level of significance is approximately 80%. That is, if the true mean in this population is 130, and we repeatedly took samples of size 25 and tested the hypothesis above at the 5% 1-sided level of significance, we would reject the null hypothesis 80% of the time. We would fail to reject the null hypothesis 20% of the time even though the true mean was 130.

The probability of rejecting H_0 when the true mean is 125 lb is approximately 35%. If, for example, it had been important to reject H_0 when the true mean was 125, perhaps the study should not have been done (at least with this sample size). For this study, with this sample size, we really only have reasonable power (0.8 or higher) for rejecting H_0 when the true mean is 130 and greater. If you are stuck with a sample size (perhaps because you are doing a retrospective study and that is the number of patients with a particular condition that have been seen in your clinic or because of cost constraints), you should generate a power curve to explore the characteristics of your study.

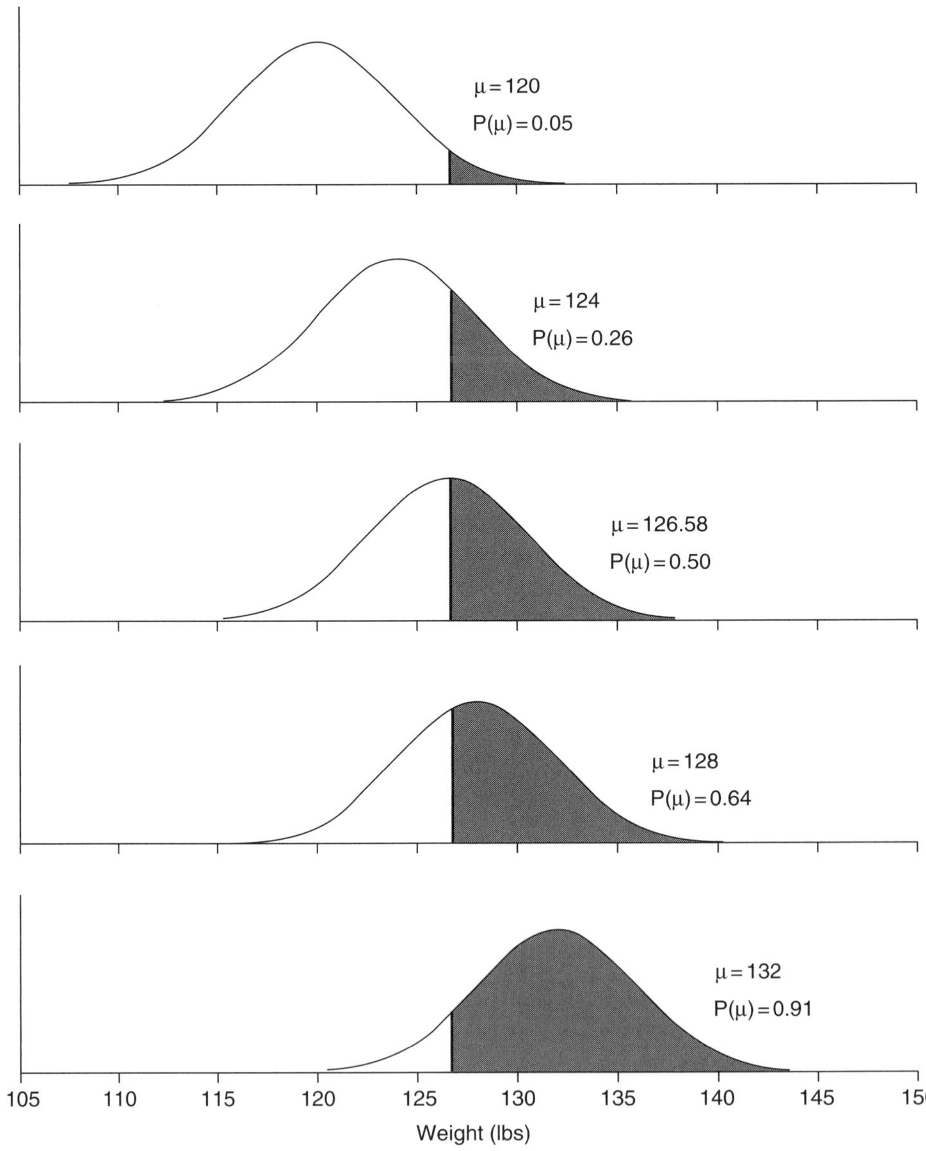

Fig. 3. Probability of rejecting H_0 for various choices of the true mean weight.

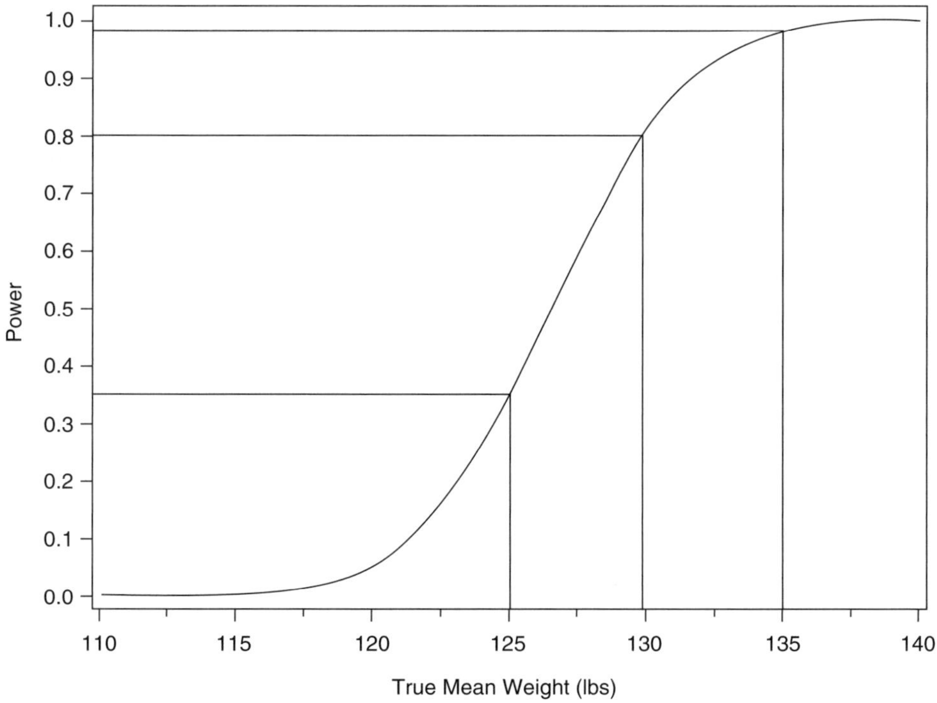

Fig. 4. Probability of rejecting H_0 as a function of the true mean weight for 1-sided test.

2.1. Two-Sided Hypothesis Tests

Consider again the following null and alternative hypotheses.

$$H_0: \mu_{wgt} = 120 \text{ versus } H_1: \mu_{wgt} \neq 120.$$

Sample means substantially less than or substantially greater than 120 support the alternative hypothesis, so the researcher would choose small or large sample means (or, correspondingly, small or large values of *Z*) as her critical region. If she wants to maintain an α size for this 2-sided hypothesis test, she could divide α equally into the lower and upper areas of the distribution of the test statistic under the null hypothesis. If the mean is to be used as the test statistic, then the lower and upper critical values would be given by

$$C_L = \mu_0 - z_{1-\alpha/2}\sigma/\sqrt{n}$$

and

$$C_U = \mu_0 + z_{1-\alpha/2}\sigma/\sqrt{n},$$

respectively, which equal 112.16 and 127.84 for our example. Of course, the corresponding values for the Z statistic are −1.96 and 1.96, respectively, because 1.96 is the normal deviate corresponding with an area of 0.025. These choices are illustrated in **Figure 5**.

The power of the test against a specific alternative is the area to the left of C_L plus the area to the right of C_U for a specific value of μ_{wgt} (note again that this is α when $\mu_{wgt} = \mu_0$). This is given by:

$$\begin{aligned}\text{Power} = P(\mu) &= 1-\Phi\left(\frac{C_U-\mu}{\sigma/\sqrt{n}}\right)+\Phi\left(\frac{C_L-\mu}{\sigma/\sqrt{n}}\right)\\ &= 1-\Phi\left(z_{1-\alpha/2}-\frac{\mu-\mu_0}{\sigma/\sqrt{n}}\right)+\Phi\left(z_{\alpha/2}-\frac{\mu-\mu_0}{\sigma/\sqrt{n}}\right).\end{aligned}$$

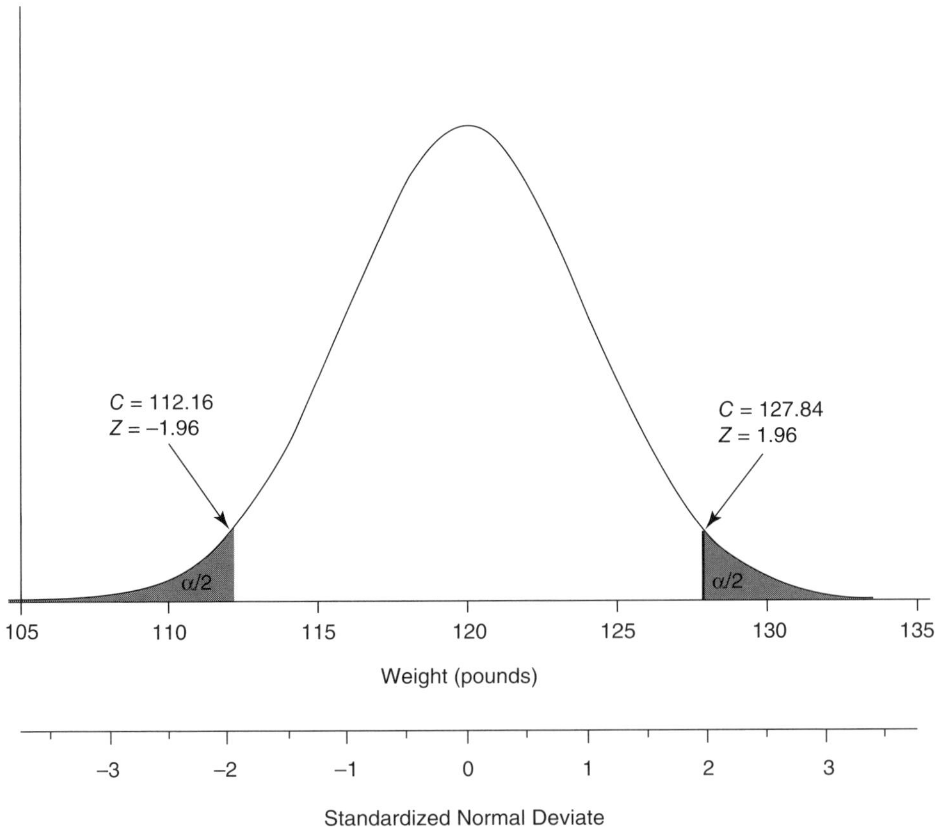

Fig. 5. Sampling distribution of the mean weight with 0.05 1-sided critical region highlighted on the original scale and corresponding standardized normal deviate scale.

Table 3
Areas of Lower and Upper Rejection Regions as a Function of the True Mean Weight with Power for 2-Sided (Sum of Lower and Upper Areas) and Corresponding 1-Sided Tests

μ_{wgt}	$\Phi\left(z_{\alpha/2} - \frac{\mu - \mu_0}{\sigma/\sqrt{n}}\right)$	$1 - \Phi\left(z_{1-\alpha/2} - \frac{\mu - \mu_0}{\sigma/\sqrt{n}}\right)$	Power (2-sided)	Power (1-sided)
105	0.9633	0.0000	0.9633	0.0000
110	0.7054	0.0000	0.7054	0.0000
115	0.2389	0.0007	0.2395	0.0019
120	0.0250	0.0250	0.0500	0.0500
125	0.0007	0.2389	0.2395	0.3465
130	0.0000	0.7054	0.7054	0.8038
135	0.0000	0.9633	0.9633	0.9824

When the true mean is much less than μ_0 the area corresponding with the upper critical region, $1 - \Phi\left(z_{1-\alpha/2} - \frac{\mu - \mu_0}{\sigma/\sqrt{n}}\right)$, will be negligible, and when the true mean is much greater than μ_0, the area corresponding with the lower critical region, $\Phi\left(z_{\alpha/2} - \frac{\mu - \mu_0}{\sigma/\sqrt{n}}\right)$, will be negligible. This is illustrated in **Table 3** for various choices of μ_{wgt}. The power for the 1-sided 5% test is also provided for reference.

The power curve for this 2-sided hypothesis test (again assuming an n of 25 with $\sigma = 20$) is shown in **Figure 6**. The 1-sided power curve shown in **Figure 4** is repeated here (shown as the dotted line) for reference. Both tests have a size of 5%. As easily seen, the 1-sided test has greater power against alternative means greater than 120 lb. However, the 1-sided test does not allow rejection for alternatives less than 120.

2.2. Simple versus Composite Hypotheses

As mentioned earlier, hypotheses can either be simple (i.e., the hypothesis completely specifies the distribution) or composite (i.e., the hypothesis does not completely specify the distribution). In the initial example above, both the null and alternative hypotheses were composite. In most applications, one or both of the competing hypotheses will be composite. However, the concepts discussed above apply as well when the hypotheses are not composite. Consider the following examples.

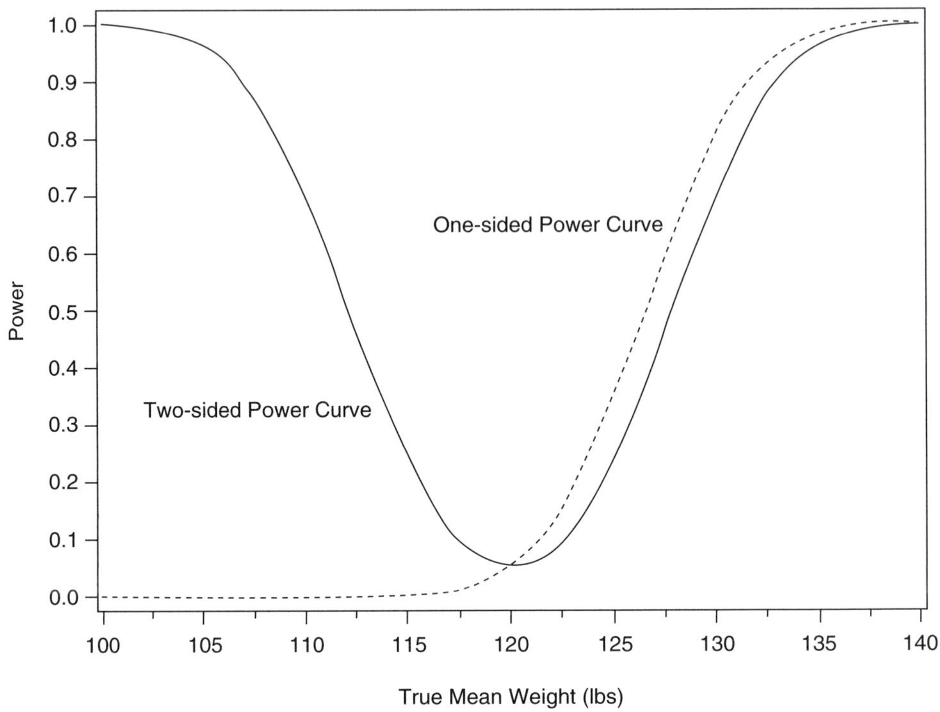

Fig. 6. Probability of rejecting H_0 as a function of the true mean weight for 1- and 2-sided tests.

Example 1

Consider a cancer researcher who is interested in the tumor response rate to a new agent. He wants to test the null hypothesis that the response rate is 0.4 against the alternative hypothesis that the response rate is 0.6. That is,

$$H_0: \theta = 0.4 \text{ versus } H_1: \theta = 0.6.$$

He treats 20 patients with the new agent. Let X denote the number of responses observed among the 20 patients. Then X has a binomial distribution with $n = 20$ and $p = 0.4$ under H_0 and $p = 0.6$ under H_1. The researcher decides to reject H_0 if $X \geq C$, where $C = 11$. What are α and β for this test?

For these simple hypotheses, the power function is only defined for 0.4 and 0.6. The power of the test for $\theta = 0.4$ is defined as α, the type I error. Cumulative binomial probabilities are shown in **Table 4**. We see that the probability of rejecting H_0 when $p = 0.4$, α, is $1 - 0.87248 = 0.12752$, whereas the probability of rejecting H_0 when $p = 0.6$, $1 - \beta$, is $1 - 0.24466 = 0.75534$. Values of α and $1 - \beta$ corresponding with several realizations of X are shown in **Table 5**.

Table 4
Cumulative Binomial Probabilities (n = 20)

X	$p = 0.4$	$p = 0.6$
0	0.00004	0.00000
1	0.00052	0.00000
2	0.00361	0.00001
3	0.01596	0.00005
4	0.05095	0.00032
5	0.12560	0.00161
6	0.25001	0.00647
7	0.41589	0.02103
8	0.59560	0.05653
9	0.75534	0.12752
10	0.87248	0.24466
11	0.94347	0.40440
12	0.97897	0.58411
13	0.99353	0.74999
14	0.99839	0.87440
15	0.99968	0.94905
16	0.99995	0.98404
17	0.99999	0.99639
18	1.0000	0.99948
19	1.0000	0.99996
20	1.0000	1.0000

Table 5
Probability of Rejecting H_0 for Various Tests

C	$P = 0.4$	$P = 0.6$
6	0.87440	0.99839
7	0.74999	0.99353
8	0.58411	0.97897
9	0.40440	0.94347
10	0.24466	0.87248
11	0.12752	0.75534
12	0.05653	0.59560
13	0.02103	0.41589
14	0.00647	0.25001
15	0.00161	0.12560

Note that because of the discreteness of the binomial distribution, it is not always possible to control the type I and II errors at specified levels. In this example, it would not be possible to have a type I probability of exactly 5%. One could choose a critical value (C) of 12 and get an α of 0.057 (with a corresponding power of 0.596 under H_1) or choose a C of 13 and get an α of 0.021 (with a corresponding power of 0.416 under H_1). Typically, if one wants to control the type I error at α, he takes the critical value corresponding with the maximum type I error $\leq \alpha$, in this case a critical value of 13. (Of course, the power of 0.416 would be unacceptable.)

Example 2

Consider **Example 1** above, except the researcher wishes to test the null hypothesis that the response rate is 0.4 against the alternative hypothesis that the response rate is not 0.4. That is,

$$H_0: \theta = 0.4 \text{ versus } H_1: \theta \neq 0.4.$$

This is a test of a simple hypothesis versus a composite hypothesis. There will only be 1 value for the power under H_0 (which is α) but multiple values under H_1. Some researchers are bothered by the simple null hypothesis, which is almost surely wrong. The researcher would now need to decide on the critical (rejection) region. Suppose he decides to reject H_0 if $C \neq 8$. Individual binomial probabilities are shown in **Table 6**. The probability of observing 8 responses when the true response rate is 0.4 is 0.1797. Thus the type I error for this test is 1 – 0.1797 = 0.8203, which is unacceptable. Note that the power of

Table 6
Binomial Probabilities (n = 20)

C	$p = 0.4$	$p = 0.6$	C	$p = 0.4$	$p = 0.6$
0	0.00004	0.00000	11	0.07099	0.15974
1	0.00049	0.00000	12	0.03550	0.17971
2	0.00309	0.00000	13	0.01456	0.16588
3	0.01235	0.00004	14	0.00485	0.12441
4	0.03499	0.00027	15	0.00129	0.07465
5	0.07465	0.00129	16	0.00027	0.03499
6	0.12441	0.00485	17	0.00004	0.01235
7	0.16588	0.01456	18	0.00000	0.00309
8	0.17971	0.03550	19	0.00000	0.00049
9	0.15974	0.07099	20	0.00000	0.00004
10	0.11714	0.11714			

Table 7
Probability of Rejecting H_0 for Various Tests

Reject H_0 if C		$p = 0.4$			$p = 0.6$		
≤ or	≥	Lower	Upper	Total	Lower	Upper	Total
0	12	0.0000	0.0565	0.0566	0.0000	0.5956	0.5956
0	13	0.0000	0.0210	0.0211	0.0000	0.4159	0.4159
1	12	0.0005	0.0565	0.0571	0.0000	0.5956	0.5956
1	13	0.0005	0.0210	0.0216	0.0000	0.4159	0.4159
2	12	0.0036	0.0565	0.0601	0.0000	0.5956	0.5956
2	13	0.0036	0.0210	0.0246	0.0000	0.4159	0.4159
3	12	0.0160	0.0565	0.0725	0.0000	0.5956	0.5956
3	13	0.0160	0.0210	0.0370	0.0000	0.4159	0.4159
4	13	0.0510	0.0210	0.0720	0.0003	0.4159	0.4162
4	14	0.0510	0.0065	0.0574	0.0003	0.2500	0.2503

this test for detecting a response rate of 0.6 is 1 – 0.0355 = 0.9645. Clearly the researcher would need to select a different critical region. **Table 7** lists several tests (choices for critical regions) that result in a type I error of 10% or less.

Note that for the binomial distribution, the areas of the lower and upper rejection regions are not necessarily equal. For example, from **Table 7** the 5% 2-sided test is seen to have a lower critical value of 3 and an upper critical value of 13. This test gives a 2-sided size of 0.037, which is the largest size less than or equal to 0.05. The area in the lower critical region under H_0 is 0.016 while the area in the upper critical region under H_0 is 0.021, both less than 0.025 but not equal. Some sample size programs require each area to be $\leq\alpha/2$. The test just mentioned satisfies this criterion. However, what if the size had been chosen to be 0.1? Several of the tests shown in **Table 7** have a total area in the rejection regions between 0.05 and 0.1 under H_0. However, none of them satisfy the criterion that each area be $\leq\alpha/2$. So the test for $\alpha = 0.1$ is the same as the test for $\alpha = 0.05$. Were α chosen to be 0.12, then several tests satisfy the $\alpha/2$ criterion, and we see that the appropriate test would have a lower critical value of 3, an upper critical value of 12, and a power of 0.5956 under the alternative hypothesis that the true response rate is 0.6. So the researcher would decide to reject H_0 if the number of responses is 3 or fewer or 13 or greater. For this test, the type I error would be (from **Table 5**) 0.03699 and the power of the test would be 0.4159.

2.3. One-Sided versus Two-Sided Hypothesis Tests

The choice between 1- and 2-sided hypothesis tests has raised considerable controversy and debate in the statistical community. Some researchers are suspicious of 1-sided studies, thinking they were used simply to reduce the sample size. Fleiss ***(5)***, in his book on rates and proportions, when discussing a situation in which a 1-sided test might be appropriate (a new treatment vs. the standard treatment), states: "If, however, the investigator intends to report the results to professional colleagues, he is ethically bound to perform a two-sided test. For if the results indicate that the new treatment is actually worse than the standard—an inference possible only with a two-tailed test—the investigator is obliged to report this as a warning to others who might plan to study the new treatment." If one believes this, then 2-sided tests would be used almost exclusively because it is almost always the objective of investigators to publish their results. Peace ***(6)***, however, argues that sidedness is an integral part of a hypothesis and 1-sided tests are appropriate if the hypotheses tests are appropriate.

It is important that researchers specify in the study design whether a 1- or 2-sided test is to be used. (It must be specified, either implicitly or explicitly, because the sample size depends on this choice.) One-sided tests should not be done at the end of a trial originally designed as 2-sided. Two-sided tests should not be done at the end of a trial originally designed as 1-sided, even if "significant" results were found in the "wrong" direction. It is this latter possibility that leads many investigators to use 2-sided designs exclusively. The power of a 1-sided test is always greater than that of a 2-sided test (of the same α) in the direction specified by the alternative hypothesis. This is due to the use of $z_{1-\alpha}$ instead of $z_{1-\alpha/2}$ in determining the critical regions. Alternatively, as we will see below, this means that the same power can be realized with a smaller sample size, which with precious subject resources, is an argument for using 1-sided designs. Indeed, it is likely that most investigators have a preconceived notion about the direction of their results. However, investigators should ask themselves how they would report results that are in the opposite direction from that predicted (significantly so had a 2-sided test been used). That is, would they be reported as significant ($P < 0.05$) or highly nonsignificant ($P > 0.975$)?

3. Sample Size Determination

An adequate power is realized by choosing an appropriate sample size. How to determine the appropriate sample size depends on the question asked, the study design, and the test statistic used. We consider in detail the case of testing hypotheses about population means.

3.1. Continuous Outcomes: One Group

Although not all continuous variables are normally distributed, the *central limit theorem* states that the mean of even nonnormal variables becomes approximately normal as the sample size gets large. Although "large" differs depending on the degree of nonnormality, sample sizes greater than 25 to 30 are usually large enough to ensure that the mean is fairly normally distributed. Thus, it is instructive to consider first the sample size formulas for normally distributed outcomes because the formulas obtained frequently serve as approximations in asymptotic applications.

We start with the simple case of a single normally distributed mean with known variance, even though in reality we will never know the population variance. Recall that if $X_i \sim N(\mu, \sigma^2)$ and $\bar{X} = \sum_{i=1}^{n} X_i/n$, the standard deviation of the sampling distribution of $\bar{X}$, which is called the standard error, is given by $\sigma/\sqrt{n}$. To see the effect of varying n, consider the example used to introduce the concept of power. When $n = 25$, we see that the standard error is $20/5 = 4$. If $n = 100$, the standard error is $20/10 = 2$.

The effect of quadrupling the sample size is shown in **Figure 7**. In the top panel, $n = 25$, and you see that there is a lot of overlap between distributions whose means are 8 units apart (120 vs. 128). In the bottom panel, $n = 100$, and the standard error is subsequently smaller so the distributions are tighter and now there is not a lot of overlap. The 5% 1-sided critical regions are also denoted in the panels. The critical values are $120 + 1.645 \times 20/\sqrt{25} = 126.58$ and $120 + 1.645 \times 20/\sqrt{10} = 123.29$, respectively. The area to the right of the critical value is 5% under the null hypothesis ($\mu_{wgt} = 120$). Under the specific alternative, $\mu_{wgt} = 128$, the power of the test when the sample size is 25 is

$$1 - \Phi\left(\frac{126.58 - 128.0}{4}\right) = 1 - \Phi(-0.355) = 0.64$$

when the true mean is 128. That is, the probability is 64% that we would reject the null hypothesis in favor of the alternative if the true mean is 128.

In the bottom panel, with a sample size of 100, the power of the test is

$$1 - \Phi\left(\frac{123.29 - 128.0}{2}\right) = 1 - \Phi(-2.355) = 0.99.$$

By increasing n from 25 to 100, we have increased the power of the test from 0.64 to 0.99 for a fixed level of $\alpha = 0.05$.

So what if we wanted the power to be 90% for detecting a mean of 128 (i.e., 90% chance of rejecting H_0 if the true mean is 128)? How do we choose n to ensure this power? We see that the power was 64% with $n = 25$ and the power was 99% with $n = 100$. So we know n will be between these values. We could use trial and error, that is, try $n = 26$ and go through the steps above, then $n = 27$, and so forth. There is an easier way.

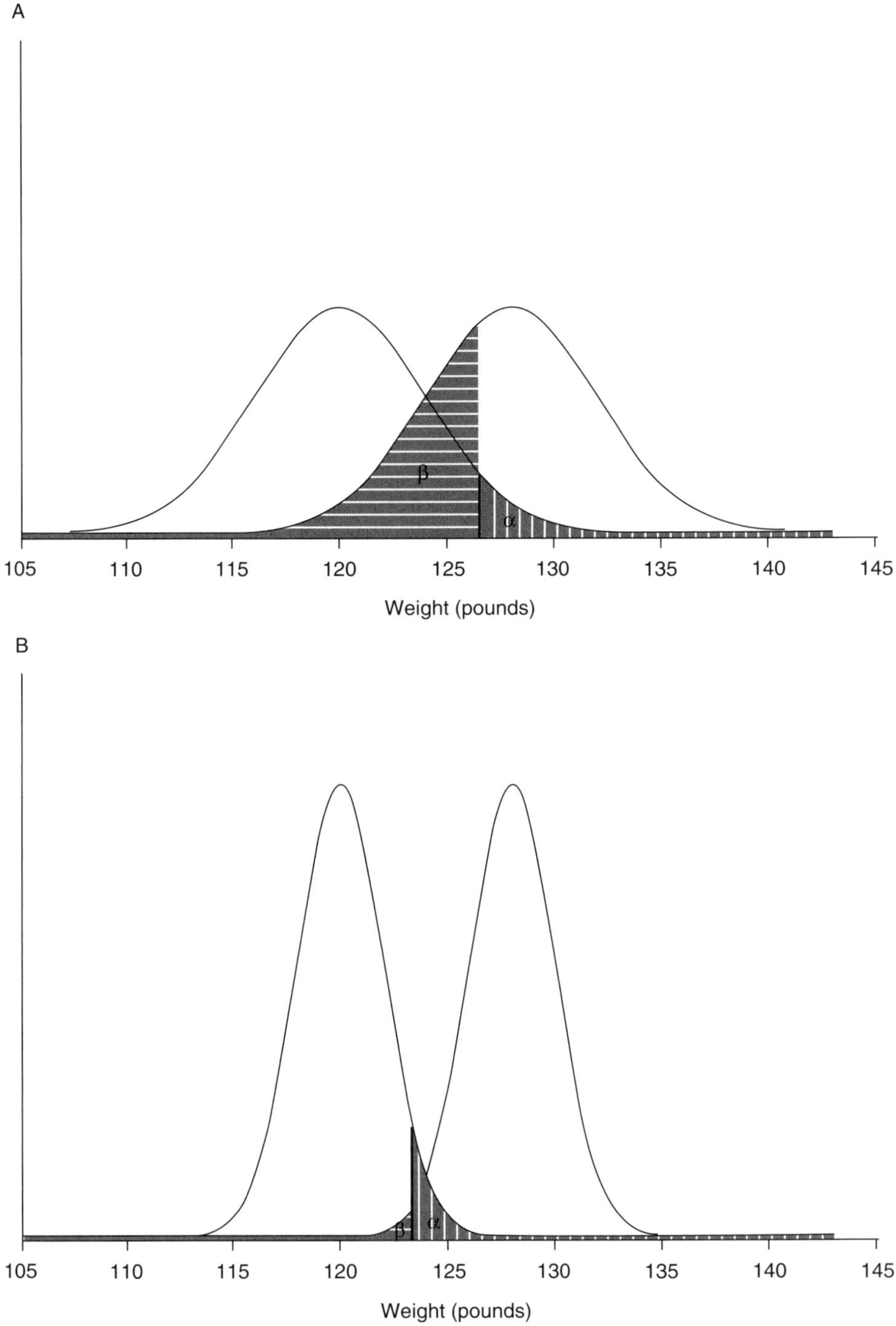

Fig. 7. Effect of increasing n from 25 (top panel) to 100 (bottom panel) on the sampling distribution of the sample mean and the decrease in the type II error for a fixed α.

Note that under the null hypothesis, the critical value (C) is given by $\mu_0 + z_{1-\alpha} \times \sigma / \sqrt{n}$. Under the alternative hypothesis, that critical value is given by $\mu_1 - z_{1-\beta} \times \sigma / \sqrt{n}$, where $z_{1-\alpha}$ and $z_{1-\beta}$ are the 100(1 – α)th and 100(1 – β)th percentiles of the standard normal distribution function, respectively. Setting these two expressions equal to one another, we get:

$$\mu_0 + z_{1-\alpha} \times \sigma / \sqrt{n} = \mu_1 - z_{1-\beta} \times s\sigma / \sqrt{n}.$$

Rearranging, we get

$$(z_{1-\alpha} + z_{1-\beta})\sigma / \sqrt{n} = (\mu_1 - \mu_0)$$

$$\frac{(z_{1-\alpha} + z_{1-\beta})\sigma}{(\mu_1 - \mu_0)} = \sqrt{n}$$

$$n = \frac{(z_{1-\alpha} + z_{1-\beta})^2 \sigma^2}{(\mu_1 - \mu_0)^2}. \qquad (1)$$

For a 2-sided alternative, replace $z_{1-\alpha}$ with $z_{1-\alpha/2}$. Note that this solution is not exact for the 2-sided problem because we are ignoring the area in the lower critical region under the alternative hypothesis (which will be negligible in most practical applications).

So for our example, had we wanted the power to be 90% at an alternative of 128 with a 1-sided $\alpha = 0.05$, our required sample size would be, using **Equation 1**,

$$n = \frac{(1.645 + 1.282)^2 \times 20^2}{(128 - 120)^2} = 53.5,$$

which we would round up to 54.

As the reader can see, calculating a sample size is fairly straightforward. The hard part is deciding on the appropriate values to include in the formulas. That is, what should the type I error be and what difference should be detected with what power? Additionally, although we are assuming we know the variance in this example, that will never be the case and we will have to come up with a reliable estimate.

The sample size formula is sometimes written as

$$n = \frac{(z_{1-\alpha} + z_{1-\beta})^2}{\Delta^2},$$

where $\Delta = (\mu_1 - \mu_0)/\sigma$. The quantity $\Delta = \delta/\sigma$, where $\delta = \mu_1 - \mu_0$, is called the effect size and is sometimes used in designing a clinical trial. While we do not believe this is a good way of designing studies, it is a good way to illustrate the effect of α, β, and the difference on the required sample size.

Table 8 and **Table 9** provide the sample sizes required for Δ ranging from 0.1 to 1.5 by 0.1. The reader will note the dramatic increase in the required sample size as Δ decreases. Because Δ is included in the formula as a squared term in the denominator, any halving in Δ results in a quadrupling in the required sample size. Consider, for example, Δ equal to 1, 0.5, and 0.25, with a 1-sided α of 0.05 and $1 - \beta$ equal to 0.95. The required sample sizes (not rounded) would be 10.82, 43.29, and 173.15, respectively. This illustrates the importance in choosing δ wisely when designing studies.

Example 3

An investigator knows from past experience that patients with syndrome X have elevated systolic blood pressure with a mean of 190 mm Hg and a standard deviation of 30 mm Hg. The investigator wants to see if a new drug will reduce the blood pressure in these patients. He hypothesizes that the blood pressure will be less than 190 for patients treated with the new drug. However, because it would be an important finding to show that the new drug actually increased the blood pressure, he decides to do a 2-sided test. He will choose between the following null and alternative hypotheses.

$$H_0: \mu_{SBP} = 190 \text{ versus } H_1: \mu_{SBP} \neq 190.$$

Table 8
Sample Sizes Required for a 1-Sided *Z*-Test as a Function of α, β, and Δ

	$\alpha = 0.05$			$\alpha = 0.1$			$\alpha = 0.2$		
$\Delta \backslash 1 - \beta$	0.80	0.90	0.95	0.80	0.90	0.95	0.80	0.90	0.95
0.1	619	857	1083	451	657	857	284	451	619
0.2	155	215	271	113	165	215	71	113	155
0.3	69	96	121	51	73	96	32	51	69
0.4	39	54	68	29	42	54	18	29	39
0.5	25	35	44	19	27	35	12	19	25
0.6	18	24	31	13	19	24	8	13	18
0.7	13	18	23	10	14	18	6	10	13
0.8	10	14	17	8	11	14	5	8	10
0.9	8	11	14	6	9	11	4	6	8
1.0	7	9	11	5	7	9	3	5	7
1.1	6	8	9	4	6	8	3	4	6
1.2	5	6	8	4	5	6	2	4	5
1.3	4	6	7	3	4	6	2	3	4
1.4	4	5	6	3	4	5	2	3	4
1.5	3	4	5	3	3	4	2	3	3

Table 9
Sample Sizes Required for a 2-Sided *Z*-Test as a Function of α, β, and Δ

	$\alpha = 0.05$			$\alpha = 0.1$			$\alpha = 0.2$		
$\Delta \backslash 1-\beta$	0.80	0.90	0.95	0.80	0.90	0.95	0.80	0.90	0.95
0.1	785	1051	1300	619	857	1083	451	657	857
0.2	197	263	325	155	215	271	113	165	215
0.3	88	117	145	69	96	121	51	73	96
0.4	50	66	82	39	54	68	29	42	54
0.5	32	43	52	25	35	44	19	27	35
0.6	22	30	37	18	24	31	13	19	24
0.7	17	22	27	13	18	23	10	14	18
0.8	13	17	21	10	14	17	8	11	14
0.9	10	13	17	8	11	14	6	9	11
1.0	8	11	13	7	9	11	5	7	9
1.1	7	9	11	6	8	9	4	6	8
1.2	6	8	10	5	6	8	4	5	6
1.3	5	7	8	4	6	7	3	4	6
1.4	5	6	7	4	5	6	3	4	5
1.5	4	5	6	3	4	5	3	3	4

Based on his clinical experience, the investigator decides that reducing the mean systolic blood pressure by 15 mm Hg would result in a clinically meaningful benefit for his patients. He designs his study to have 95% power for detecting a change in systolic blood pressure of 15 mm Hg at the 5% 2-sided level of significance. He calculates the following sample size:

$$n = \frac{(1.96 + 1.645)^2 \times 30^2}{(15)^2} = 51.98,$$

which he would round up to 52 patients.

Thus, the investigator plans to treat the next 52 patients with the new drug, after which he will measure their systolic blood pressure, compute a *Z* statistic, and compare it to the critical values of ±1.96.

As mentioned above, we never really know the variance when trying to determine a sample size for a mean. Instead we use an estimate of the variance obtained from the literature or perhaps from a pilot study, and we typically proceed with the sample size calculation as though this were the known true value. We then analyze the data using a Student's *t*-test with the variance estimated from the sample obtained.

Assuming X_i is normally distributed, the Student's t-statistic $\left(t = \frac{\bar{X} - \mu_0}{s/\sqrt{n}}\right)$ is distributed as a central t-distribution with $n - 1$ degrees of freedom. The unbiased variance estimator is given by $s^2 = \frac{\sum_{i=1}^{n}(X_i - \bar{X})^2}{n-1}$. Then the power of a 1-sided t-test is given by

$$P(\mu) = 1 - T_{\delta,n-1}(t_{1-\alpha,n-1}), \quad (2)$$

where $t_{1-\alpha}$ is the $100(1 - \alpha)$th percentile of the central t-distribution with $n - 1$ degrees of freedom, and $T_{\delta,n-1}(t_{1-\alpha,n-1})$ is the cumulative distribution function for a noncentral t-distribution with $n - 1$ degrees of freedom and noncentrality parameter $\delta = \frac{\mu - \mu_0}{\sigma/\sqrt{n}}$. For a 2-sided test, the power function will be

$$P(\mu) = 1 - T_{\delta,n-1}(t_{1-\alpha/2,n-1}) + T_{\delta,n-1}(t_{\alpha/2,n-1}). \quad (3)$$

Note how similar this is to the power function for the normal distribution. We just mentioned that σ is never actually known, and yet it appears in the noncentrality parameter. Again, the usual approach is to replace σ by s in the noncentrality parameter and proceed as though it were known. Because the critical value and the noncentrality parameter are both functions of n, the actual sample size can be obtained iteratively. Power for the example used in introducing power is shown in **Table 10** for sample sizes ranging from 52 to 58.

One sees that a sample size of 55 subjects is needed using the t-distribution, whereas 54 was the estimate based on a normal approximation. **Table 10** also

Table 10
Power for *t*-Test Using Noncentral *t*-Distribution and the Normal Approximation

n	Noncentrality parameter	Critical value	Power	Power[a]
52	2.88444	1.67528	0.88512	0.89244
53	2.91204	1.67469	0.89046	0.89746
54	2.93939	1.67412	0.89557	0.90226
55	2.96648	1.67356	0.90045	0.90685
56	2.99333	1.67303	0.90513	0.91125
57	3.01993	1.67252	0.90960	0.91545
58	3.04631	1.67203	0.91387	0.91946

[a]Normal approximation.

Table 11
Sample Sizes Required for a 1-Sided *t*-Test as a Function of α, β, and Δ

	$\alpha = 0.05$			$\alpha = 0.1$			$\alpha = 0.2$		
$\Delta \backslash 1 - \beta$	0.80	0.90	0.95	0.80	0.90	0.95	0.80	0.90	0.95
0.1	619	857	1083	451	657	857	284	451	619
0.2	155	215	271	113	165	215	71	113	155
0.3	69	96	121	51	73	96	32	51	69
0.4	39	54	68	29	42	54	18	29	39
0.5	25	35	44	19	27	35	12	19	25
0.6	18	24	31	13	19	24	8	13	18
0.7	13	18	23	10	14	18	6	10	13
0.8	10	14	17	8	11	14	5	8	10
0.9	8	11	14	6	9	11	4	6	8
1.0	7	9	11	5	7	9	3	5	7
1.1	6	8	9	4	6	8	3	4	6
1.2	5	6	8	4	5	6	2	4	5
1.3	4	6	7	3	4	6	2	3	4
1.4	4	5	6	3	4	5	2	3	4
1.5	3	4	5	3	3	4	2	3	3

gives the power based on the normal approximation. We see that the normal approximation slightly overestimates the power (thus leading to the slightly smaller estimate for *n*).

Table 11 and **Table 12** provide the sample sizes required for a 1-sample *t*-test for Δ ranging from 0.1 to 1.5 by 0.1. The reader will note how close these sample size estimates are to those provided by the normal approximation in **Tables 8** and **9**. For most practical purposes, the simple formula could be used, although modern computers make this unnecessary.

Example 4

Suppose the investigator in **Example 3** does not know the variance of systolic blood pressure in patients with syndrome X. He does a literature review and finds an article that provides an estimate of 30 mm Hg for the standard deviation. As above, the investigator wants to choose between the following null and alternative hypotheses:

$$H_0: \mu_{SBP} = 190 \text{ versus } H_1: \mu_{SBP} \neq 190.$$

As above, he designs his study to have 95% power for detecting a change in systolic blood pressure of 15 mm Hg at the 5% 2-sided level of significance.

Table 12
Sample Sizes Required for a 2-Sided *t*-Test as a Function of α, β, and Δ

	$\alpha = 0.05$			$\alpha = 0.1$			$\alpha = 0.2$		
$\Delta\backslash 1-\beta$	0.80	0.90	0.95	0.80	0.90	0.95	0.80	0.90	0.95
0.1	785	1051	1300	619	857	1083	451	657	857
0.2	197	263	325	155	215	271	113	165	215
0.3	88	117	145	69	96	121	51	73	96
0.4	50	66	82	39	54	68	29	42	54
0.5	32	43	52	25	35	44	19	27	35
0.6	22	30	37	18	24	31	13	19	24
0.7	17	22	27	13	18	23	10	14	18
0.8	13	17	21	10	14	17	8	11	14
0.9	10	13	17	8	11	14	6	9	11
1.0	8	11	13	7	9	11	5	7	9
1.1	7	9	11	6	8	9	4	6	8
1.2	6	8	10	5	6	8	4	5	6
1.3	5	7	8	4	6	7	3	4	6
1.4	5	6	7	4	5	6	3	4	5
1.5	4	5	6	3	4	5	3	3	4

Now, instead of using a *Z* statistic to test his hypothesis, he will use a 1-sample *t*-test. Using **Equation 3**, he iteratively calculates the required sample size to be 54, two more than estimated using the normal approximation in **Example 3**. Thus, the investigator would treat the next 54 patients with the new drug, after which he will measure their systolic blood pressure, compute a *t*-statistic, and compare it to the critical values of ±2.006.

3.2. Continuous Outcomes: Two Groups

As above, it is instructive to examine the case of normally distributed outcomes with known variance even though we will never know the variances. The formulas derived here will be useful for asymptotic approximations to other problems.

The test statistic for testing for a difference in 2 means when the variances are known is

$$Z = \frac{(\bar{X}_1 - \bar{X}_0) - (\mu_1 - \mu_0)}{\sigma_0^2 / n_0 + \sigma_1^2 / n_1}.$$

Note that under the null hypothesis, the critical value (C) is given by

$$\mu_0 + z_{1-\alpha}\sqrt{\sigma_0^2 / n_0 + \sigma_1^2 / n_1}.$$

Under the alternative hypothesis, that critical value is given by

$$\mu_1 - z_{1-\beta}\sqrt{\sigma_0^2 / n_0 + \sigma_1^2 / n_1}.$$

Setting these two expressions equal to one another, we get:

$$\mu_0 + z_{1-\alpha}\sqrt{\sigma_0^2 / n_0 + \sigma_1^2 / n_1} = \mu_1 - z_{1-\beta}\sqrt{\sigma_0^2 / n_0 + \sigma_1^2 / n_1}.$$

Rearranging and letting $n_1 = k_n n_0$ and $\sigma_1^2 = k_v \sigma_0^2$, we get

$$(z_{1-\alpha} + z_{1-\beta})\sqrt{\sigma_0^2 / n_0 + k_v \sigma_0^2 / k_n n_0} = \mu_1 - \mu_0$$

$$\frac{(z_{1-\alpha} + z_{1-\beta})}{(\mu_1 - \mu_0)} = \sqrt{\frac{k_n n_0}{k_n \sigma_0^2 + k_v \sigma_0^2}}$$

$$n_0 = \frac{(k_n + k_v)(z_{1-\alpha} + z_{1-\beta})^2 \sigma_0^2}{k_n(\mu_1 - \mu_0)^2}. \quad (4)$$

Unequal sample sizes and unequal variances both have an impact on required sample size. The total sample size required with unequal allocation and unequal variances relative to the total sample size required when the sample sizes and variances are equal in the two groups is given by $\frac{(k_n + k_v)(1 + k_n)}{4k_n}$, and this is shown in **Table 13** for a variety of k_n and k_v. In general, equal allocation is nearly optimal. Some efficiency is gained by allocating more subjects to the treatment with the largest variance.

Table 13
Total Required Sample Size for Various Ratios of Variances and Sample Sizes in the 2 Groups Relative to the Total Required Sample Size Assuming Equal Sample Sizes and Variances

	k_v			
k_n	1.0	1.3333	2	4
0.25	1.5625	1.9792	2.8125	5.3125
0.5	1.1250	1.3750	1.8750	3.3750
0.75	1.0208	1.2153	1.6042	2.7708
1	1.0000	1.1666	1.5000	2.5000
1.3333	1.0208	1.1666	1.4583	2.3333
2	1.1250	1.2500	1.5000	2.2500
4	1.5625	1.6666	1.8750	2.5000

Assuming equal variances and equal allocation to each treatment arm, the sample size needed in each arm is given by

$$n = \frac{2(z_{1-\alpha} + z_{1-\beta})^2 \sigma^2}{(\mu_1 - \mu_0)^2}. \tag{5}$$

As mentioned above, σ^2 will never be known, and we use an estimate in its place and proceed with the sample size calculation as though this were the known true value. We then analyze the data using a Student's *t*-test with the variance estimated from the sample obtained.

Then the power of the test is given by

$$\text{Power} = 1 - T_{\delta,N-2}(t_{1-\alpha/2,N-2}) + T_{\delta,N-2}(-t_{1-\alpha/2,N-2}), \tag{6}$$

for a 2-sided alternative, where N is the total sample size, $t_{1-\alpha,N-2}$ is the 100(1–α)th percentile of the central *t*-distribution with $N - 2$ degrees of freedom, and $T_{\delta,N-2}(t_{1-\alpha,N-2})$ is the cumulative distribution function for a noncentral *t*-distribution with $N - 2$ degrees of freedom and noncentrality parameter $\delta = \sqrt{N}\frac{\Delta}{2\sigma}$. The standard method for determining the sample size for the proposed study is to iteratively solve **Equation 6** with σ^2 replaced by s^2 such that the power provided equals the desired power. The sample size formula given by **Equation 5** with σ^2 replaced by s^2 provides a reasonable estimate.

Example 5

Suppose an investigator wants to compare a new treatment to a standard treatment for lowering systolic blood pressure (SBP). A previous study provides an estimate of 30 mm Hg for the standard deviation. The investigator wants to choose between the following null and alternative hypotheses.

$$H_0: \mu_1 - \mu_2 = 0 \text{ versus } H_1: \mu_1 - \mu_2 \neq 0,$$

where μ_1 and μ_2 are the mean SBP in the 2 groups. He designs his study to have 90% power for detecting a difference in mean systolic blood pressure of 20 mm Hg at the 5% 2-sided level of significance. Using **Equation 5** above, the investigator calculates an n of 47.3, which he would raise to 48 per group. Using **Equation 6**, he iteratively calculates the required sample size to be 49. Thus, the investigator would randomize 98 patients, half to receive the standard treatment and half to receive the new therapy.

3.3. Dichotomous Outcomes

In many cases, investigators are interested in questions regarding proportions. For example, is the toxicity experienced by patients undergoing a new experimental therapy greater than that observed historically using standard therapies or is the response rate to treatment A different from the response rate to treatment B? For single groups, the sample size formula used in testing the null hypothesis,

$$H_0: P \leq \pi_0 \text{ versus } H_1: P > \pi_0,$$

is given by

$$n = \frac{\left(z_{1-\alpha}\sqrt{\pi_0(1-\pi_0)} + z_{1-\beta}\sqrt{\pi_1(1-\pi_1)}\right)^2}{(\pi_1 - \pi_0)^2}, \tag{7}$$

where π_1 is a clinically meaningful alternative.

For comparing proportions in 2 groups, we are interested in testing the null hypothesis,

$$H_0: \pi_B - \pi_A = 0 \text{ versus } H_1: \pi_B - \pi_A \neq 0,$$

and the sample size required in each group, assuming equal sample sizes in the 2 groups, is given by

$$n = \frac{\left(z_{1-\alpha}\sqrt{2\bar{\pi}(1-\bar{\pi})} + z_{1-\beta}\sqrt{\pi_2(1-\pi_2)+\pi_1(1-\pi_1)}\right)^2}{(\pi_2 - \pi_1)^2}, \tag{8}$$

where $\bar{\pi} = (\pi_1 + \pi_2)/2$.

Example 6

Suppose an investigator wishes to compare the response rate to a new treatment for small cell lung cancer relative to that of the standard treatment. The standard treatment results in response in approximately 50% of patients. The investigator would consider the new treatment successful if it increased the response rate to 65%. She wants to have 90% power for detecting this response rate at the 5% 1-sided level of significance. The sample size needed in each group for this study would be

$$n = \frac{(1.645\sqrt{2 \times 0.55 \times 0.45} + 1.282\sqrt{0.6 \times 0.4 + 0.5 \times 0.5})^2}{(0.6 - 0.5)^2} = 422.2,$$

which she would round up to 423 per group.

3.4. Calculation of Power or Sample Size

There are many software packages devoted to the calculation of power or sample size. All require the specification of the statistical test, whether the test will be 1- or 2-sided, the significance level (α), the study design, the clinical effect of interest, and either the desired power or sample size. Power packages include nQuery Advisor (http://www.statsol.ie/), Power Analysis and Sample Size (http://www.ncss.com/), and Power and Precision (http://www.power-analysis.com/). In addition, many statistical packages have the ability to calculate power for simple designs. These include SAS (http://www.sas.com/) and S-Plus (http://www.insightful.com). Finally, there are many excellent, free online tools for the calculation of power. You can find several by doing an Internet search for "statistical power." Three good examples of Web sites offering power and sample size tools are those developed by the UCLA Department of Statistics (http://calculators.stat.ucla.edu/), Russell Lenth (http://www.math.uiowa.edu/~rlenth/Power/), and John Pezzullo (http://statpages.org/). As an example, suppose we are designing a randomized study comparing mean weight loss for 2 different regimens, one diet only and one diet and exercise. We anticipate a mean weight loss of 5 lb for the diet-only participants, and we would like to be able to detect an additional 5-lb weight loss (10-lb total) for those participants receiving the diet and exercise program with 90% power at the 5% 2-sided level of significance. Using nQuery Advisor, one specifies that means are being compared between 2 groups using a Student's *t*-test with equal variances, α (0.05, 2-sided), the difference of interest (5 lb), the standard deviation (5 lb), and the power (90%). nQuery Advisor returns a sample size of 23 per group.

Another method for power calculation is to use simulation. This approach can be used for simple power calculations such as the one just described and for much more complicated methods. In fact, many statistical methods do not have formulae for the calculation of power and power must be estimated using simulation. This approach requires that a method exist to generate data under the assumed distributions and a way to perform the statistical test. There are three main steps: (1) generate data under the specified design and alternative hypothesis, (2) perform the statistical test and determine whether it is significant at the α level of significance, and (3) repeat many times keeping track of the proportion of times that the result is significant. Recall the binomial distribution discussed in **Chapter 3**. We estimated the true proportion as the proportion in a sample. We are doing the same thing here. We do not know the true power (probability of rejecting the null hypothesis given the alternative), but we can use the proportion of times we reject the null hypothesis as an estimate of the true power. To illustrate this method, we use the same scenario described above (2-sample *t*-test, group means of 5 and 10, within-group standard deviation of 5, and a 2-sided test at the 5% level). The true power with a sample size of 23 per group is 91.250%, which is represented by the horizontal line in **Figure 8**. The jagged line is the proportion of time the null hypothesis is rejected through that simulation. Notice that there is a fair amount of variability early on but

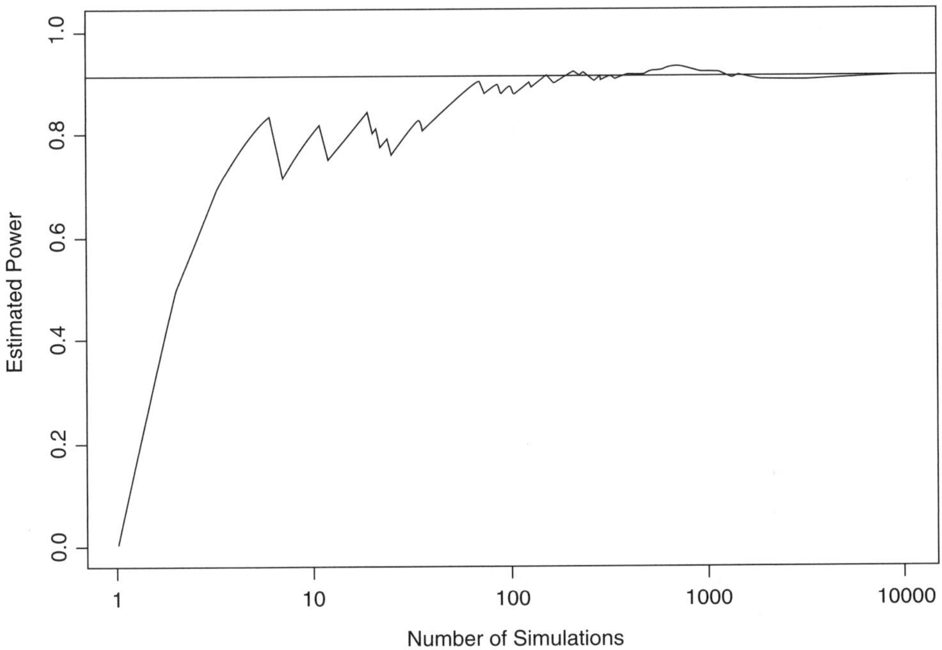

Fig. 8. The cumulative proportion of significant simulations plotted against the number of simulations. The resulting proportion estimates power. As the number of simulations increases, the estimate stabilizes at the truth of 91.25%.

that the sample proportion stabilizes at the true power as the number of simulations grows. The use of simulation is more difficult than using a formula because it takes longer to run simulations and does not allow direct estimation of sample size (you would have to fix the sample size and then estimate power and then try another sample size until you got the power you desired). On the other hand, simulation can always be used and works in many cases where formulae are unavailable.

4. Other Considerations

Typically, in calculating a sample size, the investigator must specify the variance (σ^2) of the outcome measure and a difference (δ) that is important to detect with a specified power ($1 - \beta$) and level of significance (α). Specification of the clinically meaningful difference can be difficult. What is a meaningful effect (i.e., difference in response between 2 treatments) probably differs for each investigator, and some investigators may think that any effect is meaningful. Because the difference enters the sample size formula as a squared term, halving the difference quadruples the sample size. For example, a study designed to detect a difference of 5 lb between 2 weight-loss regimens takes 4 times as many subjects as one designed to detect a 10-lb difference. However, one should resist the temptation to choose the meaningful difference as that difference which results in a feasible sample size.

Still more difficult in many cases is specification of the variance. Most sample size formulae assume that the variance is a known fixed value. Instead, estimates of the variance obtained from published studies or pilot studies performed by the investigator. Published studies rarely use exactly the same measurement assessed exactly the same way using the same patient population as the new study, so it is never clear how close the variance estimate is to the true parameter. It is just an estimate. Dudewicz ***(7)*** showed that using an estimated variance in traditional sample size calculations results in power estimates that can be "misleadingly large." Shiffler and Adams ***(8)*** and Browne ***(9)*** proposed inflating the traditional sample size estimates to ensure a desired power with a specified level of confidence. For example, if an investigator wants to be 90% confident that the sample size is large enough to provide 95% power, he would need to inflate the calculated sample size by d.f./$\chi^{-1}(1 - \gamma, \text{d.f.})$, where d.f. is the degrees of freedom of the variance estimate, γ is the desired confidence, and $\chi^{-1}(1 - \gamma, \text{d.f.})$ is the $100(1 - \gamma)$th percentile of the chi-square distribution with d.f. degrees of freedom. So if the degrees of freedom of the variance estimate was 20, the required sample size would need to be inflated by 1.61.

The type I and II errors are the easiest parameters to specify, but there is not universal agreement on what these values should be. Although it is clear that the type I and II errors should be small, there is no clear definition of what constitutes "small" for a particular study. This is not a trivial consideration because the sample size required approximately doubles in going from a design with 80% power at the 5% 1-sided level of significance to one with 95% power at the 5% 2-sided level of significance. We usually specify the type I (α) and type II errors (β) *a priori*. These are often fixed at $0.01 \le \alpha \le 0.05$ and $0.05 \le \beta \le 0.2$ regardless of the type of study. Lee and Zelen ***(10)*** believe that the selection of α and β should be based on the costs of making different types of wrong decisions. For example, in the early course of a line of research, an incorrect significant result (meaning the null hypothesis is really true) would lead to further investigation. Although this might be a waste of time, it is likely preferable than prematurely stopping a line of research. Lee and Zelen demonstrate using data from the Eastern Cooperative Oncology Group (ECOG) that in a series of clinical trials, the probability of the null hypothesis being false is about 30%. Although we all would prefer to believe that our intervention will definitely have the desired effect, we would all do well to remember that they rarely do.

O'Brien and Castelloe ***(11)*** elaborate on the ideas of Lee and Zelen in a more accessible presentation. They define γ as the probability that the null hypothesis is false. With the addition of this one parameter, the hypothesis testing framework can be recast as a diagnostic test as was described in **Chapter 6**. Two important questions are not addressed by the usual type I and type II errors: If the trial achieves statistical significance, what is the chance this will be an incorrect inference? If the trial does not achieve statistical significance, what is the chance this will be an incorrect inference? These quantities are defined by O'Brien and Castelloe as the *crucial* type I and type II errors. Specifically,

$$\alpha^* = P[H_0 \text{ true} \mid P \le \alpha] = \frac{\alpha(1-\gamma)}{\alpha(1-\gamma)+(1-\beta)\gamma}$$

and

$$\beta^* = P[H_0 \text{ false} \mid P > \alpha] = \frac{\beta\gamma}{\beta\gamma+(1-\alpha)(1-\gamma)}.$$

If, for example, we believe that $\gamma = 0.3$ and we design a study with 90% power and testing at the 5% level, our crucial type I error (α^*) = 0.1148 and our crucial type II error (β^*) = 0.0432. That is, if we reject the null hypothesis, the chance is 11.5% that the null hypothesis is true. If we accept the null hypothesis, the chance is 4.3% that the null hypothesis is false.

O'Brien and Castelloe describe the usual course of scientific enquiry (denoted as the "March of Science") where investigation may begin with small pilot studies, proceed to larger scale studies, and have these studies confirmed. Once an idea passes each stage of investigation, the belief in the idea grows. At the design of the first stage, we may be skeptical and believe that $\gamma = 0.25$. If a pilot study is positive, we may then believe that $\gamma = 0.5$. If many studies confirm the idea, we may ultimately believe that γ is close to 1. Incorporating γ can have an impact on the sample size chosen and should be considered as it is not difficult and is instructive.

5. Conclusion

Numerous other factors can impact power and should also be considered when determining the appropriate sample size. These include retention, noncompliance (dropouts and drop-ins), missing data, interim analyses, multiple end points, multiple arms, and nonindependence. Wittes ***(12)*** provides a very nice discussion of these and other problems that can occur and what can be done about them. In addition, Lenth ***(13)*** provides some excellent practical advice in determining appropriate sample sizes. In conclusion, sample size determination is a challenging but necessary and rewarding exercise.

Appropriate and early consideration of all the issues that can affect the power helps ensure a rigorous and well-designed study.

References

1. Cohen, J. (1988) *Statistical Power Analysis for the Behavioral Science*, 2nd ed. Hillsdale, Lawrence Erlbaum Associates.
2. Kraemer, H. C., and Thiemann, S. (1987) *How Many Subjects? Statistical Power Analysis in Research*. Newbury Park, SAGE Publications.
3. Murphy, K. R., and Myors, B. (1998) *Statistical Power Analysis: A Simple and General Model for Traditional and Modern Hypothesis Tests*. Mahwah, Lawrence Erlbaum Associates.
4. Chow, S-C., Shao, J., and Wang, H. (2003) *Sample Size Calculations in Clinical Research*. New York, Marcel Dekker.
5. Fleiss, J. L. (1981) *Statistical Methods for Rates and Proportions*, 2nd ed. New York, John Wiley & Sons.
6. Peace, K. E. (1988) Letter to the editor: some thoughts on one-tailed tests. *Controlled Clinical Trials* **9**, 383–384.
7. Dudewicz, E. J. (1972) Confidence intervals for power, with special reference to medical trials. *Aust. J. Stat.* **14**, 211–216.
8. Shiffler, R. E., and Adams, A. J. (1987) A correction for biasing effects of pilot sample size on sample size determination. *Journal of Marketing Research* **24**, 319–321.
9. Browne, R. H. (1995) On the use of a pilot sample for sample size determination. *Stat. Med.* **14**, 1933–1940; 1995.
10. Lee, S. J., and Zelen, M. (2000) Clinical trials and sample size considerations: another perspective. *Stat. Sci.* **15**(2), 95–103.
11. O'Brien, R. G., and Castelloe, J. (2006) *Sample-Size Analysis for Traditional Hypothesis Testing: Concepts and Issues, in Pharmaceutical Statistics*. (D'Agostino, R., Dmitrienko, A., and Chuang-Stein, C., eds.) Cary, SAS Press, SAS Institute.
12. Wittes, J. (2002) Sample size calculations for randomized controlled trials. *Epidemiol. Rev.* **24**, 39–53.
13. Lenth, R. V. (2001) Some practical guidelines for effective sample-size determination. *Am. Stat.* **55**, 187–193.

20

Microarray Analysis

Grier P. Page, Stanislav O. Zakharkin, Kyoungmi Kim, Tapan Mehta, Lang Chen, and Kui Zhang

Summary

Microarrays and related technologies have allowed investigators to ask biological questions in far greater detail than has previously been possible. Microarrays had a troubled beginning, but most of these problems resulted from the growing pains of this technology, which, like many new things, was initially more promise than delivery. Nevertheless, over the past few years, investigators have learned how to achieve optimal performance of technology, and now exciting discoveries are made using microarray-based research. Many of the advances have come from the realization that microarrays are not a magic tool but rather are like any other measurement device. Unless microarray experimentation is coupled with good experimental practices, it will not yield valid results or, worse yet, may lead to misleading results. In this chapter, we highlight some of the important steps that should be taken to successfully conduct a microarray study. These steps include a clearly stated biological question, experimental design, careful experimental conduct, complete statistical analysis, validation/verification of results, and dissemination of the data.

Key Words: Experimental design; microarray; plants; statistical analysis.

1. Introduction

Microarrays have been described as "the hottest thing in biology and medicine since the advent of the polymerase chain reaction a decade ago." This technology emerged circa 1996 *(1,2)* and had its first high-profile uses in 1998 and 1999 *(3–5)*. What is "hot" is that microarrays allow the simultaneous measurement of mRNA expression levels from thousands of genes from samples. DNA microarray studies have already been published for a variety of species, including human *(6)*, mouse *(7)*, and rat *(8)*. These studies have led to major

From: *Methods in Molecular Biology, vol. 404: Topics in Biostatistics*
Edited by: W. T. Ambrosius © Humana Press Inc., Totowa, NJ

contributions to the analyses of plants and agricultural products such as maize ***(9,10)***, *Arabidopsis* ***(10)***, tobacco ***(11)***, peppers ***(12)***, wheat ***(13)***, *Populus* ***(14)***, and sorghum ***(15)*** under a variety of experimental conditions. In the future, we envision the impact of microarrays on research increasing as they are applied in many new areas, such as comparative genomic hybridization ***(16)***, genotyping ***(17)***, phage display ***(18)***, tissue arrays ***(19)***, and tiling arrays ***(20)***.

Many researchers initially embraced microarrays with the glee of "a kid in a candy shop." Many, however, have also suffered the effects of trying to overindulge on more data than can be digested. The effects of such overindulgence are preventable. Researchers need only to remain scientifically hard-minded by employing sound principles of experimental design, analysis, and interpretation. Microarrays are like any other hot new laboratory tool that has come before; future utilization of microarrays lies in building on what has already been done to generate an ever more rigorous foundation for discovery and inference ***(21)***. Our goals in this chapter are to educate students and scientists about microarray research and to teach them about the steps that must be taken for the proper conduct of a microarray study. After a brief description of microarray technology, the following steps will be described as the sections of the chapter: clearly defining objectives of the study, appropriately designing experiments, applying statistically rigorous data analysis, verifying results, and paying close attention to data handling and management issues ***(22,23)***.

2. Materials

2.1. What Is a Microarray?

A microarray is a solid substrate (usually glass) upon which many different sense or anti-sense (which depends on the technology) cRNAs or cDNAs have been spotted in specific locations. Normally, many different cDNAs are spotted onto a single array in a grid pattern and the cDNAs are taken from genes known to code for poly A+ RNA or from EST sequences. First, RNA from a tissue sample is extracted and labeled with fluorescent dyes or radioactive nucleotides. This labeled RNA is then hybridized to cDNA immobilized on the array. The labeled RNA binds to its complementary sequence approximately proportional to the amount of each mRNA transcript in a sample. The amount of radioactivity of fluorescence can be measured, allowing estimation of the amount of RNA for each transcript in the sample.

2.2. Types of Microarrays

A variety of different microarray technologies, depending on the species being studied, may be available for use. In general, the technologies can be broken down into 3 general groups: short oligonucleotides (Affymetrix,

NimblGen), long oligonucleotides (Agilent, Illumina, Amersham), and “spotted” cDNA amplicons (National Institute on Aging (NIA), Stanford, etc.). Each particular technology has its advantages and disadvantages.

The “short oligos” have certain advantages in that they are manufactured to high standards. Because these arrays use the sum of several probes, the quantification of RNA levels appears to be robust. In addition, the probe sequences are known. There are also some factors that could be of disadvantage; specifically, they are not particularly robust to polymorphisms. This is not a large problem in humans, but across the *Arabidopsis* genus more than half the probe sets may not be fully functional. Short oligo arrays tend to be relatively expensive, with a limited number of species available. However, NimbleGen appears to be able to offer relatively inexpensive custom arrays for any species for which sequence information can be provided.

Long oligos arrays (50–80 mers) offer advantages in that they are robust to some polymorphisms, can often be used across related species, the sequence is known, the spots can be applied at a known concentration, and probes are single-stranded. However, long oligos have a high cost of synthesizing and are nonrenewable but are less expensive than short oligo platforms. The complete genome sequence is required to generate robust (unique) oligos with minimal cross-hybridization.

cDNA amplicons or spotted arrays ***(4)*** use sets of plasmids of specific cDNAs, which may be full- or partial-length sequences, to generate the sequence to be spotted upon an array. The cDNA microarrays allow investigators to print any genes from any species for which a clone can be obtained. cDNA and some long oligo arrays allow for 2 samples (usually labeled Cy3 and Cy5) to be hybridized to a single array at a time. The array’s design can be rapidly changed and the cost per array is low once the libraries have been established. Also, the cDNA is a renewable supply. There are several disadvantages, including variable amount of DNA spotted in each spot and high (10% to 20%) dropout rates due to failed polymerase chain reaction (PCR), failed spots, and possible contamination and cross-hybridization with homologues. As a result, many cDNA arrays are of poor quality.

A high-quality microarray study requires high-quality microarrays. The manufacture of high-quality spotted arrays is very time and resource intensive. Failure to generate high-quality arrays can lead to the arrays being no more than an expensive random number generator. The particulars on how to manufacture high-quality arrays are beyond the scope of this review ***(24)***.

3. Define Objectives of the Study

The first step of any experiment, microarray or otherwise, is to define a hypothesis. Although it may seem perfectly obvious to state this, the objectives of a study are likely to be met if that study is designed in such a way that it is consistent with meeting those objectives. A study can only be designed to meet the objectives if the objectives can be clearly articulated prior to initiating the study. It is common for investigators who have conducted a microarray study not to be able to clearly state the objective of their study. (For example, to identify genes that were differentially expressed in response to obesity or whose differential expression led to obesity.) To further complicate matters, even if the question is articulated, the design often does not answer the question. All too often, investigators have inadvertently allowed themselves to go down the slippery slope from hypothesis-generating studies to studies that might be termed "objective generating." By objective generating, we mean studies in which the investigator has initiated activity without a clear concept of what the exact objectives of the study are in the sole hopes that, by some mysterious process, the mass of data that will come out at the end of the microarray study will make all things clear. We do not believe that such outcomes are likely.

However, that is not to say that the experimental objectives may not include the generation of new objectives/hypotheses on interesting pathways/genes that may be highlighted as a result of the microarray study. Thus, we urge investigators to articulate clearly what they hope to get out of microarray studies at the end so that the studies can be designed to meet those objectives from the beginning.

4. Experimental Design for Microarray Studies

After stating a hypothesis, the most important step in an experiment is the experimental design. If an experiment is designed well, the choice of analytical methods will easily follow. The 3 essential principles of experimental design for microarrays are randomization, replication, and design of experiments.

4.1. Randomization

Random assignment of varieties/samples/groups and random sampling of populations is the physical basis for the validity of statistical inferences *(25)*. It is important that samples are randomly assigned to groups in any experimental design. Microarrays and most laboratory techniques are liable to nonbiological sources of variation including day, chip lot, reagent lot, and personnel. In many cases, these sources of variation are larger than the biological variation *(26)*. Thus, if nonbiological and biological differences are confounded, the experiment can be essentially useless. Microarray experiments can require multiple levels of randomization to help minimize unanticipated biases from known factors. One step to be taken is to randomly assign the samples to be run (e.g., do not run all chips of 1 experimental condition on 1 day and the rest on another) ***(27,28)***. In addition, randomization helps spread the variation from unmeasured sources of variation evenly among biological groups.

4.2. Replication

Individuals vary within a population. Measurement tools, such as microarrays, measure with some error. Thus, a single sample cannot be used to make inference about groups or a population. Replication serves the purpose of assessing the magnitude of such variation at a variety of different levels. In addition, it allows for making appropriate inference at the group or population level.

4.2.1. Types of Replication

Replication in the context of microarrays can be incorporated at several levels:

- [R1] gene-to-gene: genes can be spotted multiple times per array;
- [R2] array-to-array: mRNA samples can be used on multiple arrays, and each array is hybridized separately;
- [R3] subject-to-subject: mRNA samples can be taken from multiple individuals to account for inherent biological variability.

The first 2 types of replication are referred to as technical replication. The first measures within-array variation, whereas the second measures between-array variation. These types of replication are important for assessing the measurement error and reproducibility of microarray experiments. In addition, the first 2 types are extremely useful for improving precision of measurements. On the other hand, the third type of replication allows us to asses the biological variation within populations. Technical replicates [R1] and [R2] cannot substitute for biological replicates [R3] for inferring that the mean expression of a gene differs in two populations ***(29)***.

4.2.2. Power and Sample Size

Sample size (number of [R3] replications above) has a major impact upon how confidently genes can be declared expressed either differentially or not differentially ***(30,31)***. Choosing the best sample size for microarray research may not be well served by traditional power analysis, which is based on the control of type I errors for detection of a specifically sized difference for a single hypothesis test (e.g., 80% power at a significance level $\alpha = 0.05$ to detect a 0.5 standard deviation (SD) reduction in a metabolite due to a treatment). Although methods for calculating sample sizes have been developed along these lines ***(32,33)***, this may not be the most appropriate technique. First, every gene is differentially expressed in a different degree, thus it is difficult to specify a biologically meaningful amount of change or on which gene to base power calculations. Second, if one uses a multiple testing adjustment, such as Bonferroni or Sidak correction, the sample size required per group for a chip that covers all expressed human genes would be more than 250 per group, which is not normally achievable in microarray experimentations for budgetary reasons. Third, many genes will probably be differentially expressed and they will not be independent, which violates assumptions underlying Bonferroni and Sidak type corrections for multiple testing.

We believe it is more appropriate in microarray analysis to choose sample size based on control of the false discovery rate (FDR) ***(34)*** and the expected discovery rate (EDR). The FDR is an estimate of the expected proportion of genes declared significant that are in fact not differentially expressed; that is, that are "false discoveries" ***(35,36)***. The EDR is the expected proportion of genes where true differences between conditions exist and are found to be significantly different.

The relationship between FDR and EDR is a complicated function of the sample size per group, the significance level (α), the magnitude of the effect, and the amount of biological heterogeneity in the samples. In general, for a constant sample size, choosing a smaller significance level will decrease both the EDR (a negative result) and FDR (a positive result), while increasing the sample size will increase the EDR and decrease the FDR.

Methods have been developed based upon pilot data to extrapolate the FDR and EDR for different sample sizes ***(37,38)***, as well as to plan future studies' sample sizes. One interactive Web site, the PowerAtlas ***(39)*** (http://www.poweratlas.org), allows one to search among the published microarray data sets and to find which is most similar to one's planned study. This data can then be used to extrapolate the sample size for a planned study.

4.3. Design of the Experiment

Careful design of experiments is very important to successfully answer the primary hypothesis, as well as to minimize the cost of the experiment (number of arrays). There are 3 common experimental designs for a spotted microarray experiment: reference design, incomplete block design, and loop design. Because other types of arrays only allow 1 sample per chip, this experimental design is easier, primarily involving sufficient replication and randomization of the samples. All experiments should concentrate on generating the highest possible quality samples by carefully outlining the steps for sample preparation and labeling. If possible, nothing should be left to chance; for example, if dealing with female animals, control for cycle, and if dealing with diurnal animals, conduct experiments at a consistent time across all animals.

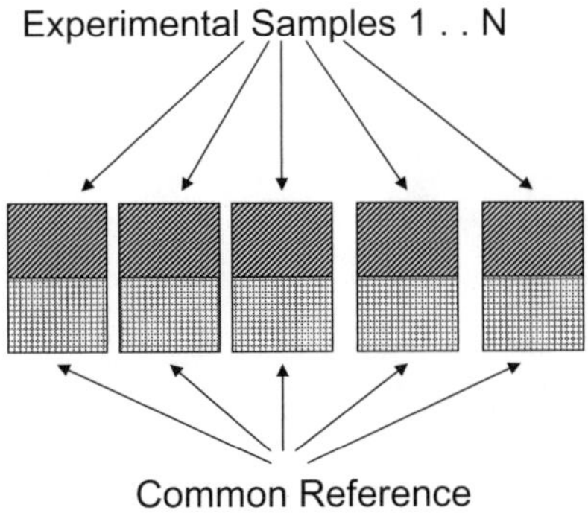

Fig. 1. Common reference with dye swap. One experimental sample and one reference sample per array. Slash = green and grid = red.

4.3.1. Reference Design

The reference design (**Fig. 1**) is commonly used for studying *n* varieties of a factor of interest by introducing an additional variety to serve as a reference. Introducing a reference as an intermediate step allows the indirect comparisons because each sample is compared directly to the reference. The reference design allows extension of the number of samples in the experiment by simply adding another array using the same reference sample. In practice, the reference sample is labeled with one dye and the other samples of interest are labeled with the other dye. This allows for all *n* varieties in the experiment to be compared with a single common reference sample and to the same experimental condition. As a side benefit, one could see that gene expression levels in some samples exhibit more variation than others. This design also has disadvantages. For example, half of the data is collected on the reference sample, although it is not usually a variety of interest. Therefore, this design is less resource efficient than some other methods.

4.3.2. Incomplete and Complete Balanced Block Design

Each spotted microarray in the experiment is probed with 2 differently labeled cDNA samples, each labeled with 2 different dyes. In other words, arrays can be treated as blocks of size 2. When only 2 varieties are compared, every pair of varieties will occur once in the same block. Designs of this type are referred to as complete block designs (**Fig. 2**). This balanced design ensures that the pairs of varieties (experimental conditions) are compared with approximately the same precision. However, complete block designs are seldom used.

In many cases, one is interested in more than 2 varieties of the factor of interest. In this situation, this design does not have the orthogonality of variety

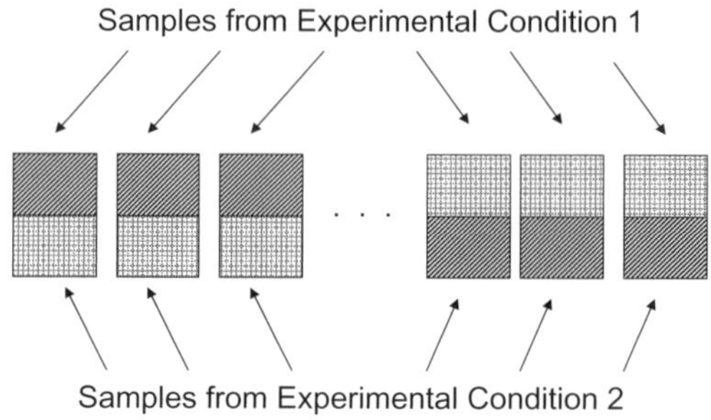

Fig. 2. Balanced block design with dye swap. Slash = green and grid = red.

and dye effects as not every variety can appear on every array *(27)*. This design is referred to as an incomplete block design. Therefore, we recommend choosing an even incomplete block design so that varieties can be balanced in terms of dyes to ensure that the effects of interest are not confounded with other sources of systematic variation.

The balanced block design has a substantial advantage over the reference design. The balanced design doubles the amount of information because half of each array is not wasted on a reference sample. However, reference designs can be set up to avoid dye biases (difference in incorporation of dyes into specific gene sequences) that really devil uncommon reference designs.

4.3.3. Loop Design

The loop design (**Fig. 3**) has received much attention lately ***(27,28,40,41)***. The loop design has several advantages over the balanced block design. In the loop design, each sample is labeled with each dye. The samples may either be type 2 or 3 replication, with type 3 being favored. In its simplest form, the green (cy3) of **sample 1** is hybridized with the red (cy5) of **sample 2**. The green of **sample 2** is hybridized with red of **sample 3**, and so forth, until the green of

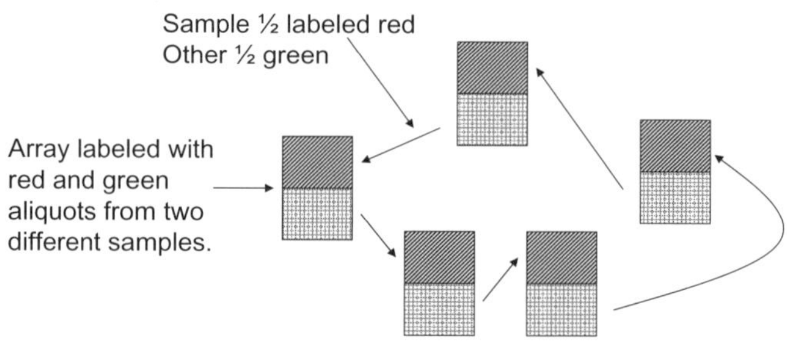

Fig. 3. Loop design with dye swap. Slash = green and grid = red.

sample *n* is hybridized with the red of **sample 1**. This allows *n* samples to be assayed on *n* arrays with dye swap. This design is very sample efficient and allows for all experimental groups to be compared. However, there can be quite a bit of error in the comparisons if the loop is large, especially if there is large spot to spot variability ***(41)***. In addition, if a chip is lost, the loop is broken, and much of the information is lost. Advanced multiple cyclic loops have been developed to reduce some of the issues while only requiring a few more than *n* chips for *n* samples.

5. Microarray Analysis

Once a microarray experiment has been conducted, the analysis of the data involves several steps. The first step is taking the image from the scanner and processing the extracted information.

5.1. Image Processing from cDNA and Long Oligo Arrays

Image processing consists of 3 steps: addressing, segmentation, and information extraction.

5.1.1. Gridding/Addressing

Typically, a microarray is an ordered array of spots, with constant separation between the row and column; grids or spots must be the same throughout the microarray. Addressing is the process of finding the location of spots on the microarray or assigning coordinates to each spot on the microarray. However, the spotting is rarely perfect; there are variations to deal with. While software usually does a good job, manual review and occasional intervention results in better data, but it is a very time consuming process.

5.1.2. Segmentation

Segmentation is the most important, and also the most difficult part, of image analysis. In this process, each image pixel is classified as either signal or the background noise. The popular methods of segmentation include fixed circles, adaptive circles, adaptive shapes, and the histogram method. The first 2 methods provide the means to separate the circular spots from the background by clearly defining the boundaries of the spots. The literature includes a variety of comparisons of the methods, with no clear winner ***(42,43)***.

5.1.3. Information Extraction

In the final step of image analysis, the mean and median values of the spot intensities and the background spot intensities are calculated. Usually, correlations between spot intensities, percentage of the spots without any signal, their distributions and signal-to-noise ratio for each spot, and variation of the pixel intensities are also calculated. The spot intensities are then measured as the sum of intensities of all the pixels inside the spot. Next, the average intensity per pixel for each spot is calculated. This information is used to identify the quality of the experiment.

5.2. Image Analysis of Affymetrix GeneChip Microarrays

Affymetrix GeneChips are the most widely used oligonucleotide arrays. Unlike the other systems, the Affymetrix system represents each gene as 11 to 20 probe pairs. Each probe pair is composed of a 25-base-pair perfect match (PM) probe, which represents the gene's known sequence, and a mismatch (MM) probe, which differs from the PM by the middle base. The expression level is some function of the averages of difference in intensities of PM and MM over the 20 sequences. Several algorithms have been developed for averaging the probe pairs to yield a final quantification. These include Dchip ***(44)***, GCRMA-EB and GCRMA-MLE ***(45)***, MAS5 ***(46)***, PDNN ***(47)***, and RMA ***(48,49)***, all of which have different measurement properties, and it is not yet clear which is best ***(50)***.

5.3. Normalization of DNA Data

One of the early and near universal steps in microarray study is the use of a technique called either normalization or transformation. The normalization technique has at least 2 purposes; the first purpose is to adjust microarray data for effects that arise from variation in the technology rather than from biological differences between the RNA samples or between the printed probes ***(51)***. The second purpose is "[to aid] in the analysis of the data by bending the data nearer the Procrustean bed of the assumptions underlying conventional analyses" ***(52)***, which will allow for reliable statistical and biological analyses. The former is considered to be an adjustment for measured covariates such as dye biases, whereas the latter is the true meaning of normalization. Either way, a wide variety of methods have been developed for all meanings of normalization, including several varieties of linear models ***(53)***, loess ***(51)***, quantile-quantile ***(54)***, $\log_2$, and others ***(55,56)***. Normalization is usually required in cDNA microarray experiments to reduce dye biases. This is still an area of active research, and it is not clear which methods are appropriate for each chip and experimental design.

5.4. Analysis

There are three types of analyses that can be conducted upon microarray data. *Class prediction* involves building models to predict which group samples should be assigned to. This type of analysis is often used in cancer research where the goal is to determine if a sample is a particular type of cancer or if it will metastasize. The second set of analyses is *class discovery*, which involves the unsupervised analysis of data in order to identify previously unknown relationships between genes or among genes and samples. The final type is *class differentiation*, which usually involves the inferential statistical analysis.

5.4.1. Class Prediction Analysis

We use the term *prediction* to define the construction of a model that involves using gene expression experiments to classify objects into preexisting known classes, to develop indexes that can serve as a biomarker, or to predict to which class a sample should be assigned. In the field of cancer research, this first type of class prediction is very common ***(57,58)***.

There are many methods for constructing such scores. The earliest was Fisher's discriminant function analysis. In the ideal situation, Fisher's method is the optimal method. However, this is rarely the case, and alternative approaches need to be utilized. These include machine learning approaches such as support vector machine (SVM), least squares approaches, and related techniques ***(59)***. It is not clear which of these techniques are best, but the goal of all methods is to find the best compromise between complexity and simplicity in the model. As the predicted model becomes more and more complex by utilizing more and more of the sample information, the predicted ability in the *sample in hand* will increase; however, the sample data contain not only information about the true structure of the data but also "noise" due to sample variation. Thus, as the model becomes more complex, it becomes increasingly likely that one is simply fitting a model to the noise. The more one overfits the data in this manner, the poorer the subsequent prediction on independent data (sometimes referred to as cross-validation) is likely to be. Thus, the push-and-pull between complexity and simplicity. To build the models that will predict new samples well, one must build cross-validated models. Cross-validating requires that one have sufficient data to hold some out of the estimation process so that one can subsequently check how good the prediction is using the held-back data. For cross validation to be accurate, the held-back data used in the cross-validation must have in no way been used in the selection of the structure of the prediction model ***(60,61)***. This has often been violated in microarray research.

5.4.2. Class Discovery Analysis

Since Eisen and others first introduced the hierarchical clustering methods in microarray data analysis in 1998, cluster analysis has emerged as a prevalent tool for the exploration and visualization of microarray data. All cluster analyses aim to construct classes such that observations within a class are more similar to each other than they are to observations in different classes.

A variety of cluster methods including hierarchical and nonhierarchical methods exist. Among hierarchical methods there are agglomerative and divisive methods. For the hierarchical methods, there are different ways to measure distance between genes (including Pearson's correlation, Euclidian distance, and Kendall's tau) as well as a variety of methods for linking genes based on their distance (including average, single, complete, and median). A number of nonhierarchical methods exist including K-nearest neighbors, self-organizing maps, support vector machines, and singular value decomposition (SVD). Each method and approach has its own pluses and minuses that should be evaluated.

Although clustering is a commonly used tool for microarray data analysis, it is unlike hypothesis testing (**Section 5.4.3**) in that there are no probabilistic foundations and no correct answers. These pitfalls are directly related to several criticisms of cluster analysis. First, the cluster algorithms are guaranteed to produce clusters from data, no matter what kind of data has been used. Second, there is no valid way to establish the number of clusters in nonhierarchical cluster analysis. Third, different methods can produce drastically different results, and the search for the best choice among them has just begun ***(62,63)***. Therefore, cautions are required for performing such analysis and interpreting the results. We view them as exploratory descriptive methods that have the potential to suggest fruitful hypotheses and to provide concise displays of complex data.

5.4.3. Class Differentiation Analysis

One of the main tasks in analyzing microarray data is to determine which genes are differentially expressed between 2 or more groups of samples. This type of analysis is a conventional hypothesis test. A variety of different methods are used for estimation. One such method is fold change, which was used in early studies to compare the expression levels of genes. As sample sizes and sophistication in microarray analysis have advanced, fold change is being replaced by the change in expression in standard deviations. For making inferences in these situations, virtually any statistical method can be used including *t*-test, ANOVA, and linear models. A variety of Bayesian methods and information borrowing approaches such as Cyber-t ***(64)*** and SAM ***(35)*** have been developed.

5.4.4. Adjusting for Multiple Testing

Because each microarray can contain more than 30,000 genes, some adjustment for multiple testing is required to avoid many false-positive results. One way is to control the family-wise error rate (FWE), which is the probability of wrongly declaring at least 1 gene as differentially expressed. A Bonferroni correction is a method to adjust *P* values from independent tests. Permutation methods can be used to control FWE in presence of nonindependent tests ***(65)***. Another approach to address the problem of multiple testing is the false discovery rate (FDR), which is the proportion of false positives among all genes initially identified as being differentially expressed ***(34,66)***. In addition, a variety of Bayesian and alternative FDR methods have been developed ***(36)***.

6. Interpretation

The gene-by-gene statistical analysis of microarray data is not the end of a study by any stretch of the imagination. Unfortunately, unless one is using a model organism, the annotation of the data set can be quite time consuming and incomplete. Affymetrix provides some annotation, but the amount of details and quality is dependent upon the scientific community doing the annotation. In addition, a number of other tools such as Dragon database ***(67)***, NetAffx ***(68)***, and Milano ***(69)*** have been developed to assist in the process of developing a biological story for the data.

Microarray data is such that if the same data is given to a physiologist, they will see physiologic pathways, and if given to a breeder, they will see genes for selection. As a result, much of the published microarray analyses are highly biased in their interpretation because investigators may see the genes that they know and tell a story around those genes. Thus, statistical methods are needed to overcome some of this subjectivity. A global test statistic is one possible solution that involves testing if classes of genes tend to be up- or downregulated more than expected or if the genes in these classes are differentially expressed more than expected. There are several methods to identify classes including annotated databases such as Gene Ontology (GO) ***(70)*** and KEGG ***(71)***. A wide variety of techniques exist for identifying significantly differentially expressed classes including vote-counting procedures ***(72)*** and rank comparisons ***(73)***. Online tools such as Onto-Express ***(72)*** and ErmineJ ***(74)*** exist to assist with these calculations. Other tools such as GoMiner ***(75)*** allow for mining Medline abstracts to assist in data interpretation as well. These types of analyses can provide relatively objective identification of the classes of genes that are differentially expressed.

Microarrays allow for the study of gene expression at a near genome-wide level, allowing a better biological picture. Combining microarrays with other types of data can provide an even better picture. The high-dimensional biological (HDB) techniques that have been combined include microarrays ***(4,76)***, proteomics ***(77)***, linkage analysis ***(78–81)***, association studies ***(82,83)***, comparative genome hybridization (CGH), metabolomics ***(84)***, *N*-ethyl *N*-nitrosourea ENU mutagenesis ***(85)***, imaging, and biochemical system theory ***(86)***. Recently, there have been a few successes in the identification of causative genes using multipronged approaches for studying a disease ***(9)***. Methods to combine data range from simple Venn diagrams ***(87)*** to more formal intersection-union tests ***(88)***.

7. Validation of Microarray Experiments

A plethora of factors, which include biological and technical variability, inherent characteristics of different array platforms, and processing and analytical steps can affect results of a typical microarray experiment ***(89)***. Thus, the findings from studies need to be independently verified. In fact, a number of journals now require some sort of validation in order for a paper to be published. Sometimes, it is possible to confirm the outcome without doing any additional laboratory-based analysis. The array results can be compared with information available in the literature or in expression databases such as GEO ***(90)***. Agreement between different groups provides general confidence in data and in novel discoveries. However, such *in silico* validation is not always possible or appropriate. One needs to confirm that expression levels of genes do change under different conditions and that the changes are real and not due to some technical problems peculiar to the method. Although RT-PCR is commonly used, other approaches such as SAGE ***(91)*** have been used. In our opinion, the verification that investigators pursue should advance their hypotheses rather than arbitrarily verifying genes by RT-PCR.

8. Microarray Informatics

For most investigators, microarrays will involve the highest informatics hurdles they have faced in their research careers. Microarrays generate huge amounts of data, and there are data reporting and quality standards that are being enforced by many journals for publication.

8.1. Data Handling

Microarray experiments generate volumes of data with which many biological researchers may not be accustomed. A single Affymetrix chip will generate about 50 MB of data. After initial processing, each chip will provide 6000 to 60,000 numbers for analysis. After analysis, summary statistics, such as changes in expression and associated significance probabilities, will be available for all of the genes on the chips. Sorting through significance tests for 60,000 genes "manually" and trying to deduce biological principles is a Herculean task. The volume of data generated by the new technology is overwhelming. Before an experiment has begun, consideration should be paid to how data are stored, viewed, and interpreted ***(92)***. A variety of statistical tools such as HDBSTat! ***(93)*** and Bioconductor (www.bioconductor.org) exist to analyze the data.

8.2. MIAME and Standards

MIAME, which stands for Minimal Information About Microarray Experiments ***(94)***, is a set of standards developed by the Microarray Gene Expression Data (MGED) Society (www.MGED.org). MGED is an international organization of biologists, computer scientists, and data analysts that aims to facilitate the sharing of microarray data generated by functional genomics and proteomics experiments. Despite the name, MIAME compliance can involve quite a bit of record keeping in many stages of the microarray studies, especially if new microarrays are being developed. Several journals, including all from the Nature Publishing Group, require compliance with MIAME standards and deposition of microarray data in a public microarray database in order to publish ***(95)***. The MIAME standard covers most aspects of microarray experiments except for the analysis of data, and investigators should familiarize themselves with the standard before a study. Although standards for analysis have not yet been fully developed, the first suggestions of standards have been put forward ***(23)***, thus analysis reporting standards will probably be coming as well.

8.3. Databases

The storage and dissemination of microarray data can be aided by the use of databases. For internal uses, several microarray databases can be installed. These include the Stanford Microarray Database (SMD), the Longhorn Database (LMD) ***(96)***, and the Yale Microarray Database (YMD) ***(92)***. The installation and maintenance of these databases can require a large amount of effort. Several public databases for the deposit and dissemination of results exist. In the United States, there is Gene Expression Omnibus (GEO) (www.ncbi.nlm.nih.gov), and in Europe, ArrayExpress (http://www.ebi.ac.uk/arrayexpress/).

9. Conclusion

Microarrays allow investigators to conduct many studies that are not possible with other technologies. In this chapter, we gave a brief overview of each step in microarray-based research. As this technology is relatively new, there are many options available to researchers. The successful application of microarrays depends upon building a good knowledge base so that the investigators can carefully plan the experiment, rigorously conduct it, correctly and completely analyze the data, verify the results, and then disseminate them to the research community.

References

1. Chee, M., Yang, R., Hubbell, E., Berno, A., Hunag, Z. C., Stern, D., Winkler, J., Lockhart, D. J., Morris, M. S., and Fodor, S. P. A. (1996) Accessing genetic information with high-density DNA arrays. *Science* **274**, 610–614.
2. Lockhart, D. J., Ding, H., Byrne, M., Follettie, M. T., Gallo, M. V., Chee, M. A. S., Mittmann, M., Wang, C., Kobayashi, M., Horton, H., and others. (1996) Expression monitoring by hybridization to high-density oligonucleotide arrays. *Nature biotechnology* **14**, 1675–1680.
3. Lee, C-K., Kloop, R. G., Weindruch, R., and Prolla, T. A. (1999) Gene expression profile of aging and its restriction by caloric restriction. *Science* **285**, 1390–1393.
4. Eisen, M. B., Spellman, P. T., Brown, P. O., and Botstein, D. (1998) Cluster analysis and display of genome-wide expression patterns. *Proc. Natl. Acad. Sci. U.S.A.* **95**, 14863–14868.
5. Perou, C. M., Jeffrey, S. S., van de Rijn, M., Rees, C. A., Eisen, M. B., Ross, D. T., Pergamenschikov, A., Williams, C. F., Zhu, S. X., Lee, J. C., and others. (1999) Distinctive gene expression patterns in human mammary epithelial cells and breast cancers. *Proc. Natl. Acad. Sci. U.S.A.* **96**, 9212–9217.
6. Ginos, M. A., Page, G. P., Michalowicz, B. S., Patel, K. J., Volker, S. E., Pambuccian, S. E., Ondrey, F. G., Adams, G. L., and Gaffney, P. M. (2004) Identification of a gene expression signature associated with recurrent disease in squamous cell carcinoma of the head and neck. *Cancer Res.* **64**, 55–63.
7. Higami, Y., Pugh, T. D., Page, G. P., Allison, D. B., Prolla, T. A., and Weindruch, R. (2003) Adipose tissue energy metabolism: altered gene expression profile of mice subjected to long-term caloric restriction. *FASEB J.* **18**, 415–417.
8. Zakharkin, S. O., Kim, K., Mehta, T., Chen, L., Barnes, S., Scheirer, K. E., Parrish, R. S., Allison, D. B., and Page, G. P. (2005) Sources of variation in Affymetrix microarray experiments. *BMC Bioinformatics* **6**, 214.
9. Schadt, E. E., Monks, S. A., Drake, T. A., Lusis, A. J., Che, N., Colinayo, V., Ruff, T. G., Milligan, S. B., Lamb, J. R., Cavet, G., and others. (2003) Genetics of gene expression surveyed in maize, mouse and man. *Nature* **422**, 297–302.
10. Ruuska, S. A., Girke, T., Benning, C., and Ohlrogge, J. B. (2002) Contrapuntal networks of gene expression during Arabidopsis seed filling. *Plant Cell* **14**, 1191–1206.
11. Chua, Y. L., Mott, E., Brown, A. P., MacLean, D., and Gray, J. C. (2004) Microarray analysis of chromatin-immunoprecipitated DNA identifies specific regions of tobacco genes associated with acetylated histones. *Plant J.* **37**, 789–800.
12. Lee, S., Kim, S. Y., Chung, E., Joung, Y. H., Pai, H. S., Hur, C. G., and Choi, D. (2004) EST and microarray analyses of pathogen-responsive genes in hot pepper (Capsicum annuum L.) non-host resistance against soybean pustule pathogen (Xanthomonas axonopodis pv. glycines). *Funct. Integr. Genomics* **4**(3), 196–205.
13. Ogihara, Y., Mochida, K., Nemoto, Y., Murai, K., Yamazaki, Y., Shin, I., and Kohara, Y. (2003) Correlated clustering and virtual display of gene expression patterns in the wheat life cycle by large-scale statistical analyses of expressed sequence tags. *Plant J.* **33**, 1001–1011.

14. Gu, R., Fonseca, S., Puskas, L. G., Hackler, Jr L., Zvara, A., Dudits, D., and Pais, M. S. (2004) Transcript identification and profiling during salt stress and recovery of Populus euphratica. *Tree Physiol.* **24**, 265–276.
15. Zhu-Salzman, K., Salzman, R. A., Ahn, J. E., and Koiwa, H. (2004) Transcriptional regulation of sorghum defense determinants against a phloem-feeding aphid. *Plant Physiol.* **134**, 420–431.
16. Zhao, J., Roth, J., Bode-Lesniewska, B., Pfaltz, M., Heitz, P. U., and Komminoth, P. (2002) Combined comparative genomic hybridization and genomic microarray for detection of gene amplifications in pulmonary artery intimal sarcomas and adrenocortical tumors. *Genes Chromosomes Cancer* **34**, 48–57.
17. Gunderson, K. L., Steemers, F. J., Lee, G., Mendoza, L. G., and Chee, M. S. (2005) A genome-wide scalable SNP genotyping assay using microarray technology. *Nat. Genet.* **37**, 549–554.
18. Cekaite, L., Haug, O., Myklebost, O., Aldrin, M., Ostenstad, B., Holden, M., Frigessi, A., Hovig, E., and Sioud, M. (2004) Analysis of the humoral immune response to immunoselected phage-displayed peptides by a microarray-based method. *Proteomics* **4**, 2572–2582.
19. Gulmann, C., Butler, D., Kay, E., Grace, A., and Leader, M. (2003) Biopsy of a biopsy: validation of immunoprofiling in gastric cancer biopsy tissue microarrays. *Histopathology* **42**, 70–76.
20. Mockler, T. C., Chan, S., Sundaresan, A., Chen, H., Jacobsen, S. E., and Ecker, J. R. (2005) Applications of DNA tiling arrays for whole-genome analysis. *Genomics* **85**, 1–15.
21. Page, G. P., Edwards, J. W., Barnes, S., Weindruch, R., and Allison, D. B. (2003) A design and statistical perspective on microarray gene expression studies in nutrition: the need for playful creativity and scientific hard-mindedness. *Nutrition* **19**, 997–1000.
22. Lee, M. L., Kuo, F. C., Whitmore, G. A., and Sklar, J. (2000) Importance of replication in microarray gene expression studies: statistical methods and evidence from repetitive cDNA hybridizations. *Proc. Natl. Acad. Sci. U.S.A.* **97**, 9834–9839.
23. Tumor Analysis Best Practices Working Group. (2004) Expression profiling—best practices for data generation and interpretation in clinical trials. *Nat. Rev. Genet.* **5**, 229–237.
24. Zhang, W., Shmulevich, I., and Astola, J. (2004) Microarray quality control. Norwell, Kluwer Academic Publishers, pp.1–66.
25. Rubin, D. B. (1991) Practical implications of modes of statistical inference for causal effects and the critical role of the assignment mechanism. *Biometrics* **47**, 1213–1234.
26. Woo, Y., Affourtit, J., Daigle, S., Viale, A., Johnson, K., Naggert, J., and Churchill, G. (2004) A comparison of cDNA, oligonucleotide, and Affymetrix GeneChip gene expression microarray platforms. *J. Biomol. Tech.* **15**, 276–284.
27. Kerr, M. K., and Churchill, G. A. (2001) Statistical design and the analysis of gene expression microarray data. *Genet. Res.* **77**, 123–128.

28. Kerr, M. K., and Churchill, G. A. (2001) Experimental design for gene expression microarrays. *Biostatistics* **2**, 183–201.
29. Oleksiak, M. F., Churchill, G. A., and Crawford, D. L. (2002) Variation in gene expression within and among natural populations. *Nat. Genet.* **32**, 261–266.
30. Lee, M. T., Kuo, F. C., Whitmore, G. A., and Sklar, J. (2000) Importance of replication in microarray gene expression studies: statistical methods and evidence from repetitive cDNA hybridizations. *Proc. Natl. Acad. Sci. U.S.A.* **97**, 9834–9839.
31. Allison, D. B., and Coffey, C. S. (2002) Two-stage testing in microarray analysis: what is gained? *J. Gerontol. A Biol. Sci. Med. Sci.* **57**, B189–B192.
32. Pan, W., Lin, J., and Le, C. T. (2002) How many replicates of arrays are required to detect gene expression changes in microarray experiments? A mixture model approach. *Genome Biol.* **3**(5), research 0022.
33. Hwang, D., Schmitt, W. A., Stephanopoulos, G., and Stephanopoulos, G. (2002) Determination of minimum sample size and discriminatory expression patterns in microarray data. *Bioinformatics* **18**, 1184–1193.
34. Benjamini, Y., and Hochberg, Y. (1995) Controlling the false discovery rate: a practical and powerful approach to multiple testing. *J. Roy. Stat. Soc. Ser. B Methodol.* **57**, 289–300.
35. Tusher, V. G., Tibshirani, R., and Chu, G. (2001) Significance analysis of microarrays applied to the ionizing radiation response. *Proc. Natl. Acad. Sci. U.S.A.* **98**, 5116–5121.
36. Allison, D. B., Gadbury, G. L., Heo, M., Fernandez, J. R., Lee, C. K., Prolla, T. A., and Weindruch, R. (2002) A mixture model approach for the analysis of microarray gene expression data. *Computat. Stat. Data Anal.* **39**, 1–20.
37. Gadbury, G. L., Xiang, G., Edwards, J. W., Page, G. P., and Allison, D. B. (2005) The role of sample size on measures of uncertainty and power. In: Allison, Page, Edwards, and Beasley, eds. *DNA Microarrays and Related Genomic Technologies*. Boca Raton, CRC Press, 3, pp. 51–60.
38. Gadbury, G. L., Page, G. P., Edwards, J., Kayo, T., Weindruch, R., Permana, P. A., Mountz, J., and Allison, D. B. (2004) Power analysis and sample size estimation in the age of high dimensional biology. *Stat. Methods Med. Res.* **13**, 325–338.
39. Page, G. P., Edwards, J. W., Gadbury, G. L., Yelisetti, P., Wang, J., Trivedi, P., and Allison, D. B. (2006) The PowerAtlas: a power and sample size atlas for microarray experimental design and research. *BMC Bioinformatics* **7**, 84.
40. Kerr, M. K., and Churchill, G. A. (2001) Statistical design and the analysis of gene expression microarray data. *Genet. Res.* **77**, 123–128.
41. Kerr, M. K. (2003) Experimental design to make the most of microarray studies. *Methods Mol. Biol.* **224**, 137–147.
42. Nagarajan, R. (2003) Intensity-based segmentation of microarray images. *IEEE Trans. Med. Imaging* **22**, 882–889.
43. Li, Q., Fraley, C., Bumgarner, R. E., Yeung, K. Y., and Raftery, A. E. (2005) Donuts, scratches and blanks: robust model-based segmentation of microarray images. *Bioinformatics* **21**(12), 2875–2882.

44. Li, C., and Hung, W. W. (2001) Model-based analysis of oligonucleotide arrays: model validation, design issues and standard error application. *Genome Biol.* **2**, 32–35.
45. Wu, Z., and Irizarry, R. A. (2005) Stochastic models inspired by hybridization theory for short oligonucleotide arrays. *J. Comput. Biol.* **12**, 882–893.
46. Hubbell, E., Liu, W. M., and Mei, R. (2002) Robust estimators for expression analysis. *Bioinformatics* **18**, 1585–1592.
47. Zhang, L., Wang, L., Ravindranathan, A., and Miles, M. F. (2002) A new algorithm for analysis of oligonucleotide arrays: application to expression profiling in mouse brain regions. *J. Mol. Biol.* **317**, 225–235.
48. Irizarry, R. A., Bolstad, B. M., Collin, F., Cope, L. M., Hobbs, B., and Speed, T. P. (2003) Summaries of Affymetrix GeneChip probe level data. *Nucleic Acids Res.* **31**, e15.
49. Irizarry, R. A., Hobbs, B., Collin, F., Beazer-Barclay, Y. D., Antonellis, K. J., Scherf, U., and Speed, T. P. (2003) Exploration, normalization, and summaries of high density oligonucleotide array probe level data. *Biostatistics* **4**, 249–264.
50. Shedden, K., Chen, W., Kuick, R., Ghosh, D., Macdonald, J., Cho, K. R., Giordano, T. J., Gruber, S. B., Fearon, E. R., Taylor, J. M., and Hanash, S. (2005) Comparison of seven methods for producing Affymetrix expression scores based on false discovery rates in disease profiling data. *BMC Bioinformatics* **6**, 26.
51. Smyth, G. K., and Speed, T. (2003) Normalization of cDNA microarray data. *Methods.* **31**, 265–273.
52. Tukey, J. W. (1964) On the comparative anatomy of transformation. *Ann. Math. Stat.* **28**, 602–632.
53. Wolfinger, R. D., Gibson, G., Wolfinger, E. D., Bennett, L., Hamadeh, H., Bushel, P., Afshari, C., and Paules, R. S. (2001) Assessing gene significance from cDNA microarray expression data via mixed models. *J. Comput. Biol.* **8**, 625–637.
54. Bolstad, B. M., Irizarry, R. A., Astrand, M., and Speed, T. P. (2003) A comparison of normalization methods for high density oligonucleotide array data based on variance and bias. *Bioinformatics* **19**, 185–193.
55. Durbin, B. P., Hardin, J. S., Hawkins, D. M., and Rocke, D. M. (2002) A variance-stabilizing transformation for gene-expression microarray data. *Bioinformatics* **18**(Suppl. 1), S105–S110.
56. Durbin, B. P., and Rocke, D. M. (2004) Variance-stabilizing transformations for two-color microarrays. *Bioinformatics* **20**, 660–667.
57. Somorjai, R. L., Dolenko, B., and Baumgartner, R. (2003) Class prediction and discovery using gene microarray and proteomics mass spectroscopy data: curses, caveats, cautions. *Bioinformatics* **19**, 1484–1491.
58. Radmacher, M. D., McShane, L. M., and Simon, R. (2002) A paradigm for class prediction using gene expression profiles. *J. Comput. Biol.* **9**, 505–511.
59. Ringner, M., and Peterson, C. (2003) Microarray-based cancer diagnosis with artificial neural networks. *Biotechniques* Suppl. 30–35.

60. Ambroise, C., and McLachlan, G. J. (2002) Selection bias in gene extraction on the basis of microarray gene-expression data. *Proc. Natl. Acad. Sci. U.S.A.* **99**, 6562–6566.
61. Simon, R., Radmacher, M. D., Dobbin, K., and McShane, L. M. (2003) Pitfalls in the use of DNA microarray data for diagnostic and prognostic classification. *J. Natl. Cancer Inst.* **95**, 14–18.
62. Garge, N. R., Page, G. P., Sprague, A. P., Gorman, B. S., and Allison, D. B. (2005) Reproducible clusters from microarray research: whither? *BMC Bioinformatics* **6**(Suppl. 2), S10.
63. Datta, S., and Datta, S. (2003) Comparisons and validation of statistical clustering techniques for microarray gene expression data. *Bioinformatics* **19**, 459–466.
64. Baldi, P., and Long, A. D. (2001) A Bayesian framework for the analysis of microarray expression data: regularized t-test and statistical inferences of gene changes. *Bioinformatics* **17**, 509–519.
65. Westfall, P. H., Zaykin, D. V., and Young, S. S. (2002) Multiple tests for genetic effects in association studies. *Methods Mol. Biol.* **184**, 143–168.
66. Benjamini, Y., Drai, D., Elmer, G., Kafkafi, N., and Golani, I. (2001) Controlling the false discovery rate in behavior genetics research. *Behav. Brain Res.* **125**, 279–284.
67. Bouton, C. M., and Pevsner, J. (2000) DRAGON: Database Referencing of Array Genes Online. *Bioinformatics* **16**, 1038–1039.
68. Liu, G., Loraine, A. E., Shigeta, R., Cline, M., Cheng, J., Valmeekam, V., Sun, S., Kulp, D., and Siani-Rose, M. A. (2003) NetAffx: Affymetrix probesets and annotations. *Nucleic Acids Res.* **31**, 82–86.
69. Rubinstein, R., and Simon, I. (2005) MILANO—custom annotation of microarray results using automatic literature searches. *BMC Bioinformatics* **6**, 12.
70. Ashburner, M., Ball, C. A., Blake, J. A., Botstein, D., Butler, H., Cherry, J. M., Davis, A. P., Dolinski, K., Dwight, S. S., Eppig, J. T., and others. (2000) Gene ontology: tool for the unification of biology. *Nat. Genet.* **25**, 25–29.
71. Kanehisa, M., Goto, S., Kawashima, S., and Nakaya, A. (2002) The KEGG databases at GenomeNet. *Nucleic Acids Res.* **30**, 42–46.
72. Khatri, P., Draghici, S., Ostermeier, G. C., and Krawetz, S. A. (2002) Profiling gene expression using onto-express. *Genomics.* **79**, 266–270.
73. Mootha, V. K., Lindgren, C. M., Eriksson, K. F., Subramanian, A., Sihag, S., Lehar, J., Puigserver, P., Carlsson, E., Ridderstrale, M., Laurila, E., and others. (2003) PGC-1alpha-responsive genes involved in oxidative phosphorylation are coordinately downregulated in human diabetes. *Nat. Genet.* **34**, 267–273.
74. Pavlidis, P., Qin, J., Arango, V., Mann, J. J., and Sibille, E. (2004) Using the gene ontology for microarray data mining: a comparison of methods and application to age effects in human prefrontal cortex. *Neurochem. Res.* **29**, 1213–1222.
75. Zeeberg, B. R., Feng, W., Wang, G., Wang, M. D., Fojo, A. T., Sunshine, M., Narasimhan, S., Kane, D. W., Reinhold, W. C., Lababidi, S., and others. (2003) GoMiner: a resource for biological interpretation of genomic and proteomic data. *Genome Biol.* **4**, R28.

76. Lee, C. K., Weindruch, R., and Prolla, T. A. (2000) Gene-expression profile of the ageing brain in mice. *Nat. Genet.* **25**, 294–297.
77. Patterson, S. D., and Aebersold, R. H. (2003) Proteomics: the first decade and beyond. *Nat. Genet.* **33**(Suppl.), 311–323.
78. Morton, N. E. (1955) Sequential tests for the detection of linkage. *Am. J. Hum. Genet.* **7**(3), 277–318.
79. Doerge, R. W. (2002) Mapping and analysis of quantitative trait loci in experimental populations. *Nat. Rev. Genet.* **3**, 43–52.
80. Mackay, T. F. (2001) The genetic architecture of quantitative traits. *Annu. Rev. Genet.* **35**, 303–339.
81. Lander, E. S., and Schork, N. J. (1994) Genetic dissection of complex traits. *Science.* **265**, 2037–2048.
82. Allison, D. B. (1997) Transmission-disequilibrium test for quantitative traits. *Am. J. Hum. Genet.* **60**, 676–690.
83. Page, G. P., and Amos, C. I. (1999) Comparison of linkage-disequilibrium methods for localization of genes influencing quantitative traits in humans. *Am. J. Hum. Genet.* **64**, 1194–1205.
84. Taylor, J., King, R. D., Altmann, T., and Fiehn, O. (2002) Application of metabolomics to plant genotype discrimination using statistics and machine learning. *Bioinformatics* **18**(Suppl. 2), S241–S248.
85. Balling, R. (2001) ENU mutagenesis: analyzing gene function in mice. *Annu. Rev. Genomics Hum. Genet.* **2**, 463–492.
86. Voit, E. O., and Radivoyevitch, T. (2000) Biochemical systems analysis of genome-wide expression data. *Bioinformatics* **16**, 1023–1037.
87. Persson, S., Wei, H., Milne, J., Page, G. P., and Somerville, C. R. (2005) Identification of genes required for cellulose synthesis by regression analysis of public microarray data sets. *Proc. Natl. Acad. Sci. U.S.A.* **102**, 8633–8638.
88. Kim, K., and Allison, D. B. (2006) Composite hypothesis testing from multiple experiments: An approach built on intersection-union tests and Bayesian posterior probabilities. In: Guerra and Allison, eds. *Meta-analysis and Combining Information in Genetics.*
89. Murphy, D. (2002) Gene expression studies using microarrays: principles, problems, and prospects. *Adv. Physiol. Educ.* **26**, 256–270.
90. Edgar, R., Domrachev, M., and Lash, A. E. (2002) Gene Expression Omnibus: NCBI gene expression and hybridization array data repository. *Nucleic Acids Res.* **30**, 207–210.
91. Tuteja, R., and Tuteja, N. (2004) Serial analysis of gene expression (SAGE): unraveling the bioinformatics tools. *BioEssays.* **26**, 916–922.
92. Cheung, K. H., White, K., Hager, J., Gerstein, M., Reinke, V., Nelson, K., Masiar, P., Srivastava, R., Li, Y., Li, J., and others. (2002) YMD: a microarray database for large-scale gene expression analysis. *Proc. AMIA. Symp.* 140–144.
93. Trivedi, P., Edwards, J. W., Wang, J., Gadbury, G. L., Srinivasasainagendra, V., Zakharkin, S. O., Kim, K., Mehta, T., Brand, J. P., Patki, A., Page, G. P., and

Allison, D. B. (2005) HDBStat!: a platform-independent software suite for statistical analysis of high dimensional biology data. *BMC. Bioinformatics* **6**, 86.

94. Brazma, A., Hingamp, P., Quackenbush, J., Sherlock, G., Spellman, P., Stoeckert, C., Aach, J., Ansorge, W., Ball, C. A., Causton, H. C., and others. (2001) Minimum information about a microarray experiment (MIAME)-toward standards for microarray data. *Nat. Genet.* **29**, 365–371.
95. Ball, C. A., Sherlock, G., Parkinson, H., Rocca-Sera, P., Brooksbank, C., Causton, H. C., Cavalieri, D., Gaasterland, T., Hingamp, P., Holstege, F., and others. (2002) An open letter to the scientific journals. *Bioinformatics* **18**, 1409.
96. Killion, P. J., Sherlock, G., and Iyer, V. R. (2003) The Longhorn Array Database (LAD): an open-source, MIAME compliant implementation of the Stanford Microarray Database (SMD). *BMC. Bioinformatics* **4**, 32.

21

Association Methods in Human Genetics

Carl D. Langefeld and Tasha E. Fingerlin

Summary

Genetic association studies are increasingly used in the search for susceptibility variants for human traits. While many of the statistical tools available for such studies are well established, the field is advancing rapidly, as biological and technological developments allow investigators to generate vast amounts of detailed genetic data. This chapter gives an overview of the statistical evaluation of genetic data in both unrelated individuals and families. A brief introduction to fundamental population genetics concepts is followed by detailed examinations of measures of linkage disequilibrium and single-marker and haplotype association tests. Emphasis is given to the historical development of family-based tests to provide the context for more recent advancements. The chapter concludes with a discussion of design strategies for genetic association studies with dense genotyping of hundreds or thousands of markers, such as those planned for follow up of a linkage-candidate region or genome-wide association studies.

Key Words: Family-based association; genetic association; genome-wide association study; haplotypes; population stratification; statistical genetics.

1. Fundamental Concepts in Human Genetics

Homo sapiens are diploid organisms with most nonreproductive cells containing 23 pairs of chromosomes. These 23 pairs of chromosomes can be partitioned into 22 pairs of autosomes and 1 pair of sex chromosomes. Genomic sexual dimorphism exists in humans, with females having two X chromosomes and males having an X and a Y chromosome. Although chromosomes X and Y differ in size and content, there are 2 pseudoautosomal regions (i.e., PAR1, PAR2) in which the chromosomal regions are homologous. The centromere divides each chromosome into 2 arms (p for the shorter arm, q for the longer arm). The notation sequentially identifies the chromosome, short (p) or long (q)

From: *Methods in Molecular Biology, vol. 404: Topics in Biostatistics*
Edited by: W. T. Ambrosius © Humana Press Inc., Totowa, NJ

arm of the chromosome, the chromosomal region, and band within region. For example, 1q42 represents the portion of the genome about the second band within the fourth region on the long arm of chromosome 1.

A chromosome is composed of 2 strands of deoxyribonucleic acid (DNA) that are in the form of a double helix and are held together by weak hydrogen bonds. The individual strands of DNA are composed of a sequence of deoxyribose sugar. Attached to each sugar is one of four bases: adenine (A), guanine (G), cytosine (C), and thymine (T). Adenine and guanine are purines, while cytosine and thymine are pyrimidines. Adenine bonds with thymine and guanine bonds with cytosine. In this sense, the 2 strands of DNA are complementary (i.e., A is at the complementary position of T and G is at the complementary position of C), and a base with its complement base forms a base pair.

A gene is defined to be a specific sequence of bases within a specific region or locus of the genome. A gene may span a few to a several thousand kilobases (kb; where 1 kb = 1000 bases). Not all of the DNA in a gene codes for proteins. Rather, there are units of DNA called *exons* that are transcribed into RNA, then into mRNA (messenger RNA), and ultimately translated into a sequence of amino acids that form a specific protein. Interspersed between pairs of exons are noncoding regions called *introns*. A *codon* is composed of 3 bases that collectively code for a specific amino acid. Thus, a single change in a base of the DNA changes the codon and therefore may change the amino acid. Changing the amino acid sequence may change the properties of the protein the gene produces. In addition to coding for amino acids, there exist codons that start and stop translation.

There are 2 cell division processes: mitosis and meiosis. *Mitosis* is 1 of 4 phases in the cell cycle and is designed to identically reproduce the diploid somatic cell. In contrast, *meiosis* has 2 additional phases, occurs only in germ cells, results in haploid cells (i.e., egg and sperm), and increases genetic diversity through sexual reproduction. During meiosis, homologous chromosomes pair up and are connected to and aligned with each other by chiasmata. Chromosomes X and Y pair up and align only relative to the 2 pseudoautosomal regions. Each chromosomal arm usually has at least 1 chiasma. At chiasmata, the homologous chromosomes are in close physical proximity and may exchange homologous segments of the chromosome in a process called recombination. Here, specific enzymes cut the chromosomes and reconnect the aligned partner chromosomal segment at the chiasma. Thus, there is an exchange of genetic material from the maternally and paternally derived chromosomes before the formation of the germ cell is complete. The evolutionary significance of recombination is the resulting increase in genetic diversity.

Although the vast majority of DNA is the same within a population, there exists important variation. If the DNA at a specific base or region of the genome varies within the population, then we define the locus as polymorphic. For a defined locus, an allele is a particular polymorphism (i.e., sequence of DNA). A single nucleotide polymorphism (SNP) is a simple but extremely important example of a genetic polymorphism.

The inheritance of genetic factors follows 2 basic laws first postulated by Gregor Mendel *(1)*. The *law of segregation* states that alleles along chromosomes are transmitted from parent to offspring through gametes and that each gamete contains only 1 copy of the allele. The *law of independent assortment* states that if 2 loci are not physically close on the same chromosome, then they segregate independently of each other.

Genetic factors are known to influence variation in traits. However, it must be understood that genetic factors influence variation in traits in the context of many other factors. Specifically, an individual's phenotype is the result of complex interactions among environmental, behavioral, and genetic factors.

2. Estimation of Familial Aggregation

If a phenotype is influenced by genetic factors, then the phenotype should exhibit familial clustering. That is, related individuals should appear more similar to each other than unrelated individuals. Familial aggregation is the tendency for a trait to cluster within a family. Family studies are important tools in establishing a genetic component of a disease. For example, in a twin study, one might estimate and compare the probabilities of disease for monozygotic and dizygotic twins. Although twin studies are powerful tools for estimating familial aggregation, they do have weaknesses *(2)*. Alternatively, one might compute the probability of disease among relatives of cases and controls as

$$\frac{\Pr(\text{disease in relative type } R \text{ of case} \mid \text{case})}{\Pr(\text{disease in relative type } R \text{ of control} \mid \text{control})}.$$

The numerator in this equation reads, "The probability of a random relative of type R (e.g., sibling) of a case has the disease, given the case has the disease." If there is a genetic contribution to the risk for disease, then this ratio will be greater than 1. The larger this ratio, the greater the familial aggregation. Risch *(3)* defined the recurrence risk ratio for relative type R as

$$\lambda_R = \frac{\Pr(\text{disease in relative of type } R \mid \text{case})}{K}.$$

Here, K is the population prevalence (probability of disease for a random individual). The most commonly reported estimate of the recurrence risk ratio is the sibling risk ratio.

Table 1 provides estimates of the sibling risk ratio for a range of complex genetic diseases. Among these diseases, systemic lupus erythematosus, multiple

Table 1
Estimated Familial Aggregation of Some Complex Genetic Diseases

Disease	MZ/DZ[a] twin concordance ratio	Sibling recurrence risk ratio
Systemic lupus erythematosus	11.3	20
Multiple sclerosis	7.5	20
Crohn disease	5.3	20
Type 1 diabetes mellitus	3.1	15
Ulcerative colitis	3.3	12
Rheumatoid arthritis	6.9	8
Chronic fatigue syndrome	2.9	6
Psoriasis	2.7	6
Type 2 diabetes mellitus	1.7	3
Prostate cancer	3.7	2.5

[a]Concordance ratio: ratio of fraction of twin pairs with both twins affected or both twins unaffected for MZ (monozygotic) to DZ (dizygotic) twins.

sclerosis, and Crohn disease exhibit the largest estimates of familial aggregation, whereas type 2 diabetes mellitus and prostate cancer exhibit significant but smaller estimates of familial aggregation. The above estimates of familial aggregation do not remove the confounding effects of shared environment and can be influenced by genetic and nongenetic factors such as age, gender, and environmental risk exposures.

3. Estimation of Allele Frequency

A central tool in the examination of genetic polymorphisms is the estimation of allele frequencies. For a particular allelic form, its allele frequency is simply the proportion of all alleles of that type or form. In a diploid organism, the genotype is composed of the 2 observed alleles. Homozygous individuals have the same 2 alleles, and heterozygous individuals have 2 different alleles. Therefore, the allele frequency is calculated as the proportion of alleles (not people) of a given type. That is, the numerator is simply twice the number of individuals homozygous for the allele of interest plus the number of individuals that are heterozygous for that allele, and the denominator is twice the number of individuals in the sample (i.e., number of alleles). Consider N individuals sampled from a population. At a particular locus, this population has two alleles (**A**, **a**). Assume that **A** is the more common allele (often denoted the major allele), and **a** is the less common allele (often denoted as the minor allele). Let N_{AA}, N_{Aa}, and N_{aa} be the number of individuals homozygous for the major allele, heterozygous for these alleles, and homozygous for the minor allele, respectively. Let p_1 and p_2 be the major and minor allele frequencies, respectively, such that $p_1 + p_2 = 1$. Then p_1 can be estimated as

Table 2
Calculation of Allele Frequencies

	Genotype			
	A/A	**A/a**	**a/a**	Total
Number of individuals	121	55	24	200
Number of **A** alleles	242	55	0	297
Number of **a** alleles	0	55	48	103
Total number of alleles	242	110	48	400

$$\hat{p}_1 = [2(N_{AA}) + N_{Aa}]/2N$$

and p_2 is estimated as

$$\hat{p}_2 = 1 - \hat{p}_1 = [2(N_{aa}) + N_{Aa}]/2N.$$

The sampling variance for p_1 is estimated by $\mathrm{Var}(\hat{p}_1) = \hat{p}_1(1 - \hat{p}_1)/2N$. These estimates are the familiar binomial estimates (see **Chapter 3**) and are maximum likelihood estimates. Often, p_2 is denoted as q. Consider the hypothetical example in **Table 2**. Assume that 200 individuals are sampled from a population with 2 alleles observed at a particular locus. That locus is said to be diallelic. Then the allele frequency for **A** is estimated to be $\hat{p}_1 = (242 + 55)/400 = 0.74$ and for **a** is $\hat{p}_2 = 1 - \hat{p}_1 = 0.26$. If the locus is not diallelic and has more than 2 alleles (e.g., J alleles), the calculations are computed in the same manner, but we denote the estimated allele frequencies as $\hat{p}_1, \hat{p}_2, \hat{p}_3, \ldots \hat{p}_J$, where $\hat{p}_1 + \hat{p}_2 + \hat{p}_3 + \ldots + \hat{p}_J = 1$.

Heterozygosity is a measure of the allelic variation at a locus. For J alleles, heterozygosity is defined as

$$H = 1 - \sum_{i=1}^{J} \hat{p}_i^2.$$

Thus, H is the probability that a randomly sampled individual is heterozygous for any 2 alleles, and H reaches its maximum when the alleles are equally frequent. For the example in **Table 2**, $H = 1 - [(0.74)^2 + (0.26)^2] = 0.38$.

4. Hardy-Weinberg Principle

It is often important to contrast the observed versus expected genotype frequencies within and between samples. The computation of these expected genotype frequencies are simple under the Hardy-Weinberg principle ***(4,5)***. For autosomal loci, the principle makes the following 8 assumptions:

1. diploid organism;
2. sexual reproduction;

3. discrete generations;
4. random mating with respect to the locus of interest;
5. infinite population size;
6. no migration;
7. no selection; and
8. no mutation.

Under these 8 assumptions, for a diallelic autosomal locus (**A** and **a**) with allele frequencies p_1 and p_2 $(p_1 + p_2 = 1)$, the genotype frequencies of **AA**, **Aa**, and **aa** are p_1^2, $2p_1p_2$, and p_2^2, respectively. In a large population, the genotype frequencies will reach equilibrium in 1 generation of random mating. That is, all subsequent generations will have the same genotype frequencies unless there are violations in these assumptions. This equilibrium is called Hardy-Weinberg equilibrium (HWE), and it plays a critical role in many population genetics applications and statistical tests. Although the assumptions appear to be highly restrictive, in practice it appears that the Hardy-Weinberg principle or its derivatives are highly robust to departures from all of the assumptions except random mating relative to the locus of interest.

It is relatively easy to test whether the observed genotype frequencies are within sampling variation of the expected genotype frequencies under HWE. Again, consider a diallelic marker with alleles **A** and **a**. The corresponding allele frequencies are p_1 and p_2. Under HWE, the expected genotype frequencies of **AA**, **Aa**, and **aa** are p_1^2, $2p_1p_2$, and p_2^2, respectively. The corresponding expected genotype counts are Np_1^2, $2Np_1p_2$, and Np_2^2, respectively, where N is the number of individuals in the sample. The observed and expected genotype counts can be compared via a standard chi-squared goodness-of-fit test. Specifically, let N_{AA}, N_{Aa}, and N_{aa} be the observed genotype counts $(N = N_{AA} + N_{Aa} + N_{aa})$. Then

$$
\begin{aligned}
X_{obs}^2 &= \sum \frac{(\text{observed}_i - \text{expected}_i)^2}{\text{expected}_i} \\
&= \frac{(N_{AA} - Np_1^2)^2}{Np_1^2} + \frac{(N_{Aa} - Np_1p_2)^2}{Np_1p_2} + \frac{(N_{aa} - Np_2^2)^2}{Np_2^2}.
\end{aligned}
$$

Here, X_{obs}^2 has an asymptotically chi-squared distribution with 1 degree of freedom. In practice, we replace the allele frequency parameters with their estimated values. The extension to multiple alleles is straightforward with the degrees of freedom equal to the number of possible genotypes minus the number of alleles. Recently, more computationally efficient exact tests have been developed ***(6,7)***. It is important to note that departures from HWE proportions may reflect real biologic effects or genotyping errors ***(8)***. Therefore, under a case-control design, it is often informative to compute these tests separately

Table 3
Computations for Hardy-Weinberg Equilibrium Proportions

Genotype	Observed count	Expected count	Contribution to chi-squared statistic
aa	74	107.9	10.65
Aa	554	486.2	9.45
AA	514	547.9	2.10
Total	1142	1142	22.20

for cases and controls. Extensions to X-linked markers and multiple loci are available ***(9)***.

As an example of testing for departures from HWE proportions, consider the following example (**Table 3**). A sample of 1142 individuals are sampled from a population, and a diallelic marker (**A**, **a**) is genotyped. For the observed counts in **Table 3**, the observed chi-squared statistic is $X^2_{obs} = 22.20$. With 1 degree of freedom, this corresponds to a P value equal to 0.00000246. Thus, there is evidence that the genotypes are not consistent with HWE proportions. The departure manifests as an excess number of heterozygotes and a deficit of both homozygotes. A common departure from Hardy-Weinberg equilibrium proportions results from the genotyping technology failing to consistently detect heterozygotes.

5. Linkage Disequilibrium

Linkage disequilibrium (LD) is the tendency for alleles at 2 linked loci to exhibit a statistical association. Thus, there are 2 components to LD. First, the 2 loci must be linked. Two loci are linked if they are sufficiently close together such that the probability of a recombination between the 2 loci is less than ½. We denote the probability of a recombination as $\theta (0 \le \theta \le \frac{1}{2})$. Second, the alleles show a statistical dependence, often denoted by δ. To quantify the degree of linkage disequilibrium, it is helpful to define a haplotype. A haplotype is the set of alleles at loci on the same chromosome inherited from an individual parent.

There are numerous statistics designed to measure the level of LD between 2 loci. Devlin and Risch ***(10)*** and Guo ***(11)*** provide useful reviews of several statistics under the assumption of Hardy-Weinberg equilibrium. For consistency, we use their notation. Consider 2 linked diallelic loci (**Table 4**): the disease-predisposing locus and a marker locus in close physical proximity to the disease-predisposing locus. In **Table 4**, p_{11} corresponds with the marker locus allele **A** and disease locus allele haplotype frequency. The other 3 haplotype frequency parameters are p_{12}, p_{21}, and p_{22}. The marker allele frequency for

Table 4
Parameters for Measures of Linkage Disequilibrium: Haplotype and Allele Frequencies

	Disease locus		
Marker **A**	Disease allele	Normal allele	
A	p_{11}	p_{12}	p_{1+}
a	p_{21}	p_{22}	p_{2+}
	p_{+1}	p_{+2}	

A is p_{1+}, and the disease allele frequency is p_{+1}. The classical definition of the LD coefficient is ***(12)***

$$D = p_{11} - p_{1+}\, p_{+1}.$$

Here, D measures the deviation between the haplotype frequency and its expectation under independence. Please note that in tests of association discussed later in the chapter, D is also denoted δ to be consistent with the cited literature. It is known that LD decays over generations due to recombination. The rate of decay can be shown to be $D_t = D_0(1-\theta)^t$, where $\theta(0 \le \theta \le \frac{1}{2})$ is the recombination fraction, t is the number of generations, D_0 is the level of LD in the initial generation, and D_t is the level of LD at generation t. Clearly, the more tightly linked 2 markers are (i.e., the smaller θ), the slower the decay in LD. However, as long as $\theta > 0$ (i.e., not completely linked) and the number of generations is sufficiently large, the 2 loci will effectively be in linkage equilibrium, $D_t = 0$.

The ability to cheaply genotype large numbers of single nucleotide polymorphisms (SNPs) has motivated the investigation of the patterns of LD across the genome. This investigation estimates marker-marker LD based on 2 variations of D. It is important to recognize that statistics based on D are dependent on allele frequencies, sample size, and evolutionary history (e.g., mutation rates). For example, estimates of LD tend to be biased upward in small samples. Although not completely successful, there are 2 common measures of LD that attempt to minimize allele frequency dependence. The first is due to Lewontin ***(13)*** and is defined as

$$D' = \begin{cases} \dfrac{D}{\min(p_{1+}p_{+2},\, p_{+1}p_{2+})}, & \text{for} \quad D > 0 \\[2ex] \dfrac{D}{\min(p_{1+}p_{+1},\, p_{+2}p_{2+})}, & \text{for} \quad D < 0. \end{cases}$$

The denominator of D' is the absolute maximum D could achieve given the observed allele frequencies. Thus, D' is the proportion of the maximum value of $D(0 \leq D' \leq 1)$. Large values of D' imply greater marker-marker LD. Values of $D' = 1$ imply that there are fewer observed haplotypes than theoretically predicted under independence. Conversely, $D' = 0$ implies linkage equilibrium.

Another commonly used measure of marker-marker LD is Δ, the correlation coefficient from a 2 × 2 contingency table. Assuming $D > 0$, Δ is defined as

$$\Delta = \frac{D}{\sqrt{p_{1+}p_{2+}p_{+1}p_{+2}}},$$

and ranges between 0 and 1. In practice, Δ is often squared and denoted as $r^2(0 \leq r^2 \leq 1)$. r^2 will be 1 only when $D' = 1$ and the allele frequencies at the 2 markers are identical. The r^2 statistic has a particularly useful property. Ardlie and others ***(14)*** state, "To have the same power to detect the association between the disease and the marker locus, the sample size must be increased by approximately Δ^2 when compared with the sample size for detecting association with the susceptibility locus itself." Note that here, Δ^2 is the LD between the marker locus and the susceptibility locus. As a rough rule of thumb, marker-marker LD values of $r^2 > \frac{1}{3}$ are often sited as useful for fine mapping. There is some evidence that r^2 may replicate better than D' within and between populations ***(15)***; albeit, there always exist population-specific evolutionary factors.

Numerous studies have investigated the variation and extent of LD across the genome. Ardlie and others ***(14)*** provide a nice summary. These empirical studies suggest that there is less LD over short distances and more over long distances than theoretical expectations predict ***(16)***. In fact, there appear to be islands of high levels of LD, referred to as blocks, separated by regions of low levels of LD. These blocks of LD exhibit low haplotypic diversity, and most of the common haplotypes can be defined by a relatively small number of markers (e.g., 3 to 5 SNPs). Whether block boundaries identify hot spots of recombination or merely old recombination events is of considerable debate and is likely region specific.

The Daly ***(17)*** study of a 500-kb region on chromosome 5q31 in 256 patients with Crohn disease provides an illustrative example of the paucity of haplotypes within blocks (**Table 5**). In the extreme, Daly and others ***(17)*** report that the seventh block spanned 92 kb and contained 31 SNPs, but only 4 of the 2^{31} possible haplotypes were observed in 94% of the chromosomes. Current estimates of the extent of LD appears to be in the neighborhood of 10 to 30 kb (22 kb) in individuals of European descent, less in individuals of African-American descent (~8 kb), and intermediate but more similar to individuals of European descent for other ethnicities.

Table 5
Haplotypic Diversity Across 11 Blocks at 5q31 in 256 Patients with Crohn Disease *(17)*

Block	Block size (kb)	Number of SNPs	Expected number of haplotypes	Observed number of common haplotypes	Proportion of chromosomes that match common haplotypes
1	84	8	256	2	0.96
2	3	5	32	2	0.97
3	14	9	512	3	0.92
4	30	11	2,048	4	0.94
5	25	5	32	4	0.93
6	11	5	32	4	0.97
7	92	31	2,147,483,648	4	0.93
8	21	7	128	4	0.91
9	27	6	64	4	0.92
10	55	7	128	3	0.90
11	19	5	32	3	0.98

An important application of LD is that a reduced number of genetic markers can often capture the vast majority of common genetic variation. Thus, depending on the stringency of the algorithm to choose SNPs, it is possible to genotype a reduced number of SNPs and have confidence that all but the most uncommon variants have been tested for association with a trait. As noted above, Ardlie and others ***(14)*** show that the level of LD is a critical factor in the power of these tests. The HapMap Project ***(18)*** seeks to capitalize on this phenomenon through a massive resequencing effort designed to identify block structures and a minimal number of genetic markers that capture the majority of haplotypic variation.

6. Testing for Genetic Association

Linkage disequilibrium requires the 2 loci to be linked and exhibit an association. Therefore, in genetic association studies, the 3 hypotheses of interest are $\delta \neq 0$ (association, recalling $D = \delta$ above), $\theta < \frac{1}{2}$ (linkage), and $\delta(1 - 2\theta) \neq 0$ (linkage disequilibrium). In the following discussion, consider a single diallelic disease locus with disease-predisposing and normal alleles **D** and **d**, respectively. Likewise, consider a single diallelic marker locus with codominant alleles **M** and **m**.

6.1. Association Methods for Unrelated Individuals

Association studies with unrelated individuals can test the first hypothesis, $\delta \neq 0$, but cannot test the second or third hypotheses. The key to an appropriate analysis with unrelated individuals is that all individuals come from the same genetic population, thereby avoiding the confounding effects of differential population ancestries (i.e., admixture, population stratification; see below). In the absence of population stratification, evidence of association can be interpreted as evidence for linkage disequilibrium between the marker and the disease locus.

The simplest design for unrelated individuals is the case-control design, where affected cases are compared with unaffected controls with no adjustment for covariates. Statistical tests can be based on standard contingency table tests for independence. To test for independence between the marker alleles and case or control status using N cases and N controls, the following table can be constructed (**Table 6**). Here, X^2_{Allele} is compared with a chi-squared distribution with 1 degree of freedom, χ^2_1, to determine the statistical significance. Note that this test assumes HWE so that under the null hypothesis of no association, the 2 alleles from each individual are independent.

Similar tables can be constructed to compare genotypic frequencies between cases and controls, either as the 2 d.f. test across all 3 genotypic categories from a 2 × 3 contingency table or as tests for association under a specific genetic model (i.e., dominant, additive, recessive). **Table 7** provides an example for a test under a dominant model and is merely the standard chi-squared test for a 2 × 2 contingency table. The recessive model is constructed analogously. A Cochran-Armitage trend test can be used to test for an additive genetic effect ***(19)***. These tests, plus tests for HWE and haplotype analysis (see below), are computed using the software package SNPGWA (available free of charge at http://www.phs.wfubmc.edu/public_bios/sec_gene/apps.cfm).

Table 6
Table for Case-Control Design Comparing Allele Frequencies

	Allele **M**	Allele **m**	Total
Case	w	x	$2N$
Control	y	z	$2N$
Total	$w + y$	$x + z$	$4N$

$$X^2_{\text{Allele}} = \frac{4N(wz - xy)^2}{(w+y)(x+z)4N^2}.$$

Table 7
Comparing Genotypic Frequencies Under Dominant Model for Case-Control Design

	Genotype **MM** or **Mm**	Genotype **mm**	Total
Case	w	x	N
Control	y	z	N
Total	$w + y$	$x + z$	$2N$

$$X^2_{\text{Dominant}} = \frac{2N(wz - xy)^2}{(w + y)(x + z)2N^2}.$$

Simple linear regression can be used to test for an association between a genetic marker and a quantitative trait without adjusting for covariates. Let y_i be the phenotypic measure of the ith ($i = 1, 2, \ldots n$) individual. Let $\boldsymbol{x}_i$ be the covariates corresponding with the desired genetic test of association (e.g., 2 degree of freedom genotypic test, or dominant, additive, recessive genetic model). For the dominant, additive, and recessive genetic models, a simple linear model can be fit to the data:

$$y_i = \beta_0 + \boldsymbol{\beta}_1\boldsymbol{x}_i + \varepsilon_i.$$

Here, $\boldsymbol{x}_i$ is a dummy variable for the dominant and recessive models and a 3-level ordinal categorical variable under the additive model. A test of $\beta_1 = 0$ is a test of H_0: no association between the marker locus and the trait of interest. Obviously, for the 2 d.f. test of overall genotypic association, two dummy variables are required, and a joint test of $\beta_1 = 0$ and $\beta_2 = 0$ is used to test for association.

If other phenotypic or genetic data is available, there may be an interest in computing tests of association that adjust for these factors. Adjusting for covariates in tests of genetic association when the samples are of independent individuals can be accomplished under a generalized linear model framework. Such an approach allows the simultaneous and conditional tests of the genetic factors and nongenetic factors. Two familiar examples include logistic regression and linear regression. For binary traits such as disease status, a logistic regression model can be employed. When the trait (perhaps after transformation) is approximately normally distributed conditional on the covariates and genetic factors and it exhibits homogeneity of variance, the standard linear model can be used. The use of standard model diagnostics, including influence statistics, is critical for valid statistical inference from these models. These methods are easily implemented in widely available statistical software packages. More generally, for unrelated individuals, a genetic risk factor can essentially be treated as any other categorical variable in any analysis for which the data collected are appropriate (e.g., survival analysis).

6.2. Haplotype Analysis

The detection of association via linkage disequilibrium relies on the assumption that a putative disease-predisposing allele resides on 1 or more unique haplotypes. With each generation, and with a probability that is a nondecreasing function of genetic distance, recombination events randomly shuffle alleles at marker loci near the disease-predisposing locus. Over time, the decay of LD is a function of both the genetic distance (the recombination fraction) and the number of generations since the origin of the haplotype. Thus, nearby marker alleles present on the ancestral haplotype(s) tend to be associated with the susceptibility allele (i.e., linkage disequilibrium). If the disease-predisposing locus is located within "ancient" haplotypes, the region of the genome that exhibits detectable LD with the disease will be small.

It is not always possible to determine with certainty the haplotypes for an individual based solely on the genotypic data. Specifically, if an individual is heterozygous at more than 1 marker included in the haplotype, it is not possible to unambiguously assign haplotypes for that individual. This ambiguity is referred to as unknown linkage phase. The most commonly used statistical tests for comparing the haplotype frequencies between cases and controls are the likelihood ratio and score tests. Both tests differ when phase is known versus when phase is not known.

When phase is known with certainty, a standard likelihood ratio test can be applied. Specifically, let h_i be the haplotype pair for the ith individual, where without loss of generality individuals 1, . . . , N are cases and individuals N + 1, . . . , $2N$ are controls. Haplotype frequencies in the three samples (i.e., $\hat{f}_{cases}$, $\hat{f}_{conrols}$, and $\hat{f}_{combined}$) are estimated by simply counting the number of each haplotype in the sample, which corresponds with the maximum likelihood estimator from a multinomial distribution. Thus, the likelihood ratio test for the phase-known case is

$$L_1 = 2\log\left[\frac{\prod_{i=1}^{N}\Pr(h_i \mid \hat{f}_{cases})\prod_{j=N+1}^{2N}\Pr(h_j \mid \hat{f}_{controls})}{\prod_{k=1}^{2N}\Pr(h_k \mid \hat{f}_{combined})}\right].$$

Asymptotically, L_1 has a chi-square distribution with degrees of freedom equal to 1 less than the number of distinct haplotypes in the combined case-control sample. Alternatively, L_1 can also be evaluated using a permutation test.

When phase is not known, the expectation-maximization (EM) algorithm is used to estimate haplotype frequencies for the 3 sets of samples (i.e., $\tilde{f}_{cases}$, $\tilde{f}_{controls}$, and $\tilde{f}_{combined}$) as outlined in Excoffier and Skatkin *(20)*. For the phase unknown case, the likelihood ratio statistic takes the form:

$$L_2 = 2\log\left[\frac{\prod_{i=1}^{N}\Pr(g_i|\tilde{f}_{cases})\prod_{j=N+1}^{2N}\Pr(g_j|\tilde{f}_{controls})}{\prod_{k=1}^{2N}\Pr(g_k|\tilde{f}_{combined})}\right],$$

where g_i represents the multilocus genotypes in the *i*th individual. The primary distinction between L_1 and L_2 is that L_2 is based on the conditional probability of the genotypes instead of the conditional probability of the haplotype pair. Asymptotically, L_2 has approximately a chi-square distribution with degrees of freedom approximately equal to 1 less than the number of distinct haplotypes estimated by the EM algorithm to have nonzero probability in the combined sample of cases and controls. The approximate distribution results from the fact that the haplotype frequencies, the "unobserved complete data," are estimated from the multilocus genotypes, the "observed incomplete data." The statistical significance of L_2 should be evaluated using a permutation test framework in which the EM algorithm is recomputed for each permutation of these data. In practice, the large sample test can be used as a first approximation to the statistical significance of the comparison, and this approximation can subsequently be used to determine the number of permutations that are needed in the permutation test. The program Dandelion *(21)* computes these tests and is freely available at http://www.phs.wfubmc.edu/public_bios/sec_gene/apps.cfm.

Schaid *(22)* reviews the studies that have compared the power of single-marker association tests to haplotype tests for association. The comparisons made have largely been between the omnibus test for haplotype frequency differences discussed above and the largest of the single-marker statistics after Bonferroni correction for the number of tests performed. Haplotype analyses have the potential to be *more* powerful than single-SNP analyses when the haplotype is more correlated with the underlying locus than any single marker or multiple alleles at different markers are required to confer disease risk, and those alleles need to be on the same chromosome *(23)*. Haplotype analyses have the potential to be *less* powerful than single-SNP analyses when a single SNP is in near perfect LD with the predisposing locus and the other markers fragment that association or there are many more haplotypes than single SNPs.

Similar to single-marker analyses, haplotype analyses can be extended in a modeling framework. For example, consider 2 marker loci, the first with alleles $\mathbf{M_1}$ and $\mathbf{m_1}$, the second with $\mathbf{M_2}$ and $\mathbf{m_2}$. If phase is known, a logistic model with an additive effect of each haplotype can be formulated as follows:

$$\text{Logit}(p) = \beta_0 + \beta_1 x_{M_1M_2} + \beta_2 x_{M_1m_2} + \beta_3 x_{m_1M_2}$$

with design matrix shown in **Table 8**.

Other covariates and interaction terms are easily incorporated into such a model. Implementing this model using standard statistical software requires first estimating haplotype frequencies, separately for cases and controls, and then using the most likely haplotype for each individual (based on the posterior probability of each haplotype combination given the individual's genotypes and the haplotype frequency estimates). This simple solution ignores the variability due to unknown linkage phase, however, so parameter estimates may be biased and statistical tests may have elevated type I error rates ***(24,25)***. An alternative approach is to estimate haplotype frequencies jointly with the other parameters in the regression model ***(26,27)***. Schaid and others ***(26)*** implement this strategy in the Haplo.Stats package and use score tests for association.

An association model that uses haplotypes is simply 1 specific model for multiple loci. Other models can incorporate multiple loci in the same gene or across multiple genes. For example, Conti and Gauderman ***(28)*** describe several multimarker models, some of which ignore linkage phase, so that unphased genotypic data can be used to estimate the model parameters. Estimation of these model parameters allows for a wide variety of tests for association between combinations of markers and the outcome of interest.

Table 8
Design Matrix for Additive Haplotype Model

Haplotype pair	$x_{M_1M_2}$	$x_{M_1m_2}$	$x_{m_1M_2}$
$\mathbf{M_1M_2/M_1M_2}$	2	0	0
$\mathbf{M_1M_2/M_1m_2}$	1	1	0
$\mathbf{M_1M_2/m_1M_2}$	1	0	1
$\mathbf{M_1M_2/m_1m_2}$	1	0	0
$\mathbf{M_1m_2/m_1M_2}$	0	1	1
$\mathbf{M_1m_2/M_1m_2}$	0	2	0
$\mathbf{M_1m_2/m_1m_2}$	0	1	0
$\mathbf{m_1M_2/m_1M_2}$	0	0	2
$\mathbf{m_1M_2/m_1m_2}$	0	0	1
$\mathbf{m_1m_2/m_1m_2}$	0	0	0

Table 9
Population Stratification[a]

Sample	Presence of allele **M**	No. of persons	No. affected	% affected
White	Yes	500	25	5.0
	No	500	25	5.0
	All	1000	50	5.0
Black	Yes	100	1	1.0
	No	900	9	1.0
	All	1000	10	1.0
Combined sample	Yes	600	26	4.3
	No	1400	34	2.4
	All	2000	60	3.0

[a]From Khoury *(29)*.

7. Population Stratification

The primary disadvantage of using unrelated individuals for genetic association studies is the potential confounding effects of population stratification on tests of association. If cases and controls are not carefully matched on ancestry and there are differences in the outcome (e.g., disease rates or mean levels of quantitative outcome) between ancestral populations present in the sample, then spurious associations may result. **Table 9** shows how a spurious association with allele **M** at a marker may result if the frequency of both disease and the **M** allele are increased in one sample compared with the other.

Population stratification may be due to the presence of subpopulations with differing rates of disease and allele frequencies, as in the above example. Population stratification may also arise due to admixture, where the sample is made up of individuals each with differing fractions of ancestry from distinct populations (e.g., African-American individuals with differing levels of Caucasian ancestry) and the average level of genetic admixture differs between cases and controls.

There are three basic strategies for dealing with potential population stratification: carefully match cases and controls at the design stage, detect and control for population stratification in the analysis stage, or use family-based sampling and analysis strategies. The 2 main analytic strategies used to control for spurious association due to population stratification are the structured association approach [SA *(30,31)*] and the genomic control approach [GC *(32)*]. Under the assumption that population stratification will be present at markers spread throughout the genome, each approach uses a set of unlinked, anonymous markers to estimate either subpopulation membership (or admixture fraction, SA) or the factor by which the standard chi-square statistic is inflated due to population stratification (GC). Advocates of the SA approach argue for stratifying the analysis by identified subpopulation, or adjusting for the admixture fraction in a linear model. The GC approach simply deflates each observed chi-squared statistic by the (assumed constant) estimated inflation factor. This is a very active area of research (e.g., Refs. ***33, 34, 35, 36***).

8. Family-Based Methods of Association

Family-based tests of association are those that use family members as controls. In general, these tests have also become synonymous with protection from false-positive results due to population stratification, although the earliest tests did not confer such protection. Below we develop some of the most historically important family-based tests of association and describe more recent tests.

8.1. Affected Family-Based Control Designs

The affected family-based controls design (AFBAC) ***(37,38)*** is an extension of the genotype-based haplotype relative risk (GHRR) design ***(39,40)***, which uses the parental alleles not passed to an affected child as a pseudo-genotype for a control population. Terwilliger and Ott ***(37)*** note that under the null hypothesis of no association, when there is no population stratification, the transmission of all 4 parental marker alleles is independent. Using this fact, they propose scoring each of the $4N$ alleles, instead of the genotypes, as having been transmitted or not transmitted to the affected offspring. Using the example of the parent-child trio (child **Mm**), the father (**Mm**) transmits an **m** allele and does not transmit an **M** allele, and the mother (**MM**) both transmits and does not transmit an **M** allele. Note that this approach does not require heterozygous parents. These data are organized as in **Table 10**, and the proposed statistical test is the standard contingency table test for independence. Terwilliger and Ott ***(37)*** denote the corresponding odds ratio as the haplotype-based haplotype

Table 10
Table for AFBAC Design

	Allele **M**	Allele **m**	Total
Transmitted allele	w	x	$2N$
Nontransmitted allele	y	z	$2N$
Total	$w + y$	$x + z$	$4N$

$$\chi^2_{AFBAC} = \frac{4N(wz - xy)^2}{(w+y)(x+z)4N^2}.$$

relative risk (HHRR) statistic. This design is also known as the affected family-based controls (AFBAC) ***(38)***.

For a valid test of association, the AFBAC design requires the parental genotypes to be available and in HWE. It only has power to detect an association in the presence of linkage ($\theta < \frac{1}{2}$) and uses all parental genotypes. The AFBAC design assumes that only a single affected individual, randomly selected from the affecteds in a family, is used in the analysis. The use of multiple affected siblings introduces correlation among the observations and invalidates the use of the chi-squared distribution for the contingency statistic; generalized linear mixed models [GLMM ***(41)***], generalized estimating equations [GEE ***(42)***], and permutation tests can provide alternative approaches to this problem. The use of family members as controls is an important step in the evolution of association methodology. The clear advantage of the AFBAC and related approaches is that these procedures guarantee that for each case there exists a control subject sampled from the identical genetic population. However, the AFBAC method uses an unmatched analysis scheme for an inherently matched design and is still subject to the confounding effects of population stratification.

8.2. Transmission/Disequilibrium Tests

Terwilliger and Ott ***(37)*** and Spielman and others ***(43)*** proposed applying a matched analysis to the AFBAC design as a test for *linkage* in the presence of association that is due to linkage disequilibrium. This test is known as the transmission/disequilibrium test (TDT), and it tests the joint hypothesis that $\delta(1 - 2\theta) = 0$. That is, the TDT tests whether either $\delta = 0$ (no disequilibrium) or $\theta = \frac{1}{2}$ (no linkage), because if either are true, the test statistic will have the null distribution.

Treating each affected child-parent pair as a stratum in a matched analysis leads to a conditional analysis ***(44,45)***; here we condition on the parental genotypes. The advantage of matching within an individual and computing the matched (conditional) analysis is that it destroys/eliminates the confounding between association and population structure by only tracking the joint transmission of a marker allele and a disease-predisposing allele *within* a person. Lazzeroni and Lange ***(46)*** list the genetic assumptions necessary for the validity of the TDT. Conditional upon the parental genotypes, the mother's and father's transmissions to a child are independent; alleles transmitted to different children are independent; and the probability of transmitting the first allele at a locus and not the second equals the probability of transmitting the second allele and not the first—this probability equals $\frac{1}{2}$. In addition, under $H_0{:}\theta = \frac{1}{2}$ and regardless of the value of δ, a parent will transmit the marker allele independent of the

Table 11
Possible Outcomes Within a Stratum (Parent-Child Pair)

	M	**m**	**M**	**m**	**M**	**M**	**M**	**m**	Total
Transmitted (case)	1	0	1	0	0	1	0	1	1
Nontransmitted (control)	1	0	0	1	1	0	0	1	1
Total	2	0	1	1	1	1	0	2	2

allele that is transmitted at the disease locus and independent of the disease status of the child.

To construct the TDT statistic, consider each parent-child pairing as a stratum consisting of the 2 pairs of observations: a parent's transmitted (case) and non-transmitted (control) marker alleles. For each stratum, there are 4 possible outcomes for a diallelic marker locus with alleles **M**, **m** (**Table 11**). For each stratum, the (row and column) marginal totals are fixed, and the first and last subtables contribute no information about the odds ratio, because conditional on the marginal totals their probability is 1. Note that if both parents and the child are heterozygous at the marker locus, we cannot determine which parent transmitted which allele. However, we do know that together 1 parent transmitted the **M** allele while the other parent transmitted the **m** allele. Thus, we can include the joint contribution of the 2 parents by entering one in the second subtable and the other in the third subtable. If a family has multiple affected children, each child-parent trio is entered separately because, as noted above, they are independent conditional upon the parental genotypes. We summarize these data as in **Table 12**, where the contribution of the heterozygous parents is represented by the off-diagonal cells. Conditional on the total number of heterozygous parents, this distribution is binomial. These data are classically evaluated via McNemar's test *(47)* or equivalently the sign test, which are the uniformly most powerful unbiased tests *(45)*; we compare McNemar's test statistic to a chi-squared distribution with 1 degree of freedom. The conditional

Table 12
Transmission/Disequilibrium Test

		Nontransmitted allele		
		M	**m**	Total
Transmitted allele	**M**	a	b	w
	m	c	d	x
	Total	y	z	$2N$

maximum likelihood estimate of the odds ratio can be used as a measure of the magnitude of the relationship and is estimated by the ratio *b*/*c*. Corresponding confidence intervals about the estimated odds ratio can be computed ***(44)***.

The TDT can be used as a test for association as well as linkage. Recall that the TDT tests the joint hypothesis $\delta(1 - 2\theta) = 0$. To test for linkage we require $\delta \neq 0$. Conversely, provided we use only 1 affected child per family and there exists linkage ($\theta < \frac{1}{2}$), we can use the TDT to test for an association ($\delta \neq 0$). If more than 1 affected child per family is included in a test for association using the TDT and $\theta < \frac{1}{2}$, then the resulting sample no longer consists of independent binomial variables and the null distribution of the test statistic is no longer chi-squared with 1 degree of freedom.

The TDT assumes that Mendel's first law holds: that under the null hypothesis, transmission of each allele is equally probable. If there is transmission distortion, one allele may be seen in offspring more than 50% of the time (e.g., maternal-fetal incompatibility if offspring has certain allele). This phenomenon has been demonstrated recently at several places along the genome ***(48)***. One simple check of the assumption is to use the TDT to test for linkage and association in a sample of unaffected individuals and their parents. If the same linkage and association is seen in controls as cases, then transmission distortion should be suspected.

The contrasting of transmitted and nontransmitted alleles, genotypes, or haplotypes requires the availability of parental marker data. When parental marker data are missing for one parent, it is tempting to apply the TDT to the remaining parent-child pair or to attempt to deduce the missing parental genotype using the remaining heterozygous parent, the child and siblings. These methods can be useful, but they can also lead to biased tests. Curtis and Sham ***(49)*** illustrate that if only the child's and heterozygous parent's data are available, results based on the TDT will be biased by the allele frequencies in the population, unless the parent-child pair are discarded whenever the child either has the same genotype as the heterozygous parent or is homozygous.

An alternative construction of a proper control that is not dependent on parental data uses the unaffected full siblings of the affected individuals. Clearly, these controls are also from the identical genetic population as the affecteds. However, they are not as strong a contrast with the affecteds' marker genotypes, because they do not form a complementary partition of the parental genotype. Specifically, we would expect a tendency for unaffected siblings to carry the disease-predisposing allele more often than unrelated controls or the nontransmitted parental haplotype. As noted by Curtis ***(50)***, this should reduce the power of the resulting tests. Likewise, misclassification of unaffecteds (incorrect diagnosis, reduced penetrance) will reduce power.

The sib transmission/disequilibrium test (S-TDT) ***(51)*** and Discordant Allele Test [DAT ***(52)***] were developed for the comparison of allele frequencies between affected individuals and their unaffected siblings. The S-TDT design requires that in each family sampled there exist at least 1 affected and at least 1 unaffected sibling and that not all of the siblings have the same marker genotypes. These data consist of the total number of each allele (genotype) in the affecteds and the corresponding count in the unaffecteds; they are organized in a standard contingency table. The S-TDT includes concordant observations and treats families as strata. Because the observations from members of the same family are not independent, the authors propose a permutation test in which the affection status is randomly permuted across genotypes conditional on the within-family total number of affecteds. This approach preserves the marginal totals within each stratum (family), thus eliminating the potential confounding affects of population stratification. The significance of the desired test statistic is evaluated from the resulting (empirical) distribution. Lazzeroni and Lange's ***(46)*** permutation approach for multiple loci is a natural and useful extension of this approach. Under some models, using only the sibling pairs with discordant marker genotypes (DAT) can be more powerful than using all siblings. Unless the disease is highly penetrant and the disease-predisposing polymorphism is relatively common, the DAT is more powerful than the S-TDT for 1 affected and 1 unaffected sibling.

8.3. Pedigree Disequilibrium Tests

In many situations, particularly association studies that follow linkage studies, data are available both for some parents and some discordant siblings. The pedigree disequilibrium test [PDT ***(53,54)***] combines information from the TDT and S-TDT within families. For each trio in a family, let X_T = (number of copies of allele M transmitted) – (number of copies of allele M not transmitted). Similarly, for each discordant sib pair (DSP) in a family, let X_S = (number of copies of allele M in affected sib) – (number of copies of allele M in unaffected sib). For the ith pedigree with n_T triads and n_S DSPs, define

$$D_i = \frac{1}{n_T + n_S}\left(\sum_{j=1}^{n_T} X_{T_j} + \sum_{j=1}^{n_S} X_{S_j}\right).$$

Under H_0: no linkage or association, $E[X_T] = 0$ and $E[X_S] = 0$ for each family. Therefore, $E[D_i] = 0$. Let N be the total number of families and define

$$T_{PDT-avg} = \frac{\sum_{i=1}^{N} D_i}{\sqrt{\sum_{i=1}^{N} D_i^{\,2}}}.$$

Under H_0, $T_{PDT-avg} \sim N(0,1)$.

An alternative to the PDT-avg is the PDT-sum ***(54)***, where

$$D_{i,sum} = \left(\sum_{j=1}^{n_T} X_{T_j} + \sum_{j=1}^{n_S} X_{S_j} \right).$$

The test based on D_{sum} gives more weight to larger families. The PDT tests are implemented in software distributed at http://www.chg.duke.edu/research/software.html. **Figure 1** gives a sample pedigree and the associated statistic D_{avg} for allele C.

The PDT has been extended to incorporate haplotypes via the EM algorithm, where the test statistic is a weighted average over the possible haplotype phases for individual families ***(55)***. The haplotypic PDT tests have been implemented in the UNPHASED program ***(55)***. The Family Based Association Test [FBAT ***(56)***] is a generalization of the TDT that is valid in the presence of missing

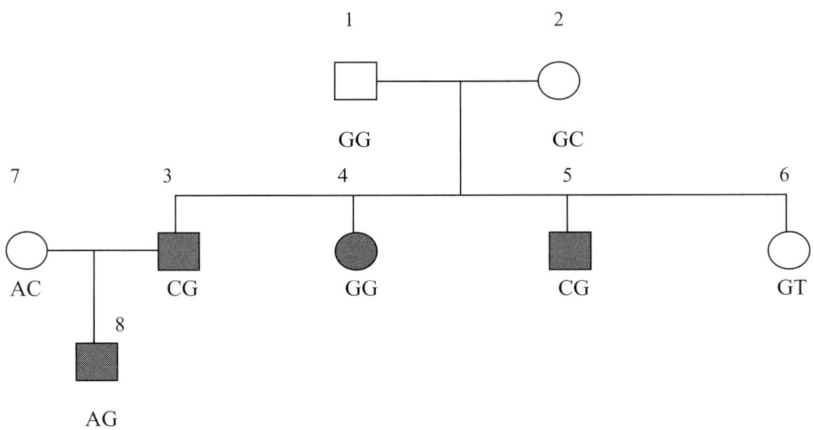

Trio	X_t	DSP	X_s
1-2-3	1	3-6	1
1-2-4	0	4-6	0
1-2-5	1	5-6	1
7-3-8	0		
Total:	2		2

Fig. 1. Calculation of the numerator for PDT_{sum} statistic: $D_{avg} = \frac{1}{4+3}(2+2) = \frac{4}{7}$

parental data and relatedness among trios. When parental information is missing, the FBAT conditions on the partially observed parental genotypes to obtain the conditional distribution of the offspring genotypes ***(57)***. The FBAT has also been extended to incorporate haplotypes ***(58)***.

9. Quantitative Measures

Quantitative transmission disequilibrium tests (QTDTs) are used to test for linkage and association between marker alleles and the alleles at a locus influencing the distribution of a quantitative trait (QTL). Allison ***(59)*** and Rabinowitz ***(60)*** first introduced such tests; they have been generalized and extended by others including Fulker and others ***(61)*** and Abecasis and others ***(62)***. Similar to a review by Whittaker and Morris ***(63)***, we use the notation and derivation as in Abecasis and others ***(62)***. Assume data for N nuclear families, each with n_i children, for a marker with 2 alleles, **M** and **m**. Define the phenotype of the jth child in the ith family as y_{ij} and the genotype as g_{ij}, where

$$g_{ij} = \begin{Bmatrix} 1 \text{ if genotype is MM} \\ 0 \text{ if genotype is Mm} \\ -1 \text{ if genotype is mm} \end{Bmatrix}.$$

A simple linear model can be fit and a likelihood ratio test of $\beta_1 = 0$ can be used to test for association using $y_{ij} = \beta_0 + \beta_1 g_{ij} + \varepsilon_{ij}$. Unfortunately, this test is not robust to population stratification. To construct a test for association that is robust to population stratification, the genotypic score is decomposed into orthogonal within- and between-family components ***(61,62)***. The within-family component is a valid estimate of the additive genetic value, a, where $E[y_{ij}] = \mu + g_{ij}a$, and μ is the population mean of the phenotypic trait.

Alternative methods for testing for association include generalized linear mixed models (GLMM) or generalized estimating equations (GEE) that explicitly account for the within-family correlation. Without additional methods such as genomic control or structural analyses, these tests are not robust to population stratification. Working correlation matrices based on independence, exchangeable correlation, and genetic relatedness can be specified. The latter is implemented in the program SOLAR and is often labeled as a variance components measured genotype (VCMG) approach. The sandwich estimator of the variance is a consistent estimator of the variance, even when the within-pedigree correlation structure is incorrectly specified. A common practice is to select a simple correlation structure that contains a minimal number of parameters and use the sandwich estimator to estimate the variance of the regression parameters for the GEE1 and GLMM models. These approaches are easily completed using standard software such as SAS (GEE1, GLMM). However, we do not recommend the use of the sandwich estimator except for additive or dominant models and large numbers of pedigrees. Because the properties of tests in the GLMM and GEE framework have not been carefully explored in this context, obtaining empirical *P* values for such tests via gene-dropping is appropriate. The VCMG approach implemented in SOLAR appears to have appropriate type I error rates.

10. High-Density SNP Mapping

Advances in genotyping technology now facilitate genotyping large numbers of markers on large numbers of individuals. Dense SNP mapping is increasingly common, especially in restricted regions such as candidate genes or linkage candidate regions. In addition, genome-wide association studies are now being planned and implemented. Dense SNP mapping of either type requires attention to both SNP-selection procedures and statistical methods to efficiently evaluate association with dozens to hundreds of thousands of markers.

10.1. Restricted Regions

In candidate gene studies, it is increasingly common (and expected) for investigators to genotype markers throughout the gene (exons and introns) and regulatory regions. These markers may include known functional variants, although most are anonymous markers with unknown function. Careful examination of the marker-marker LD between known variants can inform the SNP selection process by taking advantage of the correlation among markers. Often, a relatively small number of markers can be chosen to represent the majority of either allelic or haplotypic variation across the gene. Several algorithms have been described for both approaches (e.g., Refs. ***64,65***), and no one approach has been shown superior for all underlying genetic models. Linkage candidate regions are much larger than candidate gene regions, but the same principles apply in terms of choosing SNPs to type, whether only candidate genes within the linkage region are surveyed or the entire linkage region is considered.

10.2. Genome-wide Association

The genotyping technology for genome-wide association studies dictates that individual investigators do not have control over SNP selection, although in developing the technology, the above issues have been considered. As such, for genome-wide association studies, the primary concern has been to develop analysis strategies for studies that might include 500,000 SNPs based on relatively small sample sizes (e.g., 500 to 1000 individuals). The use of multistage designs is logical and useful in this setting. Several have advocated 2-stage designs (e.g., Refs. ***66, 67, 68, 69, 70***) where in the first stage, some proportion,

π_1, of the sample is typed on all markers. In the second stage, the remaining sample is genotyped on some proportion, π_2, of the most highly associated markers from the first stage and the data from both stages are analyzed jointly with an appropriate adjustment for the first-stage results. The 2 parameters, π_1 and π_2, can be chosen to maximize power given available samples and genotyping resources. For qualitative traits, this design is often nearly or as powerful as a 1-stage design that types all markers on all individuals in the sample, at a considerable cost savings.

11. Conclusion

Statistical methods for association mapping of complex genetic traits are rapidly evolving. This chapter has reviewed classic and proven approaches as well as emerging methods for association mapping. The critical assumption of random mating relative to the locus of interest and the consistency of these data with Hardy-Weinberg proportions is central to many tests of single locus and haplotypic association. Departures from these proportions may reflect biologic realities or genotyping problems.

The existence of linkage disequilibrium is a key biological property of the genome that motivates tests of association without complete sequence data or well-characterized functional data. This is true of both single marker analyses and haplotype analysis. In addition to departures from Hardy-Weinberg proportions, population stratification has the potential to both mask true associations and generate false associations. Fortunately, recent advances in methods such as genomic control (GC) and structured association (SA) have provided means of protecting against these errors. More classically, family-based methods of association such as the TDT have provided less powerful but more rigorous protection from these errors.

The future of association studies includes more thorough candidate gene, candidate region, and genome-wide association studies. These methods rely on the technological advances associated with high-throughput genotyping. The potential impact of this volume of data is tremendous. The basic statistical approaches for these studies are similar to that outlined for a small number of markers except the frequent need to apply more stringent methods to control for the type I error rate due to multiple testing. This is frequently done via recruiting multiple samples under a multistage design. These study designs suggest genotyping a proportion of the sample on all genetic markers in stage 1, identifying those markers that are candidates for further study, and then genotyping those markers in the remaining sample. The identification of markers for further study requires both statistical and genomic data (e.g., risk of polymorphism to yield alternative form of protein). Relatively soon, high-density genotyping will be replaced by sequence data integrated with greater functional information on observed polymorphisms. Data mining and machine learning tools may provide an important source of alternative approaches to those discussed here. Collectively, these tools will allow us to identify and interpret the importance of specific polymorphisms and integrate these data with behavioral and environmental factors to more completely understand traits and diseases of public health burden.

References

1. Mendel, J. G. (1865) Verhanlungen des naturforschenden Vereines in Brünn. *Abhandlungen* **4**, 3–47.
2. Guo, S. W. (2001) Does higher concordance in monozygotic twins than in dizygotic twins suggest a genetic component? *Hum. Hered.* **51**, 121–132.
3. Risch, N. (1990) Linkage strategies for genetically complex traits. II. The power of affected relative pairs. *Am. J. Hum. Genet.* **46**, 229–241.
4. Hardy, G. S. (1908) Mendelian proportions in a mixed population. *Science* **28**, 49–50.
5. Weinberg, W. (1908) Ueber den Nachweis der Vererbung beim Menschen. *Jahreshefte des Vereins fuer vaterlaendische Naturkunde in Wuerttemberg* **64**, 368–382.
6. Guo, S. W., and Thompson, E. A. (1992) Performing the exact test of Hardy-Weinberg proportion for multiple alleles. *Biometrics* **48**, 361–372.
7. Wigginton, J. E., Cutler, D. J., and Abecasis, G. R. (2005) A note on exact tests of Hardy-Weinberg Equilibrium. *Am. J. Hum. Genet.* **76**, 887–883.
8. Wittke-Thompson, J. K., Pluzhniko, A., and Cox, N. J. (2005) Rational inferences about departures from Hardy-Weinberg Equilibrium. *Am. J. Hum. Genet.* **76**, 967–986.
9. Hartl, D. L., and Clark, A. G. (1989) *Principles of Population Genetics*, second ed. Sunderland, Sinauer Associates.
10. Devlin, B., and Risch, N. (1995) A comparison of linkage disequilibrium measures for fine-scale mapping. *Genomics* **29**, 311–322.
11. Guo, S. W. (1997) Linkage disequilibrium measures for fine-scale mapping: a comparison. *Hum. Hered.* **47**(6), 301–314.
12. Hill, W. G. (1974) Estimation of linkage disequilibrium in randomly mating populations. *Heredity* **33**(2), 229–239.
13. Lewontin, R. (1964) The interaction of selection and linkage. I. General considerations: heterotic models. *Genetics* **49**, 49–67.
14. Ardlie, K. G., Kruglyak, L., and Seielstad, M. (2002) Patterns of linkage disequilibrium in the human genome. *Nat. Rev. Genet.* **3**, 299–309.
15. Evans, D. M., and Cardon, L. R. (2005) A comparison of linkage disequilibrium patterns and estimated population recombination rates across multiple populations. *Am. J. Hum. Genet.* **76**, 681–687.
16. Weiss, K. M., and Clark, A. G. (2002) Linkage disequilibrium and the mapping of complex human traits. *Trends Genet.* **18**(1), 19–24.

17. Daly, M. J., Rioux, J. D., Schaffner, S. F., and others (2001) High-resolution haplotype structure in the human genome. *Nat. Genet.* **29**, 229–232.
18. Couzin, J. (2002) Genomics. New mapping project splits the community. *Science* **296**, 1391–1393.
19. Agresti, A. (1990) *Categorical Data Analysis*. New York: John Wiley & Sons.
20. Excoffier, L., and Skatkin, M. (1995) Maximum-likelihood estimation of molecular haplotype frequencies in a diploid population. *Mol. Biol. Evol.* **12**(5), 921–927.
21. Green, L. E., Lange, E. M., and Langefeld, C. D. (2001) Power comparison of phase-known versus phase-unknown haplotype analyses for case-control designs. *Am. J. Hum. Genet.* **69**, 1948.
22. Schaid, D. J. (2004) Evaluating associations of haplotypes with traits. *Genet. Epidemiol.* **27**(4), 348–364.
23. Bader, J. S. (2001) The relative power of SNPs and haplotype as genetic markers for association tests. *Pharmacogenomics* **2**(1), 11–24.
24. Carroll, J. (1998) Measurement error in epidemiological studies. In: Armitage, P., and Colton, T., eds. *Encyclopedia of Biostatistics*. New York, John Wiley & Sons, pp. 2491–2519.
25. Kraft, P., Cox, D. G., Paynter, R. A., Hunter, D., and De Vivo, I. (2005) Accounting for haplotype uncertainty in matched association studies: a comparison of simple and flexible techniques. *Genet. Epidemiol.* **28**(3), 261–272.
26. Schaid, D. J., Rowland, C. M., Tines, D. E., Jacobson, R. M., and Poland, G. A. (2002) Score tests for association between traits and haplotypes when linkage phase is ambiguous. *Am. J. Hum. Genet.* **70**(2), 425–434.
27. Lake, S. L., Lyon, H., Tantisira, K., Silverman, E. K., Weiss, S. T., Laird, N. M., and Schaid, D. J. (2003) Estimation and tests of haplotype-environment interaction when linkage phase is ambiguous. *Hum. Hered.* **55**(1), 56–65.
28. Conti, D. V., and Gauderman, W. J. (2004) SNPs, haplotypes, and model selection in a candidate gene region: the SIMPle analysis for multilocus data. *Genet. Epidemiol.* **27**(4), 429–441.
29. Khoury, M. J., Beaty, T. H., and Cohen, B. H. (1993) *Fundamentals of Genetic Epidemiology*. New York, Oxford University Press.
30. Pritchard, J. K., and Rosenberg, N. A. (1999) Use of unlinked genetic markers to detect population stratification in association studies. *Am. J. Hum. Genet.* **65**(1), 220–228.
31. Pritchard, J. K., Stephens, M., and Donnelly, P. (1999) Inference of population structure using multilocus genotype data. *Genetics* **155**, 945–959.
32. Devlin, B., and Roeder, K. (1999) Genomic control for association studies. *Biometrics* **55**(4), 997–1004.
33. Roeder, K., Bacanu, S. A., Sonpar, V., Zhang, X., and Devlin, B. (2005) Analysis of single-locus tests to detect gene/disease associations. *Genet. Epidemiol.* **28**(3), 207–219.
34. Roeder, K., Bacanu, S. A., Wasserman, L., and Devlin, B. (2006) Using linkage genome scans to improve power of association in genome scans. *Am. J. Hum. Genet.* **78**, 243–252.

35. Gorroochurn, P., Heiman, G. A., Hodge, S. E., and Greenberg, D. A. (2006) Centralizing the non-central chi-square: a new method to correct for population stratification in genetic case-control association studies. *Genet. Epidemiol.* **30**(4), 277–289.
36. Heiman, G. A., Gorroochurn, P., Hodge, S. E., and Greenberg, D. A. (2005) Robustness of case-control studies to population stratification. *Cancer Epidemiol. Biomarkers Prev.* **14**(6), 1579–1580.
37. Terwilliger, J. D., and Ott, J. (1992) A haplotype-based "haplotype relative risk" approach to detecting allelic associations. *Hum. Hered.* **42**, 337–346.
38. Thomson, G. (1995) Mapping disease genes: family-based association studies. *Am. J. Hum. Genet.* **57**, 487–498.
39. Rubinstein, P., Walker, M., and Carpenter, C. (1981) Genetics of HLA disease association: the use of the haplotype relative risk (HRR) and the "Haplo-Delta" (DH) estimates in juvenile diabetes from three racial groups. *Hum. Immunol.* **3**, 384.
40. Falk, C. T., and Rubinstein, P. (1987) Haplotype relative risks: An easy reliable way to construct a proper control sample for risk calculations. *Ann. Hum. Genet.* **51**, 227–233.
41. McCulloch, C. E., and Searle, S. R. (2001) *Generalized, Linear, and Mixed Models*. New York, John Wiley & Sons.
42. Hardin, J. W., and Hilbe, J. M. (2003) *Generalized Estimating Equations*. Boca Raton, Chapman & Hall/CRC Press.
43. Spielman, R. S., McGinnis, R. E., and Ewens, W. J. (1993) Transmission test for linkage disequilibrium: the insulin gene region and insulin-dependent diabetes mellitus (IDDM). *Am. J. Hum. Genet.* **52**, 506–516.
44. Breslow, N. E., and Day, N. E. (1980) *Statistical Methods in Cancer Research, Vol 1: The Analysis of Case-Control Studies*. Lyon, France, IARC Scientific Publications.
45. Lehmann, E. L. (1986) *Testing Statistical Hypotheses*, 2nd ed. New York, John Wiley & Sons.
46. Lazzeroni, L. C., and Lange, K. (1998) A conditional inference framework for extending the transmission/disequilibrium test. *Hum. Hered.* **48**, 67–81.
47. McNemar, Q. (1947) Note on the sampling error of the difference between correlated proportions or percentages. *Psychometrika* **12**, 153–157.
48. Zöllner, S., Wen, X., Hanchard, N. A., Herbert, M. A., Ober, C., and Pritchard, J. K. (2004) Evidence of excessive transmission distortion in the human genome. *Am. J. Hum. Genet.* **74**(1), 62–72.
49. Curtis, D., and Sham, P. C. (1995) A note on the application of the transmission disequilibrium test when a parent is missing. *Am. J. Hum. Genet.* 56:811–812.
50. Curtis, D. (1997) Use of siblings as controls in case-control association studies. *Ann. Hum. Genet.* **61**, 319–333.
51. Spielman, R. S., and Ewens, W. J. (1998) A sibship test for linkage in the presence of association: the sib transmission/disequilibrium test. *Am. J. Hum. Genet.* **62**(2), 450–458.

52. Boehnke, M., and Langefeld, C. D. (1998) Genetic association mapping based on discordant sib pairs: the discordant alleles test. *Am. J. Hum. Genet.* **62**(4), 950–961.
53. Martin, E. R., Monks, S. A., Warren, L. L., and Kaplan, N. L. (2000) A test for linkage and association in general pedigrees: the pedigree disequilibrium test. *Am. J. Hum. Genet.* **67**, 146–154.
54. Martin, E. R., Bass, M. P., and Kaplan, N. L. (2001) Correcting for a potential bias in the pedigree disequilibrium test. *Am. J. Hum. Genet.* **68**(4), 1065–1067.
55. Dudbridge, F. (2003) Pedigree disequilibrium tests for multilocus haplotypes. *Genet. Epidemiol.* **25**(2), 115–121.
56. Laird, N. M., Horvath, S., and Xu, X. (2000) Implementing a unified approach to family-based tests of association. *Genet. Epidemiol.* **19**(Suppl 1), S36–S42.
57. Rabinowitz, D., and Laird, N. (2000) A unified approach to adjusting association tests for population admixture and arbitrary pedigree structure and arbitrary missing marker information. *Hum. Hered.* **50**, 211–223.
58. Horvath, S., Xu, X., Lake, S. L., Silverman, E. K., Weiss, S. T., and Laird, N. M. (2004) Family-based tests for associating haplotypes with general phenotype data: application to asthma genetics. *Genet. Epidemiol.* **26**, 61–69.
59. Allison, D. B. (1997). Transmission Disequilibrium Tests for Quantitative Traits. *Am. J. Hum. Genet.* **60**, 676–690.
60. Rabinowitz, D. (1997) Transmission disequilibrium test for quantitative trait loci. *Hum. Hered.* **47**, 342–350.
61. Fulker, D. W., Cherny, S. S., Sham, P. C., and Hewitt, J. K. (1999) Combined linkage and association sib-pair analysis for quantitative traits. *Am. J. Hum. Genet.* **64**(1), 259–267.
62. Abecasis, G. R., Cardon, L. R., and Cookson, W. O. (2000) A general test of association for quantitative traits in nuclear families. *Am. J. Hum. Genet.* **66**(1), 279–292.
63. Whittaker, J. C., and Morris, A. P. (2001) Family-based tests of association and/or linkage. *Ann. Hum. Genet.* **65**(5), 407–419.
64. Carlson, C. S., Eberle, M. A., Rieder, M. J., Yi, Q., Kruglyak, L., and Nickerson, D. A. (2004) Selecting a maximally informative set of single nucleotide polymorphisms using linkage disequilibrium. *Am. J. Hum. Genet.* **74**(1), 106–120.
65. de Bakker, P. I., Yelensky, R., Pe'er, I., Gabriel, S. B., Daly, M. J., and Altshuler, D. (2005) Efficiency and power in genetic association studies. *Nat. Genet.* **37**, 1217–1223.
66. Satagopan, J. M., Verbel, D. A., Venkatraman, E. S., Offit, K. E., and Begg, C. B. (2002) Two-stage designs for gene-disease association studies. *Biometrics* **58**, 163–170.
67. Satagopan, J. M., and Elston, R. C. (2003) Two-stage genotyping in association studies. *Genet. Epidemiol.* **25**, 149–157.
68. Thomas, D. C., Haile, R. W., and Duggan, D. (2005) Recent developments in genomewide association scans: a workshop summary and review. *Am. J. Hum. Genet.* **77**, 337–345.

69. Skol, A. D., Scott, L. J., Abecasis, G. R., and Boehnke, M. (2006) Joint analysis is more efficient than replication-based analysis for two-stage genome-wide association studies. *Nat. Genet.* **38**(2), 209–213.
70. Wang, H., Thomas, D. C., Pe'er, I., and Stram, D. O. (2006) Optimal two-stage genotyping designs for genome-wide association scans. *Genet. Epidemiol.* **30**, 356–368.

22

Genome Mapping Statistics and Bioinformatics

Josyf C. Mychaleckyj

Summary

The unprecedented availability of genome sequences, coupled with user-friendly, web-enabled search and analysis tools allows practitioners to locate interesting genome features or sequence tracts with relative ease. Although many public model organism- and genome-mapping resources offer pre-mapped genome browsing, biologists also still need to perform de novo mapping analyses. Correct interpretation of the results in genome annotation databases or the results of one's individual analyses requires at least a conceptual understanding of the statistics and mechanics of genome searches, the expected results from statistical considerations, as well as the algorithms used by different search tools. This chapter introduces the basic statistical results that underlie mapping of nucleotide sequences to genomes and briefly surveys the common programs and algorithms that are used to perform genome mapping, all available via public hosted web sites. Selection of the appropriate sequence search and mapping tool will often demand tradeoffs in sensitivity and specificity relating to the statistics of the search.

Key Words: Bioinformatics; genome annotation; genome mapping; genomics; human genome; mammalian genome; sequence alignment; sequence analysis; sequence search.

1. Introduction

The completion of the human euchromatic genome sequence in 2004, and the ever-increasing number of organisms with partial or complete genomic sequence,[1] has dramatically improved the ability of biologists to predict or confirm the occurrence of sequence tracts or patterns. The torrents of data have had the desirable effect of reducing the number of false-negative results in

[1] There are 2,469 genomes as of April 2, 2007, according to the Genomes Online Database (http://www.genomesonline.org).

From: *Methods in Molecular Biology, vol. 404: Topics in Biostatistics*
Edited by: W. T. Ambrosius © Humana Press Inc., Totowa, NJ

sequence searches, accompanied by an inevitable increase in number of positive matches, particularly for analyses of short sequence features. In this genomic era, biologists must be able to perform routine genome sequence analysis, including manipulation of sequences to design experiments, and properly interpret the results in a statistical context. The public availability of many well-supported, Web-delivered genome sequence analysis tools, especially those integrated with the major public sequence databases, frees researchers from the need to install or configure programs, properly shifting the focus to the interpretation of results.

Although many public model organism and genome mapping resources offer premapped genome browsing, there is still a need for biologists to perform their own individual mapping analyses for several reasons:

1. Map errors persist even in the most highly automated, sophisticated databases, such as those housed at NCBI (National Center for Biotechnology Information; http://www.ncbi.nlm.nih.gov), UCSC (University of California at Santa Cruz; http://genome.ucsc.edu), or Ensembl (European Bioinformatics Institute/Sanger Center; http://www.ensembl.org). Therefore, genome feature annotations should be checked independently and compared, especially if a gene or region of interest will be the primary focus of a set of experiments.
2. Differences in mapping algorithms and tools lead to differences in database genome features. One need only compare gene models constructed by NCBI (Reference Sequence Collection: RefSeq) and Sanger Center (Vertebrate Genome Annotation: Vega) to uncover differences in supposedly "known" and characterized genes. Users of these database resources should understand the differences in the way genome mapping is performed and the preponderance of evidence for feature mapping.
3. Delays in remapping the ever-increasing numbers of genome features lead to "missing" annotations as new assemblies of genomes are released.
4. A need to map lab-specific sequences *de novo*: primers, sequences, or novel gene splice variants.
5. Large-scale analysis of the occurrence of genome features across extended genome regions or multiple genes. An example is the correlation of gene expression profiles determined by microarray experiments, with the occurrence of certain regulatory features within proximal gene promoter regions. These types of analyses often require nonstandard, high-throughput genome search and mapping.

2. Genome Sequence Mapping

Genome sequence (physical) mapping refers to the process of locating a shorter query sequence in a longer segment of genome sequence. If the segment represents a whole genome sequence, the analysis maps the occurrence(s) of the query in the genome of an organism. For brevity in this discussion, the term *genome sequence* will be used, understanding that it refers to either the whole genome sequence; some specific contiguous segment of the genome, for example, a single chromosome or localized map region; or a collection of genome sequences that are to be searched in one *in silico* experiment. There are 2 fundamental modalities of genome mapping: intra- and interspecies. In the first case, the query and genome search sequences derive from a single genome, resulting in alignments of high percent identity, often >98%, that are limited only by sequence quality and intraspecies genome variation. In the second case, inference of homology accompanies the location of the orthologous query search string, requiring consideration of molecular evolutionary models and protein structure or function. This chapter focuses exclusively on the first mode with nucleotide sequence matching and alignment as the most generally applicable sequence analysis methods.

If the query sequence represents a recognized element of the genome known to control or modulate cellular processes or is of genome structural import, the genome mapping process is referred to as genome annotation, and the sequences to be mapped are genome features. Examples of genome sequence mapping are

- Identification of the region of a genome or transcript amplified by selected polymerase chain reaction (PCR) primers.
- Selection of optimal PCR primers to generate a useful amplimer.
- Location of a sequence tagged site (STS) marker.
- Location of a single nucleotide polymorphism (SNP) marker.
- Prediction of gene organization (gene model) from a transcript sequence derived from reverse transcription PCR (RT-PCR).
- Location of microarray oligonucleotide probes or small interfering RNA sequences (siRNAs) in a gene transcript mapped in a genome.
- Location of putative transcription factor binding sites (TFBS).

Users of the standard sequence analysis Basic Local Alignment Search Tool (BLAST; http://www.ncbi.nlm.nih.gov/BLAST) and BLAST-like Alignment Tool (BLAT; http://genome.ucsc.edu/cgi-bin/hgBlat) programs will recognize the query sequence as the "query" input sequence for a search and the genome sequence as the "subject." In this chapter, the subject sequence is assumed to be the (usually) longer segment of genome sequence or the entire genome sequence itself. In attempting intraspecies genome mapping of a query sequence, three questions have to be considered:

1. If the query sequence is located in the genome sequence, will it be found with a search?
2. How many matches are likely to be found from such a search, including possibly unrelated sequences in the genome?
3. Is it possible to distinguish "true" matches of interest from secondary matches that are random false positives?

These are the same general questions that are of concern whenever a sequence database is searched by alignment methods.

This chapter is mostly concerned with the first 2 of these 3 questions and provides a brief introduction into the underlying statistics and implementation of genome mapping algorithms. The treatment here employs largely heuristic derivations. Readers interested in a deeper treatment are directed to the more comprehensive and excellent texts by Waterman *(1)* and Ewens and Grant *(2)*.

3. Discrete Sequence Matching

This mapping procedure is equivalent to the well-studied computer science problem of string matching; that is, finding all occurrences of a discrete string in a second, longer string or text. The string in this case is the query nucleotide sequence, and the text is the genome sequence to be searched. The problem is to map the query sequence string $S = b_1b_2b_3b_4 \ldots b_{s-1}b_s$ of length s, to a genome sequence G of length L, with query string and genome composed of an ordered set of bases $b_j \in \{A, C, G, T\}$.

3.1. Exact Matching of Query Strings

3.1.1. Finding Sequence Matches

If the query and genome sequences contain no errors or other sources of variation, and the query sequence is a subsequence of the genome, then an exact match search algorithm is assured of finding the query string in the genome:

$$P(S \text{ is found} \mid S \in G) = 1. \tag{1}$$

Many routine lab mapping operations fall into this category, particularly if there is prior belief that the query sequence is located somewhere in the genome of interest. However, exact match searches may fail to map the query sequence because of variation in the genome sequence or the query sequence itself. Possible specific reasons for failure include:

1. *Incompleteness of the genome sequence.* The genome sequence may be incomplete because of residual sequence gaps or misassembly that arises from limited genomic sequence coverage. For example, there have been 16 released public human genome assemblies since the initial draft assembly in July 2000, the latest being March 2006, hg18, NCBI build 36 version 1 (http://genome-archive.cse.ucsc.edu/ and http://genome.ucsc.edu/goldenPath/stats.html). The human genome sequence has undergone dramatic improvement over this period with increasing depth of coverage, but gaps still remain.[2] Assemblies of other eukaryotic model organisms

[2] Two hundred thirty-four euchromatic sequence gaps, 64 non-euchromatic gaps. Also see Ref. *3*.

are less complete (http://www.genome.gov/10002154). If the site(s) in the genome where the query sequence should show an exact occurrence are missing from the reference genome sequence, neither exact nor inexact match searches will recover the correct genome sites, giving a false-negative result.

2. *Sequence errors.* Related to **item 1**, sequence errors in either genome or query sequence can arise through incomplete genome sequencing coverage.
3. *Natural variation in the genome sequence.* Genome sequence databases usually contain only a single reference exemplar genome sequence, whereas natural organism populations contain base changes and rearrangements associated with interindividual variation. In humans, germ-line single nucleotide polymorphisms (SNPs) occur with a haploid frequency of approximately 1 every 1000 bases. To further complicate the situation, dysfunction of normal DNA repair mechanisms in tumor tissues may result in abnormally high frequencies of somatic mutations and aberrant splicing compared with germ-line sequence. In a similar way, analysis of bacterial, viral, or rodent genomes should take into account interstrain sequence differences.

Figure 1 shows the results of searches using the transcript sequence for the human arginine vasopressin receptor 1B (accession NM_000707.2) as a query sequence. The arginine vasopressin receptor 1B is a G protein–coupled receptor for arginine vasopressin, expressed in the pituitary organ. Exact searches of different assemblies of the complete human genome sequence using the first 50 or 100 bases of the transcript sequence[3] yield different results, illustrating the limitations of exact sequence searching, mentioned above.

Even relatively low levels of sequence variation can have a profound impact on sequence search results. If K is the mean probability that a base occupying a specific genome sequence position is invariant:

$$P(S\ is\ found \mid S \in G) = K^S. \qquad (2)$$

Table 1 illustrates the effect of different levels of genome variation on the ability to find a query string by exact match and demonstrates the rapid dropoff in ability to locate a sequence with increasing K and length of query sequence s. It is clear that exact matching is only feasible for relatively short sequences that are the length of a typical oligonucleotide primer or less. To map longer sequences and/or sequences where there is a higher degree of variation, methods that are tolerant of inexact matching are required.

Analysis of variation in genome sequence is a key consideration in the design of algorithms for sequence searches ***(4)***.

[3] According to the GenBank file annotation, the first 100 bases of the transcript are in the 5′ untranslated region (5′ UTR) of the transcript.

3.1.2. How Many Exact Matches Are Expected?

A genome search with a query string may locate the expected site (or sites) of interest, but the tests of matches to all possible sites in the genome may also find sites that are uninteresting to the researcher, such as multiple genome priming sites or false-positive regulatory elements. Searches of genomes that

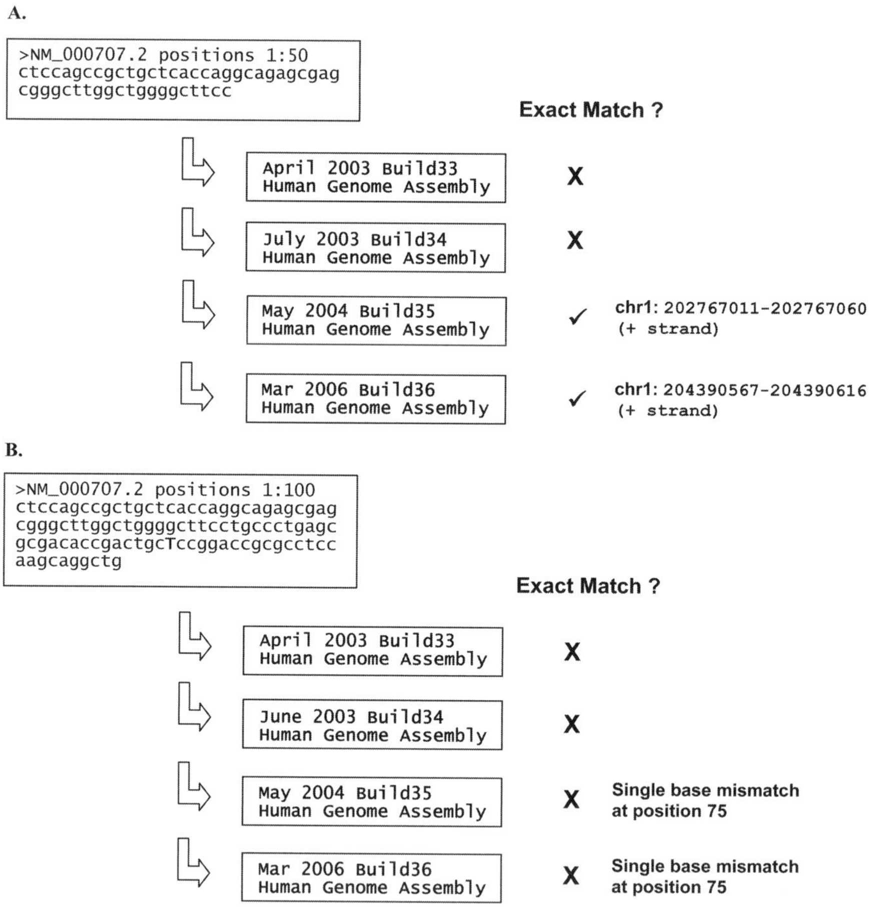

Fig. 1. The results of exact match searches using sequence from the curated NCBI mRNA RefSeq for the human arginine vasopressin receptor 1B (AVPR1B), accession NM_000707.2. In **(A)**, the sequence is found in the build 36 and 36 assemblies but not previous assemblies, presumably because of incomplete chromosomal sequence (chromosome 1 is the longest human chromosome). In **(B)**, all searches fail including the build 35 and 36 searches. There is a single mismatching nucleotide (uppercase **T** in the figure), whereas the genome sequence contains a **C** at this position. A SNP has not been previously recorded in public databases at this position, but it could be a rare SNP or a sequencing error in the mRNA query or human genome sequence.

Table 1
The Probability of Finding a Query String of Length s in an Exact Match Search of a Genome Sequence When the Query Sequence Is Drawn from an Independent Copy of the Same Genome Sequence

	$s =$				
$K = (\%)$	6	10	22	100	300
99.8	0.988	0.980	0.957	0.819	0.548
99.5	0.970	0.951	0.896	0.606	0.222
99.0	0.941	0.904	0.802	0.366	0.049
98.0	0.886	0.817	0.641	0.133	0.002
95.0	0.735	0.599	0.324	0.006	0.000
90.0	0.531	0.349	0.098	0.000	0.000

K is the mean probability that a position is invariant in the query sequence relative to the corresponding matching position in the genome sequence to be searched and hence measures the mean sequence variation between any two independent haploid copies of the same genome sequence.

$s = 6$, ~ length of a restriction enzyme recognition site; $s = 10$, ~ length of a DNA binding site motif; $s = 22$, ~ length of a typical oligonucleotide primer; $s = 100$, ~ length of a typical mammalian exon; $s = 300$, ~ length of high quality sequence from a PCR amplimer.

are millions (prokaryotes, simple eukaryotes) or billions (mammals, some plants) of bases in size will produce many random matches with small oligonucleotide strings. The expected number and distribution of random matches is important in assessing the significance and possible functional importance of sites that are discovered.

For a genome sequence of length L, let the relative frequency of each of the bases in the genome be p_A, p_C, p_G, p_T, respectively. The relative frequencies for the reference genome sequence can be derived by counting the number of occurrences of each base in the genome segment and dividing by the total number of bases in the sequence (L). (These values are nucleotide frequency parameters for the genome sequence segment but would constitute estimates for a population of genome sequences or for an incomplete sequence). If P_S is the relative frequency of occurrence of the query string S in the genome sequence and the nucleotides are distributed independently:

$$P_S = (p_A)^{nA}(p_C)^{nC}(p_G)^{nG}(p_T)^{nT}, \tag{3}$$

where $s = n_A + n_C + n_G + n_T$ and the n_b are the counts of each distinct base b in the string S. In the simplest sequence composition model, the 4 bases are assumed to be equifrequent with $p_b = 0.25$. In practice, genomes of different organisms show considerable composition bias, which affects the relative frequency of occurrence of query sequences. The estimated euchromatic base frequencies for the public build 36 assembly of the human genome are shown in **Table 2** and recapitulate known A+T biases in mammalian genomes.

If X is a random variable that represents the number of occurrences of string S in the genome sequence, the probability of finding N exact matches of S is *approximately* binomially distributed:

$$P(X = N | S, p_A, p_C, p_G, p_T) \approx \frac{L!}{(L-N)!N!} P_S^{\ N}(1-P_s)^{L-N}. \tag{4}$$

Table 2
Composition Statistics of the Euchromatin Component of the Human Nuclear Genome Sequence, Assembly Build 36, Mar 2006[a]

Hsa Chr	L	p_A	p_T	p_C	p_G
1	224999719	0.291	0.292	0.209	0.209
2	237709794	0.299	0.299	0.201	0.201
3	194704825	0.301	0.302	0.198	0.199
4	187297063	0.309	0.309	0.191	0.191
5	177702766	0.302	0.303	0.198	0.198
6	167273991	0.302	0.302	0.198	0.198
7	154952424	0.296	0.297	0.204	0.204
8	142612826	0.299	0.299	0.201	0.201
9	120143252	0.294	0.293	0.207	0.207
10	131624728	0.292	0.292	0.208	0.208
11	131130753	0.292	0.292	0.208	0.208
12	130303032	0.296	0.296	0.204	0.204
13	95559980	0.307	0.308	0.193	0.193
14	88290585	0.294	0.297	0.204	0.205
15	81341915	0.289	0.289	0.211	0.211
16	78884752	0.275	0.277	0.224	0.224
17	77800220	0.272	0.273	0.228	0.228
18	74656155	0.301	0.301	0.199	0.199
19	55785651	0.258	0.259	0.242	0.242
20	59505253	0.278	0.281	0.220	0.221
21	34170106	0.297	0.294	0.204	0.205
22	34851311	0.261	0.260	0.240	0.240
X	151058754	0.302	0.303	0.197	0.198
Y	25652954	0.299	0.302	0.199	0.201
Genome	2858012809	0.295	0.296	0.209	0.209
Promoter 1000 bases	23566887	0.233	0.231	0.268	0.267

[a]The promoter 1000 bases compositions were calculated using the UCSC upstream1000.fa sequences, derived from the 1000 bases 5′ to the transcription start site of all mapped NCBI RefSeq transcripts (http://hgdownload.cse.ucsc.edu/goldenPath/hg18/bigZips). All ambiguous nucleotides (N) denoting gaps were removed prior to calculations. Residual deviations from total $p = 1.0$ are due to rounding in the third decimal place.

Under this model, we derive the following general results for the probability of 0 matches ($X = 0$) and at least one match ($X \geq 1$):

$$P(X = 0 \mid S) = (1 - P_S)^L$$
$$P(X \geq 1 \mid S) = 1 - (1 - P_S)^L \quad (5)$$

with the base frequency parameters omitted for brevity. Using a standard result for the expected value of a binomially distributed random variable (see **Chapters 3** and **5**), the expected (mean) number of matches in a search of a genome sequence length L is given by

$$E[X] = \sum_{N=0}^{\infty} X P(X = N \mid S) = L P_S \quad (6)$$

with variance

$$\mathrm{Var}(X) = E[(X - E(X))^2] = L\,P_S(1 - P_S). \quad (7)$$

There are a number of caveats to this statistical analysis. The least problematic is that the genome sequence length L in the above equations should strictly be replaced with the end-corrected length $L' = L + s - 1$. This is not usually necessary in practice because $L \gg s - 1$ for genome-scale mapping and the correction has a minimal effect on the results. The correction necessary will also change if the analysis requires mapping of a query string to a database of multiple separate genome sequences rather than 1 contiguous sequence. The expected number for M multiple sequences is

$$E[X] = \sum_{i=1}^{M} \sum_{N=0}^{\infty} X P(X = N \mid S, i) = \sum_{i=1}^{M} L_i P_S = M L P_S \quad (8)$$

where the last identity is true if the M sequences are all the same length $L_i = L$. In this case, the end correction should be applied to each individual sequence, replacing L_i with L_i'. An example of this type of analysis is mapping a query sequence to the 5 kilobases of proximal promoter sequence for all genes in the TNF receptor superfamily. If a search were conducted in the promoter regions of 29 members of this family (http://www.gene.ucl.ac.uk/nomenclature/genefamily/tnfrec2.html), the total length of sequence searched is 5 × 29 kilobases = 145 kilobases, and the end correction of a search string length in the range $s = 10$ to 100 bases is negligible compared with the $L_i = 5$ kilobases.

A more important concern is the assumption that the L' query sequence match tests are independent (i.e., the genome sites are independent and identically distributed). In the uninteresting (!) limit of a query string of length $s = 1$, these results are exact, but in general the tests are not independent because a test of matching at genome location j in the genome interval $[j, j + s - 1]$ contains the set of genome sites $\{j, j+1, j+2, \ldots j+s-1\}$ and the probability of further query string matches in this interval will be biased by self-overlap properties of the query string. If a string match occurs at genome location j, the probability of further matches in the interval $[j, j+s-1]$ will be increased with self-overlapping query sequences and decreased if self-overlap is impossible. **Figure 2** illustrates this principle using the core prokaryotic Shine-Dalgarno sequence as an example *(5)*. This correlation between tests in colocated genome intervals alters the variance of the distribution and results in clusters of matching sites when the query is self-overlapping, and increased "repulsion" between sites when self-overlap is impossible, compared with the model binomial variance. The other assumption is that the genome sequence contains a random, independent distribution of the 4 nucleotide bases. Different feature classes in genomes have variable local base compositions, for example, exonic sequences that code for proteins or 3′ untranslated region (3′ UTR) sequences that affect mRNA stability. The assumed independent distribution of bases is confounded by local correlations between bases in certain genome features. CpG islands are distinct regions in the promoter region of genes with increased local C, G, and CG (dinucleotide) frequency that are important in the regulating gene transcription.

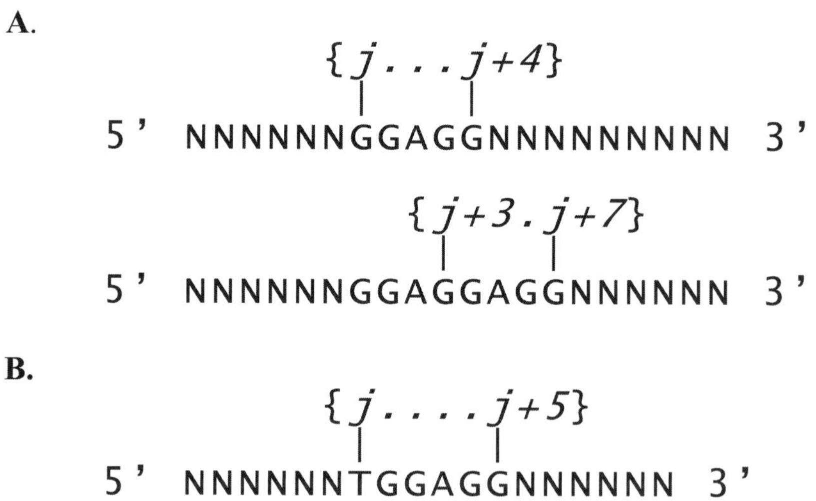

Fig. 2. Mapping of Shine-Dalgarno (SD) query sequences in prokaryotic genomes. In **(A)**, the query sequence is the most conserved GGAGG core of the SD consensus *(6)* with self-overlap possible. The query string in **(B)** extends this core by 1 base to TGGAGG and renders self-overlap impossible. Searches with either query string will locate GGAGG motifs, but the **(B)** search will discriminate further to TGGAGG sites. A match at $\{j \ldots j+4\}$ in **(A)** increases the probability of a further random match at $\{j+3 \ldots j+7\}$. The variance of random site matches will differ in the 2 cases because of the extra **T** in the search string in **(B)** and because of the self-overlap properties.

Protein coding regions also contain inherent di- and trinucleotide biases that reflect the gene organization and protein function. Other intergenic regions also contain tracts of local composition bias including homopolymer repeats.

The equations above assume that the exact match search is performed on only 1 specific strand of a genome sequence (the base frequencies are strand-specific). In many applications, the exact search is performed against *both* strands of the sequence, necessitating a slight adjustment to the model. If the query sequence is reverse complement palindromic (e.g., EcoR I recognition site GAATTC), no adjustment is necessary because searches on either strand will identify identical match sites in the genome sequence, and the $P(X = N | S)$ should be computed using the above equations. Note that an odd-length query sequence (s is odd) cannot be a reverse complement palindrome.[4] In the more general nonpalindromic case, we can compute the probabilities for the search of the second strand using the reverse complement of the query sequence searching against the identical genome sequence strand. If the forward strand search string is $S(F)$, the reverse strand string is $S(R)$, and assuming independent search events on the forward and reverse strands:

$$P(X = 0 | S(F), S(R)) \approx (1 - P_{S(F)})^L (1 - P_{S(R)})^L$$
$$P(X \geq 1 | S(F), S(R)) \approx 1 - [(1 - P_{S(F)})^L (1 - P_{S(R)})^L] \qquad (9)$$

and other results are analogous. These are approximations because the search strings have a reverse complement relationship and the searches are not truly independent. If the relative frequencies of the forward and reverse strings are approximately identical (i.e., $P_{S(F)} \cong P_{S(R)} = P_S$), then **Equations 9** reduce to the same form as **Equations 5**, but with L replaced by $2L$. The exact identity between $P_{S(F)}$ and $P_{S(R)}$ is true if either:

1. In the search string, $n_A = n_T$ and $n_C = n_G$, that is, the string contains an identical number of {A, T} and {C, G} bases; or
2. Chargaff's second rule is true for the genome, that is, $p_A = p_T$ and $p_C = p_G$.

In practice, Chargaff's second rule is a very good approximation for the genomes of prokaryotes and eukaryotes *(6–8)*, as **Table 2** shows for *Homo sapiens*.

These equations form the basis for the distribution of unique oligonucleotide primers. In mammalian genomes such as human, rat, or mouse with single-strand euchromatic lengths of approximately 3 billion bases (http://genome.ucsc.edu/cgi-bin/hgGateway), the random model of genome sequence demands a minimum length of at least 17 to 18 bases to satisfy the requirement of the

[4] Unlike English text palindromes, such as the oft-recited "Madam I'm Adam." Reverse complement palindromic sequences are sometimes referred to as containing tandem inverted repeats.

expected number of random matches $E[X] = 1$. In practice, compositional differences in the human (and other mammalian) genomes modestly affect the estimated minimum primer length, ranging from 13.7 for 100% strong bases (C, G) to 17.8 for 100% weak (A, T). Searches across both strands increase the minimum length requirement range to 14.0 to 18.4.

3.1.3. Genome Distribution of Query String Matches

In the analysis above, the expected number of matches of a query sequence in a random genome of length L was found to be $L\ P_s$ (uncorrected) or $(L - s + 1)\ P_s$ (corrected) (**Equation 6**). The distribution of distances between adjacent match sites can also be modeled as a sequence of independent Bernoulli trials, with $k - 1$ trials that fail to find a match, followed by a successful kth trial that finds the next match. A geometric distribution models this situation:

$$P(Y = k \mid P_S) = (1 - P_S)^{k-1} P_S, \tag{10}$$

where Y is a random variable representing the distance between adjacent genome match sites. This predicts a mean match site separation of

$$E[Y] = \frac{1}{P_S} \tag{11}$$

with variance

$$\mathrm{Var}(Y) = \frac{1 - P_S}{P_S^2} \tag{12}$$

as expected from the analysis leading to **Equation 6**. Hence a random genome sequence leads to randomly distributed match sites with a mean intermatch site distance parameter of $1/P_S$. The caveats that qualify the applicability of the binomial distribution to model the number of match sites in a genome search also apply here. For a query sequence of length $s > 1$, tests at contiguous genome sites in the interval $[j, j + s - 1]$ are not independent, and the variance of the intersite distances depends on self-overlap properties. In the extreme case of searches for homopolymer runs (e.g., TTTTTTT), very pronounced clustering of random match sites will occur. The clusters will be separated by a greater than expected distance compared with the value calculated in **Equation 10**, conditional on a fixed genome composition.

3.1.4. Distribution of Nonoverlapping Query String Matches

So far, the mapping analyses have allowed overlapping query string matches. In some mapping problems, the query string occurrences should not be permitted to overlap. In mapping type II restriction enzyme cut sites, a site that is cut by the enzyme could leave an unrecognizable site remnant at the end of the restriction fragment or leave insufficient flanking sequence for the enzyme to cut the second site. **Figure 3** illustrates the destruction of the second recognition site when the first site has been digested. Whether more than one of the overlapping sites is cleaved depends on the number of nonrecognition bases that flank the residual site *(9)*.

A second example where the distribution of nonoverlapping query string matches can be important is in the location of putative oligomeric regulatory

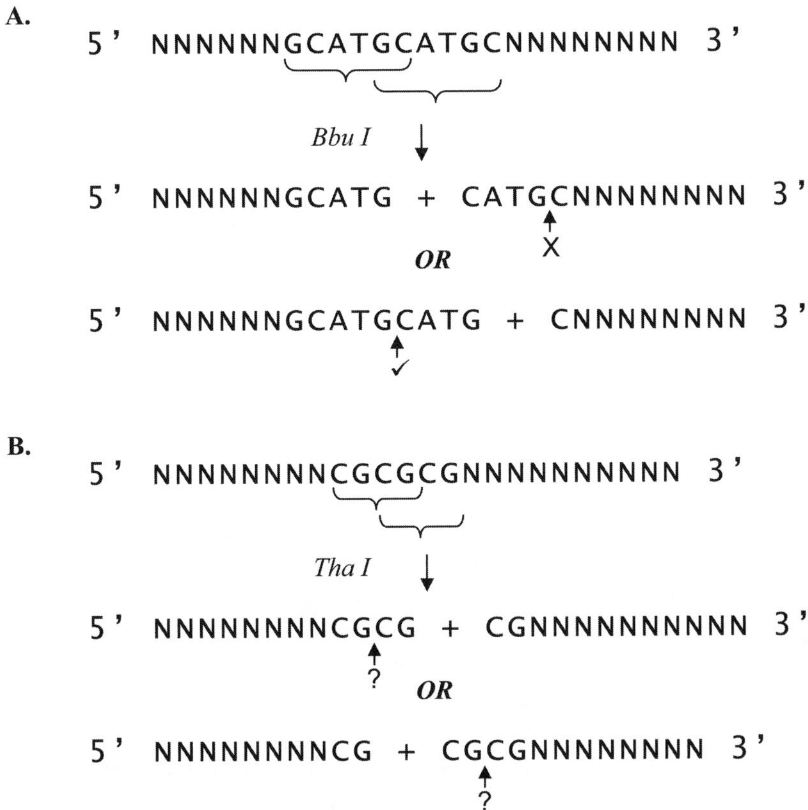

Fig. 3. Strict recognition of 2 overlapping sites in a genome could theoretically lead to cleavage at 1 or both sites depending on which site the enzyme cleaves first. In the case of *Bbu I*, which recognizes and cuts at a site GCATG∧C (**A**), cleavage at the 3′-most site leaves the 5′ site intact (✓) with 3 flanking bases, but not vice versa (✗). In the case of *Tha I*, which recognizes CG∧CG (**B**), cleavage at either site leaves no non-site flanking bases and may be less likely to cleave (?) under normal digestion conditions *(9)*.

or transcription factor binding sites, by searches of genomic sequence with consensus binding site motif sequences. Overlapping sites will presumably have single occupancy by the binding protein although binding to the alternative overlapping sites would be expected to be in dynamic equilibrium with mean occupancy of each candidate site proportional to the exponential of the binding affinity.

As illustrated with the Shine-Dalgarno (SD) sequences in **Figure 1**, the core ribosomal binding site sequence GGAGG spanning genome interval $[j, j + 4]$ can self-overlap starting at $j + 3$. Hence for the genome site match at $j + 3$ with string match interval $[j + 3, j + 7]$, the total probability of a GGAGG query string match is the sum of the direct nonoverlapping match and an overlap match conditional on the presence of a match site at j, that is,

$$\begin{aligned} P_S(j+3: j+7) &= \tilde{P}_S(j+3: j+7) + \tilde{P}_S(j: j+4)p_A(j+5)p_G(j+6)p_G(j+7) \\ &= \tilde{P}_S\{1 + p_G^2\, p_A\} \end{aligned} \tag{13}$$

where $\tilde{P}_S$ is the equivalent query string probability for *nonoverlapping* genome matches, and the term $p_G^2\, p_A$ represents $P(j + 3: j + 7 \mid j: j + 4)$ or the probability of a match at $[j + 3, j + 7]$ if there is a match at $[j, j + 4]$. The colon (:) represents the closed range in **Equation 13**, and $P(j: i)$ is shorthand for the probability of a match of query string S at the genome subsequence $[j, i]$. In general, the relationship is

$$P_S = \tilde{P}_S\{1 + \sum_{i=\text{overlap sites}} P_i(j+i : j+i+s-1 \mid j : j+s-1)\} = \tilde{P}_S\{1 + P_0\} \tag{14}$$

where the index i sums over all distinct self-overlap sites in the query string S, and P_o represents the total contribution of all self-overlap terms. Because $0 \leq P_o \leq 1$, the additional overlap terms increase the relative frequency for overlapping searches compared with nonoverlapping. Conversely, the term $\{1 + P_o\}^{-1}$ can be seen as the correction factor for the reduction of the overlapping match frequency relative to the nonoverlapping. First-order Taylor expansion of this latter correction factor gives $\{1 - P_o\}$; if query string matches in the self-overlapping case are viewed as clusters of matching genome sites, the subtracted P_o term can be interpreted as the reduction of match probability derived from direct self-overlap of the first 5′ match site in a cluster with other overlapping sites. The neglected second and higher order terms from the expansion are the n = 2nd, 3rd, . . . degree indirect genome match site contributions to the overlapping clusters. In the case of a query string without self-overlap, $P_o = 0$ and $P_S = \tilde{P}_S$. The $\tilde{P}_S$ term for nonoverlapping sites can be substituted in **Equation 11** to estimate the mean separation of nonoverlapping sites. In the case of core SD sites in *Bacillus subtilis* (% G + C = 43.5) ***(10)***, the expected core SD site separation with overlap is 1581.7 bases, which is increased to 1602.9 with no overlapping sites. The intersite distribution can be derived using the nonoverlapping $\tilde{P}_S$ substituted in **Equations 11** and **12**, although the variance estimate will be approximate, dependent on the length, composition, and self-overlap properties of the query string.

3.2. Inexact Matching of a Query Sequence

Results from **Section 3.1.1** suggest that query sequences should be as short as possible to minimize false-negative search results due to the increasing probability of at least 1 mismatching base, while **Section 3.1.2** results suggest that a longer query sequence will lead to less false-positive random genome matches. To balance these competing forces, the stringency of sequence matching must be relaxed to permit 1 or more mismatches between the query and potential-match sites in the genome sequence. If a genome search is permitted to return results that mismatch by at most a single position, the expected number of matches to a random genome sequence $E[X]$ increases dramatically. We can model the effect of this by replacing the relative frequency of occurrence of the query string (P_S) with the analogous query string frequency that permits zero or a single mismatch, $P_S^{(1)}$. This latter quantity is related to P_S by

$$P_S^{(1)} = P_S\left[1+\frac{n_A}{p_A}+\frac{n_C}{p_C}+\frac{n_G}{p_G}+\frac{n_T}{p_T}\right] \tag{15}$$

(P_S is as specified as in **Equation 1**). The expected number of matches $E[X]$ is given by the modified result of **Equation 6**, $L\ P_S^{(1)}$, and other results analogously. For a human genome primer of length $s = 21$ with $n_C + n_G = 9$ (42.8% GC content) and genome-wide $p_A = p_T = 0.295$, $p_C = p_G = 0.20$ assumed (**Table 2**), permitting a single mismatch increases the expected number of matches in a random genome search by a factor of approximately 87.

Because genome sequence variation (**Equation 2**) and tolerance for a single mismatch (**Equation 15**) both modify the relative frequency of occurrence of the query string, both effects can be incorporated into a single equation for the effective query string frequency:

$$\mathrm{P}_S^{(1)} = P_S K^s\left[1+\frac{n_A}{Kp_A}+\frac{n_C}{Kp_C}+\frac{n_G}{Kp_G}+\frac{n_T}{Kp_T}\right]. \tag{16}$$

As before, the value for P_S in **Equations 4** to **7** can be replaced by the value $\mathrm{P}_S^{(1)}$ computed using **Equation 9**, to incorporate the effects of variable genome sequence and single mismatch. The combinatorial effect of the single mismatch position in the query string more than compensates for the loss of matches due to genome sequence variation, even at 90% identity (**Table 1**), and increases the overall total number of expected random genome matches. Extension to greater numbers of mismatch positions in the query string $P_S^{(2)}$, $P_S^{(3)}$, and so forth, follow by logical extension and increase the number of matches by orders of magnitude. Hence the sensitivity of genome mapping is best enhanced through inexact matching, at the expense of lower specificity.

3.3. Joint Mapping of Two Query Sequences

A natural extension of mapping a single query sequence is to the joint mapping of 2 or more discrete sequences. One example of the use of this joint mapping of sequences is the mapping of DNA sites that are putative binding sites for zinc finger or basic helix-loop-helix (bHLH) transcription factors. Van Helden and others analyzed the yeast genome for the occurrences of pairs of trinucleotides with spacer length $n = 0, 1, \ldots, 16$ in 800 bases 5′ proximal to the open reading frames of 11 families of genes and assessed statistical significance of occurrence; that is, motifs of the form $XYZ(N)_nXYZ$. They were able to detect overrepresentation of sites bound by C_6Zn_2 binuclear cluster proteins, and other transcription factors ***(11)***.

Steroid/nuclear receptor superfamily response elements are the recognition sites for complexes of a steroid receptor bound to its cognate steroid family protein and have been identified in mammalian gene loci as well as in chicken and *Xenopus*. Binding of the receptor complex to the element, which is usually located within 1 kilobase of the gene transcription start site, leads to changes in gene expression. In the case of the class I nuclear estrogen receptor, the response element (ERE) has the core consensus GGTCANNNTGACC and the dimeric DNA contact sites are reverse complement palindromes ***(12,13)***. Because the ligand-activated estrogen receptor binds as a homodimer, this suggests that the receptor complex contacts the identical monomeric element GGTCA on opposing strands.

Spaced dimeric sequences also form the recognition sites for certain restriction enzymes. For example, *Bgl I* cuts genomic sequence at sites that match GCCNNNN∧NGGC.

The analysis in **Sections 3.1** and **3.2** is easily extended to the problem of dimeric elements with fixed spacing. The problem becomes one of matching the extended query string $S = S_1NS_2$, where S_1 and S_2 are of length s_1 and s_2, respectively, and have individual string relative frequencies given by **Equation 3**, and $(N)_n$ is a spacer sequence of length n composed of any permutation of bases. For any fixed spacer length n, the relative frequency of the extended query string S is given by:

$$P_S = P_{S_1} P_{S_2}. \qquad (17)$$

The joint relative frequency P_S can be substituted into **Equations 4** to **12** to obtain probabilities for finding the extended string, with corrected genome length $L - (s_1 + s_2 + n) + 1$. This argument also extends to situations where the spacer element can be of variable length, the most obvious example of which is the mapping of 2 individual PCR primer sequences that have been selected to amplify a particular segment of the genome, usually 200 to 1000 bases in length. Although the PCR primers map to different strands, by reverse complementing one of the primers, we convert the statistical problem to one of matching the extended query string $S = S_1NS_2$. If the range of permissible spacings between the 2 primers is $n_1, n_1 + 1, \ldots, n_2$, the expected number of genome matches is

$$E[X] = \sum_{j=n_1}^{n_2} \{L - (s_1 + s_2 + j) + 1\} P_{S_1} P_{S_2}. \tag{18}$$

Because P_S will have a much smaller relative frequency than a single primer sequence, the number of random genome matches with 2 primers with lengths s_1, s_2 in the range of 17 to 25 will in general be minuscule. As in **Section 3.1**, this is an idealized analysis, and clustering of extended query string matches can occur with self-overlapping query strings S_1, S_2.

3.4. Mapping Analysis of Overrepresented Sequences

One of the most frequent types of expression microarray experiments is the identification of sets of genes that are coregulated in response to an extracellular stimulus or perturbation. The supposition is that these gene sets are regulated by transcription factors that enhance or suppress their expression through binding to common perigenic sequence elements, particularly in the promoter regions. The sequence elements are usually small oligonucleotide consensus binding sites and are assumed to be more prevalent in the genes that have similar expression patterns. Hence, oligonucleotide sequences that are statistically overrepresented in these groups may provide clues about the transcription factor recognition sites, the factors that bind to them, and the common pathway(s) that lead to activation of the gene sets.

Analysis of the proximal promoter sequences of the genes in the set that are coregulated provides the actual count of occurrences of oligonucleotides. If the likely pathways and transcription factors that lead to the coregulation are unknown, the gene sequence analysis will require counts for a range of oligonucleotide sequence strings of different sizes. The actual counts are then compared against the expected counts for each species of oligonucleotide string S. The expected number of occurrences can be calculated using **Equation 8** using the appropriate individual base composition frequencies derived from single base composition analysis of the gene sequences. Alternatively, the expected occurrences could be modeled from the number of occurrences of the same oligonucleotide string in putatively nonfunctional intergenic regions ***(14)*** or in genes that are not dysregulated in the expression microarray experiment.

3.5. Extended Length Query Sequences

3.5.1. Sequence Alignment Algorithms Are Needed for Longer Sequences

Reference to **Table 1** shows that as the length of the query sequence increases, the probability of an expected exact match drops in proportion to the fractional identity parameter (K) to the power of the query string length. When $K = 99.8\%$ (approximately equivalent to an SNP in the diploid human genome every 500 bases), the probability of an exact match is 96% for a primer of length $s = 22$ but is only 82% for a query length typical for mammalian exons, $s = 100$. In addition, exact match searches do not permit insertions or deletions (*indels*) in query sequence relative to the genome sequence and hence are not robust to discontiguous (gapped) sequence mapping. The simple exact or nearly exact search matches must be replaced by methods that map query sequences using sequence alignment algorithms.

Optimal sequence alignment algorithms have been well studied and are used extensively in contemporary bioinformatics analyses. The two most famous are both based on dynamic programming methods: Needleman-Wunsch ***(15)*** for global pairwise alignment (alignment of all bases in 2 sequences) and Smith-Waterman ***(16)*** for local pairwise alignment (alignment of a contiguous subset of bases in 2 sequences). These algorithms can be used in direct searches of nucleotide databases and genome-scale sequences and are guaranteed to find an optimal alignment that is conditional on the scoring system used but are slower than heuristic-based algorithms such as BLAST ***(17)*** and are inefficient for mapping highly similar sequences. They explore the full matrix of sequence search space $O(N^2)$, when a very limited banded search would be sufficient to identify the sequence alignments that lie very close to one or a few diagonals in the matrix. For these reasons, several closely related programs have been developed that employ a common device to restrict the search of total pairwise sequence alignment space by seeding alignments with small oligomeric exact (or near exact) matches between subsequences in the query and subject sequences and then extend these individual matches into consistent alignments. The underlying assumption in these algorithms is that significant alignments are likely to contain local regions of perfect sequence identity between query and subject sequences. The initial search phase uses a minimum substring length or word length for the local regions of exact match. The minimum substring length is also called the word length. By searching for one or more small subsequences that exactly match, the algorithm seeds the final alignment and can dramatically increase speed of searching, but at the cost of sensitivity. In genome mapping applications, the loss of sensitivity is less of a concern compared with applications that require inference of homology, particularly between proteins or nucleotide sequences that diverged in the distant past and retain relatively low percent sequence identity.

There is a second detail in the implementation of word search–based alignment algorithms that affects search sensitivity. To increase search speed and throughput, especially repetitive searches of the same genome sequence with a batch of query sequences, it is advantageous to build an index of the positions of fixed-length words in the genome sequence and retain this entire index in computer physical memory (RAM), rather than using an index larger than available memory and having to page the index table to and from disk. To constrain RAM requirements, only a fraction of all possible $L + w - 1$ words of length w in the genome sequence are indexed, sometimes only nonoverlapping words. Although the search looks for matches between the $s - w + 1$ possible words of length s in the query sequence and the genome sequence index, the effect of the constrained index of genome word content is that the length of a query sequence that is *guaranteed* to trigger an alignment is greater than the actual word length w. A constrained genome word index is currently implemented for public BLAT and SSAHA programs as described in **Table 3**. **Figure 4** compares the algorithms for word-based genome searching.

3.5.2. Discontiguous (Gapped) Sequence Mapping

Implicit in **Sections 3.1** to **3.4** is the notion that the query string maps contiguously to the target genome segment, without permitting the introduction of gaps in either the query sequence or the target genome sequence to improve the overall match. Query strings can also map discontiguously to regions of the genome. Gapped alignments can occur because small insertions or deletions are needed to recover a higher scoring alignment of the query sequence to the genome sequence. As with mismatches in otherwise exact query string matches, these can occur because of natural variation or sequencing errors. Single nucleotide polymorphisms are the most abundant class of human genome variation, but indels of various lengths comprise ~20% of all human variation ***(18)*** and about 8% are diallelic. In other species, notably plants such as *Arabidopsis*, indels comprise a higher proportion of total variation. A search of dbSNP build 127 (http://www.ncbi.nlm.nih.gov/SNP) reveals a total of 2,995,082 variants annotated as indel class polymorphisms in all organisms and 2,156,519 in humans specifically.

In eukaryotic genome annotation work, a common example of discontiguous genome mapping is the analysis of gene organization by mapping a spliced expressed sequence tag (EST) or full-length cDNA sequence to genome sequence. In mammalian genomes, this is especially challenging, because intervening introns may reach 10 to 100 kilobases in length. In general, the alignments generated by the genome search program will usually correspond with the query sequence exons, although non–full-length query sequences will result in a truncated alignment for at least 1 exon. For the most common poly dT-primed ESTs, the alignment is biased to the 3′ end of genes, and the 5′ end of

Table 3
Publicly Available Web-Hosted Genome Mapping Alignment Programs and Their Default Parameters for Nucleotide Searches[a]

Program	Default word length (w)	Alignment trigger	Query string length matching
BLASTN @ NCBI[b]	11	1 exact word match	Guaranteed to find exact match to query string of length $s = w$
MEGABLAST @ NCBI[b]	28	1 exact word match	Guaranteed to find exact match to query string of length $s = w$[c]
BLAT @ UCSC[d]	11	2 exact word matches	Guaranteed to find exact match to query string length $s \geq 20$
SSAHA @ Ensembl[e]	13	1 exact word match	Guaranteed to find exact matches of length $s = 2w - 1$; i.e., 25 bp

[a]These programs are also available for local server installation and provide more flexible implementation options than the defaults for public-hosted sites. The listed search criteria are subject to change. The query string length matching lists guaranteed minimum, but the program implementations may be able to locate *smaller* query strings in selected cases, or may impose a minimum query length for consistency of results output.

[b]Available at http://www.ncbi.nlm.nih.gov/BLAST; human genome BLAST search is at http://www.ncbi.nlm.nih.gov/genome/seq/HsBlast.html, and other organism-specific BLAST pages are available via genome resource pages. BLAST documentation is available at http://www.ncbi.nlm.nih.gov/blast/producttable.shtml and http://www.ncbi.nlm.nih.gov/BLAST/docs/megablast.html.

[c]NCBI documentation lists the query string length match criteria as guaranteed to find exact match to query string length $s \geq 4*round(w/4) + 3$; *may* find exact match to query string length $s = w$ but not guaranteed. This is no longer true in the latest MEGABLAST engine; the word size is interpreted exactly; that is, all exact matches of length exactly w are found, whether w is divisible by 4 or not (NBCI BLAST Support, personal communication).

[d]Available at http://genome.ucsc.edu/cgi-bin/hgBlat; current Web implementation has *stepSize* = 5, which allows indexed words to overlap. For 2 exact word matches in query string, BLAT is guaranteed to find a query string of length $s \geq w + 2*stepSize - 1$. Smaller query string matches as low as $w + stepSize$ are theoretically possible but are not output based on *minMatchShown* = 20 parameter setting (UCSC Support, personal communication).

[e]Available at http://www.ensembl.org/Multi/blastview; current implementation. Overfrequent words that might be located in genome repeats are filtered from the genome word index (Ensembl Support, personal communication).

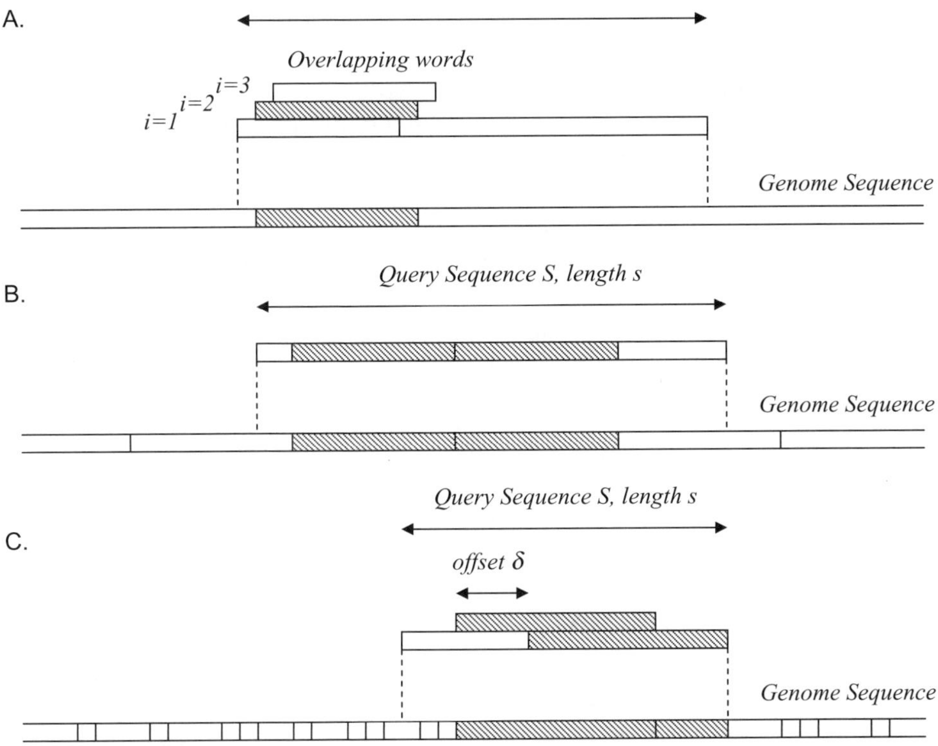

Fig. 4. Comparison of search algorithms in BLASTN **(A)** and other programs, for example, BLAT, SSAHA **(B, C)**. In **(A)**, BLASTN searches all fixed-length overlapping words in the query string (only the first 3 overlapping words in the query sequence displayed $i = 1, 2, 3$, shown stacked) against the extended target genome sequence. The cross-hatch segment in the query and genome sequence indicates that the second word ($i = 2$) has found an exact match and will trigger an extended alignment because only 1 exact word match is required. In **(B)** and **(C)**, the algorithm searches all possible overlapping words from the query sequence as in **(A)**, against an index of discrete words from the genome sequence, shown as fixed length blocks. In **(B)**, the words in the genome sequence are nonoverlapping and 2 exact word matches are required (shown in cross-hatch) to trigger an extended alignment. In **(C)**, the genome is indexed using overlapping words, and 2 genome words match in the query sequence (shown stacked). The matching words lie at the 3′ end of the query sequence, and the query sequence therefore has to be of length at least ($w + 2\delta - 1$) to guarantee finding a 2-word match irrespective of location of the query sequence in the genome (δ is the offset or spacing of words in the genome index; the *stepSize* option in BLAT).

the most distal exon will be missing. In the context of BLAST analyses, the separate alignments are referred to as *high scoring segment* pairs. Most genome search and pairwise sequence alignment programs predict exons based solely on alignment score, and there is often overhang into intronic regions by a few bases. Therefore, postprocessing of the resulting alignments is necessary to identify exons that conform to the usual GT-AG rules for mammalian splice sites. BLAT is an exception, as described below. The search word length and/or guaranteed sequence match length may cause small exons to be missed in the alignment.

3.6. Significance of Genome Search Results

Genome searches using a small oligonucleotide query string will return an expected match with probability given by **Equation 2**. Smaller query strings may return multiple results because of random string matches, with expected numbers of given by **Equations 6** to **8**, *et seq*. The number of matching sites obtained by the actual search can be tested against a null hypothesis of matching to a random genome sequence, and the 1-sided P value for the significance of the test can be computed using:

$$P(X \geq N \mid S, p_A, p_C, p_G, p_T) = \sum_{X=N}^{\infty} \frac{L!}{(L-X)!X!} P_S{}^X (1-P_S)^{L-X}. \quad (19)$$

If the expected number of matches $E[X] = LP_S \geq 5$ (**Equation 6**), the exact result from the binomial distribution can be approximated with a 1-sided test of significance using a normal distribution, with mean $\mu = LP_S$ and variance as given in **Equation 7**.

If the genome sites that match the query sequence are hypothesized to have structural or functional importance, it is not possible to distinguish "true" sites of interest from secondary random false-positive matches based solely on the sequence results, unless the search identifies genome sites with differing degrees of exactness of match. One can reasonably hypothesize that genome sites that are closer to exactness have a higher *a priori* probability of being true-positive (functional) sites, but the number of categories of "exactness" may be limited with small oligonucleotide query sequences. For exact match searches, statistical considerations alone will not distinguish true and false positives, and other information or metrics must be used to discriminate. For example, one could take into account the position of a match site relative to the nearest gene locus or the clustering of other types of sites.

As discussed above, the query string self-overlap properties affect the variance of the distribution and hence the significance of results as calculated under the model of a random genome sequence. Because the random genome sequence model is often inadequate for the compositional and correlative biases that exist in genomes, a more general approach for determining significance of results is indicated. Typically, simulation-based methods will provide an empirical null distribution that incorporates the properties of the genome sequence being searched, the nuances of the search method, and genome match site correlations. The flexibility and general applicability of these methods is usually preferred over theoretical results from parametric distributions.

For searches using extended query sequences that require sequence alignment to determine the pattern of genome matches, the alignment score of the results can be used to judge their relative merit. The highest scoring alignment(s) are the most significant and least likely to be false positives. A formal treatment of the statistics of sequence database searching was developed by Karlin and Altschul in the early 1990s to relate the significance of alignment scores to those expected under a null distribution of random sequence database matches and is based on an extreme value distribution scaled for the alignment scoring system used and the size of the query and database sequences ***(19,20)***. Discussion of this theory is beyond the scope of this chapter but is covered in many bioinformatics texts ***(2,21)*** and in online resources (e.g., http://www.ncbi.nlm.nih.gov/Education/BLASTinfo/similarity.html). The BLAST and MEGA-BLAST programs hosted at NCBI report these statistics, whereas other programs such as BLAT only report the alignment score.

4. Programs for Mapping Discrete Sequences

In principle, the exact match analysis could be performed using a text editor or word-processing program that is able to search for text strings that span multiple lines, but which also ignores end-of-line control characters that might be present in the input genomic sequence, typically in FASTA format. In the commercial Accelrys (GCG/SeqWeb) sequence analysis package, the *findpatterns* program will perform either exact match or limited mismatch searches on a particular sequence that is uploaded to the program and allows searches for small oligonucleotide sequences. The EMBOSS package (European Molecular Biology Open Software Suite) is a package of free software tools for nucleotide and protein sequence analysis. In this package, the *fuzznuc* program performs a similar search function to *findpatterns*. Publicly accessible Web-hosted sites for the EMBOSS tools are currently available at http://bioinfo.hku.hk/EMBOSS/ and http://bioweb.pasteur.fr/intro-uk.html. Because nucleotide sequence alignment programs such as BLAT and BLAST are relatively fast and publicly available, with easy-to-use Web graphical user interfaces (GUIs), the mapping of exact sequences can be performed using these similarity alignment tools. Use of these tools helps to ensure that very similar, inexact matches in the genome sequence will be detected. Use of the public GUI versions of these programs may limit the user to longer query sequences, which return fewer matches.

The major publicly available programs that implement sequence alignment by heuristic search and that have utility for genome mapping are the archetypal BLAST (BLASTN for nucleotide searches of a databases, or BL2SEQ for searches of a query sequence against a genome sequence); and the newer programs MEGABLAST *(22)*, BLAT *(4)*, and SSAHA *(23)*. Although all of these programs locate highly similar matches many times faster than BLAST (more than 10 times faster is not unusual), they all have unique features that the user must be aware of before interpreting search results. A summary of the match criteria currently implemented in the public Web versions of these programs is summarized in **Table 3**. The program parameters that most affect speed and sensitivity are

1. Subsequence word length (*w*).
2. Word spacing (offset) used when building the genome word position index.
3. Match/mismatch tolerance for a word hit between query and genome sequence.
4. Number of hits needed to trigger a full alignment.

For searches using query sequences smaller than 30 to 40 bases, different programs may result in false-negative genome search results because of the specific implementation of the program parameters and the need to balance speed with computational speed and resource usage. **Figure 4** illustrates the effect of these parameters in genome searches.

The other important consideration in selecting a program is the type of query sequence to be mapped, including whether it contains repeat sequences and/or whether it is a cDNA or EST sequence that might reflect a splicing from a gene. **Figure 5** shows an example of searches and alignment of an EST sequence from the recently discovered IL1F10 gene *(24)* against the build 36 human genome sequence. The BLASTN and MEGABLAST programs at NCBI are unintelligent of splice site consensi and predict exons based on the extension of high scoring segment pair alignments (HSPs) until the total alignment score dropoff threshold is reached. BLAT however makes an informed choice of the most likely local splice sites near the ends of the preliminary alignment using known consensus rules and assumes that intervening genomic sequence between these sites is intronic. BLAT also stitches exon alignments together to generate an overall gene model for the query sequence transcript reporting a total alignment consisting of putative exons interspersed with intronic sequence. The search word length and/or guaranteed sequence match length may cause small exons to be missed in the alignment. In the case of BLAST programs, a missing exon will be apparent from the existence of a contiguous region of nonaligned based in the query sequence. If the missing exon is sandwiched by

```
>ref|NT_022135.14|Hs2_22291 Homo sapiens chromosome 2 genomic contig
          Length = 20066235

 Score =  435 bits (235), Expect = 8e-120
 Identities = 247/252 (98%), Gaps = 3/252 (1%)
 Strand = Plus / Minus

Query: 1       tggggcacaatgtctaaaacgcagtttcctgtctccctac-agctctgtccaaagtaaa 59
                |||||||||| |  ||||||||||||||||||||||||||| |||||||| |||||||||
Sbjct: 2540900 tggggcacaagg-ctaaaacgcagtttcctgtctccctaccagctctgttcaaagtaaa 2540842

Query: 60      acttggtacgggctgagggctcactctccttggtgagctgtactggctgctggggctctg 119
                ||||||||||||||||||||||||||||||||||||||||||||||||||||||||||||
Sbjct: 2540841 acttggtacgggctgagggctcactctccttggtgagctgtactggctgctggggctctg 2540782

Query: 120     ccgggccacacaggaaccagc-aggccaggcagcagcctcaagcctgaaggcggagcctg 178
                ||||||||||||||||||||| ||||||||||||||||||||||||||||||||||||||
Sbjct: 2540781 ccgggccacacaggaaccagccaggccaggcagcagcctcaagcctgaaggcggagcctg 2540722

Query: 179     agctgctctggaagaaggtgaagcgtgtggcctcttcaccacctttgtacagttcctcaa 238
                ||||||||||||||||||||||||||||||||||||||||||||||||||||||||||||
Sbjct: 2540721 agctgctctggaagaaggtgaagcgtgtggcctcttcaccacctttgtacagttcctcaa 2540662

Query: 239     tgttcacatcct 250
                ||||||||||||
Sbjct: 2540661 tgttcacatcct 2540650


 Score = 71.3 bits (38), Expect = 3e-10
 Identities = 38/38 (100%), Gaps = 0/38 (0%)
 Strand = Plus / Minus

Query: 248     cctccagctgtagggaaggcccctcttctgtctccaca 285
                ||||||||||||||||||||||||||||||||||||||
Sbjct: 2540351 cctccagctgtagggaaggcccctcttctgtctccaca 2540314
```

A

Fig. 5. **(A)** Results of a BLASTN search of human genome build 36 at NCBI using adult placenta ORESTES library EST, accession number BI051175, which maps to the IL1F10 locus *(24)* on human chromosome 2. BLAST does not take into account known rules for consensus mammalian splice sites. Aligned bases contained within the closest intronic reverse strand donor/acceptor consensus (CT-AC) are shown in bold underline within the boxed regions and appear to be extensions of exons. **(B)** Results of a BLAT search of human genome build 36 at UCSC using the same EST as in **Figure 4A**. Only the highest scoring and most significant match to the genome is shown. The alignment of the 2 putative exons is shown. BLAT is aware of consensus splice site rules and truncates exon alignments to the best local donor and acceptor sites (shown in bold underline in boxed regions). Asterisk (*) indicates the 3′ ends of the exons as aligned (EST maps to reverse strand). The aligned query EST bases are shown in uppercase; nonaligned genomic in lowercase.

```
ACTIONS  QUERY       SCORE START  END QSIZE IDENTITY CHRO STRAND     START       END  SPAN
------------------------------------------------------------------------------------------
details BI051175     276   1      285   285  99.0%     2    -     113548862 113549448  587


   cDNA BI051175
   TGGGGGCACA AtGtCTAAAA CGCAGTTTCC TGTCTCCCTA CAGCTCTGTc  50
   CAAAGTAAAA CTTGGTACGG GCTGAGGGCT CACTCTCCTT GGTGAGCTGT  100
   ACTGGCTGCT GGGGCTCTGC CGGGCCACAC AGGAACCAGC AGGCCAGGCA  150
   GCAGCCTCAA GCCTGAAGGC GGAGCCTGAG CTGCTCTGGA AGAAGGTGAA  200
   GCGTGTGGCC TCTTCACCAC CTTTGTACAG TTCCTCAATG TTCACATCCT  250
   CCAGCTGTAG GGAAGGCCCC TCTTCTGTCT CCACA


   tgtatggtaa catatgcaca gatcttggga ttaggatgtg gacatatttc  113549499
   gggggacatt attctgccta ccatagaccc tgagcaggat gagcttggtt  113549449
  *TGGGGGCACA AgGCTAAAAC GCAGTTTCCT GTCTCCCTAC cAGCTCTGTt  113549399
   CAAAGTAAAA CTTGGTACGG GCTGAGGGCT CACTCTCCTT GGTGAGCTGT  113549349
   ACTGGCTGCT GGGGCTCTGC CGGGCCACAC AGGAACCAGc CAGGCCAGGC  113549299
   AGCAGCCTCA AGCCTGAAGG CGGAGCCTGA GCTGCTCTGG AAGAAGGTGA  113549249
   AGCGTGTGGC CTCTTCACCA CCTTTGTACA GTTCCTCAAT GTTCACATCc  113549199
   tgtgggggca aggtggatgg ctgcagaatg agagtgctga tggaggtgtt  113549149
   agccctggag gcctctgggg cttggccaaa aaaggacctg acagggctgg  113549099
   tttttaccca agggactgcc ttctgcagat tctagagggc aagggtcttg  113549049
   gtgctaggtg cagagggaga tggtggaaac ctgaggaagt aggacatgtg  113548999
   gacctggctg ggaatccaca gataggcaca aagatgccag agcatcccag  113548949
   gggtcaggcc aggtctgcag tgtcccctag aatggggaga ggcctctcac  113548899
  *CTCCAGCTGT AGGGAAGGCC CCTCTTCTGT CTCCACAcat gccaggcagc  113548849
   ggctccctcc ctggatcccc aggaaaatgg ggaccttggt gcgggccaag  113548799
B  cctctgttag gaagtatgca gatcttctct aggaggg
```

Fig. 5. (*Continued*)

longer exons that meet the word hit requirements (two word hits for BLAT@ UCSC; **Table 3**), BLAT will attempt to align the "missed" exon sequence in the query sequence to the intervening sequence between the longer exons aligned by word matches and may recover an accurate alignment for the small intervening exon.

5. Discussion

In this chapter, I have described the basic statistical results that underlie *in silico* mapping of nucleotide sequences to genomes and briefly surveyed the common programs that are used to perform genome mapping and which are available via public hosted Web sites. The binomial and geometric distributions provide useful first approximations for the distribution of the expected number of matches (or occurrences) of sequence string in a genome sequence and the intermatch site distance, but these distributions do not properly address the complication of match site clustering in the genome sequence when the query sequence has the property of self-overlap.

When mapping extended query strings, algorithms that use heuristic word-triggered alignments are publicly available through several popular, well-managed Web sites and can provide rapid genome mapping results. Although these programs—primarily BLASTN, MEGABLAST, BLAT, and SSAHA—all utilize a similar word search method to seed alignments, the nuances of the individual implementations lead to subtle differences in search sensitivity and format of results, and users of the programs should familiarize themselves with their relative capabilities and limitations. For large-scale mapping projects that require more control over program tuning parameters, it may be necessary to collaborate with a local bioinformatics specialist to install 1 or more of the programs on a local server and run the searches on a local copy of the genome sequence using command line programming to control the search parameters and submit query sequences in batches to the program(s).

The results in this chapter should help those unfamiliar with genome sequence mapping to plan computational experiments that require nucleotide-based genome sequence searching. Before embarking on any set of extensive computational experiments, the practitioner should consider the expected results from a preliminary statistical analysis, the relative significance of matches resulting from a search, and the trade-offs in choosing a program to perform searches.

References

1. Waterman, M. S. (1995) *Introduction to Computational Biology*. London, Chapman & Hall.
2. Ewens, W. J., and Grant, G. R. (2001) *Statistical Methods in Bioinformatics*. New York, Springer-Verlag.
3. International Human Genome Sequencing Consortium (2004) Finishing the euchromatic sequence of the human genome. *Nature* **431**, 931–945.
4. Kent, W. J. (2002) BLAT—the BLAST-like alignment tool. *Genome Res.* **12**, 656–664.
5. Shine, J., and Dalgarno, L. (1974) The 3′-terminal sequence of Escherichia coli 16S ribosomal RNA: complementarity to nonsense triplets and ribosome binding sites. *Proc. Natl. Acad. Sci. U. S. A.* **71**, 1342–1346.
6. Forsdyke, D. R., and Mortimer, J. R. (2000) Chargaff's legacy. *Gene* **261**, 127–137.
7. Prabhu, V. V. (1993) Symmetry observations in long nucleotide sequences. *Nucleic Acids Res.* **21**, 2797–2800.
8. Qi, D., and Cuticchia, A. J. (2001) Compositional symmetries in complete genomes. *Bioinformatics* **17**, 557–559.

9. Zimmermann, K., Schogl, D., and Mannhalter, J. W. (1998) Digestion of terminal restriction endonuclease recognition sites on PCR products. *Biotechniques* **24**, 582–584.
10. Ma, J., Campbell, A., and Karlin, S. (2002) Correlations between Shine-Dalgarno sequences and gene features such as predicted expression levels and operon structures. *J. Bacteriol.* **184**, 5733–5745.
11. van Helden, J., Rios, A. F., and Collado-Vides, J. (2000) Discovering regulatory elements in non-coding sequences by analysis of spaced dyads. *Nucleic Acids Res.* **28**, 1808–1818.
12. Klock, G., Strahle, U., and Schutz, G. (1987) Oestrogen and glucocorticoid responsive elements are closely related but distinct. *Nature* **329**, 734–736.
13. Klinge, C. M. (2001) Estrogen receptor interaction with estrogen response elements. *Nucleic Acids Res.* **29**, 2905–2919.
14. van Helden, J., Andre, B., and Collado-Vides, J. (1998) Extracting regulatory sites from the upstream region of yeast genes by computational analysis of oligonucleotide frequencies. *J. Mol. Biol.* **281**, 827–842.
15. Needleman, S. B., and Wunsch, C. D. (1970) A general method applicable to the search for similarities in the amino acid sequence of two proteins. *J. Mol. Biol.* **48**, 443–453.
16. Smith, T. F., and Waterman, M. S. (1981) Identification of common molecular subsequences. *J. Mol. Biol.* **147**, 195–197.
17. Altschul, S. F., Madden, T. L., Schaffer, A. A., Zhang, J., Zhang, Z., Miller, W., and Lipman, D. J. (1997) Gapped BLAST and PSI-BLAST: a new generation of protein database search programs. *Nucleic Acids Res.* **25**, 3389–3402.
18. Weber, J. L., David, D., Heil, J., Fan, Y., Zhao, C., and Marth, G. (2002) Human diallelic insertion/deletion polymorphisms. *Am. J. Hum. Genet.* **71**, 854–862.
19. Altschul, S. F., and Karlin, S. (1990) Methods for assessing the statistical significance of molecular sequences by using general scoring schemes. *Proc. Natl. Acad. Sci. U. S. A.* **87**, 2264–2268.
20. Karlin, S., and Altschul, S. F. (1993) Applications and statistics for multiple high-scoring segments in molecular sequences. *Proc. Natl. Acad. Sci. U. S. A.* **90**, 5873–5877.
21. Korf, I., Yandell, M., and Bedell, B. (2003) *BLAST.* Sebastopol, O'Reilly & Associates.
22. Zhang, Z., Schwartz, S., Wagner, L., and Miller, W. (2000) A greedy algorithm for aligning DNA sequences. *J. Comput. Biol.* **7**, 203–214.
23. Ning, Z., Cox, A. J., and Mullikin, J. C. (2001) SSAHA: a fast search method for large DNA databases. *Genome Res.* **11**, 1725–1729.
24. Bensen, J. T., Dawson, P. A., Mychaleckyj, J. C., and Bowden, D. W. (2001) Identification of a novel human cytokine gene in the interleukin gene cluster on chromosome 2q12-14. *J. Interferon Cytokine Res.* **21**, 899–904.

23

Working with a Statistician

Nancy Berman and Christina Gullíon

Summary

This chapter presents some guidelines for working with a statistician, beginning with when and why you should consult one. We emphasize the importance of good communication between the statistician and the client and the need for clearly defined tasks and a timetable. Other considerations, such as security, confidentiality, and business arrangements are discussed.

Key Words: Authorship; collaboration; communication; consultation; tasks; timetable.

1. Introduction

This chapter addresses some of the considerations and practical issues involved in working with a statistician. We will begin by discussing why one would want to work with a statistician, (**Section 2**) and at what point one should seek statistical help (**Section 3**). The statistician may participate in a study as either a collaborator or consultant, and we discuss the difference in **Section 4**. The last sections deal with the most practical issues, the statistician's role and tasks (**Section 5**) and business and professional arrangements (**Section 6**). **Section 7** deals with the special considerations that arise when one is working with a team of statistical and data management professionals rather than with a single individual.

2. Why Work with a Statistician?

This chapter comes at the end of a detailed book covering statistical methods. Why, with all this information at your fingertips, should you bother to work with a statistician? Surely with this book and any of the reasonably priced or

From: *Methods in Molecular Biology, vol. 404: Topics in Biostatistics*
Edited by: W. T. Ambrosius © Humana Press Inc., Totowa, NJ

free software packages that do data analysis, you could generate your own *P* values with no extra cost or trouble.

In some projects, you may not need to work with a statistician. For instance, if you have done a simple study, with 1 design factor and 1 outcome measure, and your hypothesis and data clearly match the assumptions and purpose of a statistical test that you know how to do and to interpret, you may be able to complete and publish your research without the aid of a statistician.

Most studies and data are not so simple. Methodological considerations in choice of design, measure, and analytic methods can rapidly exceed the scope of a fundamental text such as this. This does not mean that this text is not useful—a foundation in statistical thinking and methods is invaluable for working with a statistician and understanding the rationale for methodological choices. Such a foundation also provides a common language, which can greatly facilitate communication.

A variety of situations may lead you to consult a statistician. If your data do not exactly match the assumptions outlined for a statistical test, you may need to identify an alternative analytic approach. A statistician will know about a great many more methods than can be covered by a book such as this, such as more complex or specialized statistical procedures. Alternatively, the best strategy might be to transform your data—but how do you select the optimal transformation? What are the considerations in choosing to transform versus using an alternative statistical test? A statistician can help you answer these questions.

In addition to dealing with these issues in planning an analysis, a statistician can contribute to other aspects of a research project. A statistician can help you relate your research goals to the methods in this text and to understand what results can be expected. She may also be able to help you learn to use and understand a statistical software package, if you would like to do your own analysis. If you work with a statistician from the beginning of a project, as suggested below, she can help with the design of your study and offer ways to facilitate the implementation, including data collection, and final analysis.

Example 1

In a particular study, the objective is to compare means or measures of central tendency. If the distribution of the data is skewed, it might be preferable to use a nonparametric method rather than a parametric test (**Chapter 7**). How skewed a distribution justifies this shift? Would it be preferable to first transform the data and use a test on means? A variety of transformations are available, such as natural log or square root. The optimal choice may not be obvious. A statistician can help you determine if a transformation will be useful, which transformation, and ultimately which test to use.

Example 2

In a longitudinal study with measures obtained several times on the same analytic units, a simple way to look at changes over time is to do a series of paired *t*-tests (**Chapter 7**). Unfortunately, this is not the best choice, for several reasons. The repeated tests might yield a significant result or two just by chance, resulting in an inference that might not be replicable. In addition, these single-time comparisons ignore the shape of a trend over time, which may be far more interesting and important to understanding the phenomenon you are studying. Choosing an appropriate method for analyzing longitudinal (or *repeated measures*) data is not simple. A repeated measures analysis of variance (**Chapters 11** and **13**), based on least-squares computations, provides an unbiased estimate of experimental effects only when none of the outcomes are missing and when other fairly stringent assumptions about the data are met. Each analytic unit that has any missing observation(s) is dropped from the analysis. Methods for dealing with missing data are described in **Chapter 17**. Investigators have historically imputed missing values, but *commonsense* methods such as filling in means or last observation carried forward may produce misleading results. A statistician can provide guidance regarding current best practice for analyses when some data are missing. Alternatively, he might suggest using a *generalized* method, probably based on iterative maximum likelihood estimation, that relaxes requirements regarding data completeness and other features (**Chapter 11**). These are advanced techniques that require a strong theoretical understanding as well as practical experience with the software to use successfully. Frequently, the analysis will require testing several assumptions about the data, which is not a straightforward process. This is a situation in which a statistician's expertise is an important asset.

3. When to Seek Statistical Help

The best time to seek advice from a statistician is early in the planning of a project that will involve data collection and analysis. Statisticians are trained and experienced in developing study designs and can help develop a plan that will answer the research question while also making optimal use of your resources. A statistician can help translate your study goals into testable hypotheses and can define analytic methods that will successfully test them. Using power analysis (**Chapter 19**), he can determine the number of subjects or analytic units needed to obtain a definitive answer to your question. He can also advise you on various methods for randomization and can create an appropriate randomization schedule for your study.

If you do not involve a statistician in study design and implementation, you may design a futile study, that is, one in which the data cannot answer your research question, even in the hands of a skilled statistician. For instance, you might plan to enroll too few subjects to be able to answer your research question definitively. A competent power analysis will tell you if the sample size is sufficient and may lead you to rethink the whole study.

It is important to realize that input from a statistician can add value at every stage of a study, particularly if the optimal methods require more expertise than you have to implement and interpret properly. The objective of study design is to use no more time and resources than necessary to obtain a valid, useful answer to your hypothesis question. In **Example 2** above, if you know when you are planning the study what your alternatives are regarding choice of analytic strategies, it may allow you more flexibility in timing and collecting measurements.

A statistician also usually has experience with a wide variety of measurement approaches, such as questionnaires and physical measures, and is likely to offer useful suggestions during the design and implementation of the data collection plan. A statistician often can also advise you on the design of the study database, so that the information is easily retrievable for reporting and in a form that can be efficiently accessed by the statistical software.

Example 3

Many therapeutic studies require that the investigators do interim analyses; that is, evaluate the data at periodic intervals to determine whether the study should be stopped or extended. This type of decision is needed when patients exposed to an experimental treatment or on placebo are at risk of adverse events or when one treatment might be so superior to the other that no patient should be denied the better treatment. Repeated hypothesis tests increase the likelihood of obtaining a significant result by chance. When the criterion for significance (the type I error rate, α) is adjusted to reduce this likelihood, the power of the study is reduced, and the required sample size has to be increased. A statistician can help to define an appropriate plan for interim analyses, including adjustments to α and to the sample size, can develop objective criteria for deciding whether a study should be stopped, and will know how to do these analyses while still preserving the blind.

Once a study starts, a statistician can guide you through the methodological implications of alternatives, if an unexpected event requires a change in the protocol. For instance, the rates of subject recruitment and retention may not match the optimistic assumptions in the funding proposal, or the incidence of specific outcomes may be much lower or higher than estimated in the power analysis. A statistician can help you decide whether to set new targets for recruiting subjects or may suggest alternate methods to work with the data without reducing the validity of the study. For example, during the course of a longitudinal study, it may become apparent that too few subjects are willing to participate for the full study period. The statistician can help evaluate the consequences of reducing the study period, assuming it is clinically sensible, and how to account for changes in the protocol in the analysis.

It is advisable *not* to wait to consult a statistician after the data have been collected. This happens with surprising frequency. Some statisticians might not want to bother with a study at that point, as this can be a risky and frustrating undertaking—what the great statistical pioneer R. A. Fisher famously characterized as asking a statistician "to conduct a post-mortem examination. He may perhaps say what the experiment died of" ***(1)***.

If a statistician does agree to work with you at this point, she will first need to evaluate the quality of your data, such as completeness, measurement reliability, and conformity with the protocol (e.g., timing of measures). She will review in some detail the methods used to assign subjects to treatments, the handling of protocol irregularities (e.g., wrong treatment given to a subject), and whether events that threaten internal validity, such as unblinding or reassigning randomization codes, might have occurred. Finally, the characteristics of the data will be taken into account in selecting an appropriate analytic method. If she finds serious methodological flaws in the design or conduct of the study, she may conclude that there is no point in doing the requested analysis. This can result in thwarted ambitions and disputes about deliverables—one reason a good understanding of the scope of work is needed before beginning to work with a statistician (see **Section 6**).

There are other occasions when initiating work with a statistician may be too late to serve your needs; for instance, after a paper is returned by journal referees with questions about the statistics or requests for additional analyses. If the study was well done, possibly with the input of a statistician who is no longer on the scene, it may be possible for a newly involved statistician to respond satisfactorily to the referee. A more promising situation is when a grant application is rejected on statistical grounds—often there is an opportunity to resubmit, and the project can be redesigned with input from a statistician. Time and opportunity are lost as a result of the late involvement of the statistician, but a good, funded study may be still possible.

Another difficult situation (from a statistician's point of view) is a consultation that starts after investigators find that their data do not support their hypothesis. This is troubling on several counts. First, the data have already been thoroughly examined, so any further statistical tests are *post hoc* to the original findings. If the fault lies in the study design or measurement quality, a statistician will not be able to help. On the other hand, a statistician can help you determine whether it is feasible to resurrect your study. For instance, if the sample size is too small, it may be possible to plan a larger study based on the learnings from the current one, so that you do not necessarily have to give up on your original idea.

4. Collaborator versus Consultant

A statistician may work with you, the client, as either a collaborator or consultant, and may change her role over a series of projects. This distinction may seem a bit confusing. Many statisticians refer to themselves as consultants regardless of the actual nature of their working relationship with the client. We want to distinguish here between 2 levels of involvement and expectations in such relationships; they lie on a continuum, so the distinction may not always be clear.

A statistician who is a *collaborator* typically is a full partner in your research. She will become familiar with many aspects of your field, may read extensively in the literature of your field, will expect to work with you on a study from beginning to end, and will maintain an interest in your research even when she is not responsible for a specific task. If you sought NIH funding for your project, this person would be listed in the Key Personnel as a coinvestigator.

Alternatively, a statistician may limit his involvement to serving as a *consultant* on your study. A consultant typically is someone who is more distant from the project, who is responsible for a limited scope of work. He will answer questions when asked but is not involved in the study when not performing clearly specified tasks. This does not mean that the consultant does not care about the success and integrity of a study, but that the time and level of detail to which he attends is limited. A consultant could feasibly be located at some distance from your research site and is more likely to be paid a fee (for hours or days actually worked). A statistician at an institutional consulting center is likely to set boundaries that keep your association more a consultation than a collaboration, primarily because he has too many clients to serve any one of them as an in-depth research partner. Very well-known statisticians, particularly those who are recognized authorities in a particular statistical method, are more likely to provide limited consulting time on that method than to become involved in the ongoing life of a project.

A collaborative relationship takes time to develop. You may first encounter a statistician as a consultant who responds to a limited request for help (possibly that rejected paper or grant application). As you get to know this person, you may find that it is so productive to discuss ideas and plans with him, that it is worthwhile to invite the person to become more involved in your research.

5. Roles and Tasks in a Statistical Consulting Relationship

It is important that you understand the role of a statistician and the professional ethics that guide his work to enjoy a productive, lasting association.

Moses and Louis *(2)* have observed that however the consultation may begin, it is crucial that the clinician and statistician consultant "ultimately deal with the same problems." Thus, good communication between the statistician and the investigator and his team is critical. Good communication requires that both the investigator and the statistician be ready to discuss plans and problems, take care that they speak in terms that the other party can understand, and be willing to elaborate on content when requested. It is the responsibility of the investigator to

- describe the problem he is studying and indicate what he hopes to accomplish;
- avoid jargon or provide definitions of specialized terminology; and
- be willing to take a little extra time to elaborate on esoteric principles.

Similarly, the statistician should

- be ready to listen carefully, to ask questions when necessary;
- request supplemental material (to read) to consolidate his understanding;
- explain statistical issues clearly, and avoid or explain statistical jargon; and
- be willing to take a little time to elaborate on esoteric principles.

It is also necessary that both parties listen to each other and ask questions when they don't understand something.

A second critical ingredient in a successful consulting relationship is mutual respect between the investigator and the statistician. The statistician should not expect the investigator to know or understand statistical concepts and should explain his specialized knowledge in a way that respects the intelligence and professionalism of the investigator.

Mutual respect also means that the investigator recognize that the statistician is a professional, be willing to listen to his ideas, to accept him as the expert on statistical issues, and to honor his ethical boundaries. The investigator should not define what the results of an analysis should be and expect the statistician to produce them. Accepting this expectation would actually be a violation of professional ethics *(3)* for the statistician. The results of a study are largely determined by the study design and execution, with the statistician's analytic expertise serving to reveal what is there. Changes in an analysis plan that appear to be directed toward producing a more favorable result for a hypothesis should be viewed with concern.

If the statistician is to be part of the project from the beginning, then she should be informed of its progress and consulted when problems arise, even when they do not appear to directly affect the data or statistical analysis. It also means that the statistician should be willing to listen to the ideas of the investigator on statistical issues, must recognize that there are real constraints in any study that limit the use of some statistical methods, and treat all members of the study team with respect. Both the statistician and the investigator must recognize that each has standards of professional integrity that must be observed.

5.1. Introductory Meeting(s)

The first meeting with a statistician allows you to present your problem. This gives the statistician an opportunity to determine whether she is well qualified to advise you on the project and whether she is interested in working with you.

5.1.1. Describing the Problem

You should be prepared to briefly describe the problem, the general plan for a project to study it, and what you hope to accomplish. You may not be able to give all the details at this meeting, but you should be able to give enough information so that the statistician can determine the level of effort and technical expertise required. In turn, the statistician will probably ask you questions to clarify certain points or ask to see pertinent articles that describe the methods you are interested in using. You also should be ready to discuss a likely timetable for the study, allowing adequate time for analysis.

5.1.2. Supplemental Material

If possible, you should bring supplemental material for the statistician to review. This might include:

- Pertinent sections of your proposal, particularly specific aims and methods.
- Schematics or flowcharts of the process or study flow.
- Preliminary results, such as summary statistics if the study is under way.
- Published articles that can shed more light on the problem, if any, particularly if you think you might want to reproduce a published analysis or conduct a similar study on a different population.
- Articles or pilot data that can be used as a basis for power analysis, when that is required.

Frequently, you will find that material you thought would not be of interest to the statistician is actually quite useful to give a broader understanding of the important issues in your research. It is probably better to err on the side of being too inclusive—the statistician can usually sort out the material that will be useful.

Frequently, this "first" meeting may be more than 1 meeting. For example, you may give the statistician some material to examine and arrange to meet at a later date to determine if the consultation should move forward.

5.1.3. Learning About the Statistician

If you have not worked with this statistician, this is time to learn about her credentials and experience and see if it seems likely you can work well together. The more that hangs on the project (in terms of effort and cost), the more important it is that the statistician be fully qualified by training and previous experience to take responsibility for the statistical aspects of the study. A metaphor that we like is to think of the data analysis as a funnel through which everything you've put into the study must pass in order to attain the dual goals of finding out what the study showed and communicating this information to the scientific community. If you do not have good statistical advice or services, you may find that the study does not answer your question or is unpublishable.

If you are working with an independent consultant, you should obtain the statistician's CV and look for evidence of appropriate technical training (e.g., advanced degree(s) in statistics, biostatistics, or a related quantitative discipline). When working in an academic environment, you can generally be assured that the statistician has these qualifications. Regardless of context, you should look for relevant consulting training or experience, because statisticians vary substantially in areas of expertise. Finally, look for evidence of successful collaborations and consultations—current employment is no guarantee that these occur regularly. An excellent indicator of successful collaboration is a series of journal articles with a team of investigators, obvious products of research they have worked on together. If you are considering collaborating with a less experienced statistician, investigate who the person's mentors are and whether they will be a resource for this person during the project.

5.2. Specific Tasks

If you and the statistician decide to work together, this is the point at which tasks and expectations for the project and for your professional association should be agreed upon. Sometimes the actual tasks that you want the statistician to complete can be defined at the first meeting. Subsequent meetings will usually be necessary to finalize the plans, review progress, and discuss results.

Tasks a collaborating statistician is likely to accept responsibility for are described in **Table 1**. A *consulting* statistician (as opposed to *collaborating*), might be involved in only 1 or 2 of these tasks; for instance, advising on a particular aspect of the data analysis. It should not be taken for granted that she would be available to participate in paper writing or to respond to reviewers. The extent of involvement for which a consultant will be paid should be specified.

5.3. Ongoing Process

As the study progresses, the statistician will probably attend a number of project meetings, particularly during planning and implementation. When the statistician is involved during the study, it is important to review milestones and discuss problems and unforeseen events. Sometimes the realities of study implementation may require changes in the design that will require modification

Table 1
Activities and Tasks That a Statistician Might Do

Activity	Tasks
Study design and proposal development	• Negotiate wording of testable hypotheses and associated primary aims. • Suggest a study design that is optimal in usefulness and efficiency in obtaining data for testing the hypothesis. • Evaluate literature on planned measures for adequate evidence of reliability and validity. • Determine the needed sample size or estimated power (if sample size is fixed) for the planned design and primary measure(s), and write this up. • Write a description of the randomization scheme. • Write the statistical analysis plan for the protocol. • Respond to relevant parts of a critique by reviewers of the protocol.
Implementation and study conduct	• Create the randomization scheme, which may involve a static sequence of codes or a software routine for dynamic allocation. • Collaborate on implementing a process of enrolling and randomizing study participants that meets feasibility and integrity needs. • Review proposed data collection instruments and other measures for reliability, validity, and suitability for the planned data analysis. • Establish coding rules with data entry staff, such as handling of missing data and invalid responses on questionnaires. • Participate in meetings of an advisory committee or data safety monitoring board. • Respond to questions about methodology, including protocol irregularities and changes.
Data management (if there is no programmer associated with the project)	• Specify/design the data management system (for manual data entry and/or storage of data). • Train database personnel in use of the system (unless this is a supporting department task). • Set up a tracking/audit system for monitoring participant flow and data collection.
Data analysis	• Audit data for completeness and validity. • Plan, direct (or carry out), interpret, and report any interim analysis, and advise on needed project changes. • Plan, direct (or carry out), interpret, and report the final data analysis. • Present and explain analytic results to coinvestigators and project team.
Presentation/ publication	• Design and direct (or carry out) preparation of tables and graphs. • Collaborate in writing papers, abstracts, presentations. • Review data accuracy and interpretation of inferential statistics in all reports of study methods and results. • Respond to journal referees with written comments and/or additional analyses or data.

of the analysis plan. It is not unusual for the study to take longer than expected, and this should be addressed as soon as possible to allow everyone to plan his or her time accordingly. As with the first meeting, good communication and trust is key to the success of the consultation.

Toward the end of a study, the statistician is likely to want meetings to develop publication plans and establish priorities for completion of planned analyses. As results are obtained, he will naturally want to talk about interpretation of the results and potential *post hoc* questions raised by the findings.

6. Business and Professional Arrangements

6.1. Expectations Regarding Payment for Statistical Services

In most cases, a statistician is paid for his services on a project. Some statisticians will not require or cannot accept payment, such as a postdoctoral fellow, a government employee, one who is salaried to provide statistical advice (e.g., in a university consulting center), or one who is willing to do *pro bono* work. Some institutions provide some support for study design and protocol development through funded centers, such as a General Clinical Research Center (GCRC) or a Cancer Center, through the statistics department or other institutional resources. Other statisticians may be willing to participate in proposal development "on spec" and share the risk of seeking funding.

6.2. What Costs Are Included?

What is covered by a financial agreement depends both on the setting in which you are both working and on negotiations about tasks the statistician will be responsible for. If you are both salaried and working in the same institution, this arrangement may be as simple as agreeing on a percent of full-time salary that your project will cover in the institution's payroll budget. It is still useful to be explicit about what aspects of the project will be the direct responsibility of the statistician and what might be covered by other staff in the institution, such as a programmer or a statistical analyst. **Section 5.2** lists the most common tasks that the statistician or an alternate staff member might be expected to perform and that will be the basis for estimating the time needed to complete project tasks.

If the statistician is a freelance consultant or will be working with you under a subcontract with another institution (say, a statistical consulting firm), you should ask the statistician to develop a cost proposal. Statisticians vary quite a bit in what they include in such a proposal ***(4)***. Some may propose a strictly time-based method of charging, that is, hours or days of direct project activity, whereas others may propose time plus costs. What is covered by a time-based rate also can vary quite a bit. More sophisticated or experienced statistical consultants are aware that their consulting income must cover not only direct time to complete an analysis but also indirect costs such as benefits, operating costs (equipment depreciation, software, office supplies), "downtime" between projects, continuing education, professional books and journals, and travel time. These indirect costs may or may not be explicit in the rate quoted and may be included even when the statistician is in the same institution.

It is important to have a list of tasks ("deliverables") in the cost proposal, so that you can be sure you and the statistician have the same understanding about what will be done and what contingencies are covered by the cost proposal. For instance, sometimes data have quality problems that take the statistician's time to figure out and consequently delay or even prevent completing the analysis. Will the statistician be paid if the analysis cannot be completed because of a flaw in the study; what are the limits on this? Are there reasonable terms for extending or terminating an agreement? Other things that a statistician might include in project time are learning about the science of the study, researching new methodologies, or learning to use new software required for this project.

It is important to be realistic about the time involved, including time for interpretation and other follow-up such as response to journal referees. We are aware of investigators who have consulted a statistician, paid him to do very useful but complex analysis, but then been unable to understand the results or present them to others because they have no more funds for the statistician. We believe this is also a responsibility of the statistician to see that the client allows enough time for this.

If funding is coming from a grant, and the funding agency approves the grant but reduces the funding, the impact on the consulting agreement needs to be discussed with the statistician.

6.3. The Timetable

In addition to a cost agreement, you and the statistician need to have an understanding about the timetable for the work. You should set up a schedule of approximate milestones and deadlines and an understanding of the payment process. This can be flexible but both you and the statistician need some idea of the approximate time frame for the tasks to be completed.

For example, assume a study involves a large data analysis effort after a period of data collection. The statistician will have allotted time to work with your data when it is expected to be ready, but if you are very late, she may be committed to other projects when you are ready. Similarly, if you have completed your data collection as expected and the timeline includes submitting an abstract for a major meeting, then it is reasonable to expect the statistician to be available to complete the analysis. One purpose of ongoing meetings during the implementation of the protocol is to make adjustments to the timetable, so there are no great surprises when milestones occur early or are missed.

6.4. Flexibility

How much flexibility is there in a consulting arrangement? Is the estimate of time needed an estimate of *reasonable* time or a firm agreement for maximum time? There is likely to be more flexibility in a percentage-of-salary arrangement within a single institution than in a contract with another organization or a freelance consultant. Also, the more closely you and the statistician have collaborated over time, the more likely she is to allow some elasticity in the time estimates.

6.5. Authorship

If you seek a statistician's assistance in scientific research that you intend to publish, a collaborating statistician will usually expect to participate in production of papers and presentations at the end of a study and to receive coauthorship credit commensurate with this effort. The fact that the statistician was paid to provide services to the project is no more relevant in his case than for that of any other potential coauthor, such as scientists who make a living doing research. It is worthwhile to discuss and agree on authorship expectations at the beginning of the consulting relationship. If the statistician contributes substantively to the intellectual work of the project (design, analysis, interpretation) and helps develop, review, and revise publications, then she has met normal criteria for authorship. Frequently, if the statistician is a coauthor on a paper, then she will help with responding to reviewers without further charges. The other side of this arrangement is that a statistician must agree with the presentation and interpretation of results, because there will be a presumption in the scientific community that she is responsible for these. A statistician may refuse authorship if she feels a paper does not represent her professional views.

You may want to offer acknowledgment rather than coauthorship to a statistician that has been only minimally involved. This is usually a welcome gesture, but you must ask the statistician for permission first. Not all statisticians are comfortable with acknowledgments, so you should not be offended if it is refused. The problem is that the statistician who is minimally involved usually has not had an opportunity to review the analysis and manuscript in detail and so is not willing to be responsible for analysis that was done or modified by others.

6.6. Confidentiality and Security

Every patient's privacy and confidentiality of the data must be protected. All of the data files that are given to the statistician should be stripped of identifying information, such as name, initials, or Social Security number. If the data involve health information, HIPAA (Health Insurance Portability and Accountability Act of 1996) regulations may apply. Data that are going outside of the institutional environment where they were collected may be subject to a data use agreement with the receiving institution.

The statistician must agree also to guard the confidentiality of the study, including if necessary the details of the protocol and the results, at least until it is published. The statistician should take the necessary precautions to ensure that the data she is using are not accessible to others in her workplace. In addition, she should take appropriate precautions to ensure the integrity of data, such as a firewall, password protection, good data-set manipulation practices, and regular backups of computer files.

After completion of the study, the statistician may ask permission to use some of the data or the study design for teaching purposes or as an illustration in a paper in the statistical literature. This practice can strengthen education of other statisticians in practical aspects of consulting and data analysis, but you have the right to refuse permission or put restrictions on the use of the data.

7. Consulting with More Than One Individual

Sometimes a consulting arrangement involves more that 1 consultant. You may require 2 statisticians at the same level to do all the work or a junior statistician to perform most of the tasks, with a senior person involved only in the more complex analyses. The consulting arrangement for a large project may include data management specialists to set up and manage the database and clerical staff to manage paper forms and enter the data. This arrangement is different from that with a single statistician, but the need for understanding, clear goals, and well-defined business arrangements as described above do not change.

When there are multiple consultants under a single contract, usually 1 individual is designated as the team leader, who is responsible for working with the client and developing the details of the contract, such as the responsibilities of each team member, the deliverables, and the timetable. Usually this person is part of the project team (e.g., the senior statistician), but for a large study this may be someone whose major role is management of the group rather than technical work. The team leader should specify which members of the team will be involved in each activity, either by name or by professional category, even when only a few individuals compose the team.

The team leader should have the primary responsibility for communication with the client. Communications between the client and members of the team should keep the team leader well informed (e.g., cc's on e-mails, joint meetings) to avoid confusion and questions about whether the contract terms are being met. We discourage independent communications unless these have the knowledge and permission of the team leader.

The contract for a group of consultants should specify exactly how many hours each team member will spend on the contract, either by name or job category. For a large team, the team leader may want to include time spent supervising staff as well as time spent by all staff on meetings to discuss the project and review progress. For a small group, this may be included in the overhead charges. These issues should be determined in advance and included in the contract.

8. Conclusion

A statistician can contribute many things to a study. Expectations on both sides, both general and specific, should be defined at the beginning of the relationship. Clear communication and mutual respect on both sides is critical to the success of the consulting relationship.

References

1. Fisher, R. A. (1938) Presidential Address to the First Indian Statistical Congress. *Sankhya*, **4**, 789–802.
2. Moses, L., and Louis, T. A. (1984) Statistical consulting in clinical research: the two-way street. *Stat. Med.* **3**(1), 1–5.
3. American Statistical Association. Ethical Guidelines for Statistical Practice. Available at http://www.amstat.org/profession/index.cfm?fuseaction=ethicalstatistics.
4. Gullion, C., and Berman, N. (2006) What statistical consultants do: report of a survey. *Am. Stat.* **60**(2), 130–138.

Index

B

C

N

S